Physiological Systems in Insects

Physiological Systems in Insects

Marc J. Klowden
Division of Entomology
University of Idaho
Moscow, Idaho

San Diego San Francisco New York Boston London Sydney Tokyo

Cover Photo: *Enoderus eximius* © Dennis J. Schotzko.

This book is printed on acid-free paper.

Academic Press
An Elsevier Science Imprint
525 B Street, Suite 1900, San Diego, California 92101-4495, USA
http://www.academicpress.com

Academic Press
32 Jamestown Road, London NW1 7BY, UK
http://www.academicpress.com

Library of Congress Catalog Card Number: 2001094640

International Standard Book Number: 0-12-416264-9

PRINTED IN THE UNITED STATES OF AMERICA
01 02 03 04 05 06 EB 9 8 7 6 5 4 3 2 1

Contents

3 *Developmental Systems*

4 *Reproductive Systems*

5 *Behavioral Systems*

6 *Metabolic Systems*

7 *Circulatory Systems*

8 *Excretory Systems*

Preface

Insects constitute the most diverse group of living things on, under, and above our planet. Their successful domination of every ecosystem except the ocean depths can be attributed to the evolutionary fine-tuning of their physiological systems over the hundreds of millions of years since the first primitive insects appeared. Physiology is the study of function: what makes living things work. Insects are good at what they do because of the nature of their physiological systems working together, and their small size that gives these systems distinct advantages.

Much of our knowledge of the physiological systems of insects is based on species that are the easiest to rear in the laboratory. The lab rats of insect physiology consist of a few species of moths, cockroaches, grasshoppers, blood-sucking bugs, and flies. There is probably no greater diversity than that in the group of animals systematists have placed within the class Insecta, yet our understanding is limited to a handful of species that are studied because of convenience. This is not meant to be a criticism, but is merely the reality of scientific investigation and a caution about the state and depth of our knowledge. Given the tremendous diversity of insects, it is impossible to generalize in most ways about them. Whenever anyone makes any generalization concerning insects, someone else will be sure to point out the many species that break that rule. There are many rules and many exceptions in the insect world, and in writing this book I have tried to focus on the basic principles and not on those frequent exceptions. Perhaps the one rule that has no exceptions is that every ecological peculiarity in the insect world has a physiological basis. Physiological systems are as diverse as the insects themselves. Readers are urged to consult the numerous references that follow each chapter if they desire a more detailed understanding.

Entomology and insect physiology are no longer the sole provinces of departments of entomology and entomology majors. Insects not only are important to study in their own right, but they are also exceptional models that can be used to study the physiological systems of other animals. The similarities of these systems and the ease with which insects can be manipulated to better understand how they work have made insects a favorite tool for biological research. Much of what we know about human genetics is based on early experiments with *Drosophila* by biologists who had no formal training in entomology, and our current understanding of many aspects of molecular biology is based on model systems that were developed in insects by scientists who do not think of themselves as entomologists. It is important for all biologists to understand the systems that contribute to the success of the most successful animals that ever lived. This book was written as a text for our insect physiology course that is taught cooperatively at the University of Idaho and Washington State University. It is directed toward those biologists who work with insects and must understand how they function but may not have the time or talent to become insect physiologists themselves.

To Anne

CHAPTER 1

Endocrine Systems

HORMONES IN INSECTS

Hormones are pervasive in insect systems, affecting a wide variety of physiological processes including embryogenesis, postembryonic development, behavior, water balance, metabolism, caste determination, polymorphism, mating, reproduction, and diapause. Along with the nervous system, hormones provide the necessary communication between all the cells that constitute a multicellular animal.

There is a good reason to have two communications systems existing side by side. The nervous system is certainly capable of sending messages rapidly via nerves, but a network of nerves that reached every cell and coordinated their activities would take an enormous amount of space. In endocrine systems, where chemical messengers are transported in the blood, all tissues can receive the message as long as they have the proper receptors that enable them to recognize it. Hormones allow a sustained message to be sent to all cells, but only the cells that are programmed to receive the message can respond. For example, the molting process in many insects requires hours for its full completion. It could occur faster if it were coordinated by the nervous system, but that would mean that every epidermal cell involved would have to receive a nervous message, hopelessly complicating the internal environment with neurons and leaving little room in the body for other organs. Some processes, such as feeding and escape, cannot rely on the slowness of the endocrine system and are regulated by the nervous system. If information regarding some threat in the environment were to be relayed by the endocrine system in order to initiate escape behavior, the insect would probably be eaten before the message arrived. By selecting hormones as a messenger for some systems, insects have made a trade-off between the speed of the response and the complexity of the system that would be required.

The classical definition of a **hormone,** a word coined from the Greek for "I excite," includes those substances secreted by glands and transported by the circulatory system to other parts of the body, where they evoke physiological responses in target tissues in very minute quantities. Although the term "endocrine" originally implied that multicellular glands were the sources of the chemical messengers, it is now recognized that hormones may also be produced by single cells that are not necessarily clustered into a distinct gland. In addition to these more discrete endocrine glands, there are a number of neurosecretory cells found throughout the body that also produce hormones.

Types of Hormone Release Sites in Insects

It is difficult to precisely define what an endocrine gland is because the integration of the endocrine and nervous systems blurs any distinctions. Insects have conventional **endocrine glands,** which are tissues that specialize in the secretion of chemical messengers that are transported by the blood and act on receptor-bearing target tissues elsewhere in the body (Fig. 1.1E). Examples of endocrine glands in insects are the prothoracic glands, which produce ecdysteroids, and the corpus allatum, which produces juvenile hormones. Insects, like vertebrates, also have **nerve cells** that produce chemical messages at the synapse that excite other neurons to which they are connected. In this case, the messenger, or **neurotransmitter,** binds to receptors on the postsynaptic neuron, remaining compartmentalized within the synapse and not entering the bloodstream. The neurotransmitter can thus be considered as a hormone that is acting locally within the synapse (Figs. 1.1A and 1.1D). Insects and vertebrates also have functional hybrids of these two types of cells, called **neurosecretory cells.** Neurosecretory cells are specialized neurons that produce chemical messengers that are released into the bloodstream and affect distant target tissues. Rather than doing this at the synapse between two neurons, the chemicals are released into circulation or delivered to cells at a specialized structure called a **neurohemal organ** (Fig. 1.1C). Thus, the utilization of chemical messengers lies on a spectrum, with neurons at one end that provide a local release of neurotransmitter that affects other neurons only, neurosecretory cells in the middle with their modified neurons releasing neurohormones into general circulation, and conventional endocrine glands at the other end releasing hormones into general circulation. The chemical products released from these various sites are referred to as **hormones** if they are produced by endocrine glands, **neurotransmitters** if produced by neurons, and **neurohormones** if produced by neurosecretory cells. **Neuromodulators** may be released by neurons at the synapse and modify the conditions under which other nerve impulses are transmitted and received (Fig. 1.1B). Receptors present on the postsynaptic membrane (Figs. 1.1A, 1.1B, and 1.1D) and on target cells (Fig. 1.1F) specifically bind the molecules and produce a biological effect, but nontarget cells that lack these receptors are unable to receive the message (Fig. 1.1G).

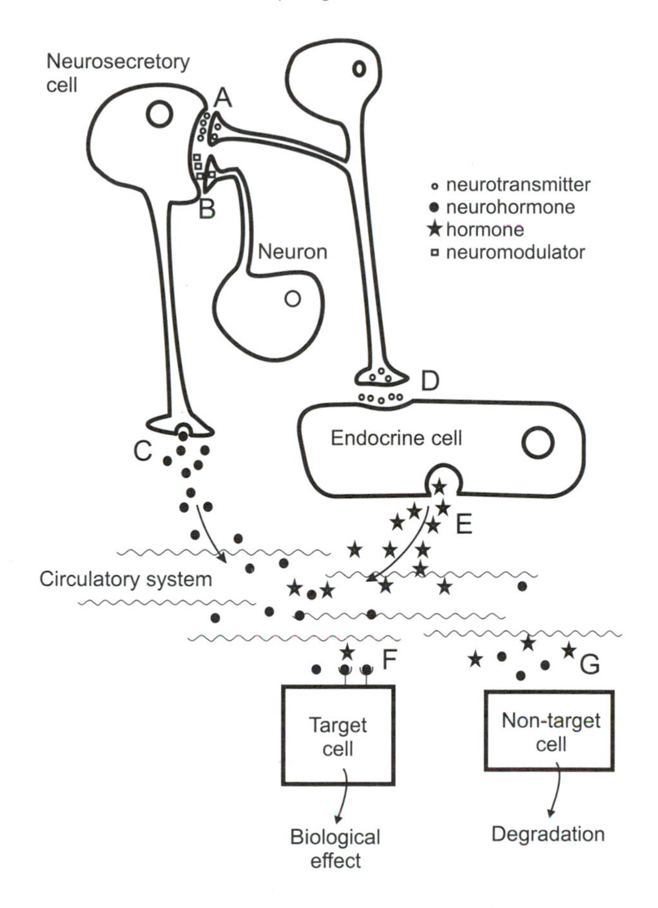

FIGURE 1.1 Some examples of neurotransmitter release. (A) A neuron synapsing with a neurosecretory cell, releasing a neurotransmitter at the synapse. (B) An inhibitory neuron synapsing with a neurosecretory cell, releasing a neuromodulator at the synapse. (C) A neurosecretory cell releasing a neurohormone into the circulatory system. (D) A neuron synapsing with an endocrine cell, releasing a neurotransmitter. (E) An endocrine cell releasing a hormone into the circulatory system. (F) Receptors on target cells recognize a neurohormone in circulation, resulting in a biological effect. (G) The absence of receptors on nontarget cells result in the cell not being able to respond to the circulating chemical messages, and any molecules taken up nonspecifically are degraded.

EARLY EXPERIMENTS THAT SET THE STAGE FOR OUR CURRENT UNDERSTANDING

The first evidence for the existence of hormones in insects is attributed to **Bataillon** in 1894, although at the time the actual involvement of chemical messengers was not recognized. When silkworm larvae were ligated, separating the anterior and posterior halves with a tightly knotted thread that restricted the flow of hemolymph between the two halves, only the anterior portions of the larvae successfully pupated. However, the result was attributed to differences in internal pressure and not to any hormones. It was not until the experiments of **Kopeć** in 1917 that the presence of hormones in insects was confirmed.

When Kopeć ligated the last instar larva of the gypsy moth just behind the head, the insects pupated normally except, of course, for abnormalities of the head. In contrast, if an earlier instar was ligated in the same way, pupation failed to occur at all. When the ligature was applied to the middle portion of the last instar larva before a critical period had passed, only the anterior half pupated, with the critical period believed to be the time at which hormone was released into circulation from the anterior portion. However, if the ligature was applied after the critical period, both halves pupated (Fig. 1.2). Removal of the brain itself before the critical period prevented pupation, but if the removal occurred after the critical period it had no effect. This was the first demonstration of an endocrine function for nervous tissue in any animal. Unfortunately, this conclusion was not well accepted at the time because the brain was not believed to have the capacity to produce hormones. It was not until the 1930s that the vertebrate brain was finally shown to also have an endocrine function. At about this same time, **Wigglesworth** repeated the experiments of Kopeć but using the blood-sucking bug, *Rhodnius prolixus*. *Rhodnius* has five larval instars, each of which requires a large meal of blood in order to molt. When fourth instar larvae were decapitated within 4 days after their blood meal, they failed to molt. However, when decapitation occurred later than 5 days following blood ingestion, the larvae did molt to the fifth instar (Fig. 1.3). Because decapitation obviously removes a number of different structures located in the head, Wigglesworth next focused on the source of the endocrine effect by excising a portion of the brain containing the neurosecretory cells. When these excised cells were implanted into the abdomens of other larvae that were decapitated early before the critical period, recipient larvae molted, demonstrating that neurosecretory cells were indeed the source of the brain's endocrine effect. The historical paths to additional insights into the existence of insect hormones will be discussed in the following sections that describe each hormone.

TYPES OF HORMONES IN INSECTS

Insects produce steroid hormones, such as **ecdysteroids,** sesquiterpenes that include all the **juvenile hormones,** and an abundance of **peptide hormones** produced by neurosecretory cells throughout the central nervous system and the midgut. There are also a number of **biogenic amines,** such as octopamine and serotonin, that are primarily neurotransmitters derived from amino acids, but that may have more widespread effects on the organism. The circulating titers of a particular hormone and its ultimate effects on target cells are precisely modulated by

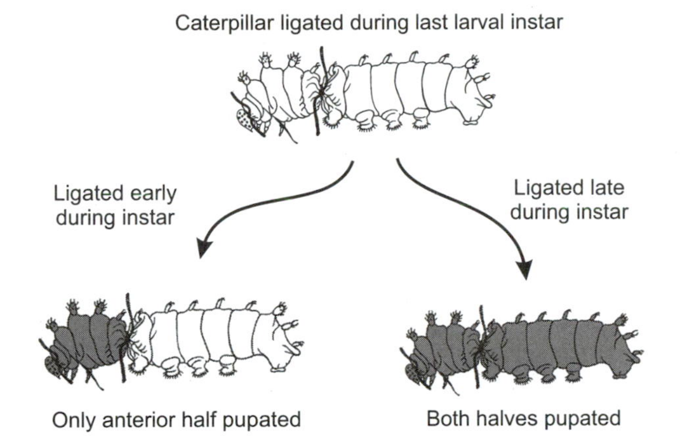

FIGURE 1.2 An experiment performed by Kopeć. When a caterpillar was ligated early during the last larval instar, only the anterior half later pupated. However, when ligated late during the last larval instar, both halves pupated. Adapted from Cymborowski (1992). Reprinted with permission.

an interplay between hormone synthesis, release, and degradation in the hemolymph once the hormone is released into circulation and by the development and specificity of receptor sites on target tissues that allow the specific hormone to be recognized (Fig. 1.4).

Modes of Action

There are fundamental differences in the manner that different hormones act on target cells. Because of their nonpolar nature, ecdysteroids and juvenile hormones are able to enter the cell and bind to cytosolic and nuclear receptors, and ultimately directly interact with DNA and its transcription (Fig. 1.5). The nuclear receptors are transcription factors that stimulate or block the synthesis of mRNA and their presence makes the cell a target for the hormone. The cell responds to the hormone by activating or inactivating specific genes.

In contrast, peptide hormones, which are much more polar, cannot pass through the cell membrane and must trigger a cellular response while remaining on the outside. The peptide hormones bind to protein receptors on the membrane's outer surface, altering the conformation of the receptor and consequently initiating the synthesis of **second messenger** molecules that carry the message inside the cell. These second messengers then act through a cascade of phosphorylations resulting in the activation or inactivation of specific enzymes. A small number of molecules of the first messenger, or hormone, can thus be amplified by the production of a larger number of these second messengers.

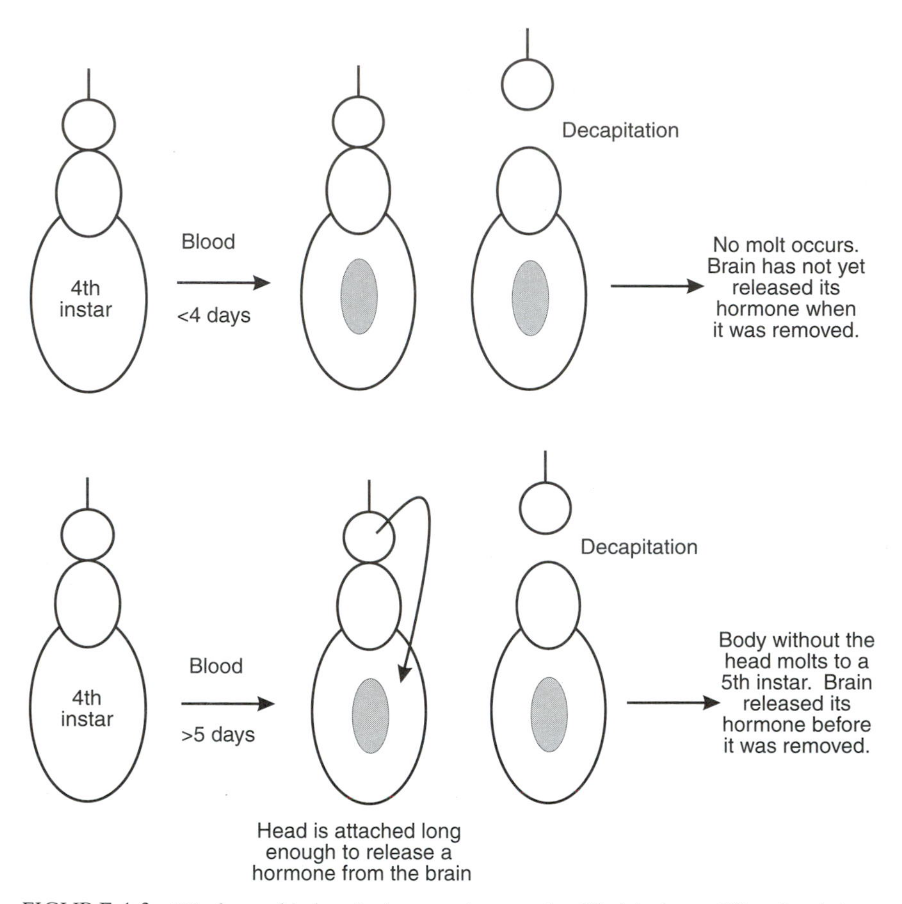

FIGURE 1.3 Wigglesworth's decapitation experiments using *Rhodnius* larvae. When fourth instar larvae were blood fed and decapitated within 4 days, they failed to molt. When they were decapitated after 5 days, the body still molted even though the head was not attached at the time.

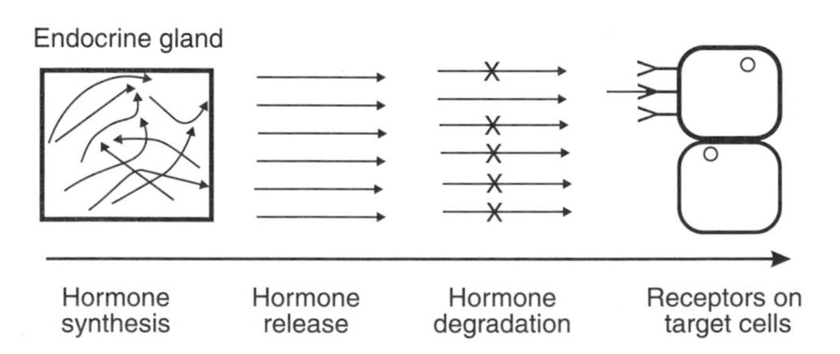

FIGURE 1.4 Factors that affect the activity of hormones. Hormonal activity in the circulatory system is regulated by its rate of synthesis by the endocrine glands, the rate of release into the blood, its degradation in the blood, and the development of hormone receptors on target cells.

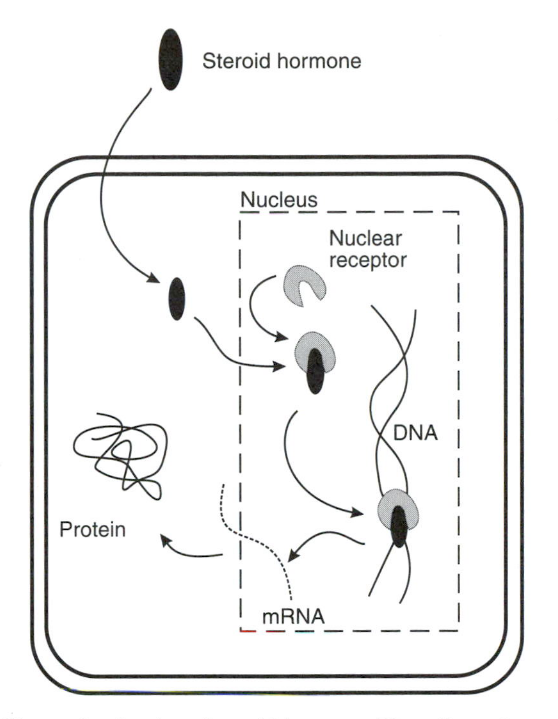

FIGURE 1.5 The mode of action of steroid hormones. The cell membranes are permeable to steroid hormones, so they pass through both the cell and nuclear membranes. They bind to receptors that serve as transcription factors, so together they directly interact with DNA and regulate transcription of mRNA and the production of proteins.

There are several different second messenger signal transduction systems, many of which involve a membrane-bound **G protein** that consists of three subunits and operates between the first and second messengers. For some hormone transduction systems, the second messenger is cyclic AMP (Fig. 1.6). When the membrane receptor for the hormone becomes bound, it changes its conformation and causes it to come in contact with the G protein. This causes the G protein subunits to dissociate, with one of the subunits activating the enzyme adenylate cyclase and forming cAMP from ATP. The cAMP then stimulates a protein kinase that phosphorylates and activates enzymes and ribosomal and nuclear proteins to elicit a biological response (Fig. 1.7A).

A second signal transduction pathway coupled to G-proteins involves the activation of a phospholipase and a subsequent increase in intracellular calcium. The hormone–receptor complex acts through a G-protein to activate a membrane-bound phospholipase that hydrolyzes the complex membrane molecule, phosphatidylinositol 4,5-diphosphate (PIP_2) to form two second messengers, triphosphoinositol (IP_3)

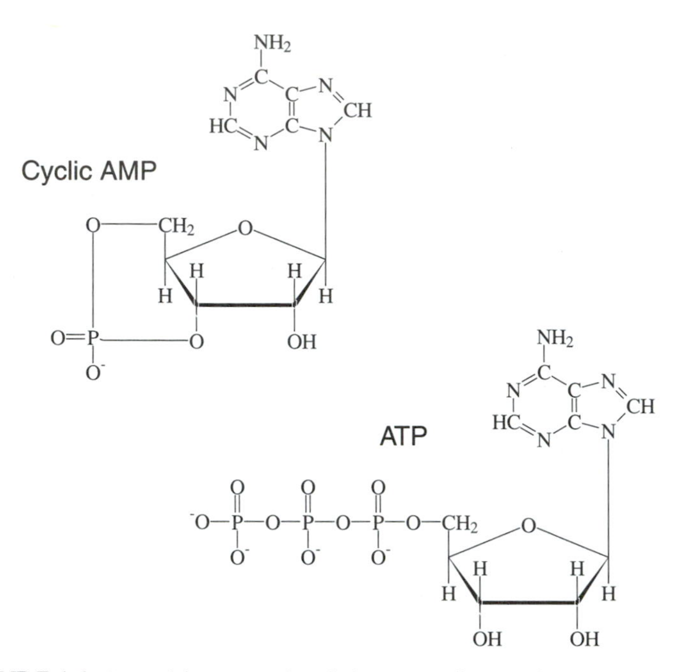

FIGURE 1.6 Two of the major roles of adenine in cells. As cyclic AMP, it acts as a second messenger in cells. As ATP, it serves as a form of energy storage and transfer.

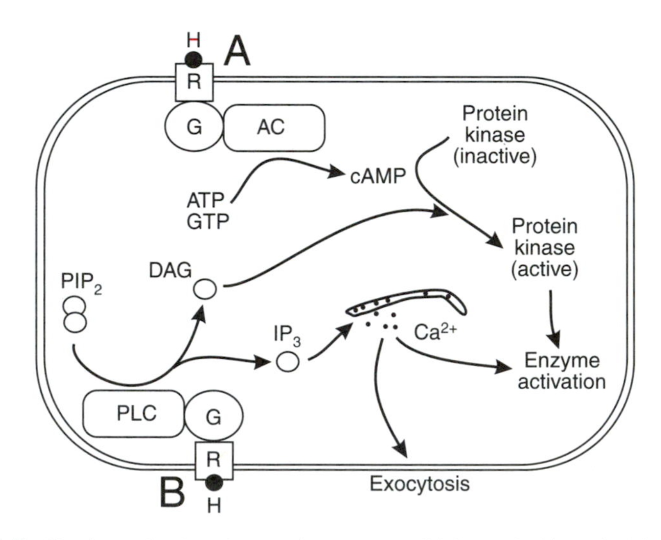

FIGURE 1.7 Signal transduction via second messengers. (A) A protein kinase is activated by the second messenger cAMP that is formed from the adenylate cyclase generated when the G-protein dissociates as the hormone binds to the membrane bound receptor. (B) The two second messengers, triphosphoinositol (IP3) or diacylglycerol (DAG), release calcium or activate a protein kinase, respectively, that can then activate enzymes. The second messengers are formed when a phospholipase is activated from the binding of the hormone to the receptor-associated G-protein.

and diacylglycerol (DAG). The IP_3 causes the release of calcium from the endoplasmic reticulum that can activate exocytosis in cell secretory mechanisms and cell enzyme cascades. The DAG activates a membrane-bound protein kinase that phosphorylates and activates other enzymes (Fig. 1.7B). In both pathways, preexisting enzymes are activated by the hormone binding, unlike the mechanism of steroid hormone action that directly involves the activation of gene transcription and the synthesis of new enzymes.

The gas, **nitric oxide,** can also serve as a second messenger in insect systems. As in vertebrates, signals from the outside of the cell activate the enzyme nitric oxide synthase that forms nitric oxide from arginine (Fig. 1.8). The nitric oxide is able to cross the cell membrane and activate a soluble guanylate cyclase that increases levels of cyclic GMP. The cyclic GMP has a wide variety of effects on

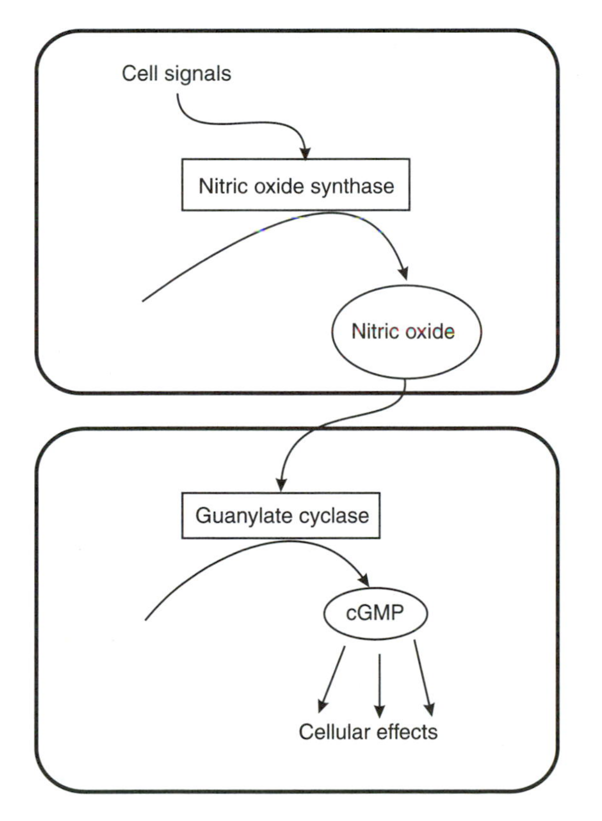

FIGURE 1.8 Nitric oxide as a second messenger. The enzyme nitric oxide synthase can be activated by many pathways. The nitric oxide formed in the cells diffuses easily through the cell membrane and is able to activate the enzyme guanylate cyclase, causing a rise in cGMP that then has cellular effects. From Davies (2000). Reprinted with permission.

the target cell, including the activation of cGMP-dependent enzymes and the permeability of membrane channels. Different isoforms of nitric acid synthase may exist in a single cell, each responsible for a different target. NO signaling in insects is associated with the Malpighian tubules, central nervous system, and the development of the compound eyes.

PROTHORACICOTROPIC HORMONE

Prothoracicotropic hormone (PTTH) was the first insect hormone to be discovered; this was the brain hormone of Kopeć's early investigations, but because the brain is known to produce so many hormones, the simple designation of "brain hormone" was no longer descriptive. Its current name underscores its ability to activate the prothoracic glands. **Williams** demonstrated this relationship in the late 1940s and early 1950s. He implanted both the prothoracic glands and a brain from a chilled pupa into a diapausing pupa and showed that both the brain and prothoracic gland were required to terminate diapause and that the brain activated the prothoracic glands. When he implanted a single chilled brain into a chain of brainless diapausing pupae connected by parabiosis, all the pupae successively underwent adult development (Fig. 1.9). PTTH was the last of the major insect hormones to be identified, perhaps because of the delay in the development of a reliable bioassay. PTTH is produced in the lateral neurosecretory cells of the brain and is released in the corpus cardiacum that terminates in the wall of the aorta, or in some insects, released by the corpus allatum. PTTH acts on the prothoracic gland to regulate the synthesis of ecdysteroids.

Early bioassays for PTTH consisted of debrained pupae, referred to as "dauer" (German: *a long time*) pupae because they could survive for 2–3 years until all the nutrients within them had been exhausted. When extracts with PTTH activity were injected, the pupae initiated metamorphosis to the adult stage. There were several problems with this bioassay, including a low reproducibility due to physiological variations between pupae and the relatively long time it took to score a response. A more direct assay for PTTH was developed by **Bollenbacher** and coworkers in 1979 using the criterion of ecdysone production by a pair of prothoracic glands that were maintained *in vitro*. The basal rate of ecdysone synthesis by a nonstimulated gland is compared to the rate of ecdysone secretion by a gland that is incubated with a suspected source of PTTH. The activation of the gland is indicated by a significant increase in ecdysone synthesis, measured by a radioimmunoassay (Fig. 1.10). It is still not a completely satisfactory bioassay because the prothoracic gland preparations that are required involve a sometimes difficult dissection and isolation.

The PTTHs from only a handful of insects have been identified, with most of the work centered on the lepidopterans *Bombyx mori* and *Manduca sexta* and the dipteran *Drosophila melanogaster*. Initial isolations of PTTH characterized them as

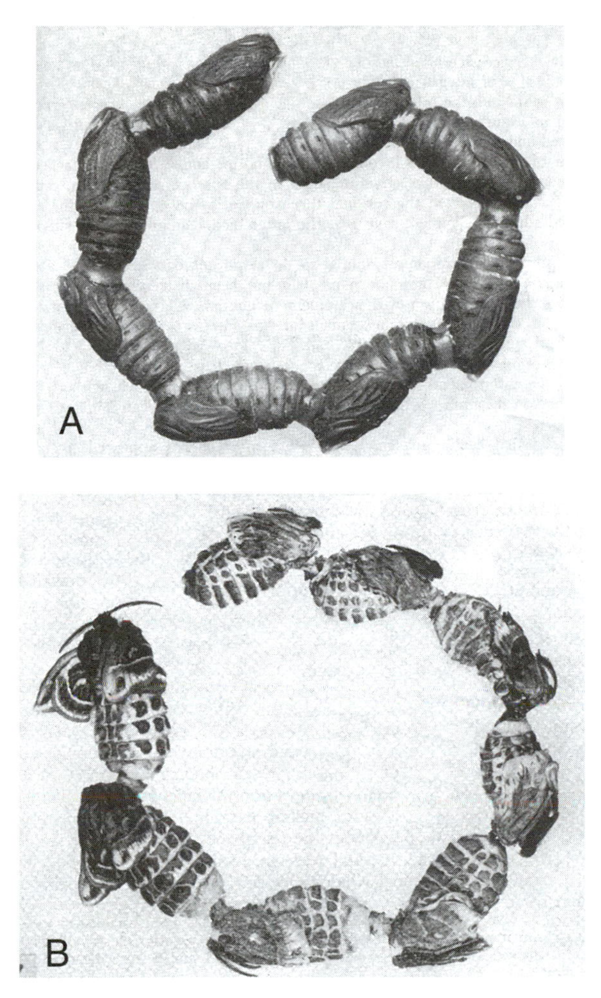

FIGURE 1.9 An experiment by Williams (1952) where a chain of brainless parabiosed pupae (A) were activated to molt by the implantation of a single brain into the first pupa (B). From Williams (1952). Reprinted with permission.

multiple forms that fell into two groups: the "big" PTTH (14–29 kDa) and the "small" PTTH (3–7 kDa). The small PTTH isolated from *Bombyx* was able to activate the prothoracic glands of the related moth, *Samia cynthia,* as well as those of the blood-sucking bug, *Rhodnius prolixus,* but curiously was unable to activate the glands of *Bombyx*. This small PTTH was renamed **bombyxin** and is no longer considered as a true PTTH, mainly because there is no relationship between its titer in the hemolymph of *Bombyx* and the levels of 20-hydroxyecdysone that result, making it a non-PTTH. The multiple molecular species of the bombyxins

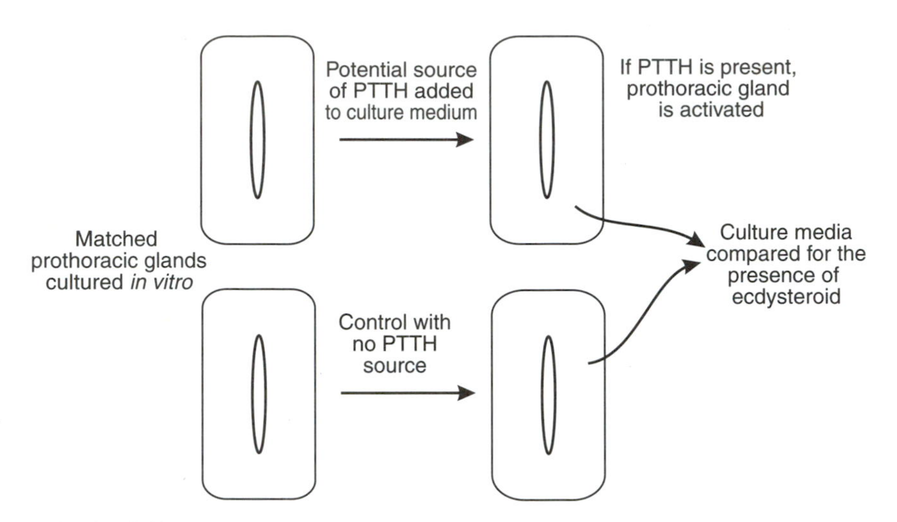

FIGURE 1.10 An assay for PTTH developed by Bollenbacher *et al.* (1979). A pair of matched prothoracic glands were removed from the insect and placed in culture. If PTTH is added to the culture, the glands produce increased amounts of ecdysteroids into the medium.

that have been identified appear to share some homology with vertebrate insulin, but their true roles in insect systems have yet to be determined. Bombyxin receptors are present on the ovaries of some lepidopterans and the hormone may be involved in ovarian development and the utilization of carbohydrate during egg maturation. The big PTTH appears to have the true PTTH activity: it stimulates the prothoracic glands to produce ecdysone. In *Bombyx,* big PTTH is synthesized as a large 224-amino-acid precursor and then cleaved to liberate a 109-amino-acid subunit. The active molecule is a homodimer, consisting of two identical chains that are held together by disulfide bonds (Fig. 1.11). The folding of the molecule is largely controlled by its intra- and intermolecular disulfide bonds.

Control of PTTH Release and Its Mode of Action

The occurrence of the molt is largely determined by the release of PTTH that activates the prothoracic glands to produce the ecdysteroid molting hormone. Most insects release PTTH from their neurohemal organ based on the receipt of environmental stimuli, which may include photoperiod, temperature, and nervous stimuli. For example, in *Rhodnius,* where molting follows blood ingestion, the abdominal distention resulting from a large blood meal triggers stretch receptors that then send a message to the brain to release PTTH. Neither a small meal nor a series of small meals is able to trigger molting; the blood volume must exceed a critical threshold to activate the stretch receptors that, through the central nervous

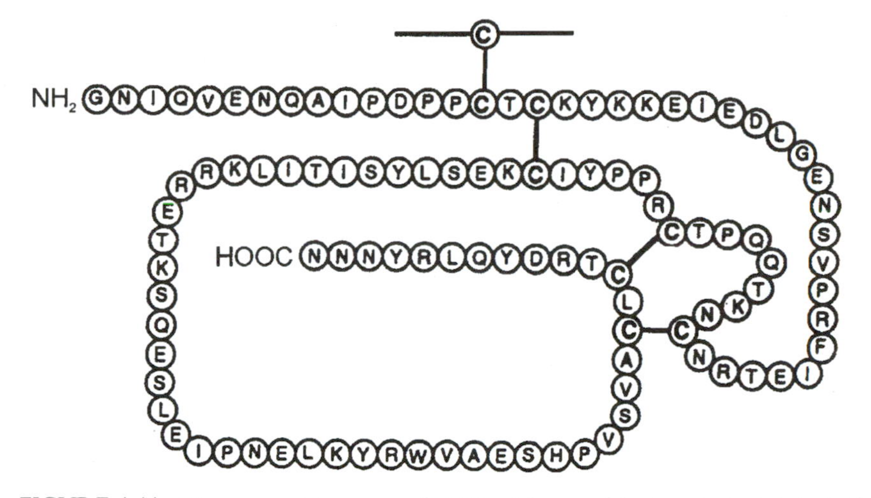

FIGURE 1.11 The amino acid structure of PTTH. Only one of the two identical chains in the homodimer is shown. Reprinted with permission from Ishbashi, J. *et al.* (1994). *Biochemistry* **33:** 5912–5919. Copyright American Chemical Society.

system, initiate PTTH release. The nutritive capacity of the blood is not important, because large meals of saline can also provoke a molt. In the lepidopteran, *M. sexta,* PTTH release is regulated by photoperiod, occurring during a circadian window. The pupal diapause in some lepidopterans results from the failure of PTTH to be released. Without PTTH and the resulting ecdysteroid, the pupa cannot develop further and molt to the adult stage. In *Hyalophora cecropia* moths, pupal diapause can be terminated by a prolonged exposure to cold temperatures followed by a warming.

Because it is a peptide hormone, PTTH is unable to enter the cell and must exert its influence from the outside through a receptor that generates cAMP as a second messenger. As described earlier, this cAMP activates a protein kinase that activates enzymes in the biosynthetic pathways that lead to a cellular response (Fig. 1.7A).

ECDYSTEROIDS

Ecdysteroids are the generic name for a group of related steroid hormones that are primarily involved in the molting process of arthropods but also have wide-ranging effects in every developmental stage. The experiments by Kopeć and Wigglesworth demonstrated the importance of the brain in the molting process, but it was **Hachlow** in 1931 who first showed that the brain does not act alone. When lepidopteran pupae were cut at different points along the body and the cut ends sealed, only the parts that contained the thorax developed adult characters. This suggested that an organ in the thorax was also necessary for molting and metamorphosis. **Fukuda,** in 1940, demonstrated that this particular organ was the prothoracic gland. Using last instar silkworm moth larvae, *B. mori,* Fukuda observed that only those portions ligated anterior to the prothoracic gland pupated, and when the prothoracic gland was implanted into the posterior portions, those also underwent pupation. Along with the experiments by Williams mentioned in the previous section, these results established that both the brain and prothoracic glands released factors necessary for a molt to occur, with the brain activating the prothoracic glands to produce the molting hormone.

Ecdysone was the first insect hormone to be structurally identified, an accomplishment that became possible only once an assay for its biological activity was developed. The *Calliphora* bioassay that was devised by **Fraenkel** in 1935 consisted of fly larvae that were ligated during their last larval instar. Because their posterior portions lacked any molting hormone, they failed to pupariate, unlike the anterior portion that contained the necessary endocrine centers. When extracts containing the molting hormone or suspected molting hormone activity were injected into the posterior portion, pupariation was induced (Fig. 1.12).

Using this bioassay, **Butenandt** and **Karlson** purified 25 mg of the hormone starting with approximately 500 kg (a half ton!) of *B. mori* pupae. Shortly afterward,

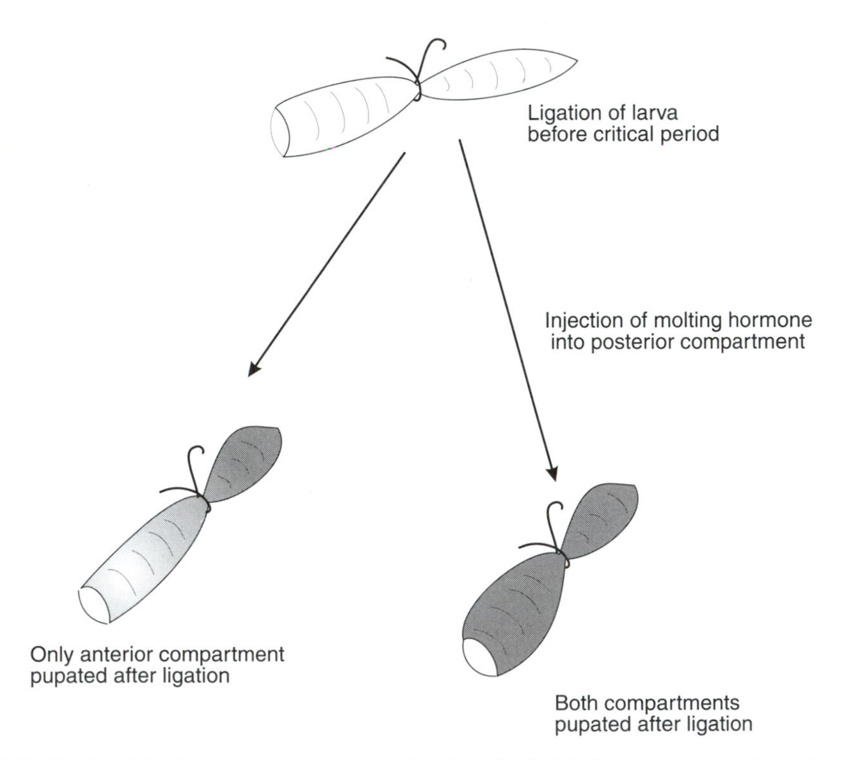

FIGURE 1.12 The *Calliphora* bioassay developed by Fraenkel (1934). When substances with molting hormone activity were injected into the posterior compartment of ligated *Calliphora* pupae, the cuticle of the posterior compartment underwent a molt along with the anterior compartment.

a second substance with molting hormone activity was isolated, and the two hormones were named α- and **β-ecdysone**, respectively. Other ecdysteroids have since been isolated, and the convention was established to use "ecdysteroid" as the generic name for the group. The first ecdysteroid to be isolated, α-ecdysone, is now referred to as **ecdysone.** The second hormone, β-ecdysone, is now referred to as **20-hydroxyecdysone** and is hydroxylated from ecdysone by target tissues (Fig. 1.13). It is the true molting hormone in that it is most active in inducing a molt. There are currently over 60 different analogs of the molting hormone isolated from insects and more than 100 **phytoecdysteroids** isolated from plants. Although they may simply represent metabolic intermediates, these phytoecdysteroids may be produced by the plants as feeding deterrents or as toxic substances that affect the survival of insect herbivores.

Identification, Synthesis, and Control of Ecdysteroid Production

The initial identification of ecdysteroids was delayed partly because they were expected to be similar to vertebrate steroid hormones in their solubility. However, given the many hydroxyl groups, one face of the ecdysteroid molecule is relatively hydrophilic and thus poorly soluble in the organic solvents that were used to extract the generally lipophilic vertebrate steroids. Ecdysone is a steroid hormone belonging to the class of substances known as terpenoids, along with the juvenile hormones. All terpenoids are synthesized by the combination of two isoprene precursors that are responsible for the production of many important biological agents in plants and animals.

The precursors for ecdysteroid synthesis by the prothoracic gland of insects are sterols, such as cholesterol (Fig. 1.14). Although most organisms can synthesize cholesterol from acetate precursors through the series of isoprene building blocks, insects cannot do this and require cholesterol in their diets. Zoophagous insects can easily ingest sufficient cholesterol, which is a major animal steroid. Phytophagous insects instead encounter campesterol, sitosterol, or stigmasterol, the major sterols in plants that contain additional methyl and ethyl groups on their side chains, and the insects must dealkylate them in order to form cholesterol. Those phytophagous insects unable to make the conversion, including many hemipterans, the honeybee, and some dipterans, produce makisterone A, or 24-methyl 20-hydroxyecdysone, as their molting hormone (Fig. 1.13). This is the only 28-carbon ecdysteroid; all the rest are 27-carbon sterols. Although the major steps in the ecdysteroid biosynthetic pathway in insects are known, the complete identification of all intermediates has yet to be described.

The primary site of ecdysteroid synthesis is the prothoracic gland, which develops during embryogenesis from ectodermal cells in the head, and in some insects remains there to be known as ventral glands. In cyclorraphan Diptera, the prothoracic gland has been incorporated into a ring gland that also consists of the

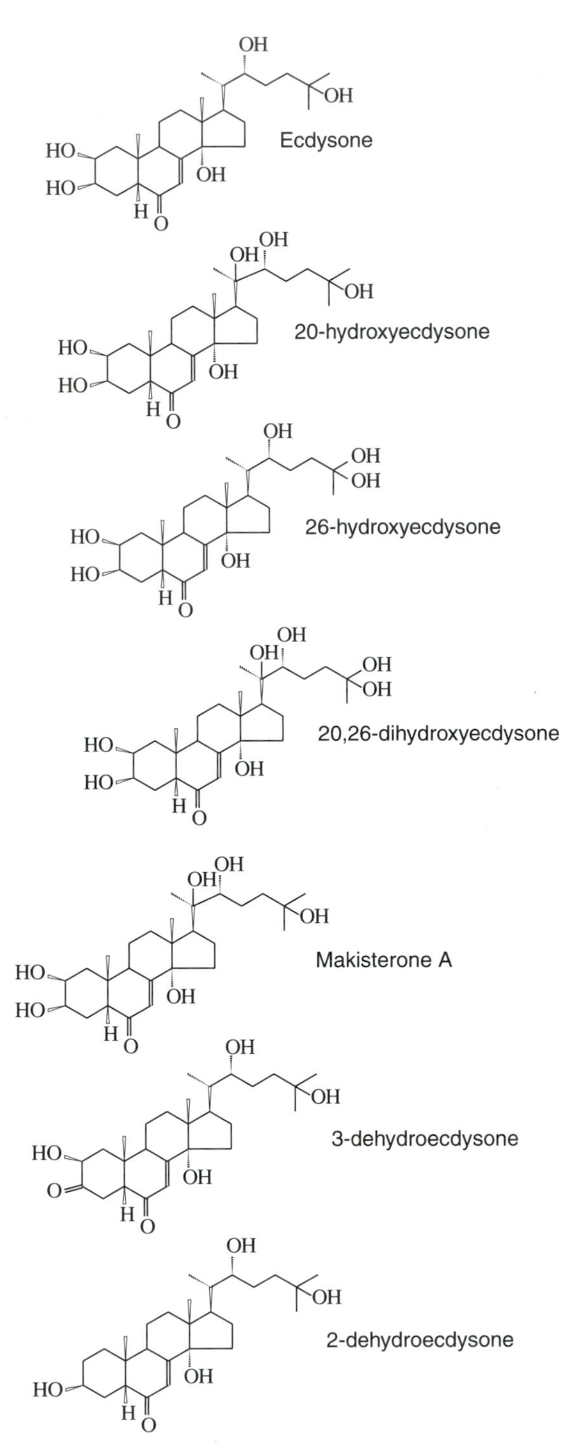

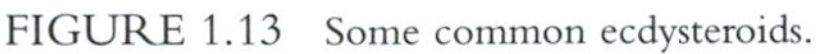
FIGURE 1.13 Some common ecdysteroids.

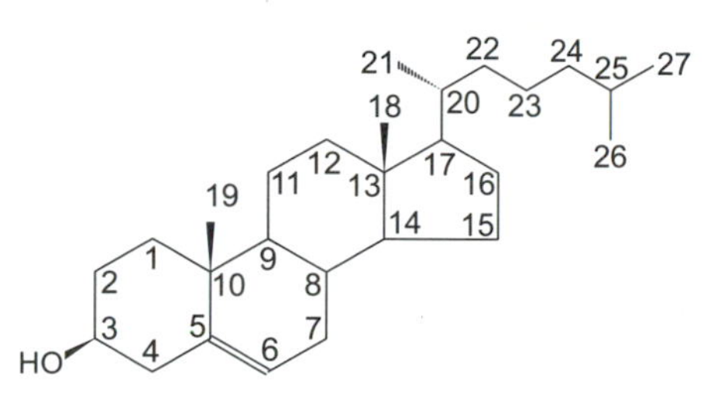

FIGURE 1.14 Cholesterol and the system of numbering its carbon atoms.

corpus allatum and corpus cardiacum (Fig. 1.15). In other insects, the glands are found in the thorax where they form loose chains of cells, with a close association with the trachea that has led to their often being called peritracheal glands (Fig. 1.16). Nervous innervations of the gland consist of a pair of nerves from the subesophageal ganglion and sometimes additional nerves that issue from the prothoracic and mesothoracic ganglia. In spite of the nervous connections, the primary mode of gland activation is hormonal.

Because adult pterygote insects no longer molt, the prothoracic gland is not necessary during this developmental stage. It degenerates in most adult pterygote insects as a consequence of the hormonal conditions present during the last molt. The stimulus for prothoracic gland degeneration is its exposure to ecdysteroids in the absence of JH, the conditions present during metamorphosis. The ecdysteroids acting alone trigger apoptosis, or programmed cell death, by the gland cells. The

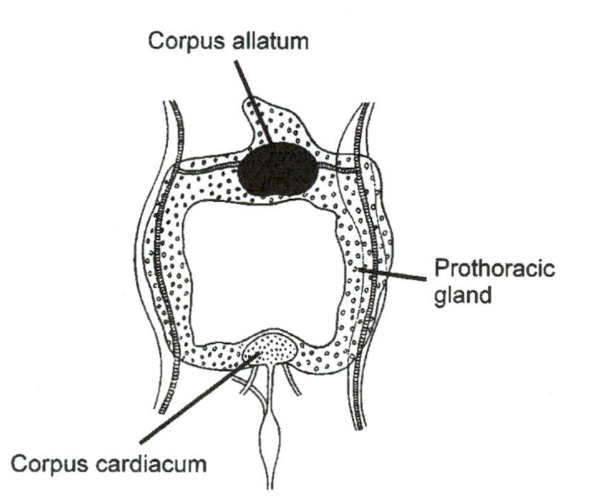

FIGURE 1.15 The ring gland of higher dipterans, consisting of the corpus cardiacum, corpus allatum, and the prothoracic gland assembled in a ring structure. From Cymborowski (1992). Reprinted with permission.

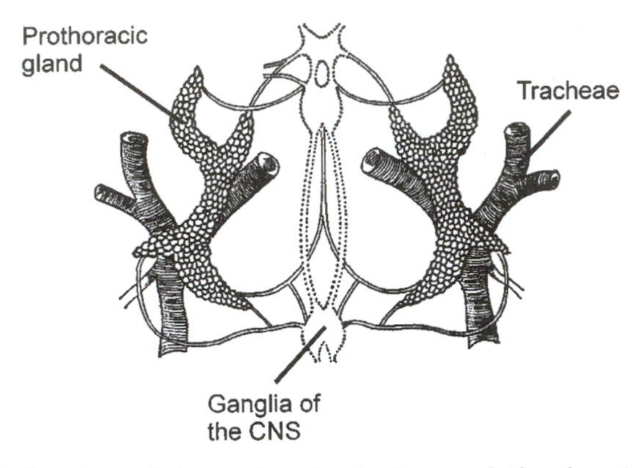

FIGURE 1.16 Location of the prothoracic glands around the thoracic tracheae. From Cymborowski (1992). Reprinted with permission.

gland persists for several days in adult *Periplaneta americana* and does not degenerate at all in gregarious female *Schistocerca* adults, although these glands can no longer produce ecdysteroids when they are cultured *in vitro*. Apertygote insects, which continue to molt as adults, retain their active prothoracic glands.

Ecdysone is the major product of the prothoracic gland, but the active form that binds to cellular receptors is 20-hydroxyecdysone, converted from ecdysone by target tissues. This raises some questions about which of these truly is considered to be the molting hormone: ecdysone released by the endocrine gland or 20-hydroxyecdysone acting on target tissues. In some lepidopterans, the ecdysteroids that are released from the prothoracic gland are a mixture of 2-dehydroecdysone and 3-dehydroecdysone, which are then converted to ecdysone in the hemolymph (Fig. 1.17). There is no evidence that the hormone is stored in the prothoracic gland and it appears to be released when it is synthesized. Once released, the hormone circulates both alone and as bound to carrier proteins. The bound form is inactive and may serve as a reservoir of the hormone.

Other Sources of Ecdysteroids

The prothoracic gland is not the only source of ecdysteroids. Even though the prothoracic gland degenerates in adult insects, ecdysteroids still occur in their hemolymph. In these adults, the site of ecdysteroid synthesis has been shifted to the ovaries and the testes. In many female insects, ecdysteroids are produced by the follicle cells of the ovaries, where they are conjugated to other molecules and incorporated into the eggs for later use during embryogenesis. Developing insect embryos contain several different ecdysteroids in both free and conjugated forms,

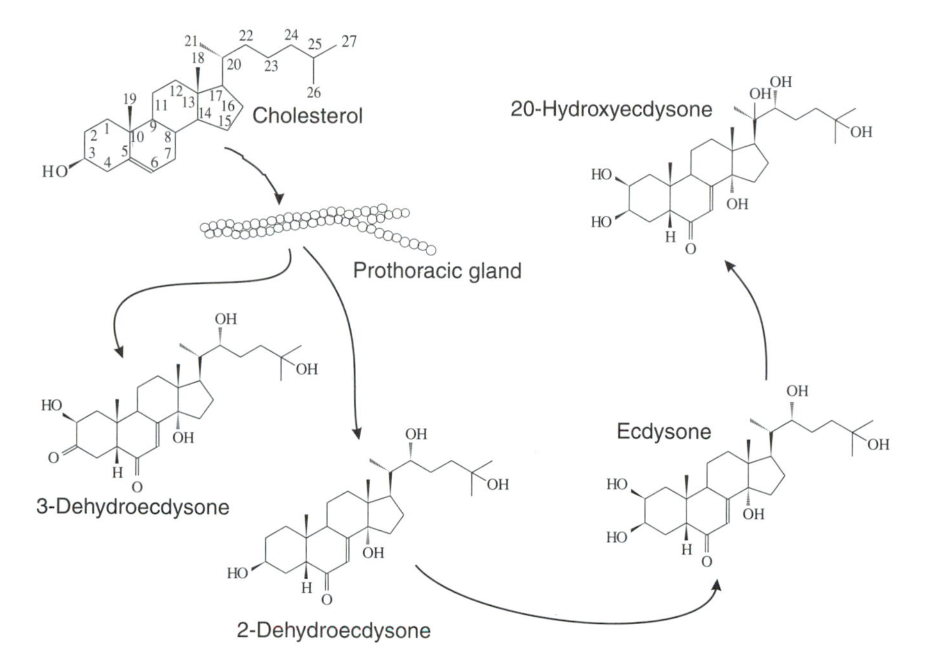

FIGURE 1.17 Synthesis of the various ecdysteroids from cholesterol.

including 20, 26-dihydroxyecdysone and 26-hydroxyecdysone (Fig. 1.13). Ovarian ecdysteroids are also released into the hemolymph and act on the fat body to activate the synthesis of yolk proteins. Ecdysteroids can be produced in males of several species by the larval and pupal sheaths of the testes. The epidermal cells may also be a source of ecdysteroids during certain developmental stages. More information on these roles of ecdysteroids will be found in the chapters on reproductive and developmental systems.

Synthesis of ecdysteroids by the prothoracic gland of an insect larva occurs by the action of PTTH. Its main sites of production are the neurosecretory cells of the brain, but PTTH activity has also been identified in the subesophageal ganglion and the ganglia of the ventral nerve cord. Synthesis of ecdysteroids during other developmental stages by additional sources such as ovaries and testes occurs in response to other ecdysiotropic neurohormones. **Ovarian ecdysiotropic hormones** and a **testes ecdysiotropin,** both produced by neurosecretory cells of the brain, modulate ecdysteroid production by these organs. In larval *Calliphora vicina* blowflies, in addition to the ecdysiotropins that stimulate ecdysteroid synthesis in the ring gland, the brain produces an **ecdysiostatic hormone** that curtails ecdysteroid synthesis. This prothoracicostatic hormone has a molecular mass of 11 kDa and can retard larval development when it is injected. The ring gland thus responds sensitively to both stimulatory and inhibitory neuropeptides. The presence of ecdysiostatins in other insects has not yet been confirmed.

Mode of Action of Ecdysteroids

Ecdysteroids, as typical steroid hormones, can easily diffuse into cells. They directly affect gene expression, causing the activation or inactivation of certain genes and the resulting synthesis or inhibition of enzymes and other regulatory peptides. The evidence that ecdysteroids influence gene transcription comes from studies of polytene chromosomes in *Drosophila,* which are chromosomes that have replicated but whose strands have not separated. Their alignment of replicated DNA makes a banding pattern visible with light microscopy. The **puffs** that are sometimes visible represent sites of active gene transcription, and puffing patterns are correlated with the developmental stage of the insect as well as the ecdysteroid titers. There is a characteristic sequence of puffs in the polytene salivary gland chromosomes of last larval instar *Drosophila* that is induced by an ecdysteroid pulse late in the instar. A few early puffs are rapidly induced by the hormone and then regress, followed by a large number of late puffs that persist through the formation of the puparium. Inhibitors of protein synthesis do not affect the formation of early puffs, suggesting that they are responding directly to the hormone, but these inhibitors do prevent their later regression. Inhibitors of protein synthesis also prevent the induction of late puffs, but these can be induced even when hormone is withdrawn once the early puffs are formed. The puffing pattern that is normally characteristic of late third instar *Drosophila* larvae can be prematurely induced by injecting early third instar larvae with ecdysteroids. The injected hormone is localized in the cell nucleus and can be identified binding to the inducible puff site.

Based on these and other observations, a model for the action of ecdysteroids on gene transcription was developed by **Ashburner** in 1974 and has been since enhanced with the identification of specific genes that are known to be activated during the puffing sequence (Fig. 1.18). In the model, ecdysteroids bind to an ecdysteroid receptor that consists of a heterodimer assembled by the products of the genes EcR and USP and that acts as a DNA binding protein. Only the EcR portion of the heterodimer binds the hormone. The binding of the hormone–receptor complex at the early puff gene sites activates them, and also represses the formation of late genes. The late genes are activated by products of the early genes that remove the repression that is induced by the ecdysteroid–receptor complex. The early gene products also repress the activity of the early genes themselves. The genes associated with the early puffs are thus regulators of late gene expression and the late genes consequently play a direct role in salivary gland morphogenesis.

Although all tissues of the insect can potentially be exposed to the ecdysteroids that are circulating in the blood, not all cells respond in the same way. The

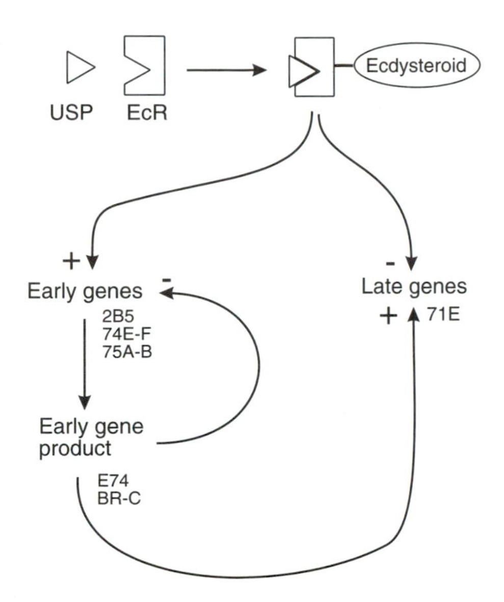

FIGURE 1.18 A model originally proposed by Ashburner (1974) for the action of ecdysteroids in the *Drosophila* salivary gland. The ecdysteroid receptor (EcR) and the product of the Ultraspiracle gene (USP) bind to the hormone. The ecdysteroid receptor complex binds to early genes, stimulating their transcription but inhibiting the transcription of the late genes. The early gene product that is produced subsequently inhibits the early genes but stimulates the late genes, demonstrating the cascade of gene activity that results in salivary gland morphogenesis. The genes that are known to be affected in each case are shown.

temporal expression of specific **isoforms** of the EcR ecdysteroid receptor that may be present in a cell accounts for the ability of these certain cells to respond to the hormone. Different EcR isoforms have been identified in different cells that show different responses to ecdysteroid during metamorphosis. For example, the isoforms EcR-A, EcR-B1, and ECR-B2, which differ in their N-terminal sequences, are expressed in varying amounts in tissues that show altered responses during metamorphosis, such as imaginal discs and neurons destined to be remodeled in going from the larval to the adult stages. In *Drosophila* larvae, the EcR-B1 isoform is predominant in larval epidermal cells that are programmed to die during metamorphosis, whereas the EcR-B2 isoform is present in imaginal discs that proliferate and differentiate during metamorphosis. USP, which is a vertebrate retinoid X receptor homolog, also is produced in at least two isoforms. The USP-EcR heterodimer binds to an ecdysone response element that lies within the promoter region of specific genes and then activates or inactivates gene expression, which can be outwardly observed as a pattern of puffs. These differences in the type of DNA binding protein found in different cells may also be responsible for the varied sensitivity and responses of tissues to the hormone.

The identification of nonsteroidal agonists, such as the substituted hydrazines RH-5849 and RH-5992, have provided additional insights into the action of ecdysteroids. These agonists bind to ecdysteroid receptors and elicit the biological effects of the true hormone. Other anti-ecdysteroids, such as certain plant brassinosteroids, can compete with ecdysteroids for their receptor-binding sites and block their action.

THE JUVENILE HORMONES

Juvenile hormone (JH) was first described by Wigglesworth as an "inhibitory hormone" that prevented the metamorphosis of the blood-sucking hemipteran *R. prolixus. Rhodnius* has five larval instars, each of which will molt after it ingests a large meal of blood. When an active corpus allatum, the source of JH, was implanted into last-instar larvae, the recipients of the gland molted to supernumerary larvae, or additional larval instars, after they fed rather than producing the adult cuticle that would normally be expected. Also, if last-instar larvae were connected to early-instar larvae by parabiosis, in which the circulatory systems of the two insects intermingled, the more mature member of the pair continued to express larval characters after it molted. This indicated that a factor circulating in the blood of the younger member was responsible for the retention of larval characters in the older member of the pair. Wigglesworth coined its present name of juvenile hormone when it became clear that the hormone acted to promote the expression of larval characters rather than to inhibit adult ones. Juvenile hormone is now the generic name for several sesquiterpenes that mediate a wide variety of functions in addition to metamorphosis. Similar to ecdysteroids, JH has multiple effects during the life of an insect, and its specific involvement in the processes of metamorphosis, diapause, reproduction, and metabolism will be described in later chapters. In fact, the name "juvenile hormone" is certainly a misnomer considering the versatility of the hormones within this group.

Major Types of JH and Their Synthesis

The identification of JH became possible when the serendipitous discovery of large quantities of the hormone became available. While Williams was performing an experiment designed to extend the life of male *H. cecropia* by parabiosing them to pupae, he noted that the parabiosed pupae molted to other pupae rather than to adults. This suggested to him that the male was supplying JH, and its accessory glands indeed contained large amounts of the hormone. With this plentiful supply of JH, sufficient quantities were finally available for its analysis.

Six major members of the juvenile hormone group are currently recognized (Fig. 1.19). The first structural identification from the male accessory gland material

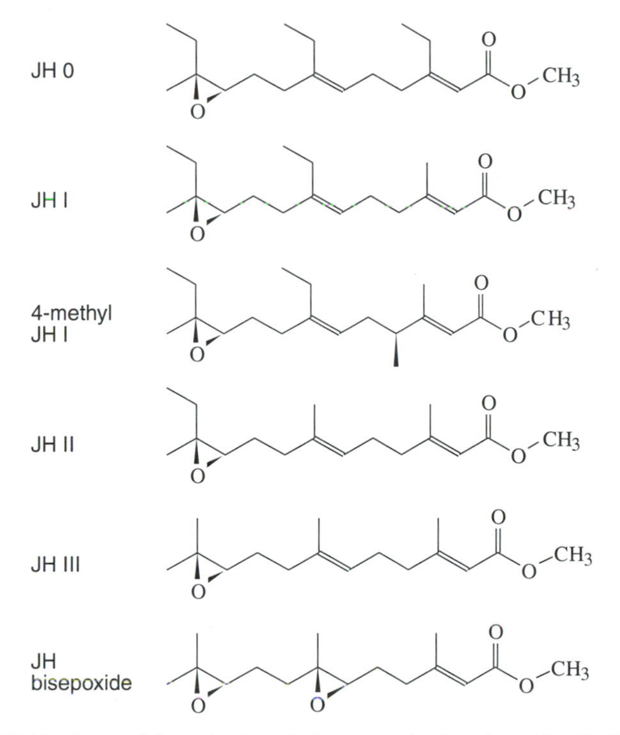

FIGURE 1.19 Some of the major juvenile hormones that have been identified in insects.

showed the hormone to be a sesquiterpenoid epoxide with an 18-carbon skeleton. After a second JH was subsequently identified from *H. cecropia* in smaller amounts, the two hormones were named **JH I** and **JH II.** As a lower homolog, the 17-carbon JH II contained a methyl group at carbon 7 instead of the ethyl group characteristic of JH I. A third homolog, **JH III,** was identified from the corpora allata of *M. sexta* cultured *in vitro*. This 17-carbon homolog contained three methyl groups and for many years was the only JH found in insect orders other than Lepidoptera. As the simplest of the juvenile hormones, it may represent the structure from which the others are derived. A fourth JH was isolated from the developing eggs of *M. sexta,* along with smaller quantities of JH I, and as the next higher 19-carbon homolog of JH I, it was named **JH 0** according to the convention that developed of naming higher homologs with lower hormone numbers. A fifth JH, another 19-carbon homolog, was identified in developing embryos of *M. sexta* as **4-methyl-JH I.**

A recently identified JH was recovered from cultured larval ring glands of *D. melanogaster*. This **JH III bisepoxide** contains a second epoxide group and has been found in several dipterans as well as in ticks. **JH acids** are produced by the

CA of *Manduca* larvae and may serve as a hormone or prohormone, because the imaginal discs may be capable of the acid methylation necessary to activate them. Several **hydroxy juvenile hormones** are produced by the CA of locusts and cockroaches (Fig. 1.20). This hydroxylation may result in molecules with greater biological activity, just as 20-hydroxyecdysone is more active than ecdysone.

We can only speculate about the possible reasons for so many different JH homologs. Advanced insects generally use the higher homologs of JH 0, I, and II, and in Lepidoptera and higher Diptera, multiple JH forms may be present in the same insect. There are differing biological activities among the homologs; JH I and II tend to be more active in morphogenesis than does JH III.

Because the JH found among the most primitive insect orders is generally JH III, the pathway to its synthesis is also believed to be the most primitive. JH III is synthesized from three molecules of acetate to form mevalonic acid that leads to the isoprenoids isopentenyl pyrophosphate (IPP) and dimethylallyl pyrophosphate (DMAP). Two molecules of IPP and one of DMAP are assembled to form farnesyl pyrophosphate, the pathway of steroid synthesis in vertebrates but one that insects cannot accomplish. In several steps, farnesyl pyrophosphate is ultimately converted into JH III. The formation of the side chains in the higher homologs JH I and II involve the combination of propionate with acetate to give rise to homoisopentyl pyrophosphate (HIPP) and ethylmethylallyl pyrophosphate (EMAP). The condensation of two IPP

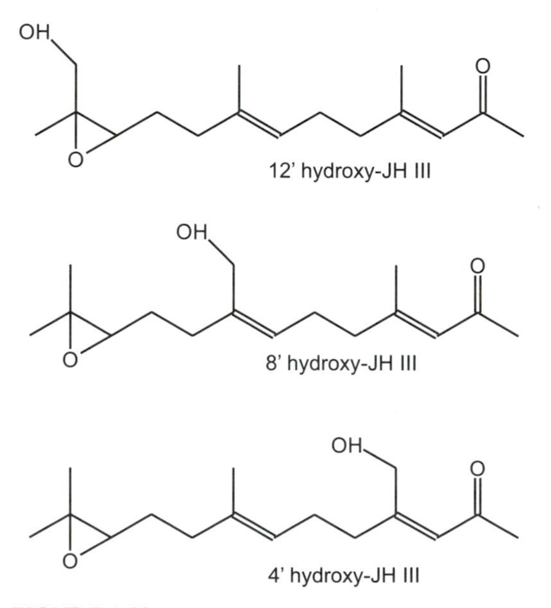

FIGURE 1.20 Examples of hydroxy juvenile hormones.

and one EMAP form JH II, and one EMAP, one HIPP, and one IPP form JH I. JH 0 originates from one EMAP and two HIPP (Fig. 1.21). More detailed steps in the synthesis of JH III are shown in Fig. 1.22. Higher JH homologs result from the incorporation of homomevalonate instead of mevalonate.

Common to all juvenile hormones are several functional groups that appear to be necessary for their morphogenetic activity. Pathways of biological degradation involve the loss or alteration of these groups. The inactive JH acid results from the loss of the methyl ester in the C-1 position. JH acid is produced at specific times by the CA of some insects and also may be a prohormone. The inactive JH diol is produced by the loss of the epoxide in the C-10,11 position. The three unsaturated positions also modulate biological activity; saturation of the 2-ene double bond drastically reduces activity, while the saturation of the 6-ene double bond can increase activity. Loss of the methyl branches along the chain reduces biological activity. An isoprenoid chain length of at least 10 $-CH_2-$ units appears to be necessary for biological activity, with 14–16 being optimal. An understanding of how modifications of the JH molecule affect activity is necessary to develop synthetic analogs that may be more effective in control. For example, the structures of some JH analogs that have been used in insect management programs are shown in Fig. 1.23.

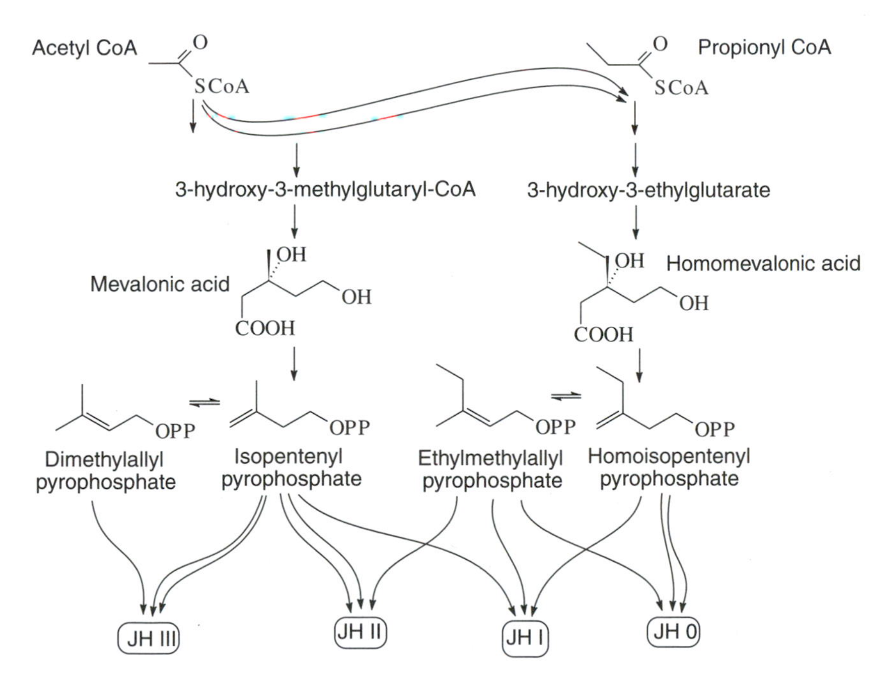

FIGURE 1.21 Initial steps in the synthesis of common juvenile hormones.

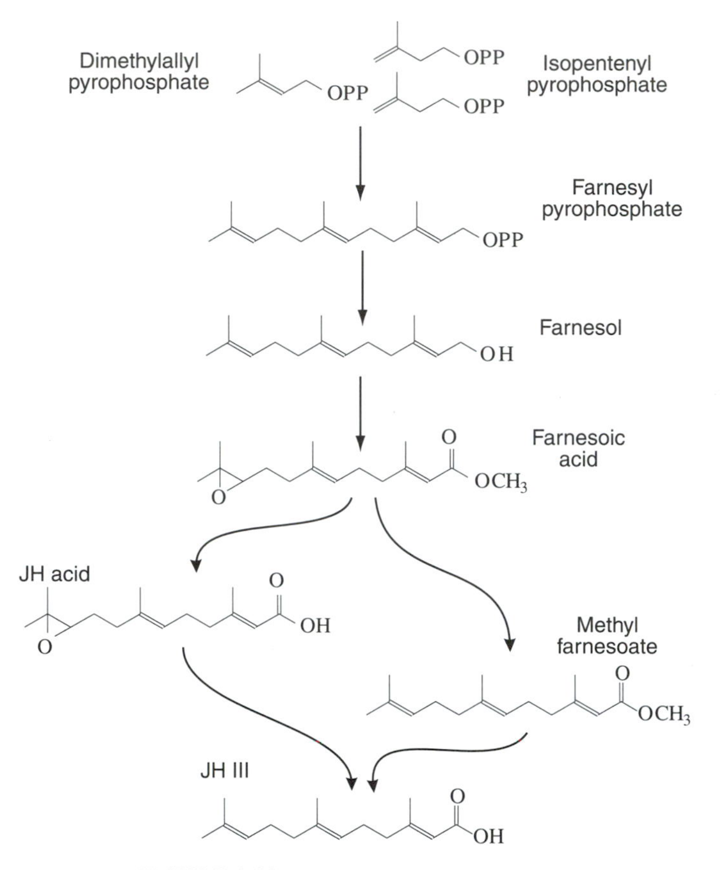

FIGURE 1.22 Final steps in the synthesis of JH III.

Sites of JH Synthesis and Control of JH Production

The corpus allatum (CA) is the major organ of JH synthesis and release, although other tissues such as the male accessory glands and imaginal discs are able to convert JH acids to JH but are unable to synthesize the basic carbon skeleton. The CA is ectodermal in origin and is usually located in the posterior regions of the head (Fig. 1.24). More primitive insect orders contain a paired CA located ventrally, but in more specialized insect orders, it has migrated to a more dorsal location. In the higher Diptera (Suborders Brachycera and Cyclorrapha), the pair is fused into one structure and is located dorsal to the aorta; in the Embioptera, Hemiptera, Dermaptera, and Pscoptera they are fused ventral to the aorta (Fig. 1.25). The CA can contain both intrinsic glandular cells as well as neurosecretory cells that have

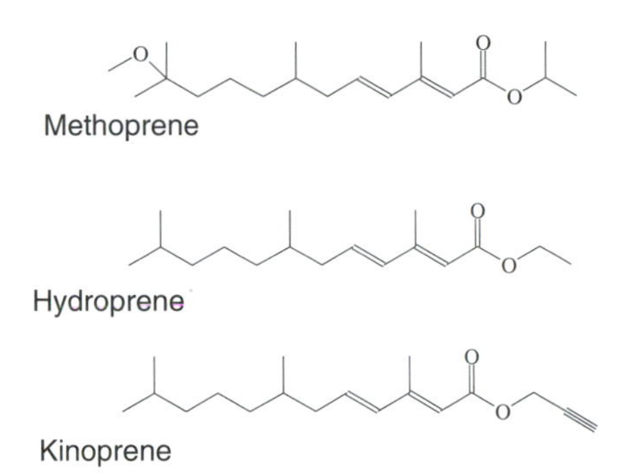

FIGURE 1.23 Examples of some JH analogs used in insect control.

their origins elsewhere. They are generally innervated from the brain by two pairs of nerves, the nervi corporis allati I (NCA I), which originate in the brain and pass through the corpus cardiacum (CC) on their way to the CA, and the NCA II, which originate in the subesophageal ganglion. Three other nerves, the nervi corporis cardiaci I, II, and III (NCC I, II, and III) originate in the brain and enter the CC and may also influence the CA (Fig. 1.26).

Given the central role of the CA in development and reproduction, sophisticated and precise mechanisms of activation and inhibition have evolved to assure

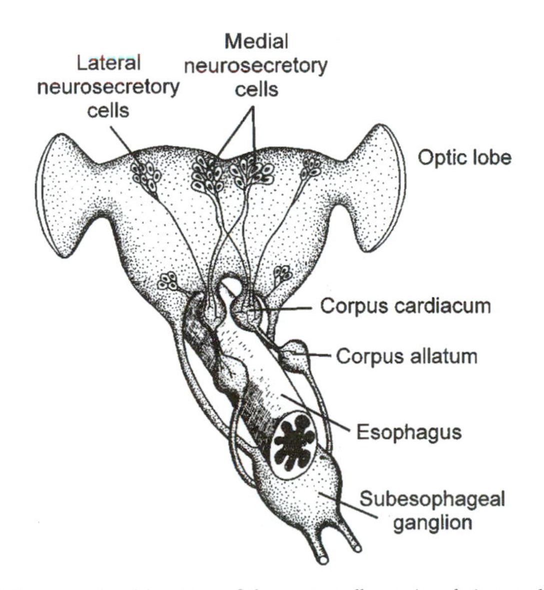

FIGURE 1.24 The generalized location of the corpus allatum in relation to the brain and other endocrine tissues. From Cymborowski (1992). Reprinted with permission.

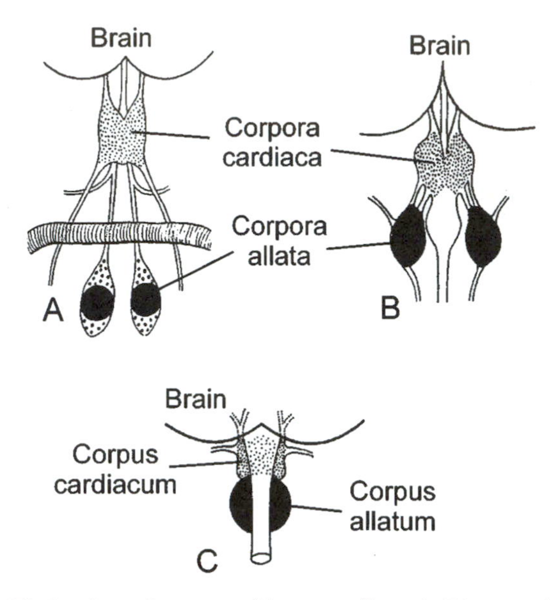

FIGURE 1.25 The location and structure of the corpus allatum in (A) a mosquito, (B) a cockroach, (C) a hemipteran. From Cymborowski (1992). Reprinted with permission.

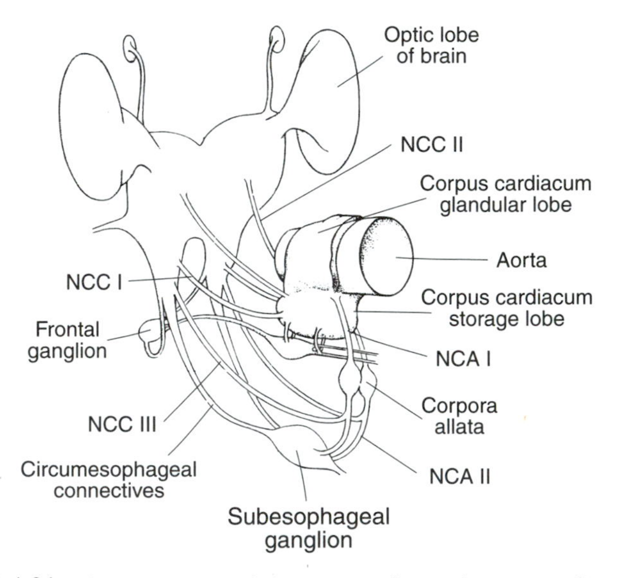

FIGURE 1.26 The innervations of the corpora allata and corpus cardiacum in the locust. Reprinted with permission from Veelaert, D., *et al.* (1998). *International Review of Cytology* **182:** 249–302. Copyright Academic Press.

both the timely production as well as the cessation and clearance of JH. Hemolymph JH titers are regulated by both biosynthesis and degradation. The most important factor determining JH titer is its synthesis by the CA. Because JH is not stored in the CA, its release is dependent upon synthesis, and this synthesis is rigidly controlled along several avenues. Environmental stimuli such as photoperiod, and endogenous factors including mating and nutritional state, are integrated by the brain and CA activity is then regulated both neurally and through the release of neurosecretory hormones, with each possibly having stimulatory or inhibitory effects. For example, in the last larval instar of most insects, JH production ceases either as a result of the absence of neural stimulation, or the presence of neural inhibition from the brain. This neural control is generally affected by the NCA I; although the NCA II innervates the CA from the subesophageal ganglion, these connections do not appear to be as important.

Neurohormones have also been identified that affect CA activity. These were initially suspected because of the stimulatory effect of brain extracts on JH production *in vitro*. These substances, called **allatotropins,** stimulate JH production. They may originate in the neurosecretory cells of the brain, but have also been found in the frontal ganglion and the terminal abdominal ganglion. The one allatotropin that has been sequenced comes from *M. sexta*. It is a tridecapeptide whose gene encodes three prohormones. Allatotropins have been associated with positive CA regulation in both larvae and adults.

In contrast, other neuropeptides produced by the brain are **allatostatins** that inhibit JH synthesis by the CA. They were originally isolated from the brain of the cockroach, *Diploptera punctata,* but more than 50 related allatostatins have since been isolated from moths, flies, and other cockroaches. The cockroach allatostatins share a remarkable sequence similarity, but an allatostatin identified in the moth, *M. sexta,* has very little similarity and has no effect on the cockroach CA. Peptides similar to the cockroach allatostatins have been isolated from dipterans, but these do not appear to have any allatostatic effects in those species. Instead, they may have other biological functions including the modulation of muscle contraction and gut motility and the inhibition of vitellogenin release from the fat body. It may be that the peptides were originally myomodulatory and assumed a secondary role of CA regulation in the cockroaches and their relatives. The true allatostatins are either delivered directly to the CA by axons of the lateral neurosecretory cells of the brain, or through the hemolymph. An **allatoinhibin** has been identified from the brain of *Manduca* that inhibits the CA nonreversibly, unlike the reversible action of allatostatins. Little is known about this factor and it has not been reported to occur in any other species.

In addition to substances from the brain, the activity of the CA in female cockroaches is affected by a humoral factor from the ovaries. The removal of the ovaries reduces JH synthesis and their reimplantation restores it. Ecdysteroid receptors have been identified in the CA of *Manduca,* suggesting that JH production may also be regulated by ecdysteroids. The physiological state of the CA itself may also determine its sensitivity to these various neuropeptides. The way that the CA responds to allatostatins may change during the reproductive cycle of the female. Finally, high levels of JH itself may suppress CA activity through feedback regulation.

The overall regulation of the CA, integrating the effects of neural and humoral pathways, is shown in Fig, 1.27. The capacity of the CA to produce JH is determined by stimulatory and inhibitory signals that arrive through both the hemolymph and the nervous system. The brain can stimulate and inhibit the CA by the nerves that innervate it and produce neuropeptides that function in a similar manner. In the cockroach, allatostatins are delivered to the CA by lateral neurosecretory cells that leave the brain via the NCCII. Thus we see that like the ecdysteroids, an important hormone like JH has a very precise mechanism of control to assure that its proper physiological concentrations are maintained when required at times that are biologically appropriate.

Postproduction Regulation of JH Activity

Once secreted by the CA, JH is transported to target tissues through the hemolymph. Because of its lipophilic nature, JH must be bound to other molecules in order to move through the aqueous hemolymph. Perhaps more importantly, binding can also protect the hormone from degradation by nonspecific tissue-bound esterases. There are a variety of such molecules that bind JH for transport and protection, collectively termed the **juvenile hormone binding proteins** (JHBP). Most of the circulating JH is bound to these JHBPs, leaving

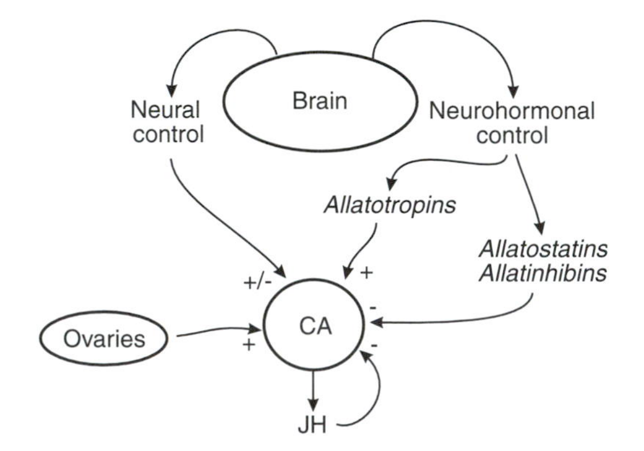

FIGURE 1.27 A summary of the ways that the corpus allatum can be regulated. The brain can stimulate or inhibit juvenile hormone synthesis by nervous or neurohormonal signals. The allatotropins stimulate JH synthesis; the allatostatins and allatinhibins inhibit it. There is some evidence for feedback regulation by circulating JH, and also stimulation by the ovaries.

little JH in free circulation. They serve not only to move the lipophilic JH throughout the hemolymph but also may be storage sites for hemolymph JH.

Three types of these binding proteins have been described in insects. The low molecular weight binding proteins consist of a single polypeptide that contains one JH binding site. A 32-kDa JHBP is produced during the larval, pupal, and adult stages of *M. sexta*. The high molecular weight binding proteins are also known as **lipophorins** that shuttle lipids between tissues within the insect. Lipophorin is composed of two apoproteins of 250 and 80 kDa and has multiple JH binding sites. The third type of binding protein is a 566 kDa hexameric protein with six JH binding sites.

The onset of metamorphosis following the last larval instar requires the clearance of JH from the hemolymph in order for epidermal cells to switch their commitment, in the presence of ecdysteroid, to the developmental pathways that lead to the production of pupal and adult cuticles. Although JH titers are reduced when the CA ceases its synthesis, the reduction of circulating JH titers is accomplished by degrading the JH that has already been produced. There are two major routes by which this degradation of JH occurs (Fig. 1.28). Specific JH esterases

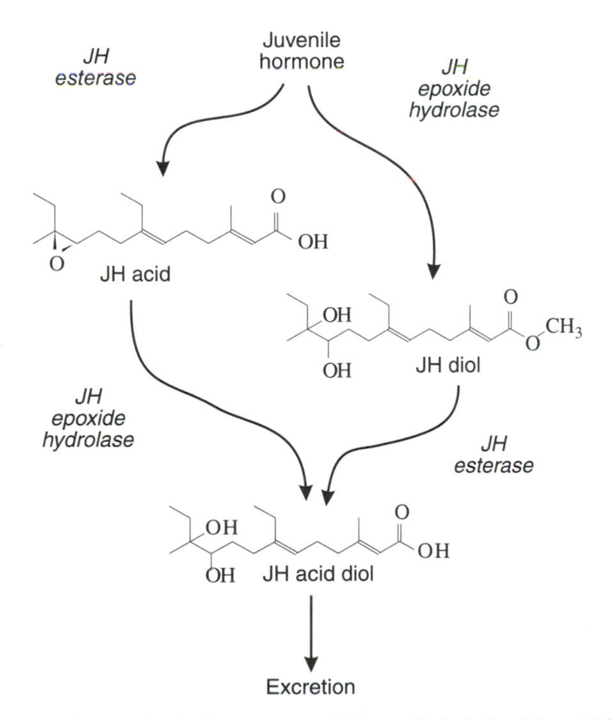

FIGURE 1.28 Degradation of JH by JH esterase and JH epoxide hydrolase. The acid diol is formed and excreted either by the initial formation of the JH acid or of the JH diol.

(JHE) recognize the JH–JHBP complex and hydrolyze the methyl ester of JH to produce the JH acid. When interacting with the JH–JHBP, JHE may institute a conformational change in the JHBP, causing the JH to be released from its binding pocket. JH epoxide hydrolase hydrates the epoxide to produce the JH diol, but its ability to degrade JH is reduced when JH is bound to the JHBP and it may be more effective against unbound JH or JH acid. High levels of JHE are correlated with low JH titers, clearing the hemolymph of any residual JH activity once the CA stops production (Fig. 1.29).

Mode of Action of JH

The major role of JH in insects is to modify the action of ecdysteroids and prevent the switch in the commitment of epidermal cells. In the presence of ecdysteroids, JH preserves the current program of gene expression. For example, the pattern of mRNAs produced by larval epidermis that is exposed to ecdysteroid will change if JH is not present. JH influences the stage-specific expression of the genome that is initiated by ecdysteroids. The search for the specific nuclear JH receptor that binds the hormone has been elusive, although there is strong evidence that in *Drosophila,* JH III and JH III acid bind to USP, one of the components of the heterodimer that also binds ecdysteroids. The hormone and its acid each induces a different conformational state when bound to USP, with the heterodimer then inducing transcription of specific genes. The relationship of both

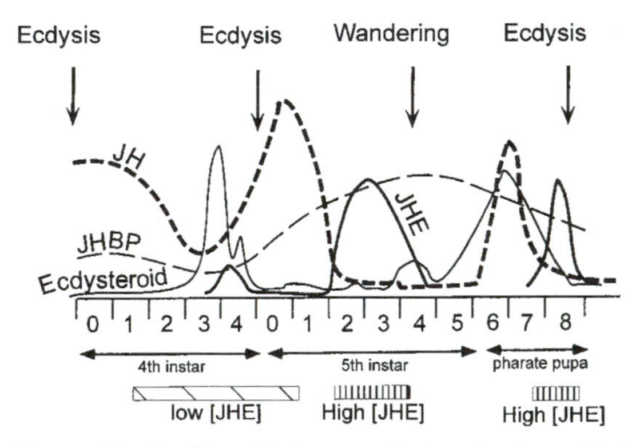

FIGURE 1.29 Proposed levels of juvenile hormone (top) and ecdysteroids (bottom) during the fourth and fifth instars of Manduca sexta. JH levels decline with increasing concentrations of JH esterase (JHE). The JH binding protein (JHBP) protects the bound JH from nonspecific hydrolytic enzymes, but not from degradation by JHE. Reprinted from *Insect Biochemistry and Molecular Biology,* Volume 25. Touhara, K., B.C. Bonning, B.D. Hammock, G.D. Prestwich. Action of juvenile hormone (JH) esterase on the JH-JH binding protein complex. An *in vitro* model of JH metabolism in a caterpillar, pp. 727–734. Copyright 1995, with permission from Elsevier Science.

JH and ecdysteroid binding to different components of the heterodimer explains how JH might modulate the effect of the cellular response to ecdysteroids.

OTHER NEUROPEPTIDES FOUND IN INSECTS

The existence of ecdysteroids, juvenile hormones, and PTTH have been known since the beginnings of insect physiology and are considered to be major hormones that regulate development and reproduction in insects, but there are many other more recently discovered hormones that are equally as important. **Adipokinetic hormone** (**AKH**) is both synthesized and released from intrinsic neurosecretory cells of the corpus cardiacum. AKH is actually a family of related peptides with varied functions in insects. The most important and well-studied function of AKH is in the regulation of lipid recruitment for flight. AKH is released in response to stimuli from prolonged flight and acts on the fat body and flight muscles to cause the mobilization and utilization of diglycerides once hemolymph carbohydrates are exhausted (see Chapter 6, Metabolic Systems). **Diuretic neuropeptides** are involved in the maintenance of water balance and as clearance factors that increase the filtration rate of the hemolymph. Diuretic peptides generally act by increasing the activity of the Malpighian tubules and inhibiting the resorption of water from the hindgut. Most diuretic neurohormones have a varying degree of homology with the corticotropin-releasing factor of vertebrates, but in the blood-sucking hemipteran, *R. prolixus*, the biogenic amine **serotonin** serves as a diuretic hormone (see Chapter 8, Excretory Systems). **Proctolin** is a small pentapeptide that causes contractions of the longitudinal muscles of the proctodeum of the hindgut. It appears to be a neuromodulator that works together with the neurotransmitter glutamate at the neuromuscular junction. Proctolin, along with several members of the AKH family, is a cardioacceleratory hormone that also affects the rate of the heartbeat. A large number of other myotropic peptides, such as the **leucokinins** and **acheatakinins,** have been identified in insects. **Pheromone-biosynthesis-activating neuropeptides** (**PBANs**) are synthesized in female lepidopterans that produce sex pheromones to attract males. PBAN regulates the enzymes that lead to pheromone biosynthesis. **Eclosion hormone** and **crustacean cardioactive peptide** release the behaviors programmed into the CNS that express the stereotypical sequence of peristalsis and other movements associated with eclosion. They are activated by an **ecdysis-triggering hormone** produced by **epitracheal glands** that consist of gland cells that are attached to a trachea near a spiracle (see Chapter 2, Integumentary Systems). **Bursicon** is a neurosecretory hormone that mediates the sclerotization of the cuticle immediately after it is synthesized after the molt. It may also be involved in the wound repair of the epidermis (see Chapter 2, Integumentary Systems).

VERTEBRATE-TYPE HORMONES IN INSECTS

Insects were once believed to be devoid of any hormones and even lack brains, and it is still surprising to learn that so many life processes in vertebrates can also be found in insects. The operation of such essential systems as protein synthesis, muscle contraction, and cell metabolism do not differ significantly between insects and us. An indication of the unity that exists between insects and vertebrates is best described by the example of the female rabbit flea that depends upon the hormones circulating in the blood of its pregnant female vertebrate host in order to reproduce. It should therefore not come as a surprise that many hormones previously isolated from vertebrates are also present in insects, although their parallel functions have yet to be fully determined and important differences in amino acid structure question their true homologies with vertebrate peptides.

Insulin controls the rate of glucose transport across cell membranes and has been identified from a number of insects. **Bombyxin,** the small PTTH, shares a significant sequence homology with the A chain of vertebrate insulin. In *B. mori*, the melanization-reddish coloration hormone shares some sequence homologies with insulin-like growth factor II. **Gastrin** and **cholecystokinin** (**CCK**) are related peptides that respectively mediate an increase in the secretion of acid in the vertebrate stomach and cause the gall bladder to contract. In insects, the **sulfakinins** show structural and functional similarities to gastrin and CCK. **Somatostatin** is a master hormone in vertebrates that controls the release of many other hormones. Somatostatin-like peptides have been found in such diverse insects as crickets, hoverflies, and locusts. Adipokinetic hormone regulates lipid mobilization from the fat body in insects and has homologies with vertebrate **glucagon.** Insect **tachykinins** stimulate visceral muscles and are structurally homologous with vertebrate tachykinins that are involved in processes as diverse as salt balance, sensory processing, and gut motility. FMRFamide-related pepides are found in many vertebrates and invertebrates. They have been isolated from insects where they stimulate muscle contraction and frequency of heartbeat, and regulate behavior in female mosquitoes. **Melatonin** is produced by the vertebrate pineal gland during the scotophase and causes drowsiness in humans. Its occurrence in the compound eyes of locusts and in a variety of other insects suggests it may also be involved in photoperiodism in invertebrates.

ADDITIONAL REFERENCES

General Insect Endocrinology

Ashburner, M. 1990. Puffs, genes, and hormones revisited. *Cell* **61:** 1–3.

Ashburner, M., Chihara, C. C., Meltzer, P. P., Richards, G. 1974. Temporal control of puffing activity in polytene chromosomes. *Cold Spring Harbor Symp.* **38:** 655–662.

Ashburner, M., Richards, G. 1976. Sequential gene activation by ecdysone in polytene chromosomes of *Drosophila melanogaster*. III. consequences of ecdysone withdrawal. *Dev. Biol.* **54:** 241–255.

Davies, S. 2000. Nitric oxide signalling in insects. *Insect Biochem. Mol. Biol.* **30:** 1123–1138.

Gade, G., Hoffmann, K. H., Spring, J. H. 1997. Hormonal regulation in insects: Facts, gaps, and future directions. *Physiol. Rev.* **77:** 963–1032.

Hall, B. 1999. Nuclear receptors and the hormonal regulation of *Drosophila* metamorphosis. *Am. Zool.* **39:** 714–721.

Henrich, V. C., Brown, N. E. 1995. Insect nuclear receptors: A developmental and comparative perspective. *Insect Biochem. Mol. Biol.* **25:** 881–897.

Henrich, V. C., Rybczynski, R., Gilbert, L. I. 1999. Peptide hormones, steroid hormones, and puffs: Mechanisms and models in insect development. *Vitam. Horm.* **55:** 73–125.

Hiruma, K., Shinoda, T., Malone, F., Riddiford, L. M. 1999. Juvenile hormone modulates 20-hydroxyecdysone-inducible ecdysone receptor and ultraspiracle gene expression in the tobacco hornworm, *Manduca sexta. Dev. Genes Evol.* **209:** 18–30.

Hodin, J., Riddiford, L. M. 1998. The ecdysone receptor and ultraspiracle regulate the timing and progression of ovarian morphogenesis during *Drosophila* metamorphosis. *Dev. Genes Evol.* **208:** 304–317.

Lafont, R. 2000. Understanding insect endocrine systems: Molecular approaches. *Entomol. Exp. Appl.* **97:** 123–136.

Linder, M. E., Gilman, A. G. 1992. G proteins. *Sci. Am.* **267:** 59–61, 64–65.

Lummis, S. C. R., Galione, A., Taylor, C. W. 1990. Transmembrane signaling in insects. *Annu. Rev. Entomol.* **35:** 345–377.

Riddiford, L. M., Hiruma, K., Lan, Q., Zhou, B. 1999. Regulation and role of nuclear receptors during larval molting and metamorphosis of Lepidoptera. *Am. Zool.* **39:** 736–746.

Roeder, T. 1994. Biogenic amines and their receptors in insects. *Comp. Biochem. Physiol. C* **107:** 1–12.

Sehnal, F., Zitnan, D. 1990. Endocrines of insect gut. *Prog. Clin. Biol. Res.* **342:** 510–515.

Siegmund, T., Korge, G. 2001. Innervation of the ring gland of *Drosophila melanogaster. J. Comp. Neurol.* **431:** 481–491.

Stowers, R. S., Garza, D., Rascle, A., Hogness, D. S. 2000. The L63 gene is necessary for the ecdysone-induced 63E late puff and encodes CDK proteins required for *Drosophila* development. *Dev. Biol.* **221:** 23–40.

Truman, J. W. 1996. Steroid receptors and nervous system metamorphosis in insects. *Dev. Neurosci.* **18:** 87–101.

Vroemen, S. F., Van der Horst, D. J., Van Marrewijk, W. J. A. 1998. New insights into adipokinetic hormone signaling. *Mol. Cell. Endocrinol.* **141:** 7–12.

Wicher, D. 2001. Peptidergic modulation of an insect Na^+ current: Role of protein kinase A and protein kinase C. *J. Neurophysiol.* **85:** 374–383.

Zitnan, D., Kingan, T. G., Beckage, N. E. 1995. Parasitism-induced accumulation of FMRFamide-like peptides in the gut innervation and endocrine cells of *Manduca sexta. Insect Biochem. Mol. Biol.* **25:** 669–678.

Classic Experiments in Insect Endocrinology

Butenandt, A., Karlson, P. 1954. Über die isolierung eines Metamorhose–Hormone der Insekten in kristallierten Form. *Z. Naturforsch.* **9b:** 389–391.

Fraenkel, G. 1934. Pupation of flies initiated by a hormone. *Nature* **133:** 834.

Fukuda, S. 1940. Induction of pupation in silkworm by transplanting the prothoracic gland. *Proc. Imp. Acad. Tokyo* **16:** 414–416.

Hachlow, V. 1931. Entwicklungsmechanik der Schmetterlinge. *Wilhelm Roux Arch. Entwicklungsmech.* **125:** 26–49.

Kopeć, S. 1917. Experiments on the metamorphosis of insects. *Bull. Int. Acad. Cracovie B.* 57–60.

Wigglesworth, V. B. 1934. The physiology of ecdysis in *Rhodnius prolixus* (Hemiptera). II. Factors controlling moulting and metamorphosis. *Q. J. Microsc. Sci.* **77:** 191–223.

Wigglesworth, V. B. 1936. The function of the corpus allatum in the growth and reproduction of *Rhodnius prolixus*. *Q. J. Microsc. Sci.* **79:** 91–123.

Wigglesworth, V. B. 1940. The determination of characters at metamorphosis in *Rhodnius prolixus* (Hemiptera). *J. Exp. Biol.* **17:** 201–223.

Wigglesworth, V. B. 1948. Functions of the corpus allatum in *Rhodnius prolixus* (Hemiptera). *J. Exp. Biol.* **25:** 1–14.

Wigglesworth, V. B. 1985. Historical perspectives. In *Comprehensive insect physiology, biochemistry and pharmacology* (G. A. Kerkut, L. I. Gilbert, Eds.), vol. 7, pp. 1–24. Pergamon Press, Oxford.

Williams, C. M. 1947. Physiology of insect diapause. II. Interaction between the pupal brain and prothoracic glands in the metamorphosis of the giant silkworm *Platysamia cecropia*. *Biol. Bull.* **92:** 89–180.

Williams, C. M. 1952. Physiology of insect diapause. IV. The brain and prothoracic glands as an endocrine system in the *cecropia* silkworm. *Biol. Bull.* **103:** 120–138.

Williams, C. M. 1956. The juvenile hormone of insects. *Nature* **178:** 212–213.

Williams, C. M., 1948. Physiology of insect diapause. III. The prothoracic glands in the *cecropia* silkworm, with special reference to their significance in embryonic and postembryonic development. *Biol. Bull.* **94:** 60–65.

Prothoracicotropic Hormone

Aizono, Y., Endo, Y., Sattelle, D. B., Shirai, Y. 1997. Prothoracicotropic-hormone producing neurosecretory cells in the silkworm, *Bombyx mori*, express a muscarinic acetylcholine receptor. *Brain Res.* **763:** 131–136.

Bollenbacher, W. E., Gray, R. S., Muehleisen, D. P., Regan, S. A., Westbrook, A. L. 1993. The biology of the prothoracicotropic hormone peptidergic neurons in an insect. *Am. Zool.* **33:** 316–323.

Bollenbacher, W. E., Agui, N., Granger, N. A., Gilbert, L. I. 1979. *In vitro* activation of insect prothoracic glands by the prothoracicotropic hormone. *Proc. Natl. Acad. Sci. USA* **76:** 5148–5152.

Dai, J. D., Gilbert, L. I. 1997. Programmed cell death of the prothoracic glands of *Manduca sexta* during pupal–adult metamorphosis. *Insect Biochem. Mol. Biol.* **27:** 69–78.

Fullbright, G., Lacy, E. R., Bullesbach, E. E. 1997. The prothoracicotropic hormone bombyxin has specific receptors on insect ovarian cells. *Eur. J. Biochem.* **245:** 774–780.

Gilbert, L. I., Rybczynski, R., Song, Q., Mizoguchi, A., Morreale, R., Smith, W. A., Matubayashi, H., Shionoya, M., Nagata, S., Kataoka, H. 2000. Dynamic regulation of prothoracic gland ecdysteroidogenesis: *Manduca sexta* recombinant prothoracicotropic hormone and brain extracts have identical effects. *Insect Biochem. Mol. Biol.* **30:** 1079–1089.

Ishibashi, J., Kataoka, H., Isogai, A., Kawakami, A., Saegusa, H., Yagi, Y., Mizoguchi, A., Ishizaki, H., Suzuki, A. 1994. Assignment of disulfide bond location in prothoracicotropic hormone of the silkworm, *Bombyx mori*: A homodimeric peptide. *Biochemistry* **33:** 5912–5919.

Kim, A. J., Cha, G. H., Kim, K., Gilbert, L. I., Lee, C. C. 1997. Purification and characterization of the prothoracicotropic hormone of *Drosophila melanogaster*. *Proc. Natl. Acad. Sci. USA* **94:** 1130–1135.

Lonard, D. M., Bhaskaran, G., Dahm, K. H. 1996. Control of prothoracic gland activity by juvenile hormone in fourth instar *Manduca sexta* larvae. *J. Insect Physiol.* **42:** 205–213.

Mizoguchi, A., Ohashi, Y., Hosoda, K., Ishibashi, J., Kataoka, H. 2001. Developmental profile of the changes in the prothoracicotropic hormone titer in hemolymph of the silkworm *Bombyx mori*: Correlation with ecdysteroid secretion. *Insect Biochem. Mol. Biol.* **31:** 349–358.

Noguti, T., Adachi-Yamada, T., Katagiri, T., Kawakami, A., Iwami, M., Ishibashi, J., Kataoka, H., Suzuki, A., Go, M., Ishizaki, H. 1995. Insect prothoracicotropic hormone: A new member of the vertebrate growth factor superfamily. *FEBS Lett.* **376:** 251–256.

Smith, W. A. 1995. Regulation and consequences of cellular changes in the prothoracic glands of *Manduca sexta* during the last larval instar: A review. *Arch. Insect Biochem. Physiol.* **30:** 271–293.

Vafopoulou, X., Steel, C. G. H. 1993. Release *in vitro* of prothoracicotropic hormone from the brain of male *Rhodnius prolixus* during larval-adult development: Identification of novel and predicted release times. *J. Insect Physiol.* **39:** 65–71.

Vafopoulou, X., Steel, C. G. H. 1996. The insect neuropeptide prothoracicotropic hormone is released with a daily rhythm: Re-evaluation of its role in development. *Proc. Natl. Acad. Sci. USA* **93:** 3368–3372.

Zachary, D., Goltzene, F., Holder, F. C., Berchtold, J. P., Nagasawa, H., Suzuki, C., Misoguchi, H., Ishizaki, H., Hoffmann, J. A. 1988. Presence of bombyxin (4K-PTTH)-like molecules in neurosecretory granules of brain-corpora cardiaca complexes of *Locusta migratoria*. Developmental aspects. *Int. J. Invertebr. Reprod. Dev.* **14:** 1–10.

Ecdysteroids

Adams, T. S., Li, Q. J. 1998. Ecdysteroidostatin from the house fly, *Musca domestica*. *Arch. Insect. Biochem. Physiol.* **38:** 166–176.

Adler, J. H., Grebenok, R. J. 1995. Biosynthesis and distribution of insect-molting hormones in plants—A review. *Lipids* **30:** 257–262.

Baehrecke, E. H. 1996. Ecdysone signaling cascade and regulation of *Drosophila* metamorphosis. *Arch. Insect Biochem. Physiol.* **33:** 231–244.

Cherbas, P., Cherbas, L. 1996. Molecular aspects of ecdysteroid hormone action. In *Metamorphosis: Postembryonic reprogramming of gene expression in amphibian and insect cells* (L. I. Gilbert, J. R. Tata, B. G. Atkinson, Eds.), pp. 175–221. Academic Press, San Diego, CA.

Dai, J. -D., Gilbert, L. I. 1991. Metamorphosis of the corpus allatum and degeneration of the prothoracic glands during the larval–pupal–adult transformation of *Drosophila melanogaster*. A cytophysiological analysis of the ring gland. *Dev. Biol.* **144:** 309–326.

Delbecque, J. -P., Weidner, K., Hoffmann, K. H. 1990. Alternative sites for ecdysteroid production in insects. *Invertebr. Reprod. Dev.* **18:** 29–42.

Fletcher, J. C., Burtis, K. C., Hogness, D. S., Thummel, C. S. 1995. The *Drosophila* E74 gene is required for metamorphosis and plays a role in the polytene chromosome puffing response to ecdysone. *Development* **121:** 1455–1465.

Hiruma, K., Bocking, D., Lafont, R., Riddiford, L. M. 1997. Action of different ecdysteroids on the regulation of mRNAs for the ecdysone receptor, MHR3, dopa decarboxylase, and a larval cuticle protein in the larval epidermis of the tobacco hornworm, *Manduca sexta*. *Gen. Comp. Endocrinol.* **107:** 84–97.

Horn, D. H. S., Bergamasco, R. 1985. Chemistry of ecdysteroids. In *Comprehensive insect physiology, biochemistry and pharmacology*, vol. 7 (G. A. Kerkut, L. I. Gilbert, Eds.), pp. 186–248. Pergamon Press, Oxford.

Hua, X. -J., Jiang, R. -J., Koolman, J. 1997. Multiple control of ecdysone biosynthesis in blowfly larvae: Interaction of ecdysiotropins and ecdysiostatins. *Arch. Insect Biochem. Physiol.* **35:** 125–134.

Hua, Y. J., Bylemans, D., De Loof, A., Koolman, J. 1994. Inhibition of ecdysone biosynthesis in flies by a hexapeptide isolated from vitellogenic ovaries. *Mol. Cell. Endocrinol.* **104:** R1–R4.

Huet, F., Ruiz, C., Richards, G. 1995. Sequential gene activation by ecdysone in *Drosophila melanogaster*: The hierarchical equivalence of early and early late genes. *Development* **121:** 1195–1204.

Kamimura, M., Tomita, S., Kiuchi, M., Fujiwara, H. 1997. Tissue-specific and stage-specific expression of two silkworm ecdysone receptor isoforms—Ecdysteroid-dependent transcription in cultured anterior silk glands. *Eur. J. Biochem.* **248:** 786–793.

Loeb, M. J., De Loof, A., Schoofs, L., Isaac, E. 1998. Angiotensin II and angiotensin-converting enzyme as candidate compounds modulating the effects of testis ecdysiotropin in testes of the gypsy moth, *Lymantria dispar. Gen. Comp. Endocrinol.* **112:** 232–239.

Rees, H. H. 1985. Biosynthesis of ecdysone. In *Comprehensive insect physiology, biochemistry and pharmacology*, (G. A. Kerkut, L. I. Gilbert, Eds.), vol. 7, pp. 249–293. Pergamon Press, Oxford.

Richter, K., Bohm, G. A. 1997. The molting gland of the cockroach *Periplaneta americana*: Secretory activity and its regulation. *Gen. Pharmacol.* **29:** 17–21.

Riehle, M. A., Brown, M. R. 1999. Insulin stimulates ecdysteroid production through a conserved signaling cascade in the mosquito, *Aedes aegypti. Insect Biochem. Mol. Biol.* **29:** 855–860.

Russell, S., Ashburner, M. 1996. Ecdysone-regulated chromosome puffing in *Drosophila melanogaster*. In *Metamorphosis. Postembryonic reprogramming of gene expression in amphibian and insect cells* (L. I. Gilbert, J. R. Tata, B. G. Atkinson, Eds.) Academic Press, San Diego, CA.

Schubiger, M., Wade, A. A., Carney, G. E., Truman, J. W., Bender, M. 1998. *Drosophila* EcR-B ecdysone receptor isoforms are required for larval molting and for neuron remodeling during metamorphosis. *Development* **125:** 2053–2062.

Song, Q., Gilbert, L. I. 1998. Alterations in ultraspiracle (USP) content and phosphorylation state accompany feedback regulation of ecdysone synthesis in the insect prothoracic gland. *Insect Biochem. Mol. Biol.* **28:** 849–860.

Wang, S. F., Li, C., Zhu, J., Miura, K., Miksicek, R. J., Raikhel, A. S. 2000. Differential expression and regulation by 20-hydroxyecdysone of mosquito ultraspiracle isoforms. *Dev. Biol.* **218:** 99–113.

Wang, S. F., Miura, K., Miksicek, R. J., Segraves, W. A., Raikhel, A. S. 1998. DNA binding and transactivation characteristics of the mosquito ecdysone receptor–Ultraspiracle complex. *J. Biol. Chem.* **273:** 27,531–27,540.

White, K. P., Hurban, P., Watanabe, T., Hogness, D. S. 1997. Coordination of *Drosophila* metamorphosis by two ecdysone-induced receptors. *Science* **276:** 114–117.

Wing, K. D., Slawecki, R., Carlson, G. R. 1988. RH 5849, a nonsteroidal ecdysone agonist: Effects on larval Lepidoptera. *Science* **241:** 470–472.

Wing , K. D., Sparks, T. C., Lovell, V. M., Levinson, S. O., Hammock, B. D. 1981. The distribution of juvenile hormone esterase and its interrelationship with other proteins influencing juvenile hormone metabolism in the cabbage looper *Trichoplusia ni. Insect Biochem.* **11:** 473–485.

Woodard, C. T., Baehrecke, E. H., Thummel, C. S. 1994. A molecular mechanism for the stage specificity of the *Drosophila* prepupal genetic response to ecdysone. *Cell* **79:** 607–615.

Juvenile Hormones

Audsley, N., Weaver, R. J., Edwards, J. P. 2000. Juvenile hormone biosynthesis by corpora allata of larval tomato moth, *Lacanobia oleracea*, and regulation by *Manduca sexta* allatostatin and allatotropin. *Insect Biochem. Mol. Biol.* **30:** 681–689.

Baker, F. C. 1990. Techniques for identification and quantification of juvenile hormones and related compounds in arthropods. In *Morphogenetic hormones of arthropods: Discovery, synthesis, metabolism, evolution, mode of action, and techniques* (A. P. Gupta, Ed.) pp. 389–453. Rutgers Univ. Press, New Brunswick.

Bendena, W. G., Donly, B. C., Tobe, S. S. 1999. Allatostatins: A growing family of neuropeptides with structural and functional diversity. *Ann. NY Acad. Sci.* **897:** 311–329.

Bhaskaran, G., Dahm, K. H., Barrera, P., Pacheco, J. L., Peck, K. E., Muszynska-Pytel, M. 1990. Allatinhibin, a neurohormonal inhibitor of juvenile hormone biosynthesis in *Manduca sexta. Gen. Comp. Endocrinol.* **78:** 123–136.

Bhatt, T. R., Horodyski, F. M. 1999. Expression of the *Manduca sexta* allatotropin gene in cells of the central and enteric nervous systems. *J. Comp. Neurol.* **403:** 407–420.

Bowser, P. R., Tobe, S. S. 2000. Partial characterization of a putative allatostatin receptor in the midgut of the cockroach *Diploptera punctata. Gen. Comp. Endocrinol.* **119:** 1–10.

Cassier, P. 1998. The corpora allata. In *Microscopic anatomy of invertebrates* (F.W. Harrison, M. Locke, Eds.), vol. 11C, pp. 1041–1058. Wiley-Liss, New York.

Darrouzet, E., Mauchamp, B., Prestwich, G. D., Kerhoas, L., Ujvary, I., Couillaud, F. 1997. Hydroxy juvenile hormones: New putative juvenile hormones biosynthesized by locust corpora allata *in vitro. Biochem. Biophys. Res. Commun.* **240:** 752–758.

Davey, K. G. 2000. The modes of action of juvenile hormones: Some questions we ought to ask. *Insect Biochem. Mol. Biol.* **30:** 663–669.

de Kort, C. A. D., Koopmanschap, A. B., Ermens, A. A. M. 1984. A new class of juvenile hormone binding proteins in insect hemolymph. *Insect Biochem.* **14:** 619–623.

de Kort, C. A. D., Granger, N. A. 1996. Regulation of JH titers: The relevance of degradative enzymes and binding proteins. *Arch. Insect Biochem. Physiol.* **33:** 1–26.

Duve, H., Thorpe, A., Yagi, K. J., Yu, C. G., Tobe, S. S. 1992. Factors affecting the biosynthesis and release of juvenile hormone bisepoxide in the adult blowfly *Calliphora vomitoria. J. Insect Physiol.* **38:** 575–585.

Gilbert, L. I., Granger, N. A., Roe, R. M. 2000. The juvenile hormones: Historical facts and speculations on future research directions. *Insect Biochem. Mol. Biol.* **30:** 617–644.

Granger, N. A., Ebersohl, R., Sparks, T. C. 2000. Pharmacological characterization of dopamine receptors in the corpus allatum of *Manduca sexta* larvae. *Insect Biochem. Mol. Biol.* **30:** 755–766.

Grieneisen, M. L., Mok, A., Kieckbusch, T. D., Schooley, D. A. 1997. The specificity of juvenile hormone esterase revisited. *Insect Biochem. Mol. Biol.* **27:** 365–376.

Gu, S. H., Chow, Y. S., Yin, C. M. 1997. Involvement of juvenile hormone in regulation of prothoracicotropic hormone transduction during the early last larval instar of *Bombyx mori. Mol. Cell. Endocrinol.* **127:** 109–116.

Hammock, B. D. 1985. Regulation of juvenile hormone titer: Degradation. In *Comprehensive insect physiology, biochemistry and pharmacology* (G. A. Kerkut, L. I. Gilbert, Eds.), vol. 7, pp. 431–472. Pergamon Press, Oxford.

Harshman, L. G., Ward, V. K., Beetham, J. K., Grant, D. F., Grahan, L. J., Zraket, C. A., Heckel, D. G., Hammock, B. D. 1994. Cloning, characterization, and the genetics of the juvenile hormone esterase gene from *Heliothis virescens. Insect Biochem. Mol. Biol.* **24:** 671–676.

Hartfelder, K. 2000. Insect juvenile hormone: From "status quo" to high society. *Braz. J. Med. Biol. Res.* **33:** 157–177.

Hidayat, P., Goodman, W. G. 1994. Juvenile hormone and hemolymph juvenile hormone binding protein titers and their interaction in the hemolymph of fourth stadium *Manduca sexta. Insect Biochem. Mol.* Biol. **24:** 709–715.

Ismail, S. M., Satyanarayana, K., Bradfield, J. Y., Dahm, K. H., Bhaskaran, G. 1998. Juvenile hormone acid: Evidence for a hormonal function in induction of vitellogenin in larvae of *Manduca sexta. Arch. Insect Biochem. Physiol.* **37:** 305–314.

Judy, K. J., Schooley, D. A., Dunham, L. L., Hall, M. S., Bergot, B. J., Siddall, J. B. 1973. Isolation, structure and absolute configuration of a new natural insect juvenile hormone from *Manduca sexta. Proc. Natl. Acad. Sci. USA* **70:** 1509–1513.

Kataoka, H., Toschi, A., Li, J. P., Carney, R. L., Schooley, D. A., Kramer, S. J. 1989. Identification of an allatotropin from adult *Manduca sexta. Science* **243:** 1481–1483.

Khan, A. M. 1988. Brain-controlled synthesis of juvenile hormone in adult insects. *Entomol. Exp. Appl.* **46:** 3–17.

Lorenz, M. W., Kellner, R., Hoffmann, K. H. 1995. A family of neuropeptides that inhibit juvenile hormone biosynthesis in the cricket, *Gryllus bimaculatus. J. Biol. Chem.* **270:** 21,103–21,108.

Mauchamp, B., Darrouzet, E., Malosse, C., Couillaud, F. 1999. 4'-OH-JH-III: An additional hydroxylated juvenile hormone produced by locust corpora allata *in vitro. Insect Biochem. Molec. Biol.* **29:**475–480.

Lorenz, M. W., Kellner, R., Hoffmann, K. H. 1995. Identification of two allatostatins from the cricket, *Gryllus bimaculatus* de Geer (Ensifera, Gryllidae): Additional members of a family of neuropeptides inhibiting juvenile hormone biosynthesis. *Regul. Pept.* **57:** 227–236.

Prestwich, G. D., Wojtasek, H., Lentz, A. J., Rabinovich, J. M. 1996. Biochemistry of proteins that bind and metabolize juvenile hormones. *Arch. Insect Biochem. Physiol.* **32:** 407–419.

Pszczolkowski, M. A., Chiang, A. 2000. Effects of chilling stress on allatal growth and juvenile hormone synthesis in the cockroach, *Diploptera punctata*. *J. Insect Physiol.* **46:** 923–931.

Rachinsky, A., Tobe, S. S., Feldlaufer, M. F. 2000. Terminal steps in JH biosynthesis in the honey bee (*Apis mellifera* L.): Developmental changes in sensitivity to JH precursor and allatotropin. *Insect Biochem. Mol. Biol.* **30:** 729–737.

Rachinsky, A., Tobe, S. S. 1996. Role of second messengers in the regulation of juvenile hormone production in insects, with particular emphasis on calcium and phosphoinositide signaling. *Arch. Insect Biochem. Physiol.* **33:** 259–282.

Ramaswamy, S. B., Shu, S., Park, Y. I., Zeng, F. 1997. Dynamics of juvenile hormone-mediated gonadotropism in the Lepidoptera. *Arch. Insect Biochem. Physiol.* **35:** 539–558.

Richard, D. S., Applebaum, S. W., Sliter, T. J., Baker, F. C., Schooley, D. A., Reuter, C. C., Henrich, V. C., Gilbert, L. I. 1989. Juvenile hormone bisepoxide biosynthesis in vitro by the ring gland of *Drosophila melanogaster*. A putative juvenile hormone in the higher *Diptera*. *Proc. Natl. Acad. Sci. USA* **86:** 1421–1425.

Richard, D. S., Applebaum, S. W., Gilbert, L. I. 1990. Allostatic regulation of juvenile hormone production *in vitro* by the ring gland of *Drosophila melanogaster*. *Mol. Cell. Endocrinol.* **68:** 153–161.

Richter, K. 2001. Daily changes in neuroendocrine control of moulting hormone secretion in the prothoracic gland of the cockroach *Periplaneta americana* (L.). *J. Insect Physiol.* **47:** 333–338.

Riddiford, L. M. 1994. Cellular and molecular actions of juvenile hormone. I. General considerations and premetamorphic actions. *Adv. Insect Physiol.* **24:** 213–274.

Riddiford, L. M., 1996. Juvenile hormone: The status of its "status quo" action. *Arch. Insect Biochem. Physiol.* **32:** 271–286.

Röller H., Dahm, K. H. 1970. The identity of juvenile hormone produced by corpora allata *in vitro*. *Naturwisse* **57:** 454–455.

Röller, H., Dahm, K. H., Sweeley, C. C., Trost, B. M. 1967. The structure of juvenile hormone. *Angew. Chem.* **6:** 179–180.

Stay, B. 2000. A review of the role of neurosecretion in the control of juvenile hormone synthesis: A tribute to Berta Scharrer. *Insect Biochem. Mol. Biol.* **30:** 653–662.

Stay, B., Fairbairn, S., Yu, C. G. 1996. Role of allatostatins in the regulation of juvenile hormone synthesis. *Arch. Insect Biochem. Physiol.* **32:** 287–297.

Stay, B., Tobe, S. S., Bendena, W. G. 1994. Allatostatins: Identification, primary structures, functions and distribution. *Adv. Insect Physiol.* **25:** 267–337.

Stoltzman, C. A., Stocker, C., Borst, D., Stay, B. 2000. Stage-specific production and release of juvenile hormone esterase from the ovary of *Diploptera punctata*. *J. Insect Physiol.* **46:** 771–782.

Tobe, S. S., Bendena, W. G. 1999. The regulation of juvenile hormone production in arthropods. Functional and evolutionary perspectives. *Ann. NY Acad. Sci.* **897:** 300–310.

Tobe, S. S., Zhang, J. R., Bowser, P. R. F., Donly, B. C., Bendena, W. G. 2000. Biological activities of the allostatin family of peptides in the cockroach, *Diploptera punctata*, and potential interactions with receptors. *J. Insect Physiol.* **46:** 231–242.

Touhara, K., Bonning, B. C., Hammock, B. D., Prestwich, G. D. 1995. Action of juvenile hormone (JH) esterase on the JH-JH binding protein complex. An *in vitro* model of JH metabolism in a caterpillar. *Insect Biochem. Mol. Biol.* **25:** 727–734.

Yin, C.-M., Zou, B.-X., Jiang, M., Li, M.-F., Qin, W., Potter, T. L., Stoffolano, J. G., Jr. 1995. Identification of juvenile hormone III bisepoxide (JHB3), juvenile hormone III and methyl farnesoate secreted by the corpus allatum of *Phormia regina* (Meigen), in vitro and function of JHB3 either applied alone or as a part of a juvenoid blend. *J. Insect Physiol.* **41:** 473–479.

Veelaert, D., Schoofs, L., De Loof, A. 1998. Peptidergic control of the corpus cardiacum–corpora allata complex of locusts. *Int. Rev. Cytol.* **182:** 249–302.
Weaver, R. J., Edwards, J. P., Bendena, W. G., Tobe, S. S. 1998. Structures, functions and occurrence of insect allostatic peptides. In *Recent Advances in Arthropod Endocrinology* (G. M. Coast, S. G. Webster, Eds.), *Soc. Exp. Biol. Seminar.* Ser. 65, pp. 3–32. Cambridge University Press, Cambridge.

Insect Neuropeptides

Brown, M. R., Raikhel, A. S., Lea, A. O. 1985. Ultrastructure of midgut endocrine cells in the adult mosquito, *Aedes aegypti*. *Tissue Cell* **17:** 709–721.
Gade, G. 1997. The explosion of structural information on insect neuropeptides. *Fortschr. Chem. Org. Naturst.* **71:** 1–128.
Winther, A. M. E., Nässel, D. R. 2001. Intestinal peptides as circulating hormones: Release of tachykinin-related peptide from the locust and cockroach midgut. *J. Exp. Biol.* **204:** 1269–1280.

Vertebrate-type Hormones in Insects

Kramer, K. J. 1985. Vertebrate hormones in insects. In *Comprehensive insect physiology, biochemistry and pharmacology, vol.* 7 (G. A. Kerkut, L. I. Gilbert, Eds.), pp. 511–536. Pergamon Press, Oxford.
Nässel, D. R. 1999. Tachykinin-related peptides in invertebrates: A review. *Peptides* **20:** 141–158.
Riehle, M. A., Brown, M. R. 1999. Insulin stimulates ecdysteroid production through a conserved signaling cascade in the mosquito *Aedes aegypti*. *Insect Biochem. Mol. Biol.* **29:** 855–860.
Satake, S., Masumura, M., Ishizaki, H., Nagata, K., Kataoka, H., Suzuki, A., Mizoguchi A. 1997. Bombyxin, an insulin-related peptide of insects, reduces the major storage carbohydrates in the silkworm *Bombyx mori*. *Comp. Biochem. Physiol.* B **118:** 349–357.
Yoshida, I., Moto, K., Sakurai, S., Iwami, M. 1998. A novel member of the bombyxin gene family: Structure and expression of bombyxin G1 gene, an insulin-related peptide gene of the silkmoth *Bombyx mori*. *Dev. Genes Evol.* **208:** 407–410.

CHAPTER 2

Integumental Systems

ADVANTAGES OF AN EXOSKELETON

Probably more than any other feature, it is the exoskeleton that has most contributed to the success of arthropods. Although it completely covers the insect like the skin of vertebrates, the exoskeleton plays a much more important role than just as a skin. Its most critical function is to serve as an interface between the insect and the environment, providing a barrier to the movement of water, ions, parasites, and environmental chemicals, including insecticides. This barrier is especially significant for small animals like insects that have a high surface-to-volume ratio and present a relatively large amount of surface area to the environment. The nature of the exoskeleton has thus had profound implications for the design of chemicals that must penetrate the integument to be used as control agents.

The exoskeleton also plays a structural role, determining the form of the insect body and making possible the dramatic changes in form that accompany metamorphosis. Insects exploit a variety of diverse habitats and diets, made possible by a developmental plasticity in body form and mouthpart structure. The rigidity that it provides allows for the insertion of muscles that can produce much more precise locomotor movements than can the soft hydrostatic exoskeletons of the annelid worms. Although being surrounded by a rigid suit of armor might limit the movement and environmental awareness of insects, the integument that makes up the exoskeleton is elastic in some areas to make flying and walking possible. Numerous sensory receptors that are concentrated in strategic areas provide windows to the outside world that allow the insect to respond appropriately to the environment.

The integument may comprise up to half the dry weight of some insects, representing a major investment of raw materials. However, because much of this is resorbed during molting and even periods of starvation, the integument could also be viewed as a food reserve. In many insects, the specific character of the exoskeleton is responsible for releasing particular behavioral sequences that are involved in mating. Specific structures required for mate recognition, as well as chemicals such as pheromones and pigments that are deposited in the exoskeleton, are releasers for the stereotyped behaviors that are necessary for mating to occur. A single layer of epidermal cells and their secretions provide the insects with all these features.

For an animal the size of an insect, an exoskeleton provides significant mechanical advantages over an endoskeleton of the same weight. Given two appendages that are of identical skeletal cross-sectional areas, the one with an exoskeleton is three times stronger than the one with an endoskeleton because of the way that the margins of the appendage, and not the center, bear the most stress when it is bent (Fig. 2.1). To be of equivalent strength, the endoskeletal appendage would have to occupy over 80% of the total appendage cross section and would contribute significantly to the weight of the animal and its investment in structural materials. Using a building analogy, the exoskeleton is similar to the lightweight scaffolding that surrounds a structure during its construction rather than the "endoskeletal" girders that are used to permanently create it.

Although the exoskeleton provides a number of advantages, it does pose a major problem for growth. In order for insects with rigid exoskeletons to undergo significant amounts of growth, a new, larger exoskeleton must be synthesized and the older one discarded. During this period of molting, the insect is relatively helpless against predators because flight or defense is difficult. Molting consumes time, energy, and metabolic resources. There is also a potential susceptibility for the loss of water because the insect can neither drink nor adjust its body to a changing environment. To reduce this susceptible period during the molting cycle, more advanced insects have evolved toward a reduced number of molts. Growth during the intermolt period is possible because the larvae of advanced holometabolous insects generally have relatively unsclerotized cuticles that can undergo a degree of stretching.

INSECT GROWTH AND DEVELOPMENT

The growth and development of insects are largely a function of the growth and development of their integuments. The cuticular molts that punctuate postembryonic growth are necessary if hard-bodied insects are to undergo any significant increases in size. Increases in body size do not always follow from molting, however. Insects that are starved during the larval stage or molt to a diapause form may actually molt to smaller individuals if they molt at all. Apertygote insects continue to molt into the adult stage, but pterygote insects are incapable of molting as adults. The inability of adult pterygote insects to molt is probably the result of the degeneration of the epidermal cells that produce the wing once it is formed. After the molt to the adult stage, the epidermal cells that make up the wing degenerate and the loss of the water contained within them makes it possible for the wing membranes to move rapidly for flight. Once they degenerate in the adult stage, the

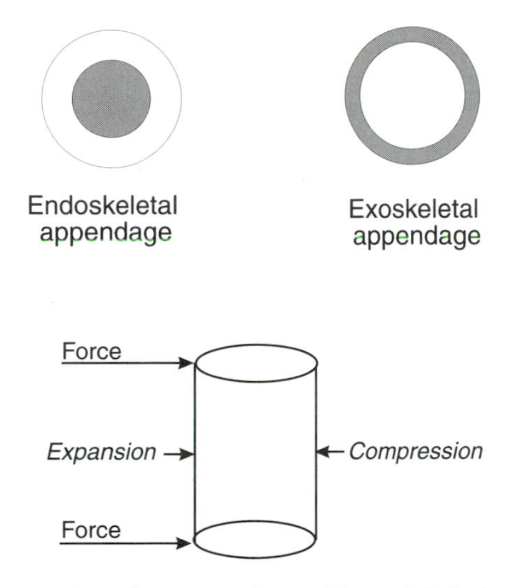

FIGURE 2.1 A cross section of two appendages with equal skeletal areas, the left being an endoskeleton and the right an exoskeleton. The forces applied to the ends of the appendage are largely borne by their walls, causing an expansion on one side of the wall and a compression on the other.

cells obviously cannot initiate a molt nor synthesize a new cuticle. Without living epidermal cells, another wing cuticle could not be formed if the insect molted again. Thus, the death of the cells that makes flight possible also makes molting as an adult impossible. Only the winged mayfly subimago is capable of a molt to another winged form, but the subimago is a poor flier because living epidermal cells must remain as part of the wing.

In many holometabolous insects, considerable growth can occur during a single larval instar in the absence of a molt because, with the exception of the head capsule, the cuticle is extensible enough to accommodate some increases in size. The last instar of *Manduca sexta* can grow from 1 g to over 9 g in weight without a molt because the pleated outer epicuticle of the exoskeleton is able to stretch to accommodate the growth that occurs within this instar. However, a molt may ultimately be necessary in order to acquire a larger head capsule and allow the sclerotized mouthparts to increase in size so the rate of food intake is increased to satisfy the demands of the larger body.

STRATEGIES FOR GROWTH

The change that occurs as an insect develops from an immature to an adult is called metamorphosis, literally a "change in form." A metamorphosis is considered as a partition in the life of an individual that separates an early feeding stage from a later reproductive stage. Insects show three major metamorphic strategies for reaching the adult stage. **Ametabolous** development is the least advanced strategy and is found in the apterygote orders Thysanura and Archeognatha, which are primitively wingless insects (Fig. 2.2). They continue to molt as sexually mature adults and the strategy is considered to be the most primitive because there are no changes and consequently no real metamorphosis. Adults develop from the gradual developmental changes that occur, and with little difference between them except for their relative sizes and the presence of genitalia, both immatures and adults occupy the same habitats.

Those insects that engage in **hemimetabolous** development, or incomplete metamorphosis, also generally occupy similar niches as larvae and adults (Fig. 2.2). Immatures lack wings and genitalia and the extent of their metamorphosis is the development of these structures in adults. Hemimetabolous larvae and adults may also show differences in cuticular structure. In hemimetabolous *Rhodnius prolixus,* numerous plaques distinguish the larval cuticle, while the adult cuticle bears a distinct ripple pattern on each segment. Several aquatic hemimetabola have larvae that bear adaptations such as gills that are lost in the molt to terrestrial adults or unusual larval mouthparts that are radically different in the adults, but retain their basic body forms. The wings of hemimetabolous insects develop externally and are often visible in the later larval instars. They are sometimes referred to as **exopterygotes** to reflect their external development of wings.

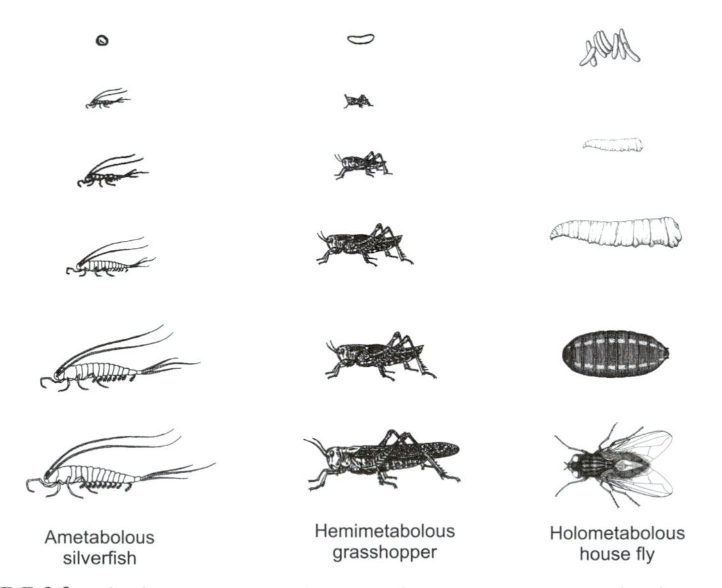

FIGURE 2.2 The three major types of metamorphosis in insects. Reprinted with permission from Ursprung, H. and R. Nöthiger (1972). *The Biology of Imaginal Discs.* Copyright Springer-Verlag GmbH & Co. KG.

Holometabolous development, in contrast, is characterized by a sometimes very radical change in form and ecological habits between immatures and adults (Fig. 2.2). These great differences between adults and their offspring presumably allow them to avoid any intraspecific competition for food and habitat. The advantages of this strategy are reflected in the relatively large number of insects, about 88% of all known insect species, that engage in holometabolous development. In addition to the differences between larvae and adults, there is also an intervening pupal stage between the two that allows the larva to be reconstructed into an adult. The histolysis of larval structures and the formation of adult structures occur during this quiescent pupal phase. The wings of holometabola arise from internal imaginal discs that evert during metamorphosis to the adult, and the insects that engage in holometabolous development are often referred to as **endopterygotes** to reflect this internal development of wings. Some argue that the designations of exopterygote and endopterygote are better classifications than are hemimetabolous and holometabolous because not all parts of the insect necessarily conform to one developmental mode. There are some hemimetabolous insects with structures that undergo radical changes characteristic of holometabolous development such as the mouthparts of the larval dragonfly that change drastically during adult development. The terms exopterygote and hemimetabola, and endopterygote and holometabola, will be used interchangeably in this text. There are some predacious or parasitic insects that engage in a developmental variation of holometaboly, **hypermetamorphosis,** in which successive larval instars have acquired different habits and a marked anatomical variation. This deviation from more conventional holometaboly occurs in some Coleoptera, Hymenoptera, Neuroptera, and all Strepsiptera, with active first-instar larvae and more inactive grub-like late-instar larvae. The active first instar is necessary to locate and enter the host or find food, but once the resource is found, the subsequent inactive stages suffice.

Endopterygote larvae generally have a relatively higher rate of growth than do exopterygotes. One reason may be that exopterygote larvae invest more material in their cuticles than do the larvae of endopterygotes. The lower growth rate of exopterygotes may reflect the increased costs involved in manufacturing the cuticle and the relatively large amounts of these materials lost in sclerotized proteins when the insects molt.

ORIGINS OF HOLOMETABOLY

Ametabolous development that lacks a metamorphosis is clearly the ancestral condition but the evolutionary route that ametabolous insects took to arrive at holometaboly has been disputed. Berlese, in 1913, suggested that the holometabolous larva was a free-living embryo resulting from a premature egg hatch, and the pupal stage represented a number of nymphal stages compressed into one. The reason for the early hatch was the reduced yolk that was supposedly contained within the egg of holometabolous insects. This interpretation considered the immature holometabola as being different than the immature hemimetabola and gave them different names. Berlese applied the term, "nymph," a French word for "pupa," to exopterygotes to indicate that the immatures of exopterygotes were equivalent to the pupa of endopterygotes, while the endopterygote immature, or "larva," was a very different, free-living embryo (Fig. 2.3B). An alternative hypothesis that is largely accepted today and was strongly supported by Hinton held that there was no difference between the immatures of exo- and endopterygotes. The difference that was claimed to exist between the amount of yolk in hemimetabolous and holometabolous eggs was not supported by any data, and there was no developmental reason for using the different terms "larva" and "nymph" for the two immature forms. The pupa was considered to represent the first of two imaginal instars that served as a mold for the development of adult musculature (Fig. 2.3C).

The most recent interpretation of the origin of holometabolous development by **Truman** and **Riddiford** returned to the original theory of Berlese. They pointed out a long-ignored stage in some terrestrial arthropods that exists much

FIGURE 2.3 (A) The sequence of development in exopterygotes, beginning with the embryo, the passage through several nymphal instars, and the adult stage. (B) The interpretation of endopterygote development by Berlese (1913), in which the larva is considered as a free-living embryo. The pupa was equivalent to the nymphal instars of the exopterygote. (C) The interpretation of endopterygote development by Hinton (1963), in which the larval instars are equivalent to the nymphal instars of exopterygotes, and the pupa was the first of two imaginal instars. (D) The most recent interpretation by Truman and Riddiford (1999), in which the pronymphal stage was believed to give rise to the larval stages of the endopterygote. The nymphal stage of exopterygotes was transformed into the transitional pupal stage.

like a free-living embryo. This stage that they refer to as a **pronymph** in some primitive hemimetabola is proposed to be the forerunner of the holometabolous larva. The pronymph lasts for only a single instar before becoming a nymph, but a shift in JH secretion during embryogenesis may have been the factor promoting its evolution into the more stable larval stage. In this scheme, the nymphal stage was reduced to a single nonfeeding instar that became the pupa, a transition between the larva and adult (Fig. 2.3D).

INSTARS, STADIA, AND HIDDEN PHASES

The molting cycle and metamorphosis present some interesting developmental conditions. The particular developmental stage of an insect is often arbitrarily referred to as an instar or stadium, but these terms actually have more precise meanings that are difficult to apply accurately but are important to understand. As will be discussed in more detail in later sections, one of the first steps in molting is apolysis, in which the old cuticle separates from the epidermis and new cuticle begins to be produced. With the old cuticle no longer directly attached to epidermis, it has effectively been discarded, although it has not been shed, and the newly formed cuticle now represents the cuticle of the next instar. For this reason, apolysis is said to mark the passage to the next instar, even though ecdysis has not yet occurred and the insect appears to still be in the skin of the earlier instar. An instar is therefore defined as the period between two apolyses and begins when the insect first becomes detached from its old skin (Fig. 2.4).

The instar that is hidden under the old, unshed cuticle before ecdysis is referred to as the **pharate instar.** This distinction is more important for some instars than others. For example, some lepidopterans undergo diapause as pharate adults that are developmentally complete adults enclosed by the detached pupal cuticle. Although based on an external examination it is easy to conclude that the insect diapauses as a pupa, but the diagnosis would not be correct. The stadium that an insect is in is defined by its ecdyses; a stadium represents the interval between one ecdysis and the next. Therefore, at apolysis, an insect passes to another instar, but does not become the next stadium until after ecdysis (Fig. 2.4). These distinctions are ordinarily not discernable or even important unless one is required to assess the actual physiological state of the insect.

STRUCTURE OF THE INTEGUMENT

The outer covering of insects is referred to both as an **exoskeleton** and an **integument.** The term exoskeleton characterizes it based on its function as the major external support for the integrity of the insect body. The term integument characterizes it more in terms of its structure. The integument consists of the

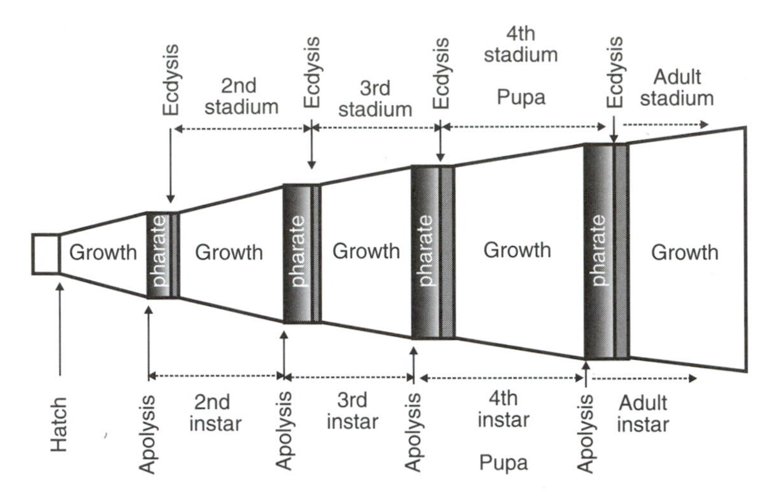

FIGURE 2.4 The molting period is punctuated by two events, apolysis and ecdysis, that define insect development. Apolysis, the separation of the epidermal cells from the cuticle, marks the beginning of the molt and the next instar. The insect is in a pharate stage until ecdysis occurs, the casting off of the old cuticle. Ecdysis marks the beginning of the next stadium. At the apolysis following the second instar, the insect enters the third instar, but is still in the second stadium until after ecdysis.

underlying basement membrane, the living epidermal cell layer, and the overlying, nonliving cuticle that is secreted by the epidermis. Given that there are over a million identified insect species and that the specific nature of the integument of each species has exquisitely evolved along with its ecological habits, it is likely that there are also over a million different varieties of integument. However, a general pattern of integumental structure has emerged, and it is this general pattern that will be described (Fig. 2.5).

We can begin on the inside and work our way to the outside. The **basement membrane** is a continuous sheet of mucopolysaccharide, as much as 0. 5 μm in thickness, that separates the epidermal cells from the body cavity. It appears to be initially secreted by hemocytes, but may also be formed by the epidermal cells during wound repair. The basement membrane is penetrated by nerves and trachea that reach into the epidermal cell layer above.

The **epidermis** that lies just above the basement membrane is the only living portion of the integument. It comprises a single layer of cells that are ectodermal in origin. Modifications of these cells produce such structural features as **dermal glands, sensory receptors** and their support cells, and **oenocytes.** Dermal glands are often multicellular, consisting of a secretory cell, a duct cell, and one or more support cells depending on their function. They secrete the cement layer that covers the epicuticle and are widely distributed over the surface of the integument. They may also produce more volatile defensive secretions and pheromones

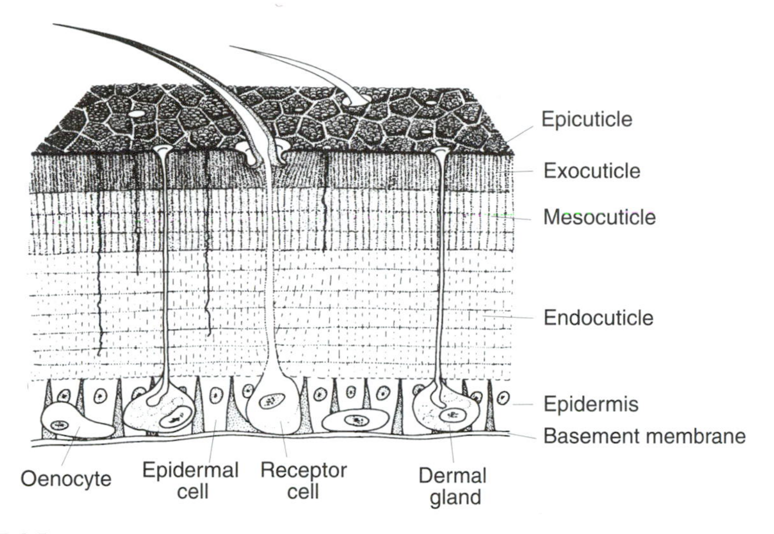

FIGURE 2.5 The generalized insect integument, consisting of the basement membrane, epidermal cells, and overlying nonliving cuticle. Reprinted with permission from Hackman, R.H. (1971). *Chemical Zoology* **6:** 1–62. Copyright Academic Press.

that are released into the environment. Sensory receptors are specialized epidermal cells that respond to environmental stimuli and consist of structural and support cells that produce the outward form of the receptor, and the dendrites of neurons that are contained inside that respond to specific stimuli. Sensory receptors will be discussed in more detail in Chapter 11. Oenocytes are large polyploid cells associated with the basement membrane. Some of these oenocytes enlarge during the molting process and appear to be secretory, suggesting that they are involved in the production of cuticular lipids that are deposited in the epicuticle. Other types of oenocytes may secrete ecdysteroid hormones. Invaginations of the epidermis during development give rise to the anterior and posterior portions of the digestive tract, the salivary glands, the trachea and tracheoles, and portions of the reproductive tracts of males and females.

The epidermal cells, along with some of the modified cells associated with them, secrete the overlying, nonliving cuticle. These secretions are largely responsible for the specific characteristics of the exoskeleton. The cuticle consists of several horizontal divisions that are produced in a certain sequence during the molting process and that may be developmentally altered both before and after the molt occurs. The cuticle is divided into two main regions, the thin outer epicuticle and the thick inner procuticle that lies just above the epidermal cells. The procuticle is the region of the cuticle that contains the chitin; no chitin has yet been found in integumental epicuticle.

The **procuticle** is secreted by the epidermal cells and consists largely of chitin and protein that are initially undifferentiated and not chemically cross-linked.

However, as the molting cycle proceeds, at least two and sometimes three layers can be identified. The first portion of the procuticle to be synthesized becomes the outer **exocuticle,** a region in which the proteins eventually become heavily cross-linked and insoluble. This insolubility prevents the exocuticle from being broken down during the molting cycle so the portion that is shed during ecdysis consists mostly of exocuticle. Various cuticular pigments such as melanin may be deposited within the exocuticle.

The **endocuticle** is the portion of the procuticle that lies just above the epidermal cells and its synthesis continues after the old cuticle is shed, often in daily layers that can be used to age-grade particular insects. It ranges from 10 to 200 μm in thickness and consists of several lamellar layers. In soft-bodied insects and in areas of flexibility such as the regions between body segments and joints, it is the endocuticle, without much developed exocuticle, that comprises most of the cuticle that is present. In contrast to the exocuticle, in the endocuticle the cross-linking of proteins is reduced, allowing it to be completely broken down by enzymes and resorbed during the molting process. These differences in chemical cross-linking of cuticular proteins give the exocuticle and endocuticle their particular properties. **Mesocuticle** has been identified in several insects and is interpreted as a transitional layer in which the proteins are untanned like the endocuticle but impregnated with lipids and proteins like the exocuticle.

The **epicuticle** is a complex consisting of several layers that are produced by both the epidermal cells and dermal glands. The epicuticle is probably the most important of the cuticular regions, largely responsible for the surface features, waterproofing properties and general impermeability of the cuticle. There are four distinct layers in the epicuticle but not all may be present in some insects (Fig. 2.6).

The outermost layer of the epicuticle is the **cement layer.** It consists mostly of lipoprotein that is secreted by different groups of dermal glands whose contents are discharged and mix on the surface, but the composition and thickness of the layer varies from species to species. In spite of its alleged importance, it is completely absent in the cuticle of honeybees. It may serve like a varnish to coat and protect the wax layer just beneath it.

Underneath the cement layer is the **wax layer,** produced by the epidermal cells and transported to the surface of the cuticle through the pore canals that permeate the procuticle. Cuticular waxes are mixtures of hydrocarbons with 25–31 carbon atoms, alcohols of 24–34 carbon atoms, and esters of fatty acids. The major role of the wax layer is in waterproofing. It is generally absent in aquatic insects, and is also lacking in other arthropods such as centipedes and millipedes that are confined to humid environments and inactive during the heat of the day. When insects are exposed to high temperatures that are above a transition point that represents the phase change of the cuticular waxes, water loss increases significantly as a result of the disorientation of the wax layer (Fig. 2.7). The transition point is a function of the composition of the cuticular waxes that are present.

The **outer epicuticle,** also referred to as the **cuticulin** layer, is the third region of the epicuticle. It is also synthesized by epidermal cells, but there is little

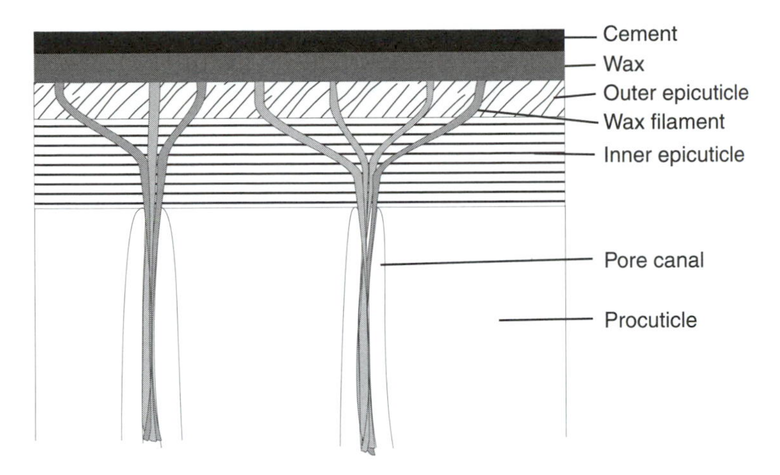

FIGURE 2.6 The layers of the epicuticle, lying above the procuticle.

known about the chemical composition of this layer. Its importance is reflected in the observation that it may be the only epicuticular layer present in all insects and among the first layers of the new cuticle to be synthesized. It is the only epicuticular layer found in the tracheoles. Although it is a very thin trilaminar membrane of 12–18 nm, it plays a significant role in cuticular permeability. It contains small

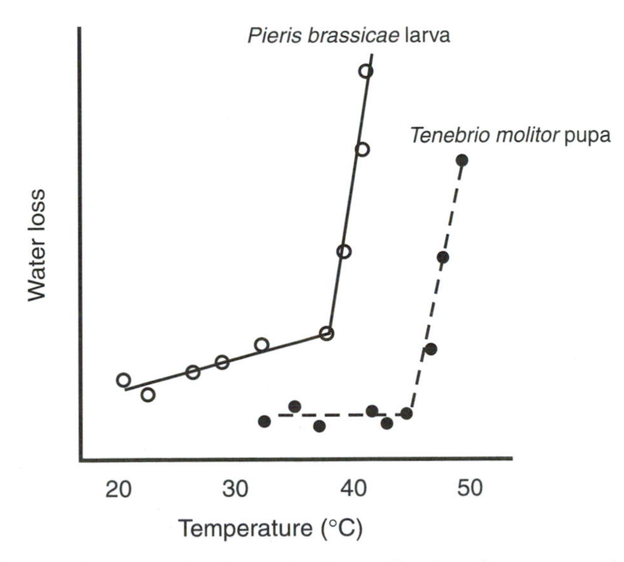

FIGURE 2.7 The relative water loss in two insects as a function of temperature. Above a critical temperature, the wax layer becomes disoriented and water loss increases substantially. From Beament (1959). Reprinted with permission.

pores that allow the molting gel and the products of cuticular hydrolysis to pass through during the molting cycle, but that restrict the passage of active molting fluid from where the old cuticle is being digested. Once the outer epicuticle is laid down, the size of the insect is fixed, and in many insects that expand after ecdysis, the epicuticle is synthesized in pleats that allow this expansion to occur.

The **inner epicuticle** contains both polyphenols and the enzyme polyphenol oxidase, which are involved in tanning the cuticle. It may also be involved in wound repair, tanning scratches that may occur. It ranges from 0. 5 to 2 μm in thickness.

MODIFIED FEATURES OF THE INTEGUMENT

The vertical organization of the integument that has been described does not necessarily apply to all areas of the insect. Depending on the functions of specific regions, there may be some layers that are either absent or more prominent. **Arthrodial membrane** refers to the flexible membranes between body segments where the exocuticle is absent (Fig. 2.8). The untanned endocuticle that remains contains special acidic proteins and the flexible protein, resilin, which provide the flexibility in the region.

Ecdysial lines are also areas of reduced exocuticle, but differ from arthrodial membrane in that they are programmed areas of weakness that serve as emergence points during ecdysis. When the endocuticle is digested during the molting cycle, only a small layer of epicuticle remains that can be disrupted when the new integument is expanded (Fig. 2.9). Soft-bodied insects such as immature lepidopterans have cuticles with reduced exocuticle, which provides the flexibility and the potential for growth during an instar that they require.

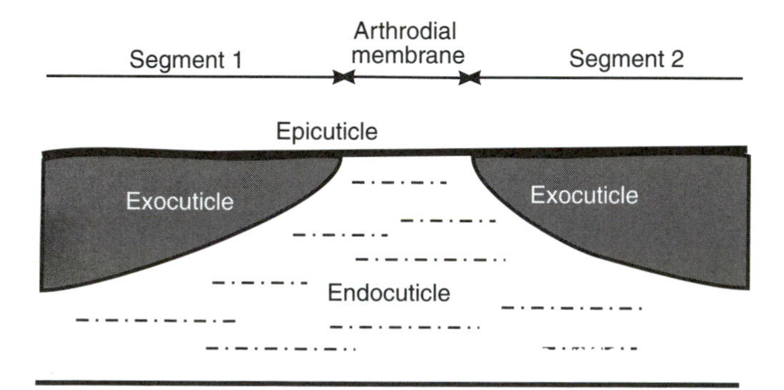

FIGURE 2.8 A cross section of the cuticle between two segments, showing the absence of exocuticle that results in the flexible arthrodial membrane.

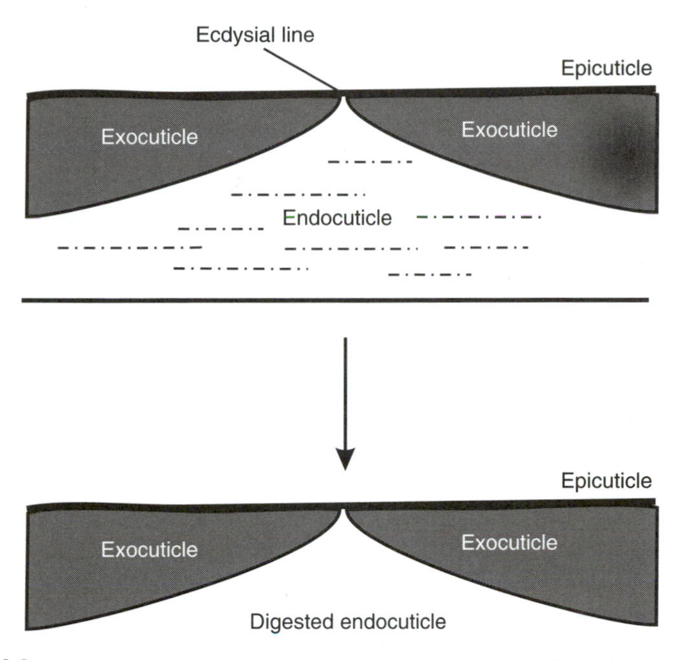

FIGURE 2.9 A cross section of the cuticle in areas that are programmed to split during ecdysis. The absence of exocuticle and the digestion of the endocuticle that remains allow the insect to easily break the remaining epicuticle for emergence.

Pore canals extend from the epidermis through the cuticle to its surface (Fig. 2.6). They are cytoplasmic extensions of the epidermal cells, rising in a helical pattern as they travel through the procuticle. Although they are not covered by the epidermal cell membrane, they do contain filaments from the cell. In the cockroach integument, approximately 200 pore canals arise from each epidermal cell, equivalent to 1. 2 million/mm^2. The pore canals transport lipids produced by the epidermal cells to the surface of the newly formed epicuticle. Pore canals are absent in transparent cuticles, such as those that cover the compound eyes.

CHEMISTRY OF THE CUTICLE

The insect cuticle is composed largely of proteins, lipids, and chitin. The proteins and chitin interact to provide the mechanical function of the cuticle, conferring the strength and hardness that is necessary for it to serve as an exoskeleton. Lipids chiefly provide a protective role, located mostly on the outermost layers of the cuticle and are secreted once the rest of the cuticle has been produced.

Proteins

Proteins can often comprise more than half the dry weight of the insect cuticle. They are primarily located within the procuticle, but the epicuticle also contains several minor proteins that are relatively difficult to extract. The stage-specific properties of insect cuticles are largely derived from the distinct stage-specific proteins that are synthesized and the manner in which they interact with the lipids that are present. Cuticular proteins are synthesized mainly by epidermal cells according to a temporal pattern during the molting process, and the sequence of their synthesis may even extend throughout the instar to change the nature of the cuticle during the instar.

Early studies of cuticular structure referred to a single protein called **arthropodin** that comprised the bulk of the cuticular proteins. However, the term arthropodin was no longer used once it became clear that a diversity of cuticular proteins was actually present. The bulk of the cuticular proteins identified to date can be placed into one of two protein familes. Proteins from areas of insect cuticles that do not become sclerotized contain a conserved amino acid sequence known as the Rebers–Riddiford consensus sequence (RR-1), named after the study that first identified them in *Manduca* and *Drosophila*. Those proteins that do not become sclerotized do not contain the RR-1 sequence, but some instead contain a variant RR-2 sequence. Both the RR-1 and the RR-2 regions of these proteins may be involved in the binding of the proteins to chitin, because they are likely to undergo a folding that places the binding regions on the same side of the molecule.

Other proteins contain repeated hydrophobic amino acid sequences (Ala-Ala-Pro-Ala/Val) that undergo frequent structural turns, resulting in a spiral molecule that is easily deformable. Areas of the cuticle that have an unusual elasticity contain the protein **resilin** (named from the Latin *resilire,* to jump back), which possesses a rubber-like property imparted by the coiling of its protein chains that are linked by di- and trityrosine residues. It also contains a high percentage of glycine residues and an amino acid sequence that minimizes cross-linking with any other proteins that could reduce its elasticity. Resilin has been called an almost perfect rubber that permits the storage of energy in leg joints, in the thoracic body wall, and in wing hinges for locomotion.

The proteins can also be classified into families based on their production and transport (Fig. 2.10). The more traditional Class C cuticular proteins are synthesized by epidermal cells and secreted into the cuticle above them. Class H proteins are secreted only into the hemolymph, in the other direction. Class BD (bidirectional) proteins are synthesized by epidermal cells and are secreted both into the cuticle into the hemolymph. The blue pigment insecticyanin is an example of a BD protein that is located in the epidermis, cuticle, and hemolymph of *Manduca*. Class T proteins are transported across the epidermal cells from the hemolymph into the cuticle and may be synthesized by hemocytes or at other nonepidermal sites. This cuticular protein classification is based entirely on their routing and not

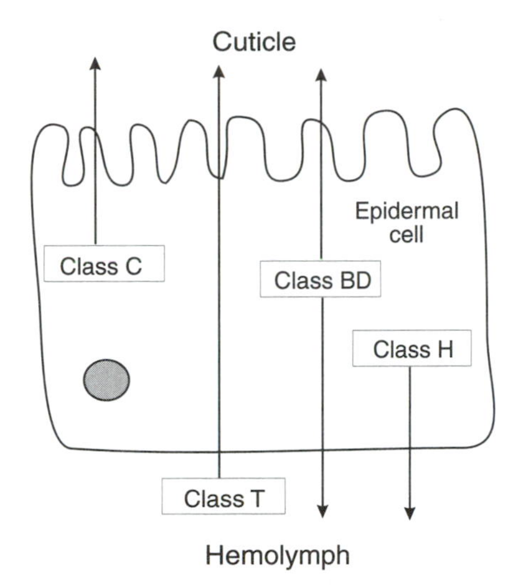

FIGURE 2.10 Families of cuticular proteins based on their transport. The usual Class C proteins are synthesized by the epidermal cells and deposited into the cuticle. Class T proteins are taken up from the hemolymph by the epidermal cells and deposited in the cuticle. Class H proteins are synthesized by the cells and secreted only into the hemolymph, while Class BD proteins are transported in both directions. Adapted from Sass *et al.* (1993). Reprinted with permission.

their function; proteins within each of these classes may have functions unrelated to others within their group. It recognizes that the epidermal cells are capable of both the synthesis as well as the transport of proteins that are destined to end up in the cuticle and hemolymph.

Although a temporal pattern of cuticular protein synthesis exists within each instar, with specific proteins synthesized and deposited in the cuticle at specific times during and after the molting cycle, the location of proteins within the cuticle does not necessarily correlate with the time of secretion. Proteins that are produced may diffuse through the cuticle as they are deposited. Many of the proteins are common within the cuticles of several instars. For example, of the 152 electrophoretic bands representing cuticular proteins of *Hyalophora cecropia,* 7% were found only in larvae, 15% were found only in pupae, and 9% were characteristic of adults. The different types of proteins present and their degree of sclerotization is related more to the type of cuticle produced in a specific area rather than to any species or instar relationships. Heavily sclerotized cuticles contain a large proportion of hydrophobic, positively charged proteins. Flexible cuticles tend to be associated with the presence of more acidic proteins that also have a higher capacity to bind water.

The character of some proteins can be altered even after the cuticle has been synthesized. Although the cuticle of the blood-sucking bug *R. prolixus* is normally stiff, when a blood meal begins to be ingested, endocrine events lower the pH of portions of the cuticle to below 6. With a more acidic pH, the conformation of the proteins is altered and they become more plastic, allowing the insect to expand its abdomen and accommodate the large blood meal. The covalent bonds between proteins produce the hard, stabilized cuticles that are found in insect mandibles. Proteins indeed may impart hardness to cuticles even in the absence of chitin. In the oothecae of cockroaches and mantids, protein cross-linking alone accounts for the hardness and tanning of the structural proteins. The asymmetrical female accessory glands produce products that sclerotize the structural proteins when they mix with enzymes in the genital vestibule (Fig. 2.11). The process of cuticular sclerotization will be discussed in more detail in a later section.

The strikingly bright cuticular coloration of insects is often due to physical colors that depend on the reflecting structural properties of the cuticles and not

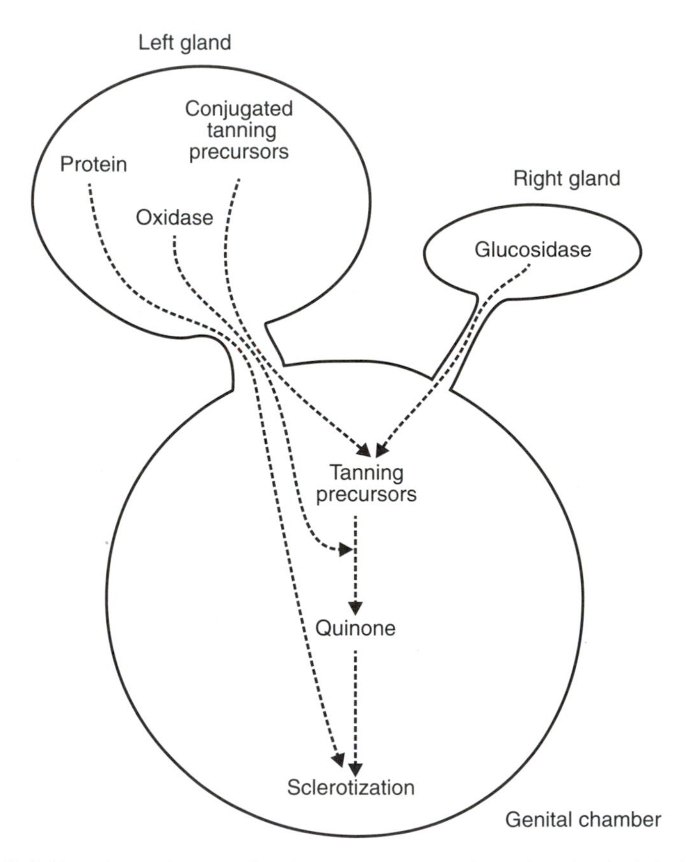

FIGURE 2.11 The mechanism of ootheca production in the cockroach. The left gland secretes oothecal proteins, tanning precursors that are conjugated and inactive, and an oxidase. The right gland secretes a glucosidase. When the contents are mixed in the genital chamber, the glucosidase removes the conjugate and makes the tanning precursors available to the action of the oxidase. The oothecal proteins are sclerotized by the quinines that result.

protein pigments. However, some coloration is the result of the absorption of certain wavelengths of light by protein pigments. Some pigments may be synthesized by the insect and deposited in the cuticle. These include the ommochromes, pteridines, melanins, porphyrins, and bile pigments. Other pigments cannot be synthesized and must be derived from plant products that are ingested. These include the carotenoid and flavenoid pigments. The common black and brown coloration in many insects is the result of either quinone sclerotization of cuticular proteins or the deposition of melanin pigments in the epidermal cells.

Chitin

Chitin is the other major component of the procuticle, consisting of 20-40% of the total dry weight of the cuticle. In addition to its presence in the cuticle of arthropods, it is present in the peritrophic membrane of the insect midgut and is also produced by some protozoans and fungi. Chitin is a polymer of *N*-acetyl-*D*-glucosamine with a few additional residues of unacetylated glucosamine, connected by unbranched 1-4 β-linkages (Fig. 2.12). Adjacent chains of chitin are cross-linked by hydrogen bonds to form chitin microfibrils about 2.8 nm in thickness, with an estimated 18 chitin chains in each microfibril. The microfibrils are known to exist in α, β, and γ crystallographic orientations that differ in the alignment of the chitin chains, but only the α-chitin form is overwhelmingly present in insects (Fig. 2.13). The microfibrils are laid down in a parallel orientation within a layer, but the orientation in successive layers is rotated by a constant angle to produce a helicoidal arrangement (Fig. 2.14). The chitin microfibrils are covalently linked to the surrounding proteins, with the relative amounts of protein and chitin varying between insect species and from one area of the cuticle to another within the same insect.

Chitin has an unusual chemical stability, being insoluble in water, dilute acids, concentrated alkali, alcohol, and organic solvents. In concentrated alkali at high temperatures the acetyl groups are detached and chitosan is formed. As a consequence of the discovery of a class of compounds that blocks chitin formation and causes

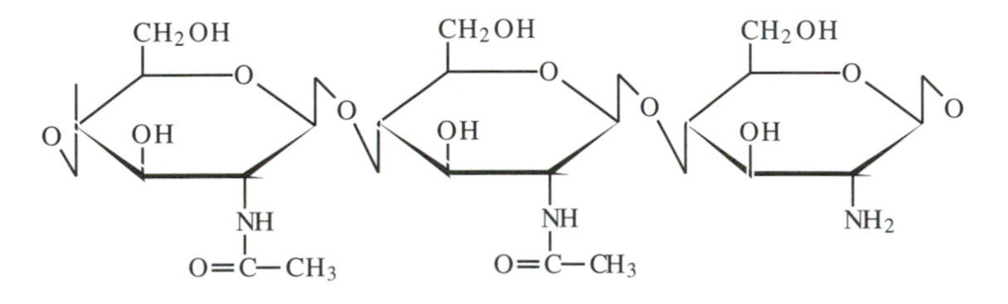

FIGURE 2.12 A portion of the chitin chain, showing two residues of *N*-acetyl-*D*-glucosamine and one of glucosamine.

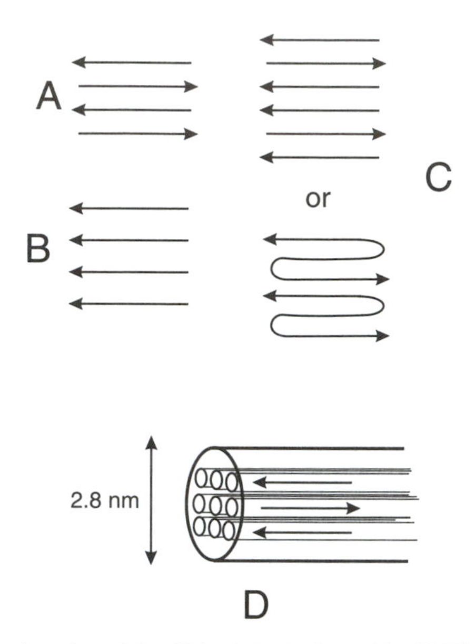

FIGURE 2.13 The orientation of the chitin chains in the cuticle. (A) The orientation of a chitin, the most common form in insects. (B) The orientation of β chitin. (C) Two possible orientations of γ chitin. Adapted from Rudall (1963). Reprinted with permission. (D) The location of the chains of a chitin in a single chitin microfibril. Adapted from Reynolds (1987). Reprinted with permission.

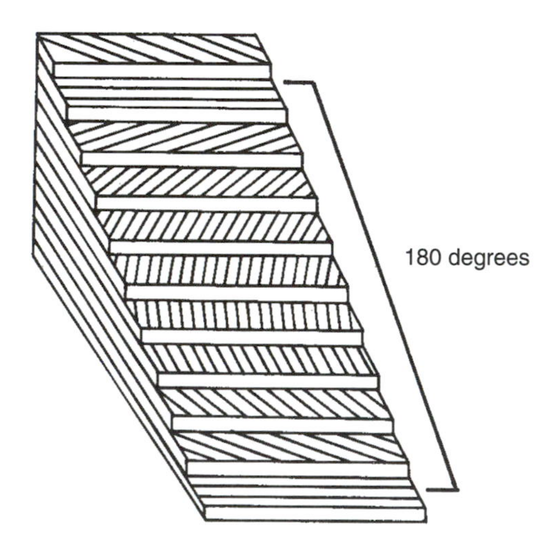

FIGURE 2.14 The helicoidal arrangement of the chitin layers as they are rotated by a constant angle during their synthesis. The bar shows the rotation of layers through 180°. From Neville (1984). Reprinted with permission.

insects to die because they fail to produce cuticles of suboptimal strength, the steps in the synthesis of chitin have become better understood. Chitin synthesis begins with the phosphorylation of glucose that is then aminated and acetylated, forming a monomer that is activated by uridine diphosphate and added to the end of an existing chitin chain (Fig. 2.15). It is degraded during the molting cycle by chitinases that are present in the molting fluid.

Lipids

The bulk of the lipids present in the cuticle are localized in the wax layer of the epicuticle, where they function to prevent desiccation and provide chemical cues

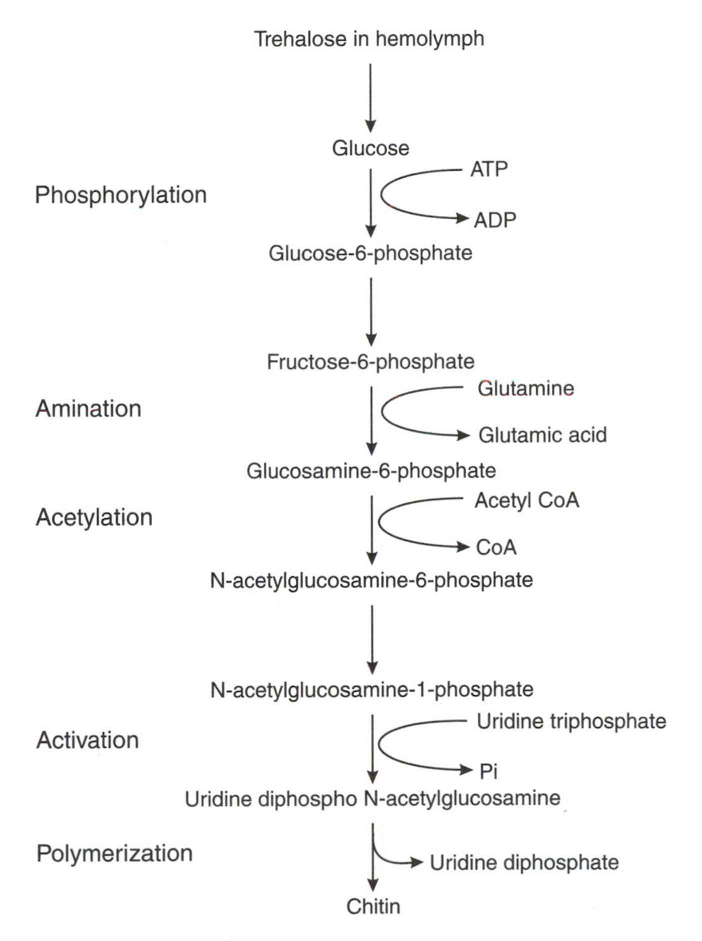

FIGURE 2.15 The steps in chitin biosynthesis.

for species recognition. As the primary barrier to the penetration of environmental chemicals, the nature of cuticular lipids must be understood in the design of effective contact insecticides that are able to cross this barrier. A number of hydrocarbons have been identified that act as sex pheromones and cues for caste recognition in social insects. Some insects secrete prodigious quantities of wax in addition to the lipid deposited in the cuticle. Honeybees produce wax to build the honeycomb of the hive, and scale insects use the wax for protection from predators and from desiccation.

The lipids in the cuticle are synthesized largely by the oenocytes and the cells of the fat body and then taken up by the epidermal cells for their distribution through the pore canals to the surface of the integument. Because the lipids are insoluble in the aqueous hemolymph, they must be transported by special protein molecules called **lipophorins** that act as reusable lipid shuttles. There are a number of different lipids found in the cuticle, commonly mixtures of *n*-alkanes, *n*-alkenes, di- and trimethylalkanes, and monomethylalkanes.

Other Components of the Cuticle

Phenols, largely derived from tyrosine metabolism, have been identified from arthropod cuticles. These are generally involved in the tanning reactions within the procuticle that stabilizes the protein matrix. A number of enzymes related to cuticular tanning are also present, including various phenoloxidases. There is usually nothing more than a trace amount of inorganic components, but some insects, including the pupae of the dipteran *Musca autumnalis,* deposit large amounts of calcium. Calcium carbonate also accumulates in the Malpighian tubules of the larvae and is incorporated into the puparium as a means of cuticular hardening.

Sclerotization

Cuticular sclerotization, also known as tanning, stabilizes the protein matrix of the cuticle to make it stiffer and harder, more insoluble, and more resistant to degradation. The process gives the integument greater strength for muscle attachment and locomotion and provides stability against hydrolytic enzymes produced by potential pathogens such as fungi. The process of sclerotization cross-links the functional groups of cuticular proteins when they react with quinones.

During sclerotization, the epidermal cells secrete several agents into the cuticle where they are transformed into more reactive compounds that are able to link the proteins. The amino acid tyrosine provides one of the precursors for sclerotization. Insects cannot synthesize the phenyl ring of tyrosine so it must be acquired in the diet and is sequestered in the hemolymph as conjugates of glucose or phosphate. Tyrosine has a very low solubility, and this conjugation may be necessary to increase its solubility in the hemolymph and protect it from being used in other metabolic reactions. The precursors for sclerotization are derived from tyrosine in three enzymatic steps (Fig. 2.16). First, the tyrosine is hydroxylated to 3,4-dihydroxyphenylalanine (DOPA). The DOPA is then decarboxylated to dopamine, followed by the acylation of the dopamine amino group with either acetate or β-alanine to form the catecholamines *N*-acetyl dopamine (NADA) or *N*-β-alanyldopamine (NBAD). The selection of which of these tanning precursors is employed may determine the character of the tanned cuticle. Areas containing high concentrations of NADA tend to be less dark than those containing NBAD. The darkening of the cuticle otherwise occurs when some of the dopamine is channeled into the pathway for melanin production.

The catecholamines are oxidized by phenoloxidases once they are released into the cuticle to form reactive quinones. In quinone sclerotization, the *o*-diphenols are oxidized to *o*-quinones. In β-sclerotization, another pathway that produces cross-linking agents, the β carbon of the aliphatic side chain is activated (Fig. 2.17). In both cases, the $-NH_2$ and $-SH$ groups on the proteins bond to the tanning agents. In addition to the association of NADA or NBAD with the type of cuticle produced, the covalent bonds that form between either the aromatic or the

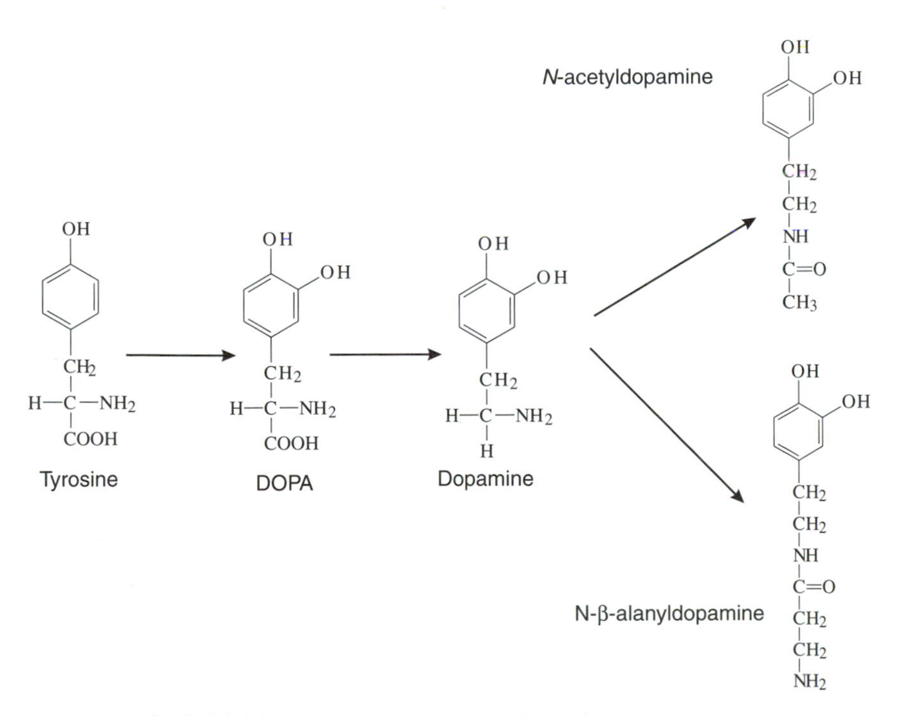

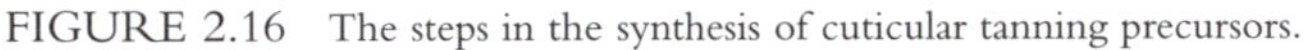
FIGURE 2.16 The steps in the synthesis of cuticular tanning precursors.

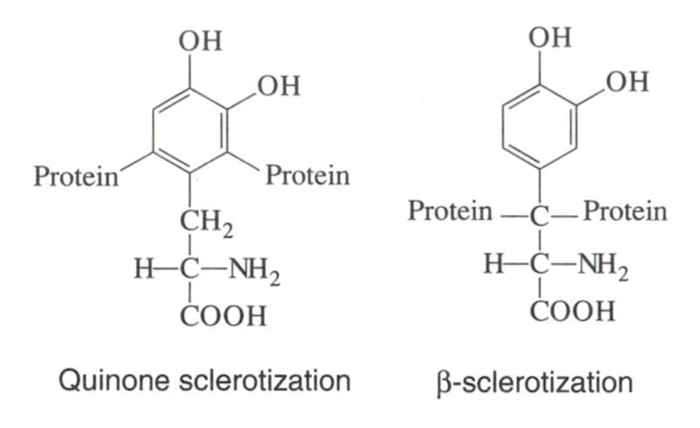

FIGURE 2.17 Differences between quinone sclerotization and β-sclerotization in where the cross-linked proteins are attached.

side-chain carbons are also believed to determine the type of sclerotized protein that results. The regulation of gene expression by the individual epidermal cells produces the enzymes, proteins, and tanning precursors that are present in localized areas of the cuticle and determine where and when the process of sclerotization occurs.

Sclerotization is regulated by at least two hormones. Ecdysteroids induce the epidermal cells to synthesize the enzyme dopa decarboxylase that is involved in the pathway toward the synthesis of NADA. The declining ecdysteroid titers that follow ecdysis then induce the release of the hormone bursicon, which is synthesized in neurosecretory cells in the brain and ganglia of the ventral nerve cord. Bursicon may increase the permeability of epidermal cells to tyrosine, allowing for the production of the tanning agents. It has also been suggested that bursicon mediates the permeability of the epidermal cells to hemolymph catecholamines.

THE MOLTING PROCESS

Although many holometabolous larvae can grow between molts, substantial growth in insects occurs only during and immediately after the molting process. Unsclerotized cuticle can accommodate some expansion for growth during an instar, but its stretching is limited by the extensibility of the folds in the epicuticle. Sclerotized cuticle is largely incapable of expansion beyond the limits set by the outer epicuticle. The molting process involves an elaborate sequence of events that produces a new cuticle capable of significant expansion before the old one is discarded. The process begins with the separation of the epidermal cells from the old cuticle, known as **apolysis,** and ends with the casting off of the old cuticle, known as **ecdysis.** It is the period between these two events when the components of the new cuticle are synthesized (Fig. 2.18).

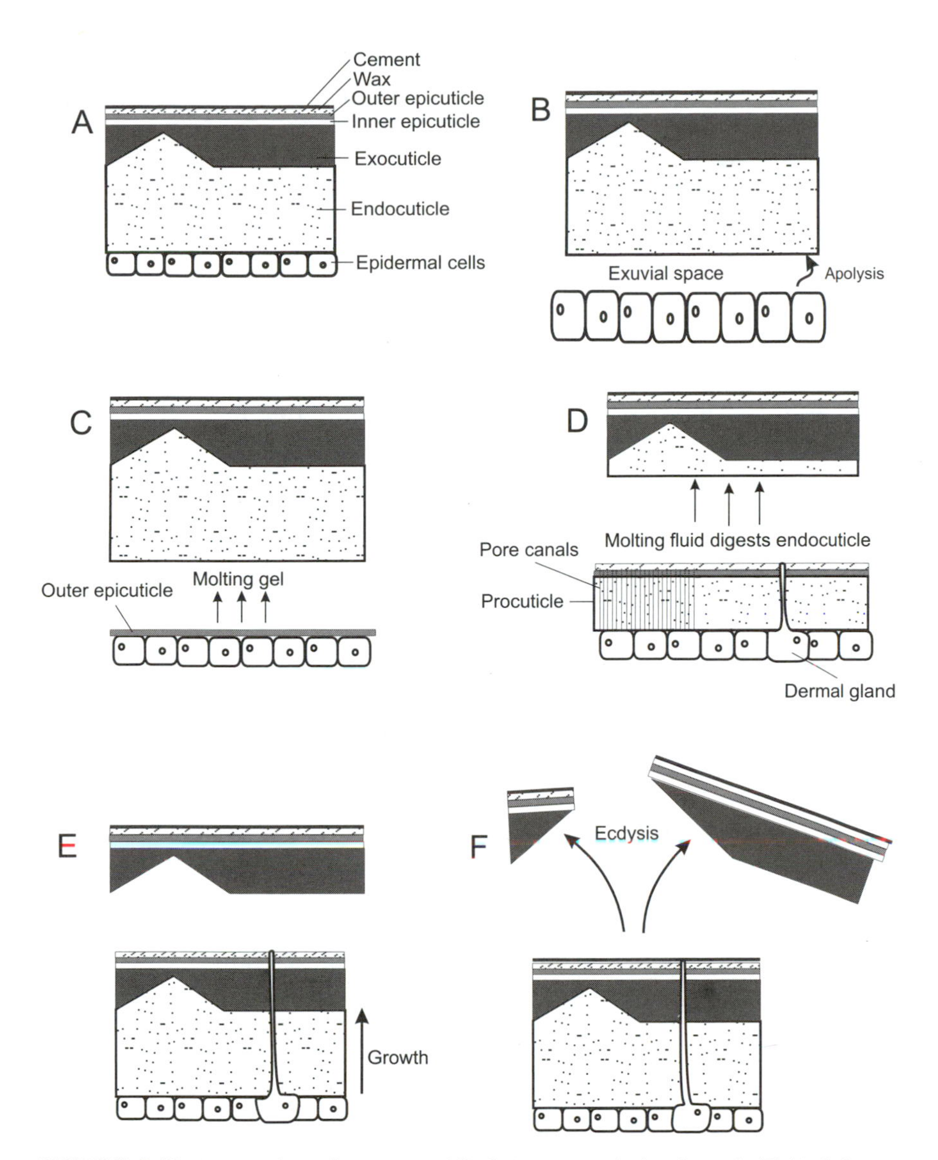

FIGURE 2.18 Steps in the molting process. (A) The integument before the molt. (B) Apolysis, separating the cuticle from the epidermis and creating an exuvial space. (C) Secretion of the molting inactive moptling gel into the exuvial space. (D) The digestion of the old endocuticle and the secretion of new procuticle. (E) Continued growth of the procuticle and epicuticle. (F) Ecdysis, the shedding of the old cuticle.

During the intermolt period, the epidermal cells are attached to the cuticle by plaques that anchor the base of the cuticle to the microvilli. At the beginning of the molting cycle, the epidermal cells undergo mitotic growth and increase their density and shape when hormones released by the endocrine system stimulate them. This causes apolysis to occur, the separation of the cuticle from the epidermis, and creates an area between the cuticle and epidermis, the **exuvial space.** The exuvial space fills with a **molting gel** that contains inactive enzymes including a chitinase and protease that will be capable of digesting the old cuticle above once they are activated. Soon afterward, the epidermal cells secrete a new outer epicuticle layer whose lipoproteins become tanned and impervious to these degradative enzymes when they later become activated by factors produced by the epidermal cells. The active enzymes, now called the **molting fluid,** are prevented from attacking the epidermal cells by the barrier provided by the outer epicuticle layer. The molting fluid is now free to begin the digestion of the old unsclerotized endocuticle, but not the sclerotized exocuticle. An estimated 90% of the materials in the old endocuticle may be resorbed and reused.

As the old cuticle is digested, the epidermal cells begin to secrete the new procuticle using the recycled raw materials. The synthesis of the new epicuticle continues with the formation of the inner epicuticle. With only the exocuticle remaining after digestion, the ecdysial lines become important points of weakness in the old cuticle. Just before ecdysis, the pore canals release the components of the wax layer on the surface of the outer epicuticle and assures the waterproofing of the new cuticle before the old cuticle is discarded. The molting fluid is then completely resorbed. A complex sequence of stereotyped behaviors follows, initiated by eclosion hormone, that causes the shedding of the old cuticle after the insect swallows air or water that expands its body and splits the old exocuticle along the ecdysial lines.

Following ecdysis, the new cuticle is soft and unsclerotized. It expands within the limits already set down by the outer epicuticle to express the increase in size that was the reason for the molt. The dermal glands release their products to the surface and create the cement layer that covers the wax layer. The tanning precursors are released into the new cuticle by the epidermal cells and the differentiation of the procuticle into the endocuticle and exocuticle is initiated. Growth of the endocuticle continues after ecdysis in many insects and does not stop until apolysis and the next molting cycle.

The epidermis is not the only portion of the insect that molts. All cells that are epidermal derivatives go through the same steps of apolysis through ecdysis. This includes the cells comprising the foregut, hindgut, and tracheal system.

Eclosion Behavior and Its Endocrine Regulation

The stereotyped behaviors that allow the insect to shed its old cuticle occur only at specific times in the life cycle. In the well-studied *M. sexta,* the behaviors associated with ecdysis have been divided into two phases: pre-ecdysis behavior and ecdysis behavior, both of which appear to be programmed into the central nervous system. During preecdysis behavior, the connections to the old cuticle are loosened through rotational movements of the abdomen. During ecdysis behavior, the loosened cuticle is shed by means of peristaltic contractions that move anteriorly along the length of the larva.

A cascade of neurohormones is responsible for eliciting eclosion behavior. At the appropriate time in the life cycle of the insect, an ecdysis-triggering hormone is released from epitracheal glands that are located on the trachea. This ecdysis-triggering hormone acts on the cells of the ventral nerve cord, causing the initial expression of pre-ecdysis behavior and the release of another hormone, eclosion hormone. The release of eclosion hormone is often gated so that it occurs only during a circadian window and accounts for the circadian molting activity of many insects. Its release forms a feedback loop with eclosion-triggering hormone and as levels of both increase, the expression of pre-ecdysis behavior continues. Eclosion hormone also initiates the release of the crustacean cardioactive peptide from the cells of the ventral ganglion that inactivates pre-ecdysis behavior and activates ecdysis behavior. The hormone is also involved in the plasticization of the cuticle that must occur in order for it to stretch immediately following ecdysis and for the release of bursicon that causes cuticular tanning (see Chapter 5, Behavioral Systems).

ENDOCRINE CONTROL OF MOLTING

The molting cycle is also initiated and regulated by the endocrine system, and molting occurs as a result of the coordinated activity of all the epidermal cells in the insect, with hormones providing the means by which this coordination occurs. The basic hormonal mechanism is identical whether the molt is larval to larval, larval to adult, larval to pupal, or pupal to adult. What does differ during metamorphosis is the nature of the cuticle that the insect molts to, and that is regulated by the presence or absence of juvenile hormone. The mechanism by which a molt results in a different kind of cuticle will be discussed in the next section.

The trigger for molting is generally correlated with some indicator of growth during the instar. In the few insects that have been studied, PTTH is secreted when a critical size is attained or when stretch receptors are triggered after a large meal is ingested. The PTTH then stimulates the release of ecdysone from the prothoracic glands, which is converted to 20-hydroxyecdysone, the active molting hormone. PTTH release is governed by a photoperiodic clock and takes place only during circadian gates when the growth-related stimulus occurs. The 20-hydroxyecdysone circulates in the hemolymph and activates the epidermal cells beginning with apolysis until ecdysis to begin the cycle of epidermal cell division and the synthesis of the new cuticle. The presence of ecdysteroid receptors and their particular varieties or isoforms in the cells at various stages of their developmental programs also determines whether and how the cells will respond to the hormone.

Insects that have been fed optimally ordinarily undergo their molts at fairly predictable times. The ultimate trigger for molting has only been established for a few insects and appears to be related to the capacity of the nervous system to monitor growth and release hormones appropriately. The blood-sucking bug, *R. prolixus,* takes a large blood meal once during each instar and then molts to the next instar within days afterward. In the absence of a blood meal or even when small blood meals are ingested that fail to produce enough abdominal distention, the insect does not molt. A large blood meal is necessary to trigger abdominal stretch receptors that cause the brain to release PTTH and initiate the molt. Larval *Oncopeltus fasciatus* milkweed bugs initiate a molt when they reach a critical weight, which is presumably monitored by stretch receptors that are activated when growth expands the abdomen.

ENDOCRINE CONTROL OF METAMORPHOSIS

The type of cuticle that is synthesized during a particular developmental stage, or that is characteristic of a certain region of the integument, depends largely on the proteins that are produced that give the cuticle its specific character. Thus, insect metamorphosis is a function of gene expression by epidermal cells and the temporal pattern of their protein synthesis. In contrast to a larval-to-larval molt, where the new cuticle produced is the same as the old one, a larval-to-adult molt in exopterygotes or the larval-to-pupal and pupal-to-adult molts in endopterygotes produce new cuticles that have different properties and contain different proteins. As in a larval-to-larval molt, ecdysteroids still trigger apolysis and the activation of the epidermal cells, but a metamorphic molt also requires the absence of juvenile hormone to reprogram the epidermal cells so they may produce the next stage-specific proteins. A molt that occurs in the presence of JH results in the same type of cuticle being formed, but in the absence of JH the epidermal cells become reprogrammed to produce the cuticle of the next instar. JH appears to act as a status quo agent, preventing epidermal cells from changing their pattern of protein synthesis in the presence of ecdysteroid. JH has an all-or-nothing effect on metamorphosis: if it is present when target cells are sensitive, the cells fail to change their developmental programs. This sensitivity results from the presence of developmentally appropriate hormone receptors.

The JH titer of holometabolous insects declines during the last larval instar, allowing PTTH to be released. This is followed by an increase in ecdysteroid titers as the PTTH activates the prothoracic glands. The last larval instar of holometabolous insects is characterized by a series of ecdysteroid releases, the first one being small and occurring in the absence of JH. This absence of JH in the presence of an ecdysteroid peak changes the commitment of the epidermal cells so that they will produce a pupal cuticle when a second increase in ecdysteroid titers occurs. The critical periods for the presence or absence of JH are associated with periods of juvenile hormone sensitivity of the epidermal cells (Fig. 2.19).

As with the decision about when to molt, the decision of when to undergo metamorphosis is ultimately determined by the insect's ability to estimate its own size and its developmental arrival in the last larval instar. In the larval lepidopteran, *M. sexta,* two indications of size are used. One indicator is the size of the sclerotized head capsule, and the other is the weight of the developing larva. The size of the head capsule tells the larva in which instar the metamorphic molt will occur. Once that recognition occurs, the weight within that final larval instar determines

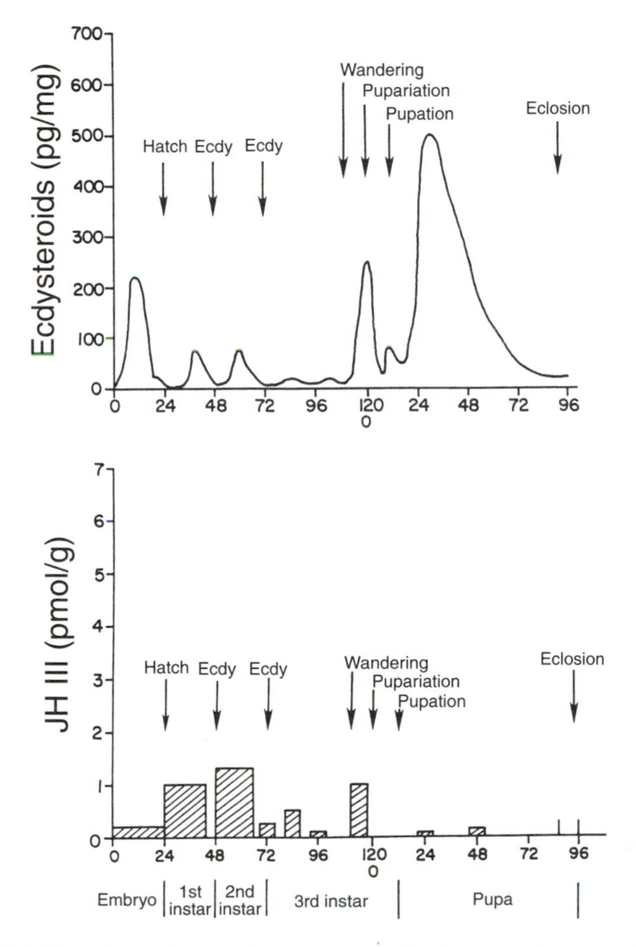

FIGURE 2.19 The correlation of hormone levels with developmental events. (Top) Ecdysteroids. (Bottom) Juvenile hormone III. Reprinted with permission from Riddiford, L.M. (1993). *The Development of Drosophila melanogaster,* Volume 2, pp. 899–939. Copyright Cold Spring Harbor Press.

the time during the instar that the metamorphic process begins. The mechanism of how the larva is able to measure these two indicators is not yet known, but it may again be related to stretch receptors.

Whether the larva behaves developmentally as if it were in its last larval instar depends on whether the width of the *Manduca* head capsule is greater than 5 mm. If this width is less than 5 mm, no matter what the developmental or nutritional history of the larva, it will not become a pupa at its next molt and will instead molt to an extra, or sixth larval instar. Thus, there is a threshold size that is required for metamorphosis in that all larvae that enter the last instar with a head capsule width of at least 5 mm will metamorphose during that instar (Fig. 2.20).

There are several physiological changes that occur once the *Manduca* larva is in what has been determined to be its last larval instar. During this last instar, as will be discussed later, the imaginal discs are committed to the production of pupal cuticle. A major change, however, involves the development of sensitivity to JH. Unlike in earlier instars, the brain is no longer able to secrete PTTH in the presence of JH. Similarly, the prothoracic glands cannot respond to PTTH if it is released in the presence of JH. Because during the last instar both the brain and the prothoracic gland develop this sensitivity to JH, neither PTTH nor ecdysteroid can be released, and a molt initiated, until all circulating JH has been cleared from the insect. The attainment of a weight of 5 g during this last instar is correlated with the development of these events and, presumably, the failure of the corpora allata to continue to secrete JH once the insect reaches this weight. After reaching a weight of 5 g, the last instar larva takes about a day to clear all JH from its sys-

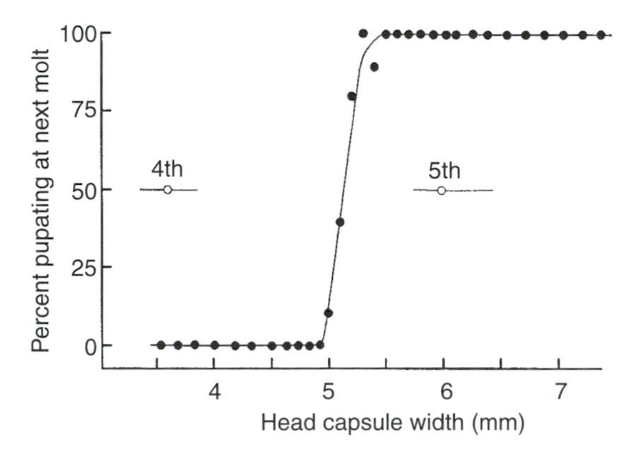

FIGURE 2.20 The relationship between the size of the *Manduca* larva and its tendency to pupate and undergo metamorphosis. By manipulating diet, larvae of a wide range of head capsule sizes can be produced. Those larvae with head capsules greater than 5 mm tend to pupate at their next molt, while those with head capsules less than 5 mm tend to first molt to another larval instar. The horizontal bars show the means and ranges of head capsule sizes for fourth and fifth (last larval instar) instar larvae. From Nijhout (1975). Reprinted with permission.

tem, and in the absence of the inhibition by JH, the brain secretes PTTH during the next allowable circadian gate that leads to a metamorphic molt.

METAMORPHOSIS AND THE RADICALLY CHANGING CUTICLE

Whether they be endopterygotes or exopterygotes, when insects molt from one larval stage to another a new cuticle is produced that is similar to the one being replaced. In contrast, the molt to an adult is associated with a cuticle of a very different nature and often completely different exoskeletal structures. In some insects, such as Lepidoptera, the epidermal cells secrete the larval cuticles as well as the different pupal and adult cuticles, all of which have characteristic proteins and overall patterns. In other insects, such as the higher Diptera, most of the larval epidermal cells are completely replaced by imaginal discs that are epidermal cells that remain undifferentiated during the larval stage but form the various structures associated with the adult, including wings, legs, mouthparts, and the cuticle of the body wall. Thus, there are two approaches to the changing pattern at metamorphosis: larval epidermal cells can change their commitment and alter their developmental secretions to produce various cuticular proteins or be replaced by new cells that form the radically different structures.

Changes in Cuticle Commitment

The ability of larval epidermal cells to change their pattern of secretion and produce a new pupal cuticle depends on the hormonal conditions during the larval–pupal molt and the sensitivity of target cells to the hormones. As discussed previously, the windows of sensitivity depend on the presence of cellular receptors that are able to recognize and bind the hormones. During the larval–pupal transformation, a small peak of ecdysteroids occurs when the levels of JH are reduced. This is the first time during the life of the insect that cells are exposed to ecdysteroids in the absence of JH. The exposure to ecdysteroid without JH changes the commitment of the cells so when they are next exposed to a higher concentration of ecdysteroid shortly afterwards, they produce a pupal cuticle. The general rule they subscribe to is this: if JH is present with ecdysteroids, continue the current developmental program. If JH is absent and ecdysteroids are present alone, go on to the next developmental state (Fig. 2.21).

We have already seen that ecdysteroids are able to induce puffing at specific sites on the chromosome. The puffing represents mRNA synthesis, and a sequence of gene activation was derived from the sequence of puffing. Many of the early genes induced by ecdysteroid produce transcription factors that can activate or inactivate other late genes. At metamorphosis, these ecdysteroid-induced transcription factors

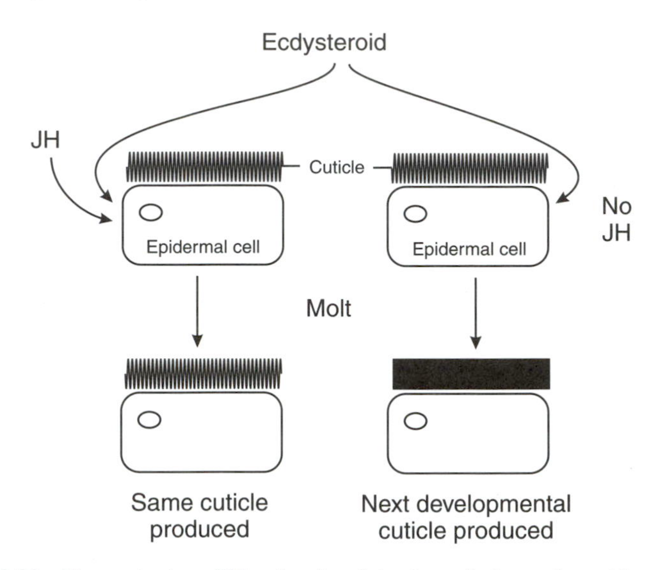

FIGURE 2.21 The mechanism of JH action. A molt is triggered when ecdysteroids act on the epidermal cells. If JH is present during a critical period during which time JH receptors are present on the cell, the epidermal cells produce the same cuticle when they molt. If JH is absent during this critical period and the receptors go unfilled, the epidermal cells produce the next developmental cuticle that they have been genetically programmed to synthesize.

appear to change, and the new transcription factors induced by ecdysteroid in the absence of JH may activate the new genes and produce the structural proteins characteristic of the next developmental stage.

Imaginal Disc Development

Metamorphosis in holometabolous insects involves an extensive remodeling of the larva to form the adult. Cells that comprise some larval structures die, while other cells are able to persist and synthesize the new products that are characteristic of the new instar. Because adult reconstruction may be beyond the ability of some cells, the new structures are formed from special epidermal cells whose growth is suppressed during larval life and that have no functions in the survival of larvae. These cells are often grouped into distinct imaginal discs or more loosely grouped clusters of histoblasts. Imaginal cells are not unique to holometabola as some hemimetabola also have imaginal histoblasts that replace larval cells. They are employed to different degrees within the holometabola. In higher Diptera and Hymenoptera, the adult epidermis is completely derived from epidermal cells, while in some Lepidoptera and Coleoptera, parts of the epidermis may be derived from discs and some from larval epidermal cells that survive metamorphosis.

Most of the information we have about imaginal discs comes from work with *Drosophila*. Different discs form the different body parts of the adult fly, each of which contributes to the formation of a different structure. The adult head and thorax are formed from 20 discs, and the abdomen is derived from about 50 histoblasts plus one imaginal disc that gives rise to the adult genitalia. Each of the discs differentiates into a variety of cell types and organs that constitute the adult structures (Fig. 2.22). The wing disc, for example, contains about 50,000 cells at maturity that give rise to the structures that comprise the wing as well as the portions of the thorax to which it is attached.

The imaginal discs arise as invaginations or thickenings of the ectoderm during late embryogenesis. Their developmental fates are determined in the late blastoderm stage. During larval development, the cells of the discs undergo division and increase in size, attaining maturity during the last larval instar. A mature disc contains columnar epithelial cells surrounded in part by a peripodial membrane and attached to the larval epidermis by a stalk (Fig. 2.23). A contractile belt of actin

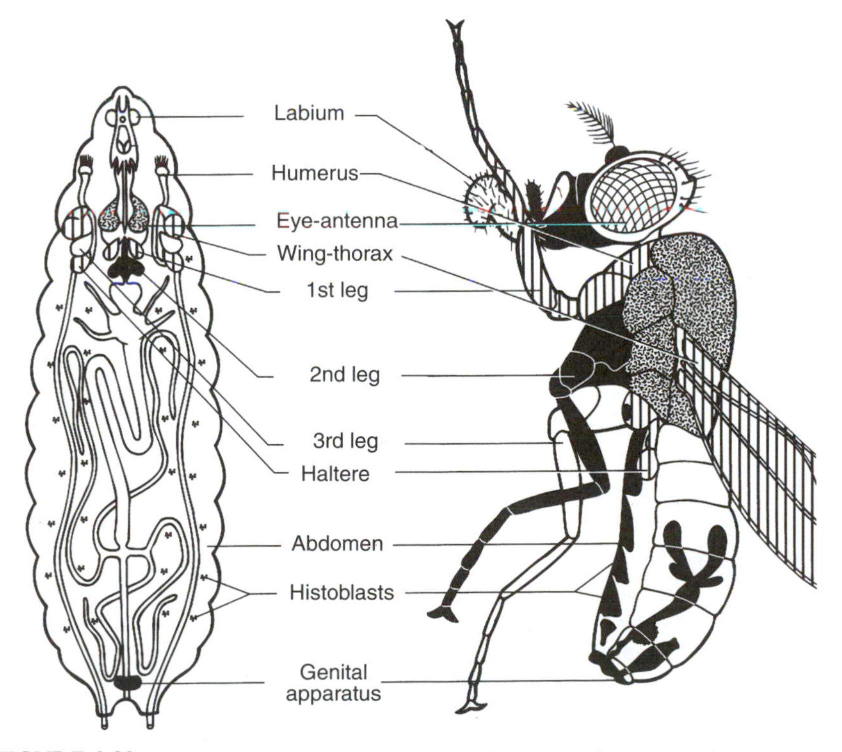

FIGURE 2.22 The imaginal discs of a larval *Drosophila* (left) and the corresponding structures in the adult (right) to which they give rise. From Nothiger (1972). Reprinted with permission.

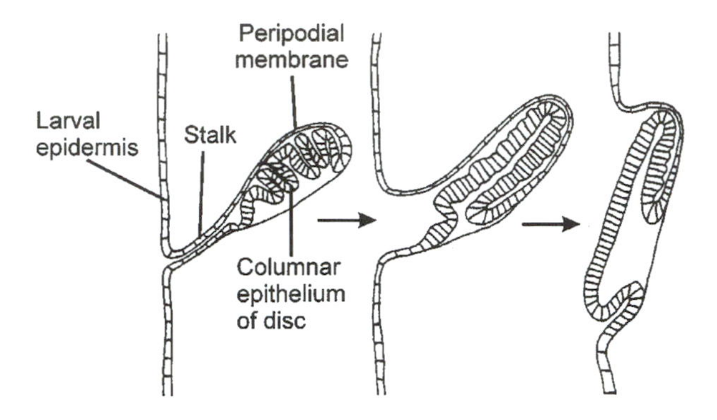

FIGURE 2.23 The evagination of a leg disc during *Drosophila* development. The columnar epithelium of the disc is folded with the remainder of the disc forming the peripodial membrane. In response to 20 hydroxyecdysone, the cells of the disc evaginate, allowing the disc surface to assume the structure of the new leg. The larval epithelial cells ultimately fuse with the disc cells to form a continuous sheet of epidermal cells. From Milner and Bleasby (1985). Reprinted with permission.

and myosin is also present. At the end of the last larval instar the cells switch from proliferation to differentiation and the discs undergo an elongation and eversion mediated by changes in cell shape and alterations in cell attachments that are driven by the actin–myosin belts. Cell rearrangement may occur with the expression of new gene products such as the cadherins that are calcium-dependent adhesion molecules that function in intracellular adhesion. The eversion causes the apical surface of the disc to become positioned on the outside of the epithelium. The cells then spread from each appendage and fuse with the cells from other discs to form the continuous epidermis of the body wall. Cuticle formation and disc morphogenesis occur at different times, with elongation and eversion possible only when the epidermal cells are not burdened with the overlying cuticle. The ungrouped imaginal histoblasts proliferate during metamorphosis, replacing the larval epidermis as they spread.

Disc development and the metamorphosis that results are regulated by both JH and ecdysteroid. JH is present in *Drosophila* during the early larval stages but decreases during the third instar, with a brief peak just before pupariation. This presence of JH during the larval stages allows the cells in the discs to proliferate without differentiating. During this time, JH prevents the discs from responding with a change in commitment when ecdysteroid is present. Differentiation of the discs occurs after pupariation and is the result of high ecdysteroid levels in the absence of JH. The evagination of the disc is followed by the deposition of new cuticle, which occurs only once ecdysteroids decline. A shift from the secretion of exocuticle to endocuticle results from a subsequent small peak of ecdysteroid that occurs slightly later.

Metamorphosis, Cell Reorganization, and Cell Death

The sometimes radical changes that occur during metamorphosis allow the life stages of the holometabolous insect to occupy radically different ecological niches. The terrestrial larval stage is mainly focused on feeding and growth and spends its time crawling. In contrast, the winged adult is occupied with reproduction and dispersal and has a completely different diet. These changes are made possible not only by the appearance of novel structures that arise from imaginal discs during development but also by the disappearance of many larval tissues and their capacity to change their structure and function in the adult.

The prothoracic glands produce the ecdysteroids that trigger the molting process and are essential for the larval stage to undergo its periodic molts. However, adult pterygotes no longer molt and the glands no longer have any function in this stage. They are absent in most adult insects due to their programmed degeneration during metamorphosis. In both *Manduca* and *Drosophila,* the glands degenerate during the pupal–adult transition as a result of their exposure to the peak of ecdysteroids that they themselves produce in the absence of JH secretion. Similarly, the silk glands that are used by Lepidopteran larvae to form the cocoon for pupation begin to degenerate after the first ecdysteroid peak that occurs in the absence of JH during the last larval instar. The larval midgut and salivary gland cells of *Drosophila* undergo a histolysis in response to the peak of ecdysteroid during the prepupal–pupal molt. The reformation of the adult midgut occurs more slowly, with the adult gut forming around the degenerating larval gut, and is preceded by a cessation of larval feeding behavior that may minimize the complications that food in the rearranging gut might create.

The cells of the larval fat body are different from that of the adult in form and function. The larval fat body of *Drosophila* is a lobed monolayer of polytene cells that separate, become depleted of glycogen, lipid, and protein, and subsequently degenerate during the early adult stages. The cell separation results from the production of the proteolytic enzyme cathepsin that is produced by hemocytes in *Sarcophaga*. The adult fat body is arranged in sheets just below the epidermis and develops from precursor cells. Another method of larval fat body restructuring occurs in the lepidopteran *Calpodes,* which is arranged in ribbons one cell thick and reforms in the adult as lobes attached to trachea.

Structural changes in appendages and the body wall also require changes in muscle and nervous innervation. During metamorphosis, the larval muscles of insects can either die or become remodeled. *Manduca* larvae bear abdominal prolegs that aid in crawling, and these structures are not present in the flying adult. The muscles that retract the larval prolegs and allow their movement to occur degenerate on the day after the wandering stage begins. There is a second wave of muscle degeneration that occurs after adult emergence when larval muscles that are also used for early adult ecdysial behaviors disappear. The larval intersegmental muscles of *Manduca*, which are also used during adult ecdysis to escape from the pupal skin, are delayed in their death until after this event occurs. The timing of the two waves of muscle degeneration in *Manduca* during metamorphosis reflects the delicate balance between the termination of larval functions and the development of adult functions during this period. The remodeling of muscles also occurs. After the degeneration of the contractile apparatus of some larval muscles and the loss of muscle cell nuclei, the remains are used to build the adult muscles. New muscle attachment sites are created and the remodeled muscle gains a function distinct from its larval predecessor.

The motor neurons that innervate these muscles also change during metamorphosis. The increase in 20-hydroxy-ecdysone with molting triggers the deaths of the neurons supplying the larval prolegs, but the delayed death of the intersegmental muscle motor neurons is a result of the declining levels of 20-hydroxy-ecdysone after adult emergence. The presence of the hormone sustains this population of neurons during larval development and the absence of the hormone after the pupal–adult molt causes their decline.

A more fundamental reorganization of the nervous system also occurs during metamorphosis. The nervous system of the newly hatched hemimetabolous insect is essentially complete and does not change substantially during its metamorphosis. Except for a few neural additions to accommodate the appearance of the adult compound eyes, the hemimetabolous larva begins life essentially with an adult nervous system. However, holometabolous larvae differ drastically from adults in locomotion, sensory reception, and behavior, and the way the nervous system couples this new sensory information to novel motor outputs must change radically to accommodate these differences. Not only must newly formed structures be innervated and the nerves providing the connections to old structures be redesignated or terminated, but the new interneurons that form the connections between the sensory input and motor output must also be created.

Many neurons of the holometabolous adult arise from embryonic neuroblasts that are arrested until metamorphosis and then proliferate to create the interneurons that link sensory and motor nerves. As a result of this developmental arrest, larval holometabola tend to have fewer interneurons in the central nervous system than do larval hemimetabola, and this relatively simplified system may explain their relatively simple behaviors and the reasons that hemimetabolous larvae are more adult-like.

The neurons that innervate larval structures destined to disappear in the adult, such as the larval prolegs, die soon after the larval–pupal molt. Other motor neurons that innervate larval muscles also used during adult emergence die soon after adult development is complete. Approximately half the neurons in the central nervous system of the larva also die shortly after adult emergence. Other larval neurons are remodeled and used for other functions in the adult.

The death of some neurons during the larval–pupal molt is also a function of the hormonal conditions that exist during the last larval instar. Two peaks of ecdysteroid occur, the first of which occurs in the absence of JH and has already been identified as the so-called commitment peak, which is the first time in the life of the organism that ecdysteroids appear without JH. When the neurons that innervate the disappearing larval prolegs are exposed to this commitment peak, they become more sensitive to ecdysteroids and degenerate several days after being exposed to the second larger ecdysteroid peak.

The temporal sequence of degeneration of larval cells and their differing responses to ecdysteroids as the trigger for degeneration may be related to the multiple isoforms of the ecdysteroid receptor, EcR. The three EcR isoforms that are currently known to combine with the other nuclear receptor, USP, to form the heterodimer that binds with ecdysteroids are EcR-A, EcR-B1, and EcR-B2. Larval neurons do not bear any EcR receptors until the last larval instar, when EcR-B1 and EcR-A predominate. High levels of EcR-A are correlated with cell death while the presence of EcR-B1 in a cell is correlated with the tendency of neurons to undergo remodeling.

ADDITIONAL REFERENCES

Structure of the Integument

Andersen, S. O. 2001. Matrix proteins from insect pliable cuticles: Are they flexible and easily deformed? *Insect Biochem. Mol. Biol.* **31:** 445–452.

Andersen, S. O., Weis-Fogh, T. 1964. Resilin. A rubberlike protein in arthropod cuticle. *Adv. Insect Physiol.* **2:** 1–66.

Beament, J. W. L. 1959. The waterproofing mechanism of arthropods I. The effect of temperature on cuticle permeability in terrestrial insects and ticks. *J. Exp. Biol.* **36:** 391–422.

Brey, P. T., Lee, W. J., Yamakawa, M., Koizumi, Y., Perrot, S., Francois, M., Ashida, M. 1993. Role of the integument in insect immunity: Epicuticular abrasion and induction of cecropin synthesis in cuticular epithelial cells. *Proc. Natl. Acad. Sci. USA* **90:** 6275–6279.

Brunet, P. C. J. 1952. The formation of the oothecae by *Periplaneta americana*. II. The structure and function of the left colleterial gland. *Q. J. Microscop. Sci.* **93:** 47–69.

Csikos, G., Molnar, K., Borhegyi, N. H., Talian, G. C., Sass, M. 1999. Insect cuticle, an *in vivo* model of protein trafficking. *J. Cell Sci.* **112:** 2113–2124.

Dotson, E. M., Cornel, A. J., Willis, J. H., Collins, F. H. 1998. A family of pupal-specific cuticular protein genes in the mosquito *Anopheles gambiae*. *Insect Biochem. Mol. Biol.* **28:** 459–472.

Fogal, W., Fraenkel, G. 1969. Melanin in the puparium and adult integument of the fleshfly, *Sarcophaga bullata*. *J. Insect Physiol.* **15:** 1437–1447.

Fuzeau-Braesch, S. 1972. Pigments and color changes. *Annu. Rev. Entomol.* **17:** 403–424.

Gibbs, A. G. 1998. Water-proofing properties of cuticular lipids. *Am. Zool.* **38:** 471–482.

Hadley, N. F. 1982. Cuticle ultrastructure with respect to the lipid waterproofing barrier. *J. Exp. Zool.* **222:** 239–248.

Hadley, N. F. 1986. The arthropod cuticle. *Sci. Am.* **255:** 104–112.

Hepburn, H. R. 1985. Structure of the integument. In *Comprehensive insect physiology, biochemistry and pharmacology* (G. A. Kerkut, L. I. Gilbert, Eds.), vol. 3, pp. 1–58. Pergamon, Oxford.

Jungreis, A. M., Ruhoy, M., Cooper, P. D. 1982. Why don't tobacco hornworms (*Manduca sexta*) become dehydrated during larval-pupal and pupal-adult development? *J. Exp. Zool.* **222:** 265–276.

Kayser, H. 1985. Pigments. In *Comprehensive insect physiology, biochemistry and pharmacology* (G.A. Kerkut, L. I. Gilbert, Eds.), vol. 10, pp. 367–415. Pergamon, Oxford.
Kramer, K. J., Koga, D. 1986. Insect chitin. Physical state synthesis degradation and metabolic regulation. *Insect Biochem.* **16:** 851–877.
Kramer, K. J., Muthukrishnan, S. 1997. Insect chitinases: Molecular biology and potential use as biopesticides. *Insect Biochem. Mol. Biol.* **27:** 887–900.
Locke, M. 1974. The structure and formation of the integument in insects. *The Physiology of Insecta* vol. 6, pp. 123–213. Academic Press, New York.
Locke, M. 1991. Insect epidermal cells. In *Physiology of the insect epidermis*, pp. 1–22. CSIRO.
Locke, M., Kiss, A., Sass, M. 1994. The cuticular localization of integument peptides from particular routing categories. *Tissue Cell* **26:** 707–734.
Lockey, K. H. 1991. Insect hydrocarbon classes: Implications for chemotaxonomy. *Insect Biochem.* **21:** 91–97.
Machin, J., Lampert, G. J., O'Donnell, M. J. 1985. Component permeabilities and water contents in *Periplaneta* integument: Role of the epidermis re-examined. *J. Exp. Biol.* **117:** 155–169.
Marcu, O., Locke, M. 1998. A cuticular protein from the moulting stages of an insect. *Insect Biochem. Mol. Biol.* **28:** 659–669.
Neville, A. C. 1983. Daily cuticular growth layers and the teneral stage in adult insects: A review. *J. Insect Physiol.* **29:** 211–219.
Neville, A. C. 1984. Cuticle: Organization. In *Biology of the integument* (J. Bereiter-Hahn, A. G. Matoltsy, K. S. Richards, Eds.), vol. 1, pp. 611–625. Springer-Verlag, Berlin.
Rebers, J. E., Niu, J., Riddiford, L. M. 1997. Structure and spatial expression of the *Manduca sexta* MSCP14. 6 cuticle gene. *Insect Biochem. Mol. Biol.* **27:** 229–240.
Rebers, J. E., Riddiford, L. M. 1988. Structure and expression of a *Manduca sexta* larval cuticle gene homologous to *Drosophila* cuticle genes. *J. Mol. Biol.* **203:** 411–423.
Rudall, K. M. 1963. The chitin/protein complexes of insect cuticles. *Adv. Insect Physiol.* **1:** 257–313.
Wolfgang, W. J., Riddiford, L. M. 1987. Cuticular mechanics during larval development of the tobacco hornworm, *Manduca sexta*. *J. Exp. Biol.* **128:** 19–34.
Yarema, C., Mclean, H., Caveney, S. 2000. L-glutamate retrieved with the moulting fluid is processed by a glutamine synthetase in the pupal midgut of *Calpodes ethlius*. *J. Insect Physiol.* **46:** 1497–1507.

Chemistry of the Integument

Andersen, S. O. 1976. Cuticular enzymes and sclerotization in insects. In *The insect integument* (H. R. Hepburn, Ed.), pp. 121–141.
Andersen, S. O. 1979. Biochemistry of insect cuticle. *Annu. Rev. Entomol.* **24:** 29–61.
Andersen, S. O. 1981. The stabilization of locust cuticle. *J. Insect Physiol.* **27:** 393–396.
Andersen, S. O. 1998. Amino acid sequence studies on endocuticular proteins from the desert locust, *Schistocerca gregaria*. *Insect Biochem. Mol. Biol.* **28:** 421–434.
Andersen, S. O. 2001. Matrix proteins from insect pliable cuticles: Are they flexible and easily deformed? *Insect Biochem. Mol. Biol.* **31:** 445–452.
Andersen, S. O., Hojrup, P., Roepstorff, P. 1995. Insect cuticular proteins. *Insect Biochem. Mol. Biol.* **25:** 153–176.
Andersen, S. O., Peter, M. G., Roepstorff, P. 1996. Cuticular sclerotization in insects. *Comp. Biochem. Physiol.* **113B:** 689–705.
Andersen, S. O., Weis-Fogh, T. 1964. Resilin. A rubberlike protein in arthropod cuticle. *Adv. Insect Physiol.* **2:** 1–66.
Bade, M. L. 1974. Localization of moulting chitinase in insect cuticle. *Biochim. Biophys. Acta* **372:** 474–477.

Brunet, P. C. J. 1980. The metabolism of the aromatic amino acids concerned in the cross-linking of insect cuticle. *Insect Biochem.* **10:** 467–500.

Cox, D. L., Willis, J. H. 1985. The cuticular proteins of *Hyalophora cecropia* from different anatomical regions and metamorphic stages. *Insect Biochem.* **15:** 349–362.

Espelie, K. E., Hermann, H. R. 1990. Surface lipids of the social wasp *Polistes annularis* (L.) and its nest and nest pedicel. *J. Chem. Ecol.* **16:** 1841–1851.

Gibbs, A. G., Louie, A. K., Ayala, J. A. 1998. Effects of temperature on cuticular lipids and water balance in a desert *Drosophila*: Is thermal acclimation beneficial? *J. Exp. Biol.* **201:** 71–80.

Hackman, R. H. 1971. The integument of arthropoda. In *Chemical zoology* (M. Florkin, B. T. Sheer, Eds.), vol. 6, pp. 1–62. Academic Press, New York.

Hackman, R. H. 1982. Structure and function in tick cuticle. *Annu. Rev. Entomol.* **27:** 75–95.

Hackmann, R. H. 1986. Chemical nature of the outer epicuticle from *Lucilia cuprina* larvae. *Insect Biochem.* **16:** 911–916.

Hackman, R. H., Goldberg, M. 1987. Comparative study of some expanding arthropod cuticles: The relation between composition structure and function. *J. Insect Physiol.* **33:** 39–50.

Hadley, N. F. 1980. Surface waxes and integumentary permeability. *Am. Sci.* **68:** 546–553.

Hopkins, T. L., Kramer, K. J. 1992. Insect cuticle sclerotization. *Annu. Rev. Entomol.* **37:** 273–302.

Hopkins, T. L., Morgan, T. D., Aso, Y., Kramer, K. J. 1982. N-β-alanyldopamine: Major role in insect cuticle tanning. *Science* **217:** 364–366.

Jensen, U. G., Rothmann, A., Skou, L., Andersen, S. O., Roepstorff, P., Hojrup, P. 1997. Cuticular proteins from the giant cockroach, *Blaberus craniifer*. *Insect Biochem. Mol. Biol.* **27:** 109–120.

Jespersen, S., Hojrup, P., Andersen, S. O., Roepstorff, P. 1994. The primary structure of an endocuticular protein from two locust species, *Locusta migratoria* and *Schistocerca gregaria*, determined by a combination of mass spectrometry and automatic Edman degradation. *Comp. Biochem. Physiol.* B **109:** 125–138.

Kramer, K. J., Koga, D. 1986. Insect chitin. Physical state, synthesis, degradation and metabolic regulation. *Insect Biochem.* **16:** 851–877.

Reynolds, S. E., Samuels, R. I. 1996. Physiology and biochemistry of insect molting fluid. *Adv. Insect Physiol.* **26:** 157–232.

Sass, M., Kiss, A., Locke, M., 1993. Classes of integument peptides. *Insect Biochem. Mol. Biol.* **23:** 845–857.

Sass, M., Kiss, A., Locke, M. 1994. Integument and hemocyte peptides. *J. Insect Physiol.* **40:** 407–421.

Sugumaran, M. 1998. Unified mechanism for sclerotization of insect cuticle. *Adv. Insect Physiol.* **27:** 229–334.

Sugumaran, M., Saul, S. J., Semensi, V. 1988. On the mechanism of formation of *N*-acetyldopamine quinone methide in insect cuticle. *Arch. Insect Biochem. Physiol.* **9:** 269–282.

Willis, J. H. 1987. Cuticular proteins: The neglected component. *Arch. Insect Biochem. Physiol.* **6:** 203–215.

Willis, J. H. 1999. Cuticular proteins in insects and crustaceans. *Am. Zool.* **39:** 600–609.

Insect Growth

Bernays, E. A. 1986. Evolutionary contrasts in insects: Nutritional advantages of holometabolous development. *Physiol. Entomol.* **11:** 377–382.

Carlson, J. R., Bentley, D. 1977. Ecdysis: Neural orchestration of a complex behavioral performance. *Science* **195:** 1006–1008.

Cho, K. O., Chern, J., Izaddoost, S., Choi, K. W. 2000. Novel signaling from the peripodial membrane is essential for eye disc patterning in *Drosophila*. *Cell* **103:** 331–342.

Cole, B. J. 1980. Growth ratios in holometabolous and hemimetabolous insects. *Ann. Entomol. Soc. Am.* **73:** 489–491.

Emlen, D. J. 2000. Integrating development with evolution: A case study with beetle horns. *BioScience* **50:** 403–418.

Emlen, D. J., Nijhout, H. F. 1999. Hormonal control of male horn length dimorphism in the dung beetle *Onthophagus taurus* (Coleoptera: Scarabaeidae). *J. Insect Physiol.* **45:** 45–53.

Emlen, D. J., Nijhout, H. F. 2000. The development and evolution of exaggerated morphologies in insects. *Annu. Rev. Entomol.* **45:** 661–708.

Evans, J. D., Wheeler, D. E. 1999. Differential gene expression between developing queens and workers in the honey bee, *Apis mellifera. Proc. Natl. Acad. Sci. USA* **96:** 5575–80.

Ewer, J., Wang, C. M., Klukas, K. A., Mesce, K. A., Truman, J. W., Fahrbach, S. E. 1998. Programmed cell death of identified peptidergic neurons involved in ecdysis behavior in the moth, *Manduca sexta. J. Neurobiol.* **37:** 265–280.

Hinton, H. E. 1948. On the origin and function of the pupal stage. *Trans. R. Entomol. Soc. London.* **99:** 395–409.

Hinton, H. E. 1963. The origin and function of the pupal stage. *Proc. R. Entomol. Soc. London. A* **38:** 77–85.

Hinton, H. E. 1971. Some neglected phases in metamorphosis. *Proc. R. Entomol. Soc. London. C* **35:** 55–64.

Hinton, H. E. 1973. Neglected phases in metamorphosis: A reply to V. B. Wigglesworth. *J. Entomol. A* **48:** 57–68.

Hinton, H. E. 1976. Notes on neglected phases in metamorphosis and a reply to J. M. Whitten. *Ann. Entomol. Soc. Am.* **69:** 560–566.

Jenkin, P. M., Hinton, H. E. 1966. Apolysis in arthropod moulting cycles. *Nature* **211:** 871.

McNabb, S. L., Baker, J. D., Agapite, J., Steller, H., Riddiford, L. M., Truman, J. W. 1997. Disruption of a behavioral sequence by targeted death of peptidergic neurons in *Drosophila. Neuron* **19:** 813–823.

Milner, M. J., Bleasby, A. J. 1985. The alignment of imaginal anlagen during the metamorphosis of *Drosophila melanogaster.* In *Metamorphosis* (M. Balls, M. Bownes, Eds.), pp. 20–35. Clarendon, Oxford.

Nijhout, H. F., Williams, C. M. 1974. Control of moulting and metamorphosis in the tobacco hornworm, *Manduca sexta* (L.): Growth of the last-instar larva and the decision to pupate. *J. Exp. Biol.* **61:** 481–491.

Nijhout, H. F. 1990. Metaphors and the role of genes in development. *Bioessays* **12:** 441–446.

Nijhout, H. F. 1999. Control mechanisms of polyphenic development in insects. *BioScience* **49:** 181–192.

Nijhout, H. F., Emlen, D. J. 1998. Competition among body parts in the development and evolution of insect morphology. *Proc. Natl. Acad. Sci. USA* **95:** 3685–3689.

Nothiger, R. 1972. The larval development of imaginal disks. In *The biology of imaginal disks* (H. Ursprung, R. Nothiger, Eds.), pp. 1–34, Springer-Verlag, Amsterdam.

Safranek, L., Williams, C. M. 1984. Critical weights for metamorphosis in the tobacco hornworm, *Manduca sexta. Biol. Bull.* **167:** 555–567.

Safranek, L., Williams, C. M. 1984. Determinants of larval molt initiation in the tobacco hornworm, *Manduca sexta. Biol. Bull.* **167:** 568–578.

Sehnal, F. 1985. Morphology of insect development. *Annu. Rev. Entomol.* **30:** 89–109.

Stern, D. L., Emlen, D. J. 1999. The developmental basis for allometry in insects. *Development* **126:** 1091–1101.

Tanaka, A. 1981. Regulation of body size during larval development in the German cockroach, *Blattella germanica. J. Insect Physiol.* **27:** 587–592.

Truman, J. W., Riddiford, L. M. 1999. The origins of insect metamorphosis. *Nature* **401:** 447–452.

Whitten, J. M. 1976. Definition of insect instars in terms of 'apolysis' or 'ecdysis'. *Ann. Entomol. Soc. Am.* **69:** 556–559.

Wigglesworth, V. B. 1973. The significance of "apolysis" in the moulting of insects. *J. Entomol.* **47:** 141–149.

Endocrine Control of Molting and Metamorphosis

Bayer, C., Von Kalm, L., Fristrom, J. W. 1996. Gene regulation in imaginal disc and salivary gland development during *Drosophila* metamorphosis. In *Metamorphosis: Postembryonic reprogramming of gene expression in amphibian and insect cells* (L. I. Gilbert, J. R. Tata, B. G. Atkinson, Eds.), pp. 321–361. Academic Press, San Diego.

Cohen, S. M. 1993. Imaginal disc development. In *The Development of Drosophila* (M. Bate, A. Martinez-Arias, Eds.), vol II, pp. 747–841. Cold Spring Harbor Laboratory Press, Cold Spring Harbor, NY.

Consoulas, C., Levine, R. B. 1997. Accumulation and proliferation of adult leg muscle precursors in *Manduca* are dependent on innervation. *J. Neurobiol.* **32:** 531–553.

Dai, J.-D., Gilbert, L. I. 1997. Programmed cell death of the prothoracic glands of *Manduca sexta* during pupal-adult metamorphosis. *Insect Biochem. Mol. Biol.* **27:** 69–78.

Dyer, K., Thornhill, W. B., Riddiford, L. M. 1981. DNA synthesis during the change to pupal commitment of *Manduca* epidermis. *Dev. Biol.* **84:** 425–431.

Ewer, J., Gammie, S. C., Truman, J. W. 1997. Control of insect ecdysis by a positive-feedback endocrine system: Roles of eclosion hormone and ecdysis triggering hormone. *J. Exp. Biol.* **200:** 869–881.

Fahrbach, S. E. 1997. The regulation of neuronal death during insect metamorphosis. *BioScience* **47:** 77–85.

Fechtel, K., Natzle, J. E., Brown, E. E., Fristrom, J. W. 1988. Prepupal differentiation of *Drosophila* imaginal discs: Identification of four genes whose transcripts accumulate in response to a pulse of 20-hydroxyecdysone. *Genetics* **120:** 465–474.

Fristrom, D. 1976. The mechanism of evagination of imaginal discs of *Drosophila melanogaster.* III. Evidence for cell rearrangement. *Dev. Biol.* **54:** 163–171.

Fristrom, D., Fristrom, J. W. 1993. The metamorphic development of the adult epidermis. In *The development of Drosophila* (M. Bate, A. Martinez-Arias, Eds.), vol II, pp. 843–897. Cold Spring Harbor Laboratory Press, Cold Spring Harbor, NY.

Gammie, S. C., Truman, J. W. 1999. Eclosion hormone provides a link between ecdysis-triggering hormone and crustacean cardioactive peptide in the neuroendocrine cascade that controls ecdysis behavior. *J. Exp. Biol.* **202:** 343–352.

Hegstrom, C. D., Truman, J. W. 1996. Steroid control of muscle remodeling during metamorphosis in *Manduca sexta. J. Neurobiol.* **29:** 535–550.

Hiruma, K., Hardie, J., Riddiford, L. M. 1991. Hormonal regulation of epidermal metamorphosis in vitro: Control of expression of a larval-specific cuticle gene. *Dev. Biol.* **144:** 369.

Hori, M., Riddiford, L. M. 1982. Regulation of ommochrome biosynthesis in the tobacco hornworm *Manduca sexta* by juvenile hormone. *J. Comp. Physiol.* **147:** 1–9.

Horodyski, F. M., Riddiford, L. M. 1989. Expression and hormonal control of a new larval cuticular multigene family at the onset of metamorphosis of the tobacco hornworm. *Dev. Biol.* **132:** 292–303.

Jones, G., Schelling, D., Chhokar, V. 1996. Overview of the regulation of metamorphosis-associated genes in *Trichoplusia ni. Arch. Insect Biochem. Physiol.* **32:** 429–437.

Jungreis, A. M. 1979. Physiology of moulting in insects. *Adv. Insect Physiol.* **14:** 109–184.

Kimbrell, D. A., Berger, E., King, D., Wolfgang, W. J., Fristrom, J. W. 1988. Cuticle protein gene expression during the third instar of *Drosophila melanogaster. Insect Biochem.* **18:** 229–235.

Kimbrell, D. A., Tojo, S. J., Alexander, S., Brown, E. E., Tobin, S. L., Frisrom, J. W. 1989. Regulation of larval cuticle protein gene expression in *Drosophila melanogaster. Dev. Genet.* **10:** 198–209.

Kingan, T. G., Adams, M. E. 2000. Ecdysteroids regulate secretory competence in inka cells. *J. Exp. Biol.* **203:** 3011–3018.

Krämer, B., Wolbert, P. 1996. Hormonal control of expression of a pupal cuticular protein gene during metamorphosis in *Galleria. Arch. Insect Biochem. Physiol.* **32:** 467–474.

Mathi, S. K., Larsen, E. 1988. Patterns of cell division in imaginal discs of *Drosophila. Tissue Cell.* **20:** 461–472.

Miner, A. L., Rosenberg, A. J., Nijhout, H. F. 2000. Control of growth and differentiation of the wing imaginal disk of *Precis coenia* (Lepidoptera: Nymphalidae). *J. Insect Physiol.* **46:** 251–258.

Nijhout, H. F. 1975. A threshold size for metamorphosis in the tobacco hornworm, *Manduca sexta*. *Biol. Bull.* **149:** 214–225.

Nijhout, H. F. 1981. Physiological control of molting in insects. *Am. Zool.* **21:** 631–640.

Nijhout, H. F., Wheeler, D. E. 1982. Juvenile hormone and the physiological basis of insect polymorphisms. *Q. Rev. Biol.* **57:** 109–133.

Nijhout, H. F., Williams, C. M. 1974. Control of moulting and metamorphosis in the tobacco hornworm, *Manduca sexta* (L.): Cessation of juvenile hormone secretion as a trigger for pupation. *J. Exp. Biol.* **61:** 493–501.

Rebers, J. E., Riddiford, L. M. 1988. Structure and expression of a *Manduca sexta* larval cuticle gene homologous to *Drosophila* cuticle genes. *J. Mol. Biol.* **203:** 411–423.

Riddiford, L. M. 1978. Ecdysone-induced change in cellular commitment of the epidermis of the tobacco hornworm, *Manduca sexta,* at the initiation of metamorphosis. *Gen. Comp. Endocrinol.* **34:** 438–446.

Riddiford, L. M. 1981. Hormonal control of epidermal cell development. *Am. Zool.* **21:** 751–762.

Riddiford, L. M. 1993. Hormones and *Drosophila* development. In *The development of Drosophila melanogaster* (M. Bate, A. Martinez-Arias, Eds.), vol. 2, pp. 899–939. Cold Spring Harbor Laboratory Press, Cold Spring Harbor, NY.

Riddiford, L. M. 1996. Molecular aspects of juvenile hormone action in insect metamorphosis. In *Metamorphosis: Postembryonic reprogramming of gene expression in amphibian and insect cells* (L. I. Gilbert, J. R. Tata, B. G. Atkinson, Eds.), pp. 223–251. Academic Press, San Diego, CA.

Riddiford, L. M., Chen, A. C., Graves, B. J., Curtis, A. T. 1981. RNA and protein synthesis during the change to pupal commitment of *Manduca sexta* epidermis. *Insect Biochem.* **11:** 121–127.

Riddiford, L. M., Curtis, A. T. 1978. Hormonal control of epidermal detachment during the final feeding stage of the tobacco hornworm larva. *J. Insect Physiol.* **24:** 561–568.

Riddiford, L. M., Hiruma, K., Lan, Q., Zhou, B. 1999. Regulation and role of nuclear receptors during larval molting and metamorphosis of Lepidoptera. *Am. Zool.* **39:** 736–746.

Riddiford, L. M., Palli, S. R., Hiruma, K., Li, W., Green, J., Hice, R. H., Wolfgang, W. J., Webb, B. A. 1990. Developmental expression, synthesis, and secretion of insecticyanin by the epidermis of the tobacco hornworm, *Manduca sexta*. *Arch. Insect Biochem. Physiol.* **14:** 171–190.

Riddiford, L. M., Truman, J. W. 1993. Hormone receptors and the regulation of insect metamorphosis. *Am. Zool.* **33:** 340–347.

Schubiger, M., Wade, A. A., Carney, G. E., Truman, J. W., Bender, M. 1998. *Drosophila* EcR-B ecdysone receptor isoforms are required for larval molting and for neuron remodeling during metamorphosis. *Development* **125:** 2053–2062.

Sehnal, F., Svacha, P., Zrzavy, J. 1996. Evolution of insect metamorphosis. In *Metamorphosis. Postembryonic reprogramming of gene expression in amphibian and insect cells.* (L. I. Gilbert, J. R. Tata, B. G. Atkinson, Eds.), pp. 3–58. Academic Press, San Diego.

Svacha, P. 1992. What are and what are not imaginal discs: Reevaluation of some basic concepts (Insecta, Holometabola). *Dev. Biol.* **154:** 101–117.

Truman, J. W. 1996. Metamorphosis of the insect nervous system. In *Metamorphosis: Postembryonic reprogramming of gene expression in amphibian and insect cells* (L. I. Gilbert, J. R. Tata, B. G. Atkinson, Eds.), pp. 283–320. Academic Press, San Diego, CA.

Truman, J. W., Reiss, S. 1995. Neuromuscular metamorphosis in the moth *Manduca sexta*: Hormonal regulation of synapse elimination and sprouting. *J. Neurosci.* **15:** 4815–4826.

Truman, J. W., Sokolove., P. G. 1972. Silk moth eclosion: Hormonal triggering of a centrally programmed pattern of behavior. *Science* **175:** 1491–1493.

Willis, J. H. 1986. The paradigm of stage-specific gene sets in insect metamorphosis: Time for revision!! *Arch. Insect Biochem. Physiol. (Suppl.)* **1:** 47–57.

Willis, J. H. 1996. Metamorphosis of the cuticle, its proteins, and their genes. In *Metamorphosis: Postembryonic reprogramming of gene expression in amphibian and insect cells* (L. I. Gilbert, J. R. Tata, B. G. Atkinson, Eds.), pp. 253–282. Academic Press, San Diego.

Wolfgang , W. J., Riddiford, L. M. 1986. Larval cuticular morphogenesis in the tobacco hornworm *Manduca sexta* and its hormonal regulation. *Dev. Biol.* **113:** 305–316.

Yund, M. A. 1978. Ecdysteroid receptors in imaginal discs of *Drosophila melanogaster. Proc. Natl. Acad. Sci. USA* **75:** 6039–6043.

Yund, M. A. 1989. Imaginal discs as a model for studying ecdysteroid action. In *Ecdysone: From chemistry to mode of action.* (J. Koolman, Ed.), pp. 384–392. Thieme Medical Publishers, Stuttgart.

CHAPTER 3

Developmental Systems

PREEMBRYONIC DEVELOPMENT

All multicellular organisms begin their lives as single-celled zygotes that divide to form groups of cells that show progressive morphological changes. Although they are genetically identical, successive daughter cells display increasing differentiation to form all the diverse cell types that will make up the multicellular insect. The development of these morphological differences and the generation of cellular diversity is the process of **differentiation.** However, long before the actual differentiation is evident, the developmental fate of the cells is fixed. Although they may not yet show any morphological differences, they have become genetically committed to a particular course of development. The commitment process that establishes the later differentiated state is **determination.** Thus, the nonvisible process of determination is followed by the detectable state of differentiation. These processes begin in the egg and continue during the life of the organism.

INSECT EGGS

The insect eggshell has a number of critical responsibilities. Most importantly, it must serve as a two-way barrier to prevent the loss of egg contents to the environment and also minimize the disturbance of those contents by environment hazards. It does this while remaining flexible enough to pass through a narrow ovipositor, permitting sperm to enter for fertilization, and facilitating the escape of the larva at the termination of embryogenesis. In many aquatic insects, the egg contains special external structures that hold them upright on the surface of the water. Specialized respiratory structures may also branch off the main body of the egg. The embryo has all its metabolic needs prepackaged inside the egg, but one critical component, oxygen, is lacking and must be acquired from outside. This also presents a challenge in design, because given the small size of insect eggs and the resulting high surface-to-volume ratio, any gaseous exchange by the egg brings with it an enormous potential for water loss. However, the complex structure comprising the insect eggshell allows this gas exchange to occur without a significant loss of water.

Holometabolous development was a major advance that is considered to be a key to the success of insects, allowing the adults and immatures of a species to occupy divergent habitats and thus reduce the competition between them for resources. This divergence of habitat preferences also reduced the opportunities for parental investment and made its development more costly, as adults no longer routinely occupied the environments of their offspring. With an increased need for eggs to survive on their own under these circumstances, the design of the egg was of considerable importance. With the evolution of ovipositors, females were able to deposit their eggs in concealed locations as well as novel habitats that were unavailable to other animals. It might be said that holometabolous development could have occurred only once the egg and egg-related structures had been sufficiently altered to allow survival under varied aquatic and terrestrial conditions.

Egg Membranes

The eggshell, or **chorion,** is a complex of several layers (Fig. 3.1). It is synthesized within the ovariole by the **follicular epithelium** that surrounds the oocyte, and begins once **vitellogenesis,** the uptake of yolk proteins, has been completed. In the silkworm moth, *Antheraea polyphemus,* there are approximately 10,000 follicle cells that surround the oocyte in a monolayer and secrete the complex layers of the chorion, only to degenerate once the chorion has been completely formed. The follicle cells often leave a species-specific imprint on the surface of the egg. Because these follicle cells surround the oocyte and direct their products inward, the inner layers of the chorion are secreted first with subsequent layers formed either by **apposition** of newly synthesized layers or by **intercalation** of components into previously formed layers.

Choriogenesis is a period of intense protein synthesis, and prior to chorion gene expression the follicle cells undergo several rounds of DNA replication without cell division to increase the DNA content and their synthetic capability. Some of the chorion genes may be additionally amplified to 60–80 times above the usual copy number to enhance the biosynthetic rate of eggshell protein production. The formation of the chorion has been studied only for a handful of insects, and most generalizations are based on *Drosophila melanogaster, Antheraea polyphemus,* or *Bombyx mori* models.

The first layer of the chorion to be synthesized by the follicle cells is the **vitelline envelope,** an inner noncellular membrane with a thickness of about 0.3 μm that completely surrounds the oocyte, except at the **micropyle,** the site of future sperm penetration. The envelope is first deposited as plaques that ultimately

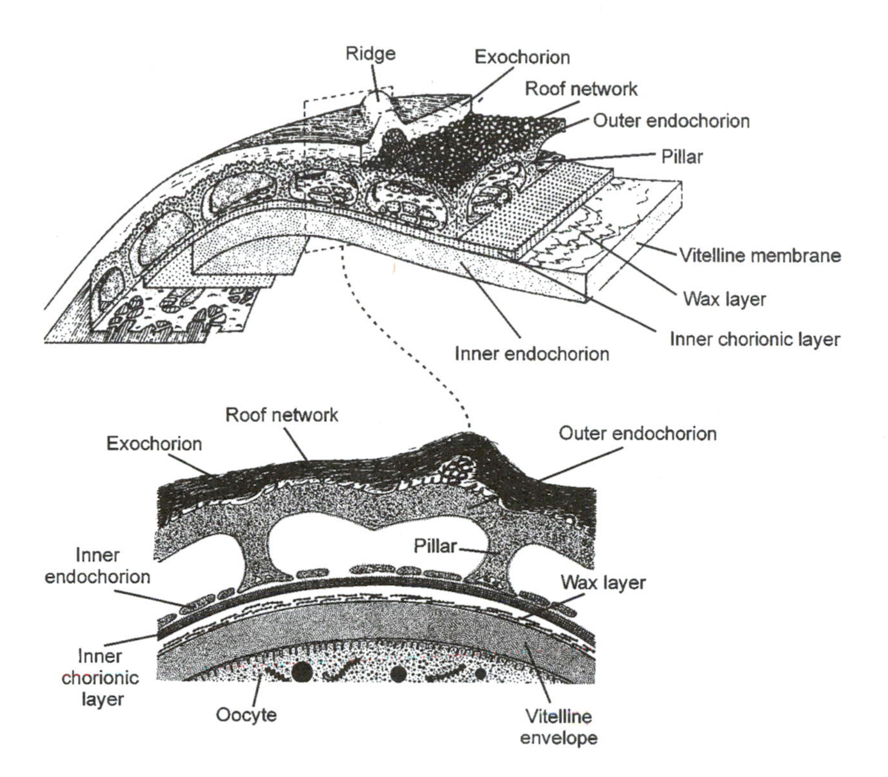

FIGURE 3.1 Cross-sectional representations of the chorion of *Drosophila melanogaster*. The figure below is a two-dimensional cross section of the region shown by the dashed square in the figure above. Reprinted from Margaritis, L.H. and M. Mazzini (1998). *Microscopic Anatomy of Invertebrates* **11C:** 995–1037. This material is used by permission of Wiley-Liss, Inc., a subsidiary of John Wiley & Sons, Inc.

coalesce over the oocyte membrane. In many insects a thin **wax layer** ranging in thickness from 5 nm to 2 μm may then be deposited between the vitelline envelope and the follicle cells to provide protection against desiccation. This layer is the primary waterproofing layer of the egg. The presence of a wax layer has not been identified in all insect chorions.

Above the wax layer are several other chorionic layers that, depending on species, may be formed by the apposition of newly secreted layers or the intercalation of secretions into existing layers. A thin **inner chorionic layer** that consists largely of protein may be secreted above the vitelline envelope. The early proteins of the **endochorion** begin to be synthesized after the wax layer is laid down and form a scaffold upon which middle and late proteins intercalate. There has been a remarkable conservation of early proteins among insect genera, with a greater divergence in the structures of the middle and late proteins. The endochorion is divided into an **inner endochorion** of about 400 Å in thickness, a network of pillars, and the **outer endochorion** of approximately 0. 2 μm in thickness that creates a roof network. Between the pillars is a meshwork that can trap air, with pores or **aeropyles** that open to the outside. The aeropyles may be grouped in specific areas, such as in the vicinity of hatching lines, and allow the passage of air into the meshwork of the endochorion while restricting the passage of water. The **exochorion** is a 0. 3-μm-thick outermost layer that may contain carbohydrate in addition to protein. Indeed, except for the wax layer, the chorion consists primarily of protein that can be cross-linked by the process of sclerotization to stabilize its structure.

There is a regional differentiation of specialized structures. Because the eggshell is so impermeable, there must be a special provision to allow sperm to enter. The **micropylar apparatus** is synthesized by the border cells, special cells dispersed among the follicle cells, early during oogenesis. It consists of an opening, the **micropyle,** and a canal that leads to the oocyte (Fig. 3.2). The micropyle is sufficiently wide to allow only a single sperm to enter. It may be a protrusion or a depression, containing small holes that penetrate through the chorion and the vitelline membrane to the surface of the oocyte. There may be a single micropyle or as many as 70, and although they are usually located at the anterior end of the egg, they are found on the ventral side in crickets and at the posterior end in termites and scorpion flies.

There are other specialized regions of the chorion. In addition to the aeropyles that channel air into the chorionic meshwork, respiratory appendages may be present that can serve as a plastron to extract oxygen from water (Fig. 3.3). These are found both in eggs of aquatic species as well as those terrestrial eggs that may become periodically inundated by water. There may be regions of programmed weakness in the chorion to allow the larva to more easily escape during hatching. The **operculum** is a cap that is surrounded by these hatching regions and opens to allow the larva to exit. In some species, first instar larvae have a special spine, or egg-burster, to facilitate the breaking of the chorion.

Pattern Formation within the Oocyte

Given the limited amount of space inside the egg, it is somewhat surprising that the internal composition of the egg is far from uniform. There are distinct regions at the anterior and posterior poles, and the cytoplasm immediately beneath the plasma membrane, the **periplasm,** is morphologically different from that of the interior of the oocyte (Fig. 3.2). The egg contains not only the oocyte and the raw materials necessary for its growth, but also a complex of materials that specify how that growth should occur. In addition to the protein yolk, lipid, messenger RNA, and maternal organelles such as mitochondria and ribosomes, there are also gradients of protein that result from the transcription of specific genes, all of which are spatially organized to create a pattern that provides the developing embryo with

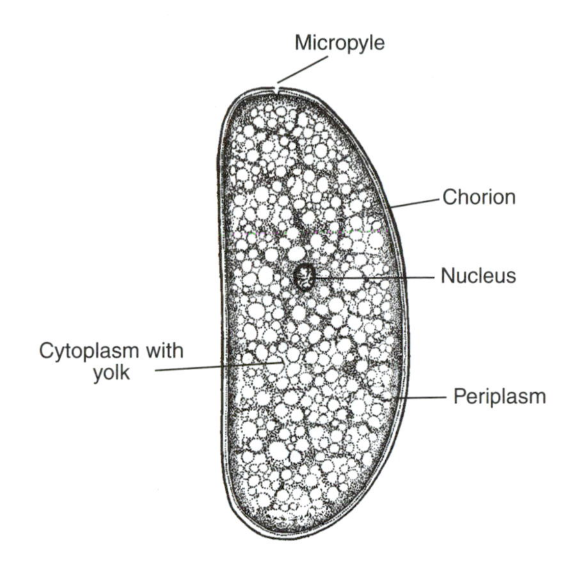

FIGURE 3.2 Cross section of a generalized insect egg prior to fertilization. From Johannsen, O. and F. Butt (1941). *Embryology of Insects and Myriapods.* Copyright McGraw-Hill Education. Reprinted with permission.

positional information within the egg. Proper development involves a sequential activation or repression of specific genes by gradients of these **morphogens,** and this positional information is crucial for the expression of genetic information at the appropriate time and place during development. Thus, although much of the information that allows the organism to develop correctly resides in the genome, considerable information is also derived from this nonuniform placement of materials in the egg.

Polarity gradients of these materials are established early during oogenesis, while the oocyte resides in the female even before fertilization and oviposition occur. An anterior–posterior polarity in the egg is established early by **maternal effect genes** in the nurse cells of the female parent that produce the mRNAs of *bicoid, nanos, hunchback,* and *caudal* that are the first to encode various regulatory proteins that diffuse throughout the egg and specify positional information that activates or represses the expression of certain genes during development.

The first of the morphogenetic gradients in the developing *Drosophila* embryo establishes the cytoskeletal framework upon which other spatial patterns are based. Soon after the oocyte differentiates from nurse cells, it migrates to the posterior of the nurse cells where it induces the follicle cells there to adopt a posterior fate rather than the anterior fate that is their default state. The oocyte nucleus serves as a collection site for a maternally derived *gurken* RNA that is translated into a Gurken protein. The Gurken protein passes into the posterior follicle cells and, in response, these cells produce a *torpedo* gene product that binds to Gurken and

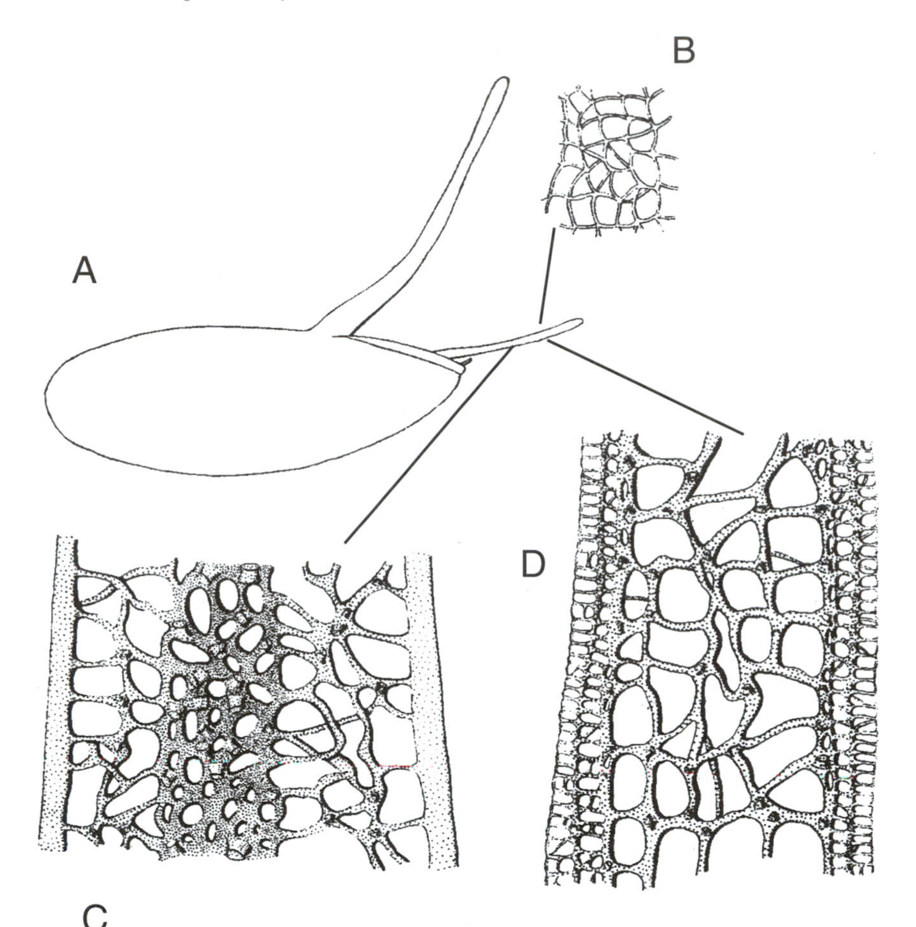

FIGURE 3.3 (A) The egg of *Drosophila gibberosa* and its respiratory horns. (B) The surface network of the horn at the region indicated. (C and D) The meshwork of the respiratory horn seen in optical section. Reprinted from Richards O.W. and R.G. Davies. *Imm's General Textbook of Entomology*, 10th edition. With kind permission from Kluwer Academic Publishers.

establishes their posterior cell identity. These posterior follicle cells then send a signal back to the oocyte that polarizes its microtubular and directs the localization of the *oskar* and *bicoid* RNAs to either end of the oocyte. The anterior development is directed by *bicoid* and the formation of the posterior pole plasm where the germ cells are localized and develop is established by *oskar* (Fig. 3.4).

This cyoskeletal repolarization induces the oocyte nucleus to move to the anterior, dorsal portion of the cytoplasm where its development continues. The Gurken signal carried by the nucleus causes the cells in this area to adopt a dorsal

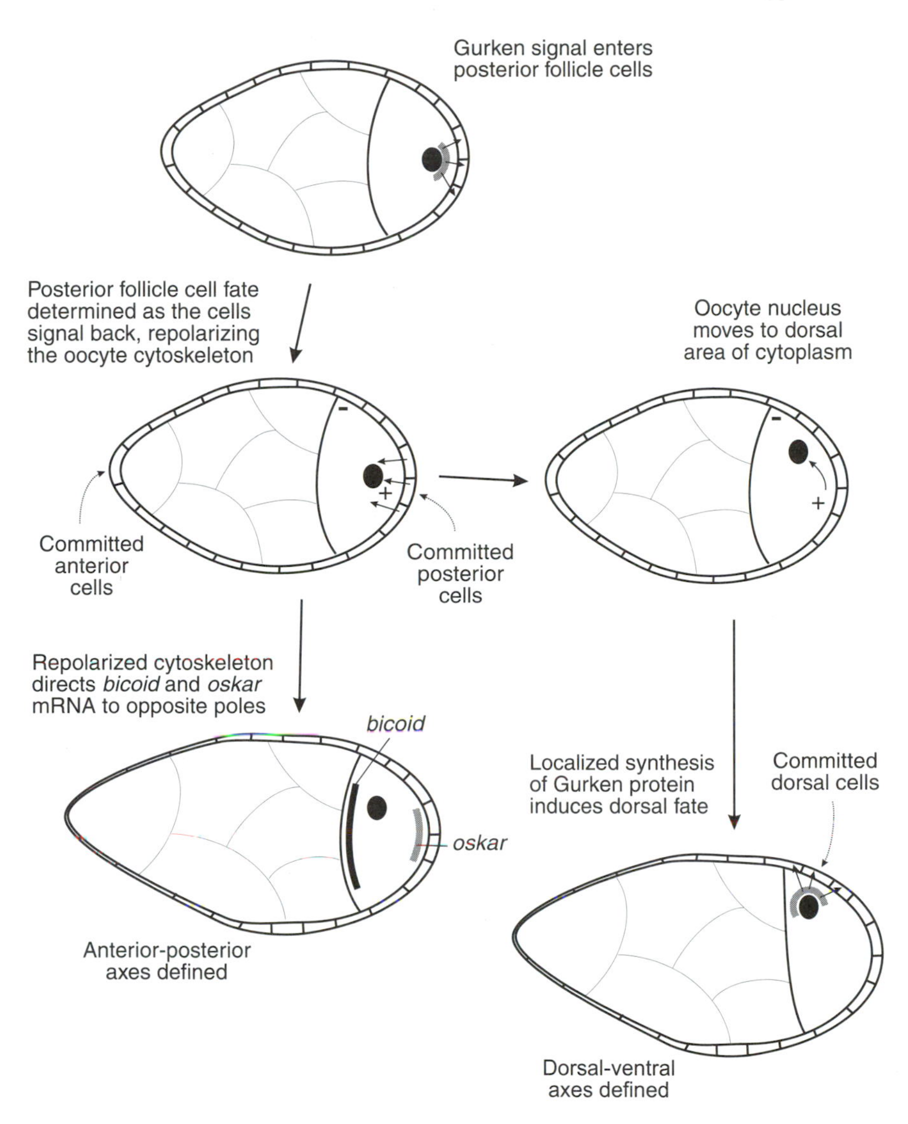

FIGURE 3.4 The mechanism of Gurken signaling to establish anterior/posterior and dorsal/ventral commitment of follicle cells. Reprinted with permission from Gonzalez-Reyes, A. *et al.* (1995). *Nature* **375:** 654–658. Copyright Nature.

fate (Fig. 3.4). So, the *gurken* message, originating from maternal mRNA, sequentially induces both posterior and dorsal cell fates even before the egg is laid.

What little is known about how the morphogenic gradients regulate the determination of cells paints a picture that is already incredibly complex. There are at least four maternal mRNAs that are involved in the subsequent determination of anterior/posterior patterning: *bicoid* and *hunchback* mRNAs regulate the development of structures characteristic of the anterior of the egg, while *nanos* and *caudal* regulate posterior structures. The Bicoid protein specifies the anterior of the embryo in two ways. It prevents the posterior determination in anterior regions by binding to and suppressing the anterior translation of the *caudal* mRNA that is normally found dispersed throughout the egg. When present anteriorly, the *caudal* message prevents the proper formation of the head and thorax. Bicoid protein also activates the *hunchback* gene that specifies anterior structures and they together specify the anterior pattern.

The maternal *nanos* gene produces mRNA that is localized at the posterior pole of the egg. The Nanos protein represses the translation of any *hunchback* RNA that is present in the posterior region at the same time the *hunchback* is being activated at the anterior by Bicoid protein. If the *nanos* gene product were to be absent, Hunchback protein would be synthesized in other areas and affect the expression of other genes that are required for differentiation of structures in the abdominal region. The distribution of these mRNAs according to this model for development is shown in Figure 3.5.

Next, the genes of the zygote that are in turn regulated by these maternal factors are expressed in certain broad regions. The first genes to be expressed in the embryo are **gap genes,** named because mutations in them produce gaps in the segmentation pattern. The concentrations of the gap gene proteins subsequently cause the transcription of **pair-rule genes** that divide the embryo into periodic units. The transcription of each of the pair-rule genes results in a striped pattern

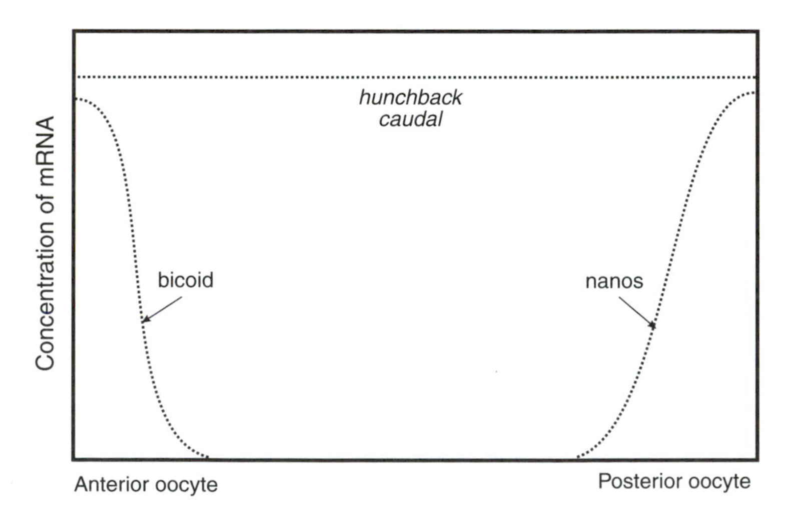

FIGURE 3.5 The gradients of *bicoid*, *nanos*, *hunchback*, and *caudal* mRNA that establish position in the *Drosophila* oocyte.

of vertical bands, and in turn activate the transcription of the **segment polarity genes.** The mRNA and proteins of the segment polarity genes divide the *Drosophila* embryo into 14 segments and establish further periodicity. At the same time, the proteins of the gap, pair-rule, and segment polarity genes interact to regulate the **homeotic genes,** or ***Hox* genes,** whose transcription determines the developmental fate of each of the segments. *Hox* genes specify relative position within an animal and not particular structures, controlling segment identity rather than specifying a particular structure within the segment. They appear to be universal among multicellular organisms, having been identified in a number of animals from nematodes to chordates. The genes all contain a highly conserved sequence of 180 base pairs called the **homeobox** that encodes a transcription factor that binds to specific DNA regions. Without the *Hox* genes, all segments in the insect would look alike. *Drosophila* has at least eight *Hox* genes in two complexes, the antennapedia complex and the bithorax complex. The *bicoid* gene mentioned earlier contains a homeobox and is therefore one of these hometic genes.

EMBRYONIC DEVELOPMENT

The oocyte, which has been arrested in metaphase of the first meiotic division, continues to mature after oviposition occurs. Shortly after the sperm penetrates the egg and oviposition takes place, the oocyte completes meiosis. This results in a haploid oocyte nucleus and three polar nuclei that inhabit the periplasm at the periphery of the egg (Fig. 3.6). The oocyte nucleus, surrounded by an island of

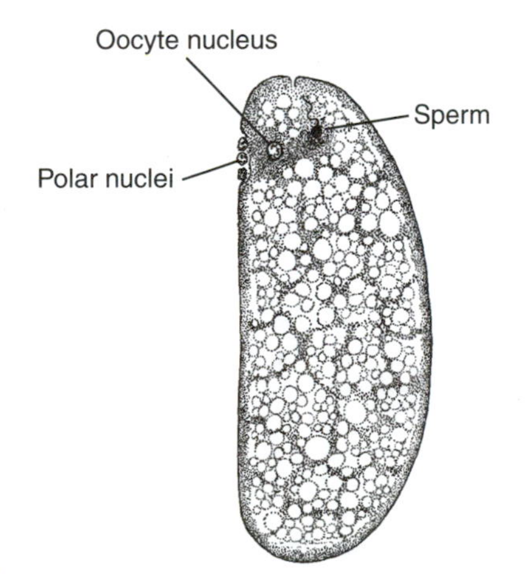

FIGURE 3.6 The generalized insect egg after sperm entry but prior to ferilization. Reprinted with permission from Johannsen, O. and F. Butt (1941). *Embryology of Insects and Myriapods.* Copyright McGraw-Hill Education.

cytoplasm, then moves to the interior of the egg where it meets the sperm that has already entered. **Syngamy,** the union of sperm and egg, occurs at the interior. The polar nuclei may accompany the oocyte or degenerate at the periphery. In some insects that reproduce without sperm involvement by **parthenogenesis,** a haploid polar nucleus combines with the haploid oocyte nucleus to restore the diploid number without requiring fertilization by male gametes.

Following the union of the sperm and the egg, the newly formed zygote undergoes **cleavage** within the patterned environment that is present in the egg. Cleavage is the process by which the zygote divides mitotically to parcel out the cytoplasm into other smaller daughter cells. The pattern of cleavage in a particular animal species is determined largely by the amount and distribution of the yolk within the egg cytoplasm. The presence of yolk in the egg tends to inhibit cleavage, so animals that have relatively little yolk are able to undergo complete or **holoblastic** cleavage (Fig. 3.7). However, the relatively large amount of yolk in most insect eggs prevents the first cleavage divisions from cutting through the entire egg, and their cleavage is more superficial, or **meroblastic.** Rather than dividing into the separate cells that would result from holoblastic cleavage, the nuclei divide without the formation of new cell membranes. Not all insect eggs are meroblastic; collembolan eggs have little yolk and do undergo holoblastic cleavage.

Blastoderm Formation

During the meroblastic cleavage of most insects, the zygote nucleus undergoes mitotic division in the center of the egg cytoplasm but the resulting daughter nuclei or **energids,** each surrounded by an island of cytoplasm, are not incorporated into new complete cells. Instead, after a series of mitoses, the energids migrate to the egg periphery with their islands of cytoplasm and continue to divide there (Fig. 3.8B). The initial cleavage divisions are synchronous, but they become more asynchronous as they continue to occur in different regions of the egg. At the periphery the energids first form a **syncytial blastoderm** that lacks any membranes, with all the cleavage nuclei contained within the common cytoplasm of the egg (Fig. 3.8C). In *Drosophila,* the egg remains a syncytium for the first 2 h of development, and the lack of any cell membranes undoubtedly facilitates the diffusion of morphogens. Some of the energids that migrate to the posterior of the egg form the **pole cells** that will give rise to the germ cells of the future adult, establishing the continuity of the germ line early during embryogenesis. Their early removal from the normal course of cell divisions assures that their genetic integrity is maintained. The pole cells differentiate in the microenvironment present at the posterior pole largely due to the *oskar* gene product that appears to be necessary for their formation.

Other energids stop dividing mitotically and, rather than migrate to the periphery, remain in the yolk to form yolk cells, or **vitellophages,** which will be involved later in the digestion of yolk and the formation of the midgut epithelium. At about

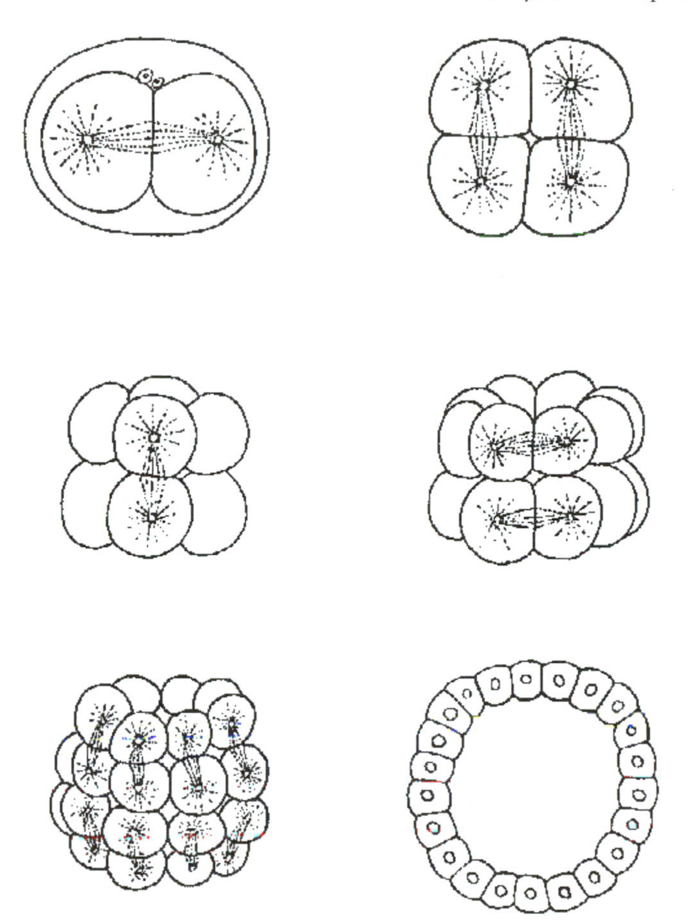

FIGURE 3.7 Typical holoblastic cleavage in a noninsect egg. At the bottom right, the figure shows a saggital section. Reprinted with permission from Balinsky, B.I. (1970). *An Introduction to Embryology,* Third edition. Copyright W.B. Saunders Company.

the time the pole cells have become distinct, membranes begin to develop around each of the energid nuclei of the syncytial blastoderm, creating individual cells that now form the **cellular blastoderm** (Fig. 3.9). The cells of the blastoderm appear to be determined at this stage, because the experimental destruction of groups of cells at this time results in specific defects later in development.

Formation of the Germ Band

As the division of the blastoderm continues, the rate of division changes in the ventral region, causing it to thicken as the cells become more columnar in shape. This thickened portion of the blastoderm becomes the **embryonic primordium,**

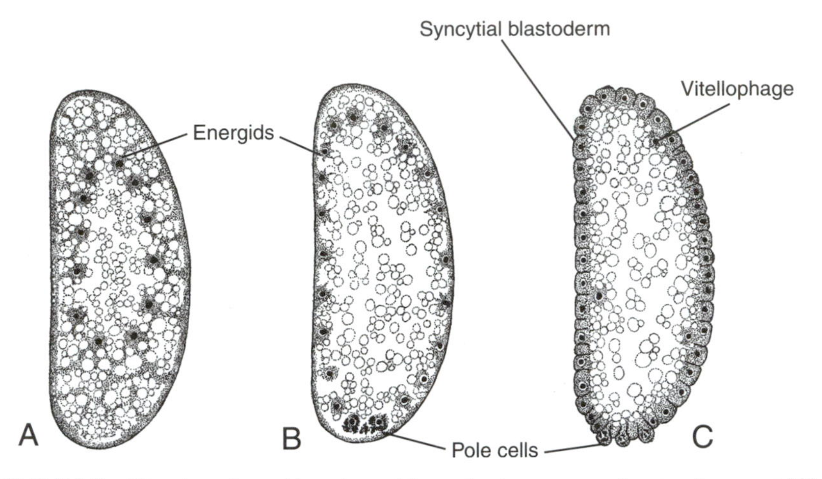

FIGURE 3.8 Migration of energids to the periplasm of an insect oocyte, forming the syncytial blastoderm. The vitellophages, agents of yolk digestion, and the pole cells, destined to form the germ cells, are derived from the energids. Reprinted with permission from Johannsen, O. and F. Butt (1941). *Embryology of Insects and Myriapods.* Copyright McGraw-Hill Education.

which will develop into the embryo, while the other cells of the blastoderm are designated as the **extraembryonic ectoderm** (Fig. 3.10). As the embryonic primordium increases in length, it forms the **germ band** that represents the ventral region of the future body. Continued proliferation of the germ band causes it to

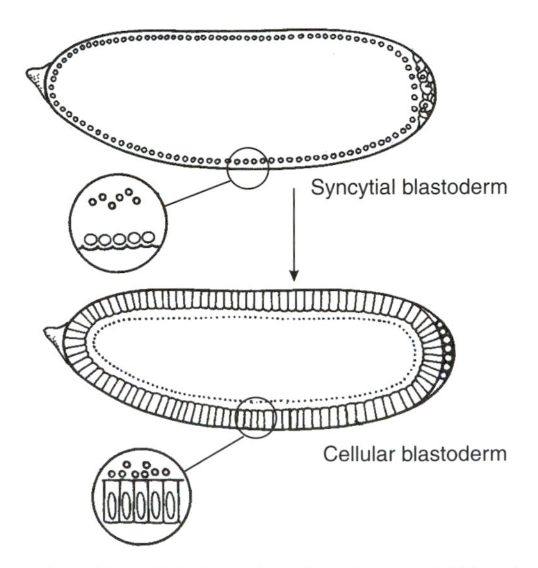

FIGURE 3.9 Formation of the cellular blastoderm from the syncytial blastoderm. From Lawrence (1992). Reprinted with permission.

penetrate into the interior of the yolk mass and the presence of morphogenetic gradients govern its continued determination and differentiation. As mentioned previously, gradients of morphogens such as Bicoid protein at the anterior pole, and the Nanos protein at the posterior pole, activate or inactivate the transcription of specific genes that are associated with anterior or posterior structures.

At this stage, the germ band is still a single layer of cells. It takes an important step in development when it becomes a double layer by the process of **gastrulation.** In other animals, gastrulation involves the invagination of the ball of cells that has formed, but in insects, the invagination is prevented by the large amount of yolk that is present. Instead, the formation of a multicellular layer occurs when some of the cells of the germ band migrate inward into the yolk (Fig. 3.11). When the cells along the ventral midline of the germ band elongate and migrate upward, a longitudinal furrow, the **gastral groove,** is created. The gastral groove is soon filled in by other cells of the germ band, with this outer layer of cells remaining as the embryonic ectoderm, while the invaginated cells proliferate to form the **mesoderm.** By virtue of their position, these cells acquire a different developmental fate and will form most of the internal organs. The germ band constitutes the trunk of the developing insect with the ectoderm on the outside and the mesoderm on the inside.

The products of the segmentation genes divide the germ band of *Drosophila* into a series of 14 regions. Segmentation begins as the maternal effect genes activate or

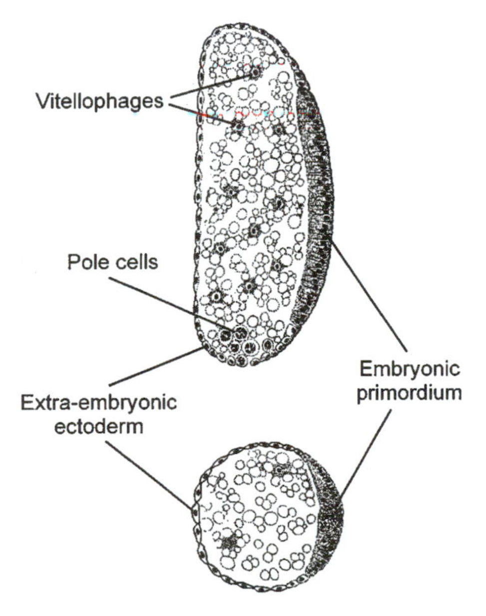

FIGURE 3.10 Development of the embryonic primordium and extraembryonic ectoderm from the blastodermal cells. Reprinted with permission from Johannsen, O. and F. Butt (1941). *Embryology of Insects and Myriapods.* Copyright McGraw-Hill Education.

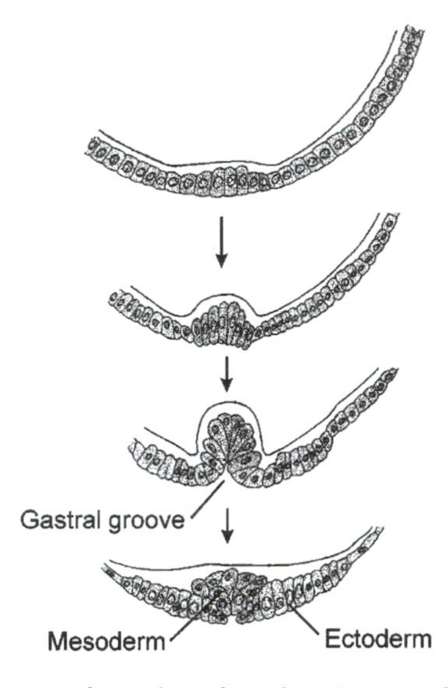

FIGURE 3.11 Formation of mesoderm from the migration of ectodermal cells. From Anderson (1972). Reprinted with permission.

repress the gap genes to establish regions of segment identity. The concentrations of Bicoid, Hunchback, and Caudal proteins determine the transcription patterns of the gap genes. The products of the gap genes then regulate the expression of the pair-rule genes, which, when transcribed, further divide the broad gap gene regions into individual **parasegments.** The parasegments partially correspond to the three mouthpart, three thoracic, and eight abdominal segments and will ultimately produce the anterior compartment of one segment and the posterior compartment of the next. The segment polarity genes assure that certain repeated structures appear in each segment, establishing the cell fates within each of the parasegments. Then, subsequent interactions of the gap and pair-rule genes regulate the homeotic genes that establish the identity of each segment and its characteristic structures. Homeotic mutants cause the misidentification of segments, often producing bizarre characters such as legs growing out of the head.

Blastokinesis and Dorsal Closure

With the subsequent formation of membranes, the germ band becomes separated from the extraembryonic ectoderm. The germ band elongates and widens, carrying the margin of the extraembryonic ectoderm with it. These **amniotic folds** extend to the ventral midline and merge to form a double layer of extraembryonic cells that are ventral to the germ band surface. As the folds merge, a com-

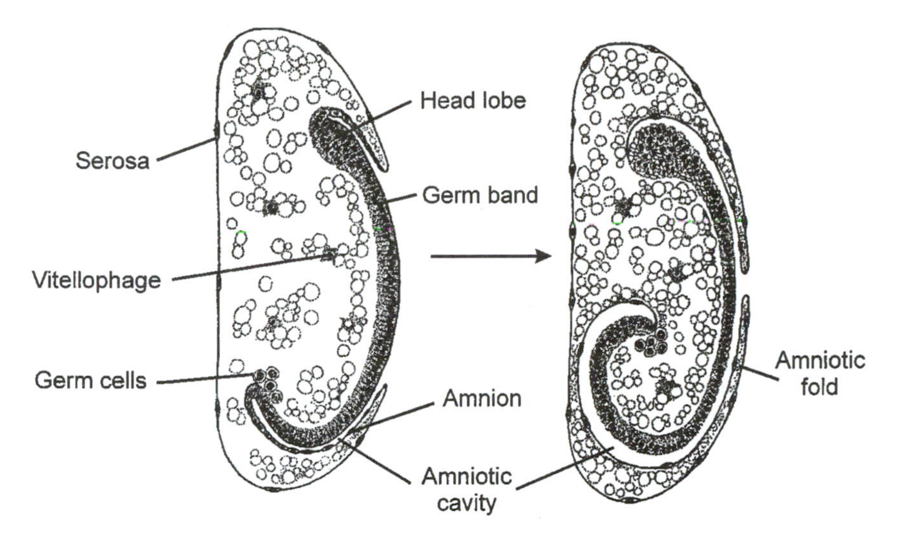

FIGURE 3.12 The formation of the amnion and the invagination of the germ band to free it from the serosa. Reprinted with permission from Johannsen, O. and F. Butt (1941). *Embryology of Insects and Myriapods.* Copyright McGraw-Hill Education.

pletely cellular membrane covers both the yolk mass and the germ band (Figs. 3.12 and 3.13). The inner walls of the amniotic folds merge to form an internal cellular membrane, the **amnion,** continuous with the margin of the embryonic ectoderm. The amnion encloses a fluid-filled **amniotic cavity.** Once detached from the germ band, the extraembryonic ectoderm is referred to as the **serosa.** The formation of the amnion separates the germ band from the extraembryonic ectoderm and allows the germ band to engage in **blastokinesis,** movements that take place within the yolk. These movements tend to be much more pronounced in hemimetabolous insects, which have a smaller germ band. The movements of the germ band reverse the relative positions of the yolk and the embryo. Earlier, the

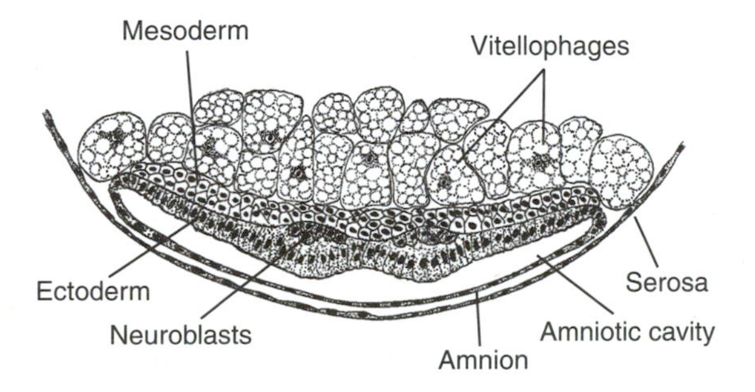

FIGURE 3.13 The formation of neuroblasts from ectodermal tissue and the proliferation of mesoderm. Reprinted with permission from Johannsen, O. and F. Butt (1941). *Embryology of Insects and Myriapods.* Copyright McGraw-Hill Education.

embryo was positioned within the yolk, but after blastokinesis and growth of the embryonic ectoderm over the dorsal portion to complete **dorsal closure,** the yolk is contained within the embryo (Fig. 3.14).

Formation of the Gut

Cells at the anterior and posterior ends of the embryo migrate inward to form the foregut, hindgut, and midgut. The foregut and hindgut arise from ectodermal cells, with some of the invaginating epithelia acquiring a new fate as the **endoderm** that will constitute the midgut. A reorganization of these endodermal cells forms sheets that enclose the yolk in a tube that creates the midgut epithelium. Vitellophages contained within the yolk that is enclosed by the midgut are integrated into the endoderm to form the definitive midgut epithelium (Fig. 3.15). Shortly before hatching occurs, the blind ends of the foregut and hindgut break down and the continuity of the digestive tract is established.

Formation of the Nervous System

The nervous system arises from ectodermal cells in the ventral region of the germ band. Proliferation of the neuroblasts in portions of the embryonic ectoderm of the germ band create longitudinal thickenings on either side of the ventral mid-

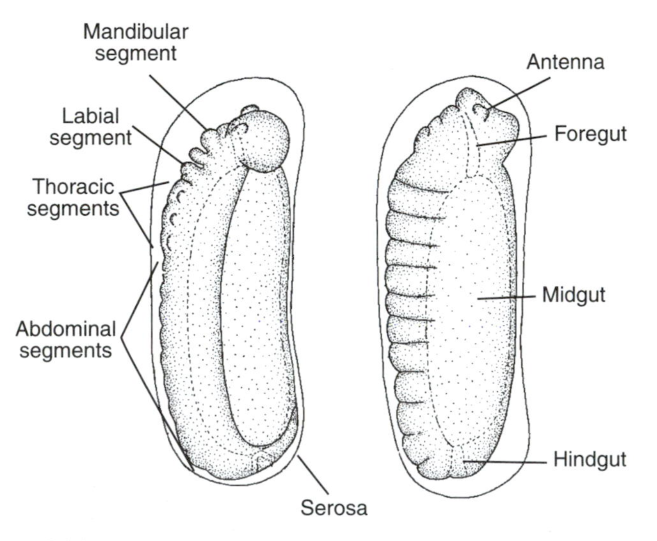

FIGURE 3.14 Dorsal closure of the embryo. From Anderson (1972). Reprinted with permission.

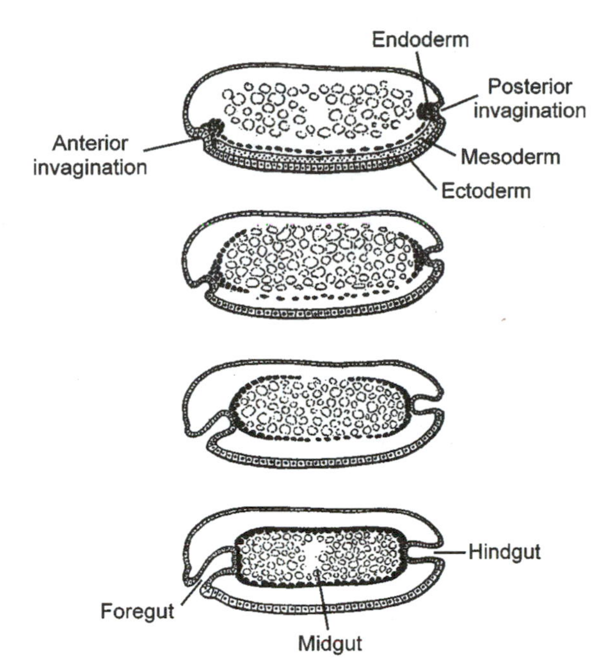

FIGURE 3.15 Formation of the foregut and hindgut from ectodermal invaginations and the development of endoderm that forms the midgut. From Johannsen, O. and F. Butt (1941). *Embryology of Insects and Myriapods.* Copyright McGraw-Hill Education. Reprinted with permission.

line, producing a **neural groove** and **neural ridges** (Fig. 3.16). The neural ridges result from the proliferation of **neuroblasts** that differentiate from the ectoderm and give rise to the nervous system. The commitment of the ectodermal cells to become neuroblasts rather than epidermal cells results from a group of **proneural genes.** The subsequent expression of **neurogenic genes** inhibits any neighboring cells from becoming neuroblasts. Three groups of neuroblasts that proliferate in the anterior region ultimately produce the protocerebrum, deutocerebrum, and tritocerebrum of the brain. Those in other segments give rise to the subsesophageal and abdominal ganglia.

Formation of Internal Organs

Mesodermal tissue gives rise to most of the internal organs of the insect (Fig. 3.17). Formed during gastrulation when the cells of the germ band migrate and give rise to two cell layers, the mesoderm forms two strands running the length of the body. Later in development, a pair of coelomic cavities develops in each segment, forming a tube through the thoracic and abdominal segments. The mesoderm is differentiated

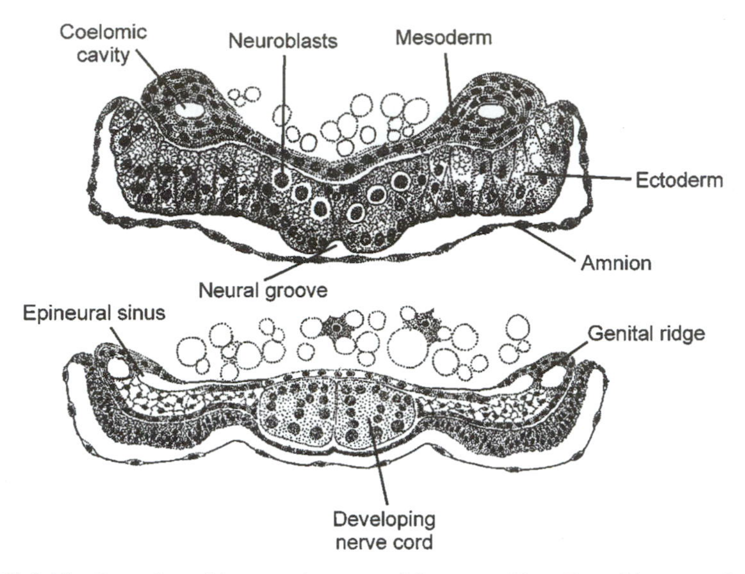

FIGURE 3.16 Formation of the ventral nerve cord from neuroblasts. From Johannsen, O. and F. Butt (1941). *Embryology of Insects and Myriapods.* Copyright McGraw-Hill Education. Reprinted with permission.

into two layers, the inner **splanchic mesoderm** that forms the visceral muscles, and the **somatic mesoderm** that gives rise to the skeletal muscles. These form from the fusion of myoblasts. The fat body is derived from the somatic layer that is not in contact with the ectodermal layer. At the junction of the splanchic and somatic layers,

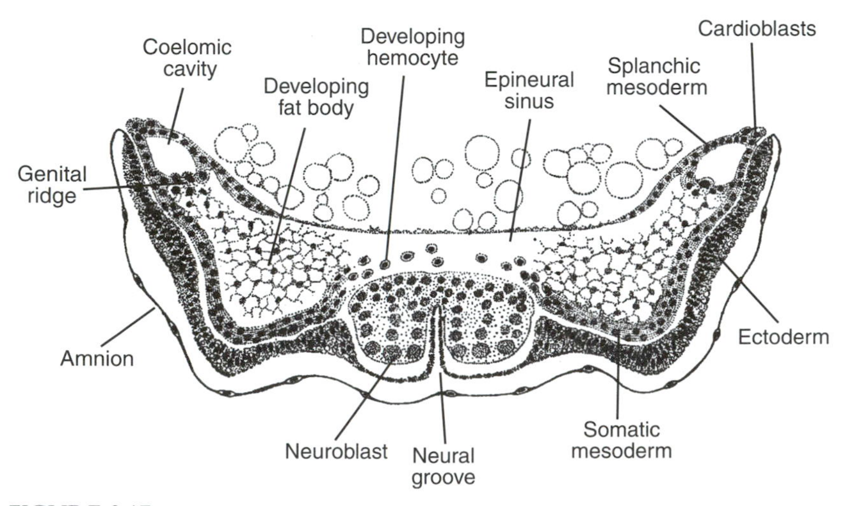

FIGURE 3.17 Mesodermal tissues that give rise to the internal organs. Reprinted with permission from Johannsen, O. and F. Butt (1941). *Embryology of Insects and Myriapods.* Copyright McGraw-Hill Education.

cardioblasts differentiate that ultimately give rise to the heart. Mesodermal tissue also forms a **genital ridge** that encloses the germ cells and forms the reproductive organs. When the median strand of mesoderm breaks down, some of its cells become hemocytes. Mesodermal somites acquire a lumen in each lateral half that eventually becomes the body cavity, or hemocele. The **epineural sinus** in which the hemocytes lie merges with the lumen to become the hemocele.

Formation of the Reproductive System

The reproductive system is constructed around the **pole cells** that had differentiated early during embryogenesis. The pole cells become the germ cells of the future adult, with surrounding mesodermal tissue enclosing those cells to form the germaria of the ovarioles and the follicles of the testes. Mesoderm also gives rise to the lateral oviducts and vasa deferentia. Invaginations of the ectoderm form the median oviduct and ejaculatory duct. A summary of all these cell lineages is shown in Fig. 3.18.

Endocrinology of Embryonic Development

It has been much more difficult to characterize the role of hormones during embryogenesis than during postembryonic development. Hormonal functions in the larval and adult stage are normally demonstrated by gland extirpation and hormone replacement therapy, procedures that are not possible for the egg. Consequently, endocrinology of the embryo has largely involved the isolation and identification of the hormones and their correlation with significant events during embryogenesis. The sources of these hormones and their fluctuations during embryogenesis were at first somewhat of a puzzle, because in the absence of embryonic glands there was no obvious place they could be synthesized. Their presence well before the embryonic endocrine organs have been formed is a result of their maternal synthesis during vitellogenesis. They are packaged into the egg largely as inactive hormone conjugates and released from the conjugates as active hormones when they are required.

The origin of the ecdysteroids in the adult female destined to be packaged in the egg was also somewhat puzzling, because the source of ecdysteroids for larval molting, the prothoracic glands, degenerate shortly after the molt to the adult and no prothoracic glands are present in the female to produce the hormone during vitellogenesis. However, an important discovery was that the follicle cells in the developing ovaries of many adult insects could also synthesize ecdysteroids and demonstrated how the hormones were produced to be incorporated into the egg.

The fluctuations of ecdysteroid titers appear to occur when the hormones are released from their inactive conjugates. Only later, when embryonic endocrine

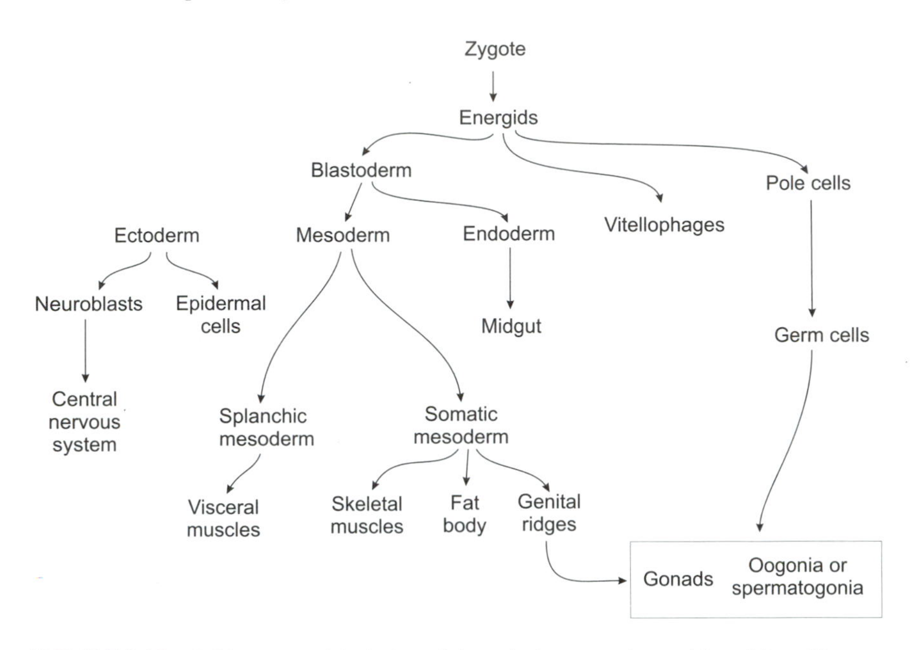

FIGURE 3.18 Cell lineages and derivation of tissues in the mature insect. Adapted from Chapman (1982). Reprinted with permission.

organs are formed, are the embryos capable of synthesizing new hormones. A large number of ecdysteroids have been isolated from insect eggs, primarily ecdysone, 20-hydroxyecdysone, 26-hydroxyecdysone, 2-deoxyecdysone, and 20,26-dihydroxyecdysone. There have also been several metabolic intermediates between cholesterol and ecdysone that have been isolated.

Molting also occurs during embryonic development in many insects, just as it does during postembryonic life. For example, the embryo of *Locusta migratoria* produces four distinct cuticles, including the serosal cuticle and three embryonic cuticles, the last of which is the cuticle of the first instar larva. All of these cuticles undergo typical molts while in the egg. The ecdysteroids are packaged as conjugates of phosphate or yolk proteins, and increases in the titers of active hormone result from the release of the conjugates. The levels of ecdysteroids typically peak with the deposition of these cuticles, and the coincidence of ecdysteroid peaks with these molts strongly suggests their involvement in embryonic molting. The embryonic prothoracic glands are formed after blastokinesis and may then begin to synthesize ecdysteroids.

There is even less known about the contributions of JH during embryogenesis. At least five different juvenile hormones have been isolated from insect eggs, and topical treatment with JH or its analogs often affects embryonic development.

There are correlations between the titers of JH and significant events during embryogenesis, with increased levels reported before and during blastokinesis and at the end of embryonic development.

Endocrine events in the egg influence whether the embryo undergoes a diapause before it hatches and can continue its postembryonic development. In the silkworm, *B. mori,* there is a maternal determination of whether an egg diapause occurs in the next generation. In effect, the embryo is relying on its mother to make a decision to diapause. When larvae are exposed to a short photoperiod, those adults later produce a diapause hormone from their subesophageal ganglion that is incorporated into their eggs and causes the resulting embryos to halt their development at gastrulation. The diapause hormone in *Bombyx* is a 24-amino-acid peptide that not only regulates the accumulation of glycogen in the developing ovary, but also determines the metabolic fate of the glycogen that is deposited in the egg. In nondiapausing eggs, glycogen levels remain high, but in diapausing eggs, levels of glycogen decrease significantly at the same time that the polyols sorbitol and glycerol increase. Sorbitol not only may provide a degree of freeze-tolerance but it also directly inhibits further embryonic development. The stimulus for the termination of embryonic diapause is prolonged chilling that allows the sorbitol and glycerol to be reconverted to glycogen, which eliminates the inhibition by, and provides an energy source for, postdiapause development and hatching.

Bombyx eggs that enter diapause also contain maternal ecdysteroids conjugated as phosphoric esters, but these are hydrolyzed to the free forms of active ecdysteroids in nondiapausing eggs. Levels of free 20-hydroxyecdysone increase significantly in eggs that will develop, but remain low in diapause eggs. Because of the strong correlation between 20-hydroxyecdysone and embryonic molting, these increases in free ecdysteroids may also be involved in the continuation of embryonic development.

ADDITIONAL REFERENCES

Anderson, D. T. 1972. The development of holometabolous insects. In *Developmental systems: Insects* (S. J. Counce, C. H. Waddington, Eds.), vol. 1, pp. 165–242. Academic Press, London.

Anderson, D. T. 1973. *Embryology and phylogeny in annelids and arthropods.* Pergamon Press, Oxford.

Beament, J. W. L. 1947. The formation and structure of the micropylar complex in the egg-shell of *Rhodnius prolixus* Stahl. (Heteroptera Reduviidae). *J. Exp. Biol.* **23:** 213–233.

Bownes, M., Shirras, A., Blair, M., Collins, J., Coulson, A. 1988. Evidence that insect embryogenesis is regulated by ecdysteroids released from yolk proteins. *Proc. Natl. Acad. Sci. USA* **84:** 1554–1157.

Calvi, B. R., Lilly, M. A., Spradling, A. C. 1998. Cell cycle control of chorion gene amplification. *Genes Dev.* **12:** 734–744.

Calvi, B. R., Spradling, A. C. 1999. Chorion gene amplification in *Drosophila*: A model for metazoan origins of DNA replication and S-phase control. *Methods* **18:** 407–417.

Carroll, S. 1995. Homeotic genes and the evolution of arthropods and chordates. *Nature* **376:** 479–485.

Deng, W., Lin, H. 2001. Asymmetric germ cell division and oocyte determination during *Drosophila* oogenesis. *Int. Rev. Cytol.* **203:** 93–138.

Dobens, L. L., Raftery, L. A. 2000. Integration of epithelial patterning and morphogenesis in *Drosophila* ovarian follicle cells. *Dev. Dyn.* **218:** 80–93.

Duchek, P., Rorth, P. 2001. Guidance of cell migration by egf receptor signaling during *Drosophila* oogenesis. *Science* **291:** 131–133.

Fotaki, M. E., Iatrou, K. 1993. Silk moth chorion pseudogenes: Hallmarks of genomic evolution by sequence duplication and gene conversion. *J. Mol. Evol.* **37:** 211–220.

Fullilove, S. L., Jacobson, A. G. 1971. Nuclear elongation and cytokinesis in *Drosophila montana*. *Dev. Biol.* **26:** 560–577.

Gehring, W. J. 1998. *Master control genes in development and evolution: The homeobox story*. Yale, New Haven, CT.

Gonzalez-Reyes, A., St. Johnston, D. 1994. Role of oocyte position in establishment of anterior–posterior polarity in *Drosophila*. *Science* **266:** 639–642.

Gonzalez-Reyes, A., Elliott, H., St. Johnston, D. 1995. Polarization of both major body axes in *Drosophila* by gurken-torpedo signalling. *Nature* **375:** 654–658.

Gonzalez-Reyes, A., Elliott, H., St. Johnston, D. 1997. Oocyte determination and the origin of polarity in *Drosophila*: The role of the spindle genes. *Development* **124:** 4927–4937.

Gonzalez-Reyes, A., St. Johnston, D. 1998. Patterning of the follicle cell epithelium along the anterior–posterior axis during *Drosophila* oogenesis. *Development* **125:** 2837–2846.

Hagedorn, H. H., O'Connor, J. D., Fuchs, M. S., Sage, B., D. Schlaeger, A., Bohm, M. K. 1975. The ovary as a source of α-ecdysone in an adult mosquito. *Proc. Natl. Acad. Sci. USA* **72:** 3255–3259.

Hinton, H. E. 1960. The structure and function of the respiratory horns of the eggs of some flies. *Phil. Trans. R. Soc. London. B* **243:** 45–73.

Hinton, H. E. 1969. Respiratory systems of insect egg shells. *Annu. Rev. Entomol.* **14:** 343–368.

Hinton, H. E. 1970. Insect eggshells. *Sci. Am.* **223:** 84–91.

Horie, Y., Kanda, T., Mochida, Y. 2000. Sorbitol as an arrestor of embryonic development in diapausing eggs of the silkworm, *Bombyx mori*. *J. Insect Physiol.* **46:** 1009–1016.

Horike, N., Sonobe, H. 1999. Ecdysone 20-monooxygenase in eggs of the silkworm, *Bombyx mori*: Enzymatic properties and developmental changes. *Arch. Insect Biochem. Physiol.* **41:** 9–17.

Illmensee, K., Mahowald, A. P. 1974. Transplantation of posterior pole plasm in *Drosophila*: Induction of germ cells at the anterior pole of the egg. *Proc. Natl. Acad. Sci. USA* **71:** 1016–1020.

Illmensee, K., Mahowald, A. P., Loomis, M. R. 1976. The ontogeny of germ plasm during oogenesis in *Drosophila*. *Dev. Biol.* **49:** 40–65.

Johannsen, O. A., Butt, F. H. 1941. *Embryology of insects and myriapods*. McGraw-Hill, New York.

Kafatos, F. C. 1981. Structure, evolution and developmental expression of the silkmoth chorion multigene families. *Am. Zool.* **21:** 707–714.

Kai, H., Kawai, T. 1981. Diapause hormone in *Bombyx* eggs and adult ovaries. *J. Insect Physiol.* **27:** 623–627.

Kai, H., Kotani, Y., Miao, Y., Azuma, M. 1995. Time interval measuring enzyme for resumption of embryonic development in the silkworm, *Bombyx mori*. *J. Insect Physiol.* **41:** 905–910.

Kuhn, D. T., Chaverri, J. M., Persaud, D. A., Madjidi, A. 2000. Pair-rule genes cooperate to activate en stripe 15 and refine its margins during germ band elongation in the *D. melanogaster* embryo. *Mech. Dev.* **95:** 297–300.

Lagueux, M., Hoffmann, J. A., Goltzené, F., Kappler, C., Tsoupras, G., Hetru, C., Luu, B. 1984. Ecdysteroids in ovaries and embryos of *Locusta migratoria*. In *Biosynthesis, metabolism and mode of action of invertebrate hormones* (J. Hoffmann, M. Porchet, Eds.), pp. 168–180. Springer-Verlag, Berlin.

Lawrence, P. A. 1992. *The making of a fly. The genetics of animal design*. Blackwell, Oxford.

Leclerc, R. F., Regier, J. C. 1993. Choriogenesis in the Lepidoptera: Morphogenesis, protein synthesis, specific mRNA accumulation, and primary structure of a chorion cDNA from the gypsy moth. *Dev. Biol.* **160:** 28–38.

Leclerc, R. F., Regier, J. C. 1994. Evolution of chorion gene families in Lepidoptera: Characterization of 15 cDNAs from the gypsy moth. *J. Mol. Evol.* **39:** 244–254.

Li, J., Hodgeman, B. A., Christensen, B. M. 1996. Involvement of peroxidase in chorion hardening in *Aedes aegypti*. *Insect Biochem. Mol. Biol.* **26:** 309–317.

Mahowald, A. P. 2001. Assembly of the *Drosophila* germ plasm. *Int. Rev. Cytol.* **203:** 187–213.

Marchini, D., Marri, L., Rosetto, M., Manetti, A. G., Dallai, R. 1997. Presence of antibacterial peptides on the laid egg chorion of the medfly *Ceratitis capitata*. *Biochem. Biophys. Res. Commun.* **240:** 657–663.

Marchiondo, A. A., Meola, S. M., Palma, K. G., Slusser, J. H., Meola, R. W. 1999. Chorion formation and ultrastructure of the egg of the cat flea (Siphonaptera: Pulicidae). *J. Med. Entomol.* **36:** 149–157.

Margaritis, L. H. 1984. Microtubules during formation of the micropylar canal in *Drosophila melanogaster*. *Cell Biol. Int. Rep.* **8:** 317–321.

Margaritis, L. H. 1985. Structure and physiology of the eggshell. In *Comprehensive insect physiology biochemistry and pharmacology* (G. A. Kerkut, L. I. Gilbert, Eds.), vol. 1, pp. 153–230. Pergamon Press, Oxford.

Margaritis, L. H., Kafatos, F. C., Petri, W. H. 1980. The eggshell of *Drosophila melanogaster*. I. Fine structure of the layers and regions of the wild-type eggshell. *J. Cell. Sci.* **43:** 1–35.

Margaritis, L. H., Mazzini, M. 1998. Structure of the egg. In *Microscopic anatomy of invertebrates* (F. W. Harrison, M. Locke, Eds.), vol. 11C, pp. 995–1037. Wiley-Liss, New York.

Mazur, G. D. 1989. Morphogenesis of silkmoth chorion: Sequential modification of an early helicoidal framework through expansion and densification. *Tissue Cell* **21:** 227–242.

Mazur, G. D., Regier, J. C., Kafatos., F. C. 1980. The silkmoth chorion: Morphogenesis of surface structures and its relation to synthesis of specific proteins. *Dev. Biol.* **76:** 305–321.

Moreno, E., Morata, G. 1999. Caudal is the Hox gene that specifies the most posterior *Drosophila* segment. *Nature* **400:** 873–877.

Munn, K., Steward, R. 1995. The anterior–posterior and dorsal–ventral axes have a common origin in *Drosophila melanogaster*. *Bioessays* **17:** 920–922.

Munn, K., Steward, R. 2000. The shut-down gene of *Drosophila melanogaster* encodes a novel fk506-binding protein essential for the formation of germline cysts during oogenesis. *Genetics* **156:** 245–256.

Nakamura, A., Amikura, R., Mukai, M., Kobayashi, S., Lasko, P. F. 1996. Requirement for noncoding RNA in *Drosophila* polar granules for germ cell establishment. *Science* **274:** 2075–2079.

Neumann, C., Cohen, S. 1997. Morphogens and pattern formation. *Bioessays* **19:** 721–729.

Pascucci, T., Perrino, J., Mahowald, A. P., Waring, G. L. 1996. Eggshell assembly in *Drosophila*: Processing and localization of vitelline membrane and chorion proteins. *Dev. Biol.* **177:** 590–598.

Patel, N. H. 2000. It's a bug's life. *Proc. Natl. Acad. Sci. USA* **97:** 4442–4444.

Pederson, J. A., Lafollette, J. W., Gross, C., Veraksa, A., Mcginnis, W., Mahaffey, J. W. 2000. Regulation by homeoproteins: A comparison of deformed-responsive elements. *Genetics* **156:** 677–686.

Pitnick, S., Karr, T. L. 1998. Paternal products and by-products in *Drosophila* development. *Proc. R. Soc. London B* **265:** 821–826.

Schwalm, F. E. 1988. *Insect morphogenesis*. Basel, Karger.

Shirk, P. D., Broza, R., Hemphill, M., Perera, O. P. 1998. α-Crystallin protein cognates in eggs of the moth, *Plodia interpunctella*: Possible chaperones for the follicular epithelium yolk protein. *Insect Biochem. Mol. Biol.* **28:** 151–161.

Spoerel, N. A., Nguyen, H. T., Towne, S., Kafatos, F. C. 1993. Negative and positive regulators modulate the activity of a silkmoth chorion gene during choriogenesis. *J. Mol. Biol.* **230:** 151–160.

Thompson, M. J., Weirich, G. F., Rees, H. H., Svoboda, J. A., Feldlaufer, M. F., Wilzer, K. R. 1985. New ecdysteroid conjugate: Isolation and identification of 26-hydroxyecdysone 26-phosphate from eggs of the tobacco hornworm *Manduca sexta* (L.). *Arch. Insect Biochem. Physiol.* **2:** 227–236.

Tolias, P. P., Konsolaki, M., Komitopoulou, K., Kafatos, F. C. 1990. The chorion genes of the medfly, *Ceratitis capitata*. II. Characterization of three novel cDNA clones obtained by differential screening of an ovarian library. *Dev. Biol.* **140:** 105–112.

Trougakos, I. P., Margaritis, L. H. 1998. The formation of the functional chorion structure of *Drosophila virilis* involves intercalation of the "middle" and "late" major chorion proteins: A general model for chorion assembly in Drosophilidae. *J. Struct. Biol.* **123:** 97–110.

Tsuzuki, S., Iwami, M., Sakurai, S. 2001. Ecdysteroid-inducible genes in the programmed cell death during insect metamorphosis. *Insect Biochem. Mol. Biol.* **31:** 321–331.

Whiting, P., Sparks, S., Dinan, L. 1993. Ecdysteroids during embryogenesis of the house cricket, *Acheta domesticus*: Occurance of novel ecdysteroid conjugates in developing eggs. *Insect Biochem. Mol. Biol.* **23:** 319–329.

Yamashita, O. 1996. Diapause hormone of the silkworm, *Bombyx mori*: Gene expression and function. *J. Insect Physiol.* **42:** 669–679.

Zeh, D.W., Zeh, J.A. 1989. Ovipositors, amnions and eggshell architecture in the diversification of terrestrial arthropods. *Q. Rev. Biol.* **64:** 147–168.

CHAPTER 4

Reproductive Systems

MAKING MORE INSECTS

One of the reasons for the huge success of insects and their domination of our planet is their enormous reproductive potential. Although relatively short-lived, they can produce prodigious numbers of offspring as a result of their extraordinary reproductive physiology. Consider, for example, the reproductive potential of the female mosquito, *Aedes aegypti*. She can produce about 125 eggs from each blood meal she ingests, and those eggs can develop to adults in about 10 days after she feeds. If one female mosquito began to reproduce in the spring with a single blood meal and all her offspring survived and likewise reproduced, by the end of the summer there would be over 7×10^{21} mosquitoes. Their biomass at the end of these 3 months would be over 1×10^{19} g, about 33,000 times the weight of the total human population. Of course, insect populations never reach those tremendous numbers because of an unusually high mortality from the many predators that depend on insects as food. An extremely high reproductive potential is the price they must pay for occupying their particular ecological niches, and the design of their reproductive systems is the result of selective pressures for meeting this potential.

The germ cells of the reproductive organs originate from the **pole cells,** which are among the first to differentiate during embryogenesis. Together with mesodermal tissue, they form the reproductive organs of the adult. A feature common to the reproductive systems of all insects is the precise coordination of these cells by hormones and transcription factors, which is ultimately regulated by physiological and environmental factors.

FEMALE REPRODUCTIVE SYSTEMS

The reproductive system of female insects generally consists of a pair of **ovaries** that are connected to a **median oviduct** by a pair of **lateral oviducts** (Fig. 4.1).

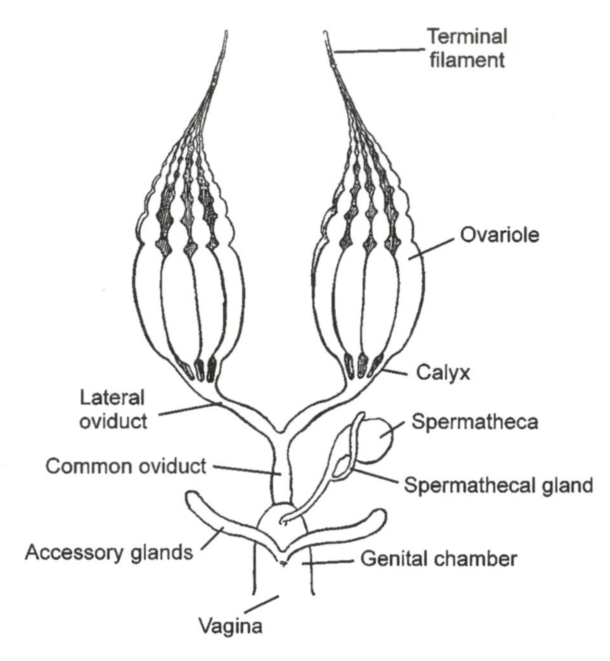

FIGURE 4.1 The generalized female reproductive system. From Snodgrass (1935). Reprinted with permission.

The ovaries are suspended in the hemocele by a **terminal filament** at their anterior end and the lateral oviducts at their posterior ends. Comprising each ovary is a series of tapering egg tubes, the **ovarioles,** the functional units that contain a progression of developing **oocytes.** The number of ovarioles in each ovary varies tremendously depending on the size and reproductive strategies of the particular insect species, and although it is genetically determined, it can be regulated by the diet of the immature stages. Ovariole number largely determines fecundity and typically ranges from 4-8 per ovary, although many dipterans have about 50 ovarioles in each ovary. The tsetse fly, *Glossina,* has only two ovarioles in each ovary, and the queen termite, *Eutermes,* has over 2,000.

Each ovariole produces oocytes that develop and grow within it. Most of the ovary develops from the splanchic mesoderm during embryogenesis. The oocytes begin their growth at the anterior end of the ovariole just beneath the terminal filament where the ovarioles are clustered (Fig. 4.2). Here, the **germarium** contains the germ cells, or **oogonia,** as well as **prefollicular cells.** The oogonia arise from the pole cells of the embryo that differentiate very early during development. The oocytes are produced by the meiotic division of the oogonia in the germarium and begin to be surrounded by the follicle cells that will later be involved in the incorporation of yolk and that will produce the eggshell, or chorion. As the

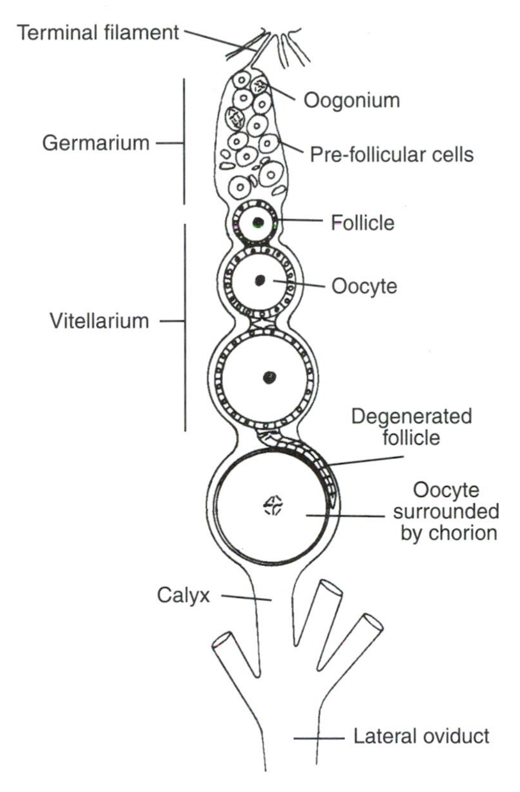

FIGURE 4.2 A single ovariole from the ovary. Oocytes arise from the oogonia in the germarium and descend into the vitellarium surrounded by follicle cells. The follicle cells produce the chorion around the oocyte and then degenerate. From Telfer (1975). Reprinted with permission.

oocytes move down the ovariole, they enter the region called the **vitellarium** where they are completely surrounded by follicle cells and increase in size as they accumulate the yolk. At the posterior end of the ovariole, the **calyx** joins it to the **lateral oviduct.** The **common oviduct** is ectodermal in origin and is lined with cuticle. The lateral oviducts are generally considered to be mesodermal along with the follicle cells and ovariole wall. A thin acellular membrane, the **tunica propria,** covers the ovariole from the terminal filament to the calyx.

Specialized structures are associated with the common oviduct. The **female accessory glands,** also known as **collaterial glands,** are modified dermal glands. These glands produce cement that allows the deposited eggs to be attached to the substrate or glued together. In some insects that retain their eggs after they hatch, such as the tsetse, *Glossina,* the accessory glands produce a nutritive secretion that nourishes the larvae during their entire larval period. In cockroaches and mantids, the female accessory glands produce the hardened egg cases, or **oothe-**

cae. The **spermathecae** are used for the storage of sperm once the female is inseminated. Also ectodermal in origin, the spermatheca generally opens into the common oviduct and releases sperm as the fully formed egg passes by. The spermathecal duct may contain glycogen deposits that can serve as an energy source for the sperm as they pass through to the egg. Nearest its opening to the outside, the common oviduct may be modified into a **genital chamber** that is capable of incubating eggs internally. The **bursa copulatrix** is an additional pouch within the chamber into which sperm are first deposited after mating. The sperm leave the bursa and then move into the spermatheca where they are stored more permanently. The bursa may have a series of toothlike structures that disrupt the spermatophore in which the sperm are contained in more primitive insects and facilitate their release. It may also secrete chemical signals into the hemolymph when it is filled with sperm to signal to the female that mating has occurred.

Types of Ovarioles

The ovariole is the basic unit of egg production. Ovarioles are classified into two major types that differ in the way that RNA and other nutrients are supplied to the oocyte. In the **panoistic** type, the germarium contains only oogonia, primary oocytes, and mesodermal prefollicular tissue (Fig. 4.3A). Most nourishment for the oocyte comes from the follicular epithelium, and the genes for ribosomal RNA of the oocyte nucleus are selectively amplified to supply the large number of ribosomes that are necessary for protein synthesis during development. Panoistic ovarioles are believed to be the most primitive of the two types, found in the more primitive insect orders, including the Thysanura, Odonata, Orthoptera, Plecoptera, and Isoptera.

More advanced is the **meroistic** type of ovariole that contains oogonia, primary oocytes, prefollicular tissue, and **nurse cells,** or **trophocytes** within the germarium. Meroistic ovarioles can be further divided into **telotrophic meroistic** and **polytrophic meroistic** (Fig. 4.3B,C). In both these types of ovarioles the trophocytes provide the oocyte with RNA, proteins, and ribosomes through much of their development that are otherwise provided only by the oocyte itself in panoistic ovarioles. In the telotrophic meroistic ovariole, the nurse cells remain in the apex of the ovariole and feed the oocyte through nutritive cords as it descends alone. The difference in electrical potential between the trophocytes and the oocyte may be responsible for the flow of materials into the cytoplasm of the oocyte. Toward the end of oocyte maturation, the cytoplasmic cord is broken. Telotrophic meroistic ovarioles are found in most Ephemeroptera, Hemiptera, and some Coleoptera.

Most holometabolous orders have the more advanced type of polytrophic meroistic ovariole, in which the oocyte is one of a group of interconnected cells, or **cystocytes,** that all descend together within the follicle (Fig. 4.3C). The nurse cells are located at the anterior of the follicle and the oocyte at its posterior end,

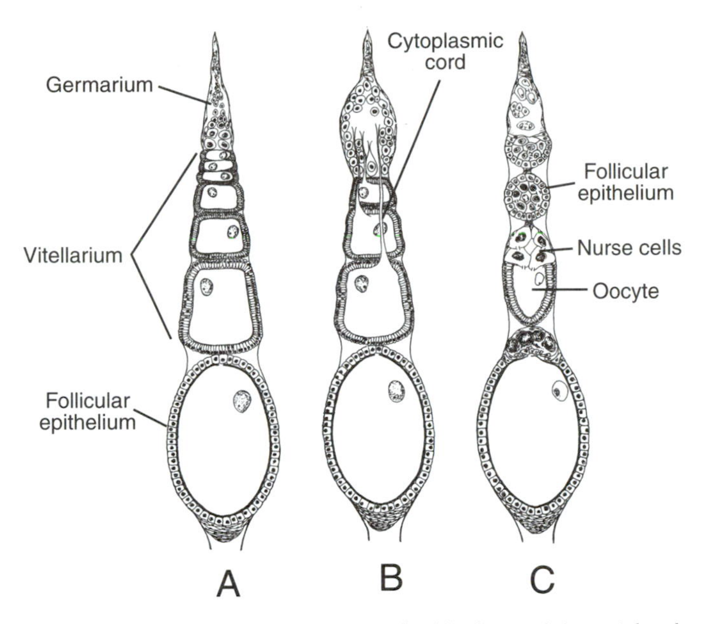

FIGURE 4.3 The three major types of insect ovarioles. (A) The panoistic ovariole, where nourishment for the oocyte comes only from the follicular epithelium. (B) The telotrophic meroistic ovariole, with a cytoplasmic cord that connects nurse cells that remain in the germarium with the descending oocyte. (C) The polytrophic meroistic ovariole, with nurse cells contained within the follicle. From Schwalm (1988). Reprinted with permission.

all surrounded by the follicular epithelium. The nurse cells are sister cells of the oocyte that arise by mitotic divisions and remain connected to the oocyte and each other by cytoplasmic bridges called **ring canals,** forming a cystocyte complex. The number of nurse cells per oocyte is $2^n - 1$, where n is the number of species-specific mitotic divisions that occur during development (Fig. 4.4). The nurse cells also undergo endomitotic chromosome replications and become polyploid, enhancing their ability to produce rRNA. This differs from panoistic ovarioles where the oocyte selectively amplifies its own rRNA. When their mitotic divisions are complete, oocyte-specific factors begin to move by polarized transport to the one cystocyte that will ultimately differentiate into the oocyte. In *Drosophila,* the oocyte always arises from one of the two oldest cystocytes that have the largest number of ring canals. When the nurse cells have passed all their contents into the oocyte, they degenerate, leaving the quiescent oocyte alone in the follicle before vitellogenesis takes place.

Vitellogenesis

From the time the egg is laid until the first instar larva emerges, the embryo must be completely self-reliant. Because it is unable to acquire nutrients or water, the

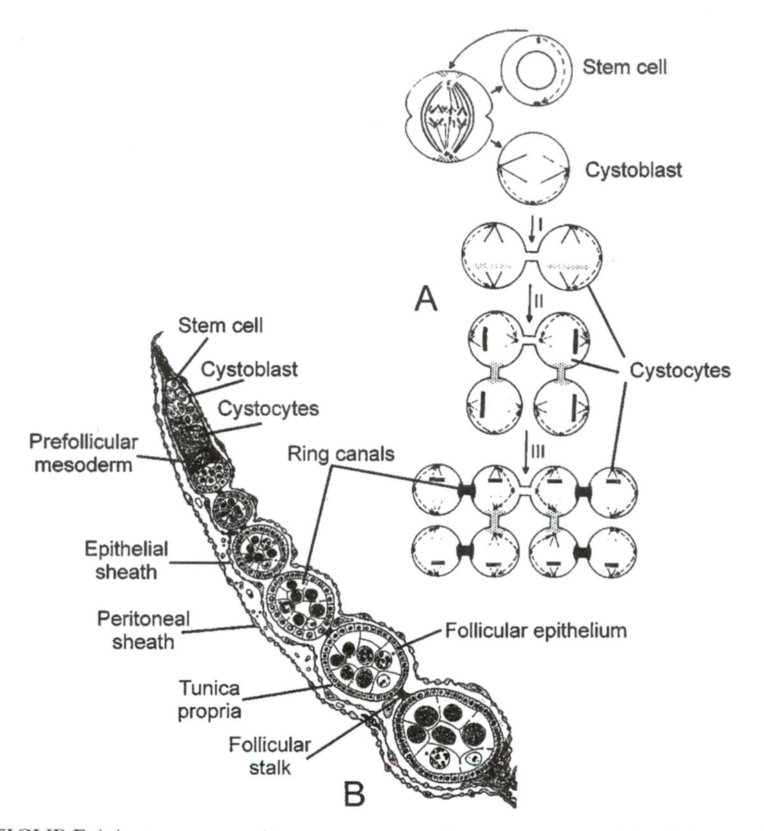

FIGURE 4.4 Maturation of the oocyte–nurse cell complex in *Drosophila*. (A) The division of the cystoblast to form cystocytes, and their connection by ring canals. From Engelmann (1970). (B) Their location and development in the ovariole, From Cummings and King (1969). Reprinted with permission.

embryo must have within the egg everything it needs for embryogenesis. During embryonic development, a large number of new proteins must be synthesized on a large number of new ribosomes. The ribosomes are acquired from either the increased synthetic capacity of the panoistic oocyte, or from the trophocytes in the meroistic oocyte. The major source of nutrients is the abundant yolk, consisting largely of proteins and lipids, that is deposited in the oocyte cytoplasm. The yolk is broken down and made available during embryogenesis by the vitellophages, those energids that differentiated early during development.

The bulk of the yolk protein consists of **vitellogenins** that are synthesized during the process of **vitellogenesis.** In **autosynthetic vitellogenesis,** found in the primitive apterygotes, the egg synthesizes its own yolk from hemolymph proteins. More advanced insects, as well as vertebrates, engage in **heterosynthetic vitellogenesis,** where the yolk is synthesized elsewhere and transported to the oocyte in the blood. In vertebrates, the liver is stimulated to release vitellogenins into the bloodsteam by hormones. In the insects, the fat body is stimulated by hormones to secrete vitellogenins into the hemolymph.

Insect vitellogenins are large glycolipophosphoproteins made in the female fat body that generally consist of two or more subunits. They are selectively taken up by the terminal oocytes and may comprise as much as 60-90% of the total soluble yolk proteins that are present. The yolk is deposited in the oocyte as it descends through the **vitellarium,** the lower portion of the ovariole, which considerably increases the volume of the oocyte cell. In *Drosophila,* the oocyte volume increases by 100,000 times during vitellogenesis. In *Periplaneta,* this increase in volume is over 2,000,000 times.

The passage of yolk proteins into the oocyte is regulated by the follicle cells that surround it. With **patency,** intracellular spaces appear between the follicle cells to regulate the uptake of vitellogenin and allow it to be taken up at the oocyte membrane by **receptor-mediated endocytosis** during vitellogenesis (Fig. 4.5). Juvenile hormone regulates this patency, as well as the endocytotic uptake of vitellogenin by the oocyte.

The vitellogenin that is produced by the fat body is further processed to form subunits. Alterations may involve the cleavage of the vitellogenin molecule into smaller molecules or addition of subunits to form the mature vitellogenin molecule. The vitellogenin that is ultimately taken up by the oocyte can be further processed by the oocyte itself to create a crystalline form of the vitellogenin called **vitellin** that is stored within yolk spheres. The vitellin is immunologically indistinguishable from vitellogenin, but may be different in its solubility and electrophoretic mobility. Other proteins can also be packaged into yolk spheres, including yolk processing enzymes, hemolymph proteins that are taken up nonspecifically, and proteins produced by follicle cells. The oocyte cytoplasm also includes a number of lipid droplets that contain triglycerides and phospholipids, and granules of glycogen.

Toward the end of vitellogenesis, the follicle cells produce the **vitelline envelope,** an acellular membrane that surrounds the oocyte. There is evidence that levels of 20-hydroxyecdysone regulate its formation. The last synthetic act of the follicle cells is **choriogenesis,** the secretion of the chorion, the complex lamellar array that was described in Chapter 3. The chorion is an intricate assemblage of proteins that are laid down sequentially to form an semipermeable barrier to the environment. The imprint of each follicle cell is patterned on the surface of the chorion and provides a species-specific signature. The versatility of the follicle cells in producing so many of the products that are necessary for reproduction is truly remarkable. Once the chorion is secreted, this layer forms an impermeable envelope that restricts the movement of substances in or out, with the exception of sperm that enter through the specialized micropyle. Once the synthesis of the chorion is completed, the follicle cells degenerate and leave the chorion as the outer surface of the egg.

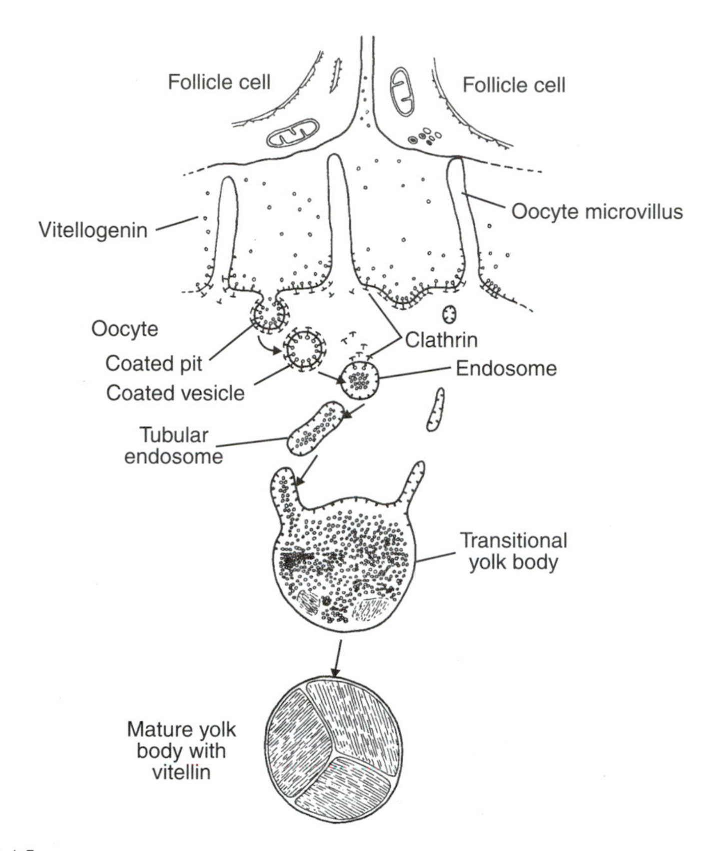

FIGURE 4.5 The mechanism of receptor-mediated endocytosis in the mosquito oocyte. The follicle cells are shown just above the microvilli on the oocyte cell membrane. Vitellogenin moves between the follicle cells and binds to receptors on the oocyte cell membrane that are concentrated in areas known as coated pits. The pits also contain the membrane protein, clathrin. The region where vitellogenin is bound to the receptor invaginates and pinches off to form a coated vesicle, which carries the vitellogenin-receptor complex to an endosome. The endosome loses its clathrin, which is recycled to the oocyte membrane. The endosomes are combined into a transitional yolk body, which recycles the vitellogenin receptors when it forms a mature yolk body that contains the vitellin. Reprinted with permission from Raikhel, A. S. (1984). *J. Ultrastructure Res.* **87:** 285–302. Copyright Academic Press.

Endocrinology of Female Reproduction

Given the tremendous amount of ecological variability among insect species, it is not surprising to find a comparable variability and complexity in the systems that regulate reproduction. There are only a few generalizations that can be made regarding the endocrine control of female reproduction. In short-lived insects that do not feed as adults, the yolk must be derived from reserves acquired during the larval stage. In these insects, vitellogenesis occurs while the pharate adult is still within the pupal skin and the adult emerges with a full complement of eggs. In the lepidopteran *Hyalophora cecropia,* this preemergence egg development appears to occur in the absence of any identifiable hormonal controls and may simply be a developmental program that is followed by the fat body. In longer-lived insects that undergo multiple cycles of reproduction, vitellogenesis occurs when the fat body is activated by hormones that allow the vitellogenins to be produced cyclically.

The specific hormones that are involved in cyclical vitellogenesis vary considerably among insects, and there is no single mechanism of hormonal control that can be described. However, there are two general approaches to egg production. In some insects, the production of vitellogenins is dependent on JH alone, such as in grasshoppers, cockroaches, some lepidopterans, and *Rhodnius.* JH acts directly on the fat body cells, causing them to initiate the translation and secretion of vitellogenin. In the gypsy moth, *Lymantria dispar,* low or declining titers of JH in the last larval instar are necessary for vitellogenin production by the fat body.

In other insects, including most dipterans, both JH and 20-hydroxyecdysone are involved. JH regulates the formation of new endoplasmic reticulum in the fat body and the sequestration of the vitellogenin produced, while 20-hydroxyecdysone regulates the rate of its production. Variations in the control of vitellogenesis by 20-hydroxyecdysone are common. Vitellogenin production in the Indian meal moth, *Plodia interpunctella,* coincides with a decline in the ecdysteroid titer of the pharate adult. Vitellogenin synthesis in the silkworm, *Bombyx mori,* coincides with a rise of 20-hydroxyecdysone. The reproductive endocrinology of relatively few insects has been examined, and there are many deviations from these generalizations within this small group.

The control of vitellogenesis in the female mosquito is a good example of the complexity of the mechanisms that has evolved to coordinate reproduction with nutritional state. Because vitellogenesis only occurs in female mosquitoes after a blood meal has been periodically acquired, blood ingestion consequently serves as a method of synchronizing the reproduction of many individuals so that the coordinated events can be better observed. The blood meal provides the precursors for yolk synthesis that are lacking after larval development is completed. After the adult female emerges, the JH that is released by the paired CA during the first few days of life prepare both the fat body and the ovary for vitellogenesis. In response to an early peak of JH, the fat body cells become polyploid to provide more templates for DNA synthesis during vitellogenesis. JH also stimulates the induction of mRNA in fat body cells, the proliferation of ribosomes, and a development of its responsiveness to 20-hydroxyecdysone. The follicle cells that surround the oocyte-nurse cell syncytium are relatively undifferentiated at emergence, but in response to JH, they begin to differentiate and increase in size. Mediated by endocrine cells dispersed throughout the midgut epithelium, a blood meal releases a pulse of the neurohormone **ovarian ecdysteroidogenic hormone** (**OEH**) from the brain, which acts on the ovaries to increase their synthesis of protein and stimulate their production of ecdysone. The 20-hydroxyecdysone then activates the transcription of vitellogenin genes in the fat body. The most abundant fat body transcript is a 6.5-kb vitellogenin mRNA that is translated into a 224-kDA provitellogenin and subsequently cleaved and then repackaged into a 380-kDa vitellogenin (Fig. 4.6). The 20-hydroxyecdysone peak also stimulates the follicle cells of the ovary to synthesize the vitelline envelope, the inner layer of the chorion, and acts on the germarium to cause the creation of a new, secondary follicle. During the later stages of egg development, an **oostatic hormone** is produced that prevents the maturation of any secondary follicles until the maturing eggs have been laid, and avoids the burden of developing so many eggs that the female could no longer fly.

Ovulation, Fertilization, and Oviposition

Once egg development has been completed, the mature eggs are moved into the oviduct by the process of **ovulation.** This is only possible once the follicle cells

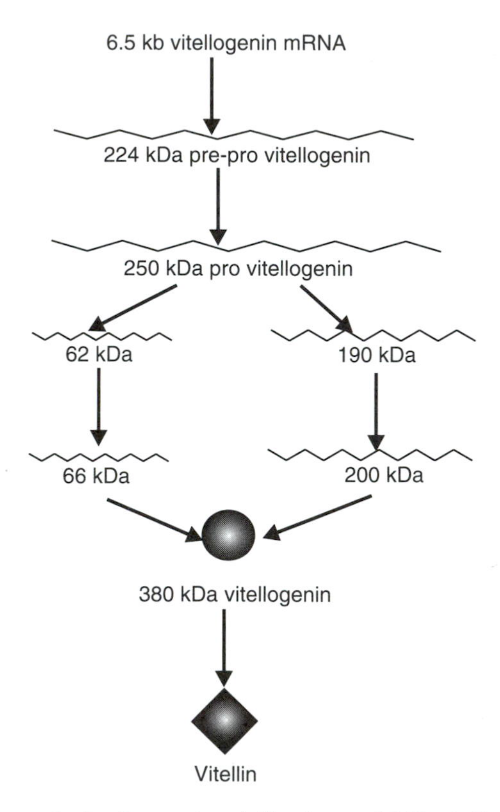

FIGURE 4.6 The synthesis of mosquito vitellin from a 6.5-kb vitellogenin mRNA. From Sappington and Raikhel (1998). Reprinted with permission.

that surround the oocyte degenerate, leaving the egg free to move out of the ovariole. The muscular contractions of the ovariole and oviduct that propel the egg through the reproductive tract are coordinated by **myotropins** that are secreted by neurosecretory cells when the central nervous system receives a confirmation that mating has occurred and that eggs are indeed mature. For example, in the blood-sucking bug, *Rhodnius prolixus,* ovulation is initiated by an FMRFamide-like myotropin that is released only when the spermatheca produces a factor once it is filled with sperm and male accessory gland substances and the maturing eggs produce 20-hydroxyecdysone. In the female tsetse, *Glossina,* the message that mating has occurred and that stimulates ovulation is not chemical, but comes from prolonged mechanical stimulation during copulation.

The presence of the egg in the oviduct triggers stretch receptors that stimulate the spermathecal muscles to contract and release the sperm that are stored within the spermatheca. As the egg is passed from the lateral oviduct into the common oviduct, it travels past the spermathecal duct and the released sperm enter the egg through the micropyle. Insect eggs have a variable number of these micropyles on the chorion. Dipteran eggs have only a single micropyle, and those of orthopterans may have as many as 40.

After they are ovulated and fertilized, the eggs are usually deposited outside of the female's body in the process of **oviposition.** The eggs move down the common oviduct by peristaltic waves of muscular contractions and out of the body through the ovipositor. The movement of the egg downward through the oviducts is facilitated by backwardly directed scales that act like a ratchet mechanism so the egg can only move in one direction, downward toward the genital opening. In many insects, oviposition follows immediately after ovulation, but in some insects the eggs may be retained for variable periods until the eggs hatch or the larvae mature. Some of these variations will be discussed later in the section on unusual methods of reproduction.

The control of oviposition in the cockroach, *Spodromantis,* is a good example of the integration of environmental and physiological information during egg-laying (Fig. 4.7). This insect normally lays its eggs at the beginning of the photophase. The brain integrates the information it receives about the photoperiod and the presence of mature oocytes and triggers the release of an oviposition stimulating hormone that activates the ovipositor, ovariole and oviduct muscles, and the terminal abdominal ganglion. As the insect probes the substrate with its ovipositor, tactile sensations from sensilla are sent to the terminal abdominal ganglion, which controls further movements of the egg and secretions by the accessory glands.

MALE REPRODUCTIVE SYSTEMS

Spermatozoa are produced within the paired **testes** of the male (Fig. 4.8). Each testis is composed of a series of tubular **follicles,** which can vary in number from

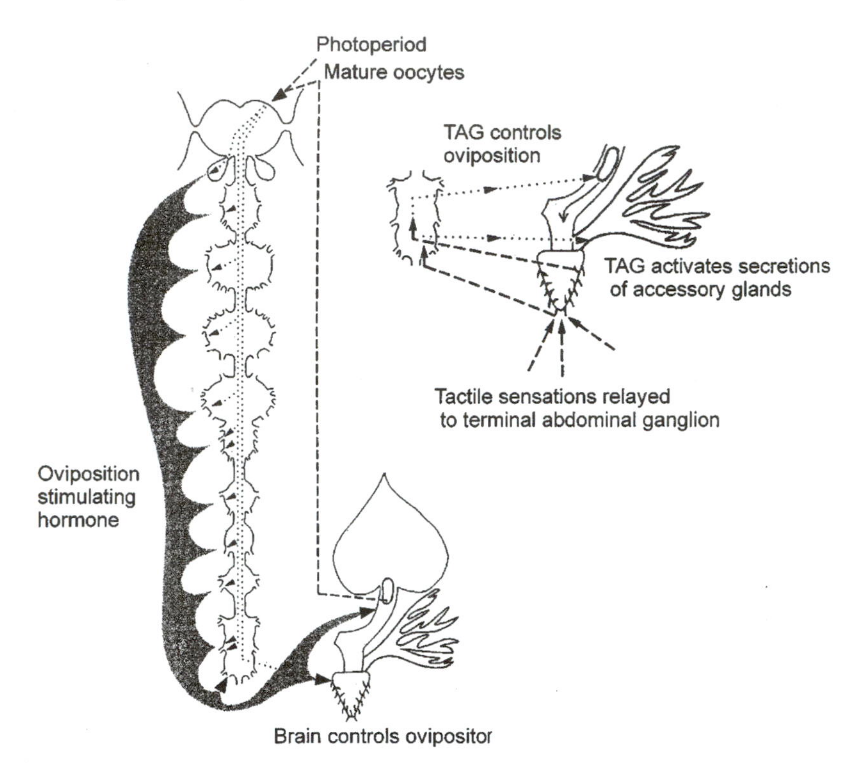

FIGURE 4.7 The control of oviposition in the cockroach, *Sphodromantis lineola*. The proper photoperiod and the presence of mature oocytes are perceived by the brain, which triggers the release of an oviposition stimulating hormone from the ventral ganglia that causes the oviduct muscles to contract. The brain also controls the searching movements of the ovipositor. Tactile sensations of the ovipositor are relayed to the terminal abdominal ganglion, which activates the secretions by the accessory glands. The perception of all these components by genital receptors causes further contractions of the oviducts and oviposition. From Mesnier (1984). Reprinted with permission.

1 in Coleoptera to over 100 in Orthoptera. The follicles are in turn enclosed by a **peritoneal sheath.** More primitive insects have a single testis, and in some lepidopterans the two maturing testes are secondarily fused into one structure during the later stages of larval development, although the ducts still remain separate. The follicles connect to a main duct, the **vas deferens,** through individual **vas efferens** tubes. A portion of the vas deferens may be enlarged as the **seminal vesicle** that serves as a storage reservoir for sperm before they are transferred to the female. The two vasa deferentia connect to the **ejaculatory duct,** which is composed of cells that are ectodermal in origin and produce a lining of cuticle. The terminal portion of the ejaculatory duct may be sclerotized to form the intromittent organ, the **aedeagus.**

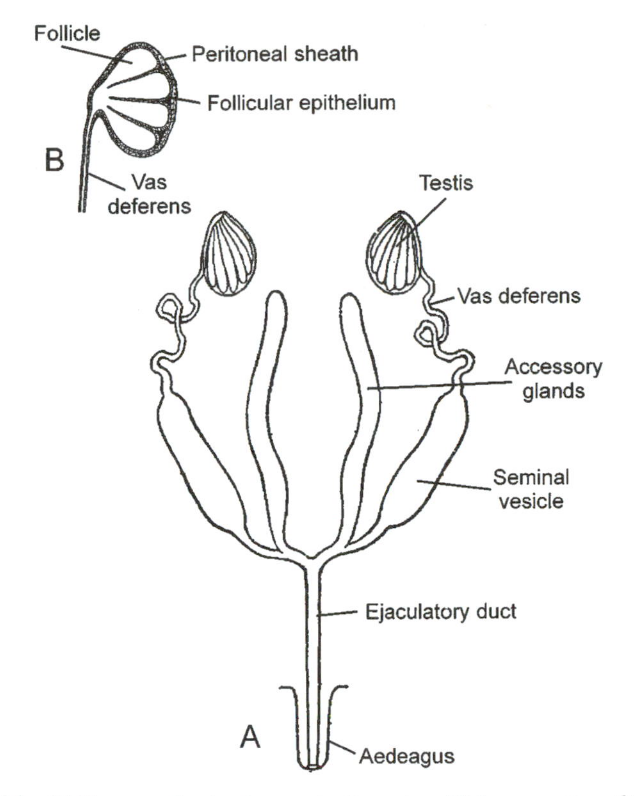

FIGURE 4.8 (A) The generalized male reproductive system. (B) A cross section of the testis. From Snodgrass (1935). Reprinted with permission.

A pair of male accessory glands is also typically present. These glands can open either into the vas deferens or the ejaculatory duct. Those arising from the vas deferens during development are mesodermal in origin, whereas those originating from the ejaculatory duct have an ectodermal origin. Male accessory glands serve a variety of functions, including the production of the **seminal fluid,** which serves as a transport and activation medium for sperm, the **vaginal mating plug** that temporarily blocks sperm from another male from entering, and the formation of **spermatophores** that are proteinaceous secretions of the male accessory glands that enclose the sperm. Insects are believed to have descended from aquatic ancestors, and the synthesis of a spermatophore appears to be associated with the evolutionary transition to the terrestrial habitats they have come to occupy. Internal fertilization is most appropriate for animals that live on land, and the spermatophore may be a transitional form. The spermatophore thus represents an initial adaptation for life on land that protects the male gametes from desiccation until they are in the female reproductive tract. Apterygote males produce a spermatophore that is deposited on the moist ground and then taken up by the female. In more advanced insects, the sperm are transferred directly in seminal fluid by internal fertilization and spermatophores are not produced.

Peptides that are produced by the male accessory glands and transferred to the female during mating can also affect several physiological systems of the female. A common effect is the prevention of subsequent mating by the female, either temporarily or permanently. Much like a nuptial gift, male accessory gland substances can also supplement the nutritional reserves of the female and allow her to increase her egg production when mated. The unmated females of many insects are incapable of laying the eggs that may develop, and the components in male accessory glands are able to remove the physiological block that prevents oviposition until mating takes place. The circadian rhythmicity of females can be altered by mating, and this alteration is commonly mediated by male accessory gland substances.

Spermatogenesis

The production of spermatozoa occurs within the follicles of the testes. The anterior **germarium** of the follicle contains apical stem cells that divide mitotically to form **spermatogonia** (Fig. 4.9). As these spermatogonia move downward in the follicle they become enclosed by somatic cells that form the **cysts** that surround them. The remainder of each follicle can be divided into zones containing cysts that display successive regions of sperm development. In zone I, the zone of growth, the spermatogonia divide mitotically 6-8 times within each cyst to form **spermatocytes,** each of which remains connected by cytoplasmic bridges, or ring canals. The number of divisions within a cyst is species specific; in *Rhodnius* the spermatocyte stage is reached at eight divisions, or 128 cells per cyst. Zone II is the zone of maturation in which the spermatocytes divide meiotically to form haploid **spermatids.** As the cysts move toward the vas deferens within zone III, the zone of transformation, the haploid spermatids within them together differentiate into flagellated spermatozoa. As this differentiation proceeds, the cysts elongate and eventually rupture, releasing the spermatozoa into the vas efferens. These spermatozoa migrate to the seminal vesicles where they remain until mating takes place. In Lepidoptera, this migration occurs before adult emergence and is dependent on a circadian clock located in the reproductive tissues that produces a two-step release, first to the vas deferens and then to the seminal vesicles.

Lepidopteran males are unusual in that they produce two types of sperm in their fused testes. Both are transferred to the spermatheca of the female, but only the conventional **eupyrene** sperm are involved in normal fertilization of the oocyte. The **apyrene** sperm have no nuclei and thus cannot have any genetic function. Instead, their function may be to assist the eupyrene sperm to move through the female reproductive tract or provide the eupyrene sperm with nutrients. Apyrene spermatozoa are initially more motile than eupyrene sperm and are

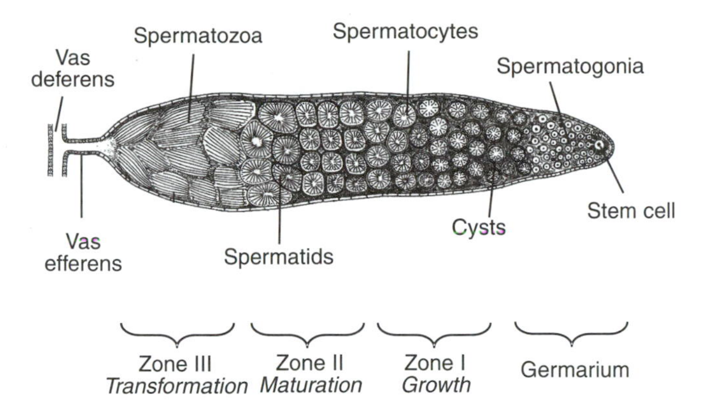

FIGURE 4.9 A cross section of the generalized male testis. Stem cells at the tip give rise to spermatogonia by mitotic division. The spermatogonia become enclosed by somatic cells that produce the cysts found in zone I. Incomplete mitotic divisions of the spermatogonia form the spermatocytes that are connected by ring canals. In zone II, the spermatocytes divide meiotically to form the haploid spermatids. The spermatids differentiate into flagellated spermatozoa in zone III; as they mature, they rupture the cyst wall and escape to the seminal vesicles. From Chapman (1982). Reprinted with permission.

present singly in the seminal vesicles when the eupyrene sperm are still in bundles, gaining their motility when they are placed into the spermatophore (Fig. 4.10). They also may displace the eupyrene sperm that are already present in the spermatheca from the previous matings of the female. As many as 90% of the spermatozoa transferred may be apyrene.

Spermatozoa

Insect sperm are morphologically very similar to those of vertebrates, containing a head region and a long flagellum that is used for locomotion (Fig. 4.11). Sperm are the only cells of the insect that bear flagella. Within the head are a haploid **nucleus** and the **acrosomal complex** at the tip that arises from the Golgi apparatus during differentiation. The acrosome contains the enzyme **acrosin,** a trypsin-like enzyme that dissolves the egg membranes for fertilization. The motor portion of the flagellum is called the **axoneme** and is composed of microtubules originating from the centriole at the base of the sperm nucleus. Although the sperm of some primitive apterygotes have no flagellum at all, most pterygote sperm have a flagellum with a 9 + 9 + 2 arrangement of microtubules (9 outer accessory tubules, 9 doublets, and 2 central tubules). The microtubules are made of the dimeric protein **tubulin,** with arms consisting of the contractile protein **dynein** that forms cross-bridges between the microtubule fibers and allows them to slide past each other to effect bending of the flagellum. The tail length is variable; in *D. melanogaster* the

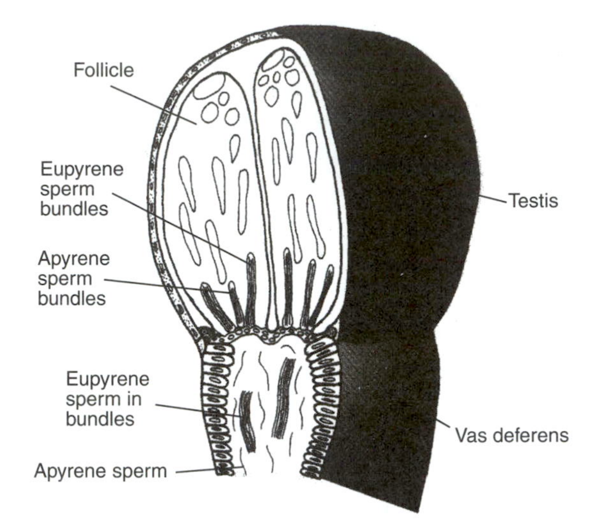

FIGURE 4.10 The fused testis of lepidopterans and their production of bundles of apyrene and eupyrene sperm. Reprinted from *Journal of Insect Physiology* Volume 43. Giebultowicz, J. M., F. Weyda, E. F. Erbe, and W. P. Wergin. Circadian rhythm of sperm release in the gypsy moth, *Lymantria dispar*: ultrastructural study of transepithelial penetration of sperm bundles, pp. 1133–1147. Copyright 1997, with permission from Elsevier Science.

sperm range in length from 1. 6 to 2 mm, but *D. bifurca* males produce sperm with a total length of over 58 mm, the longest described for any animal and many times the length of the adult fly that produces them.

Endocrine Control of the Male Reproductive System and Spermatogenesis

Relatively little is known about the endocrine regulation of male reproduction compared to what is known about the female. In short-lived insects that may not feed as adults, spermatogenesis occurs early during the larval and pupal stages. In longer-lived males, spermatogenesis continues throughout adult life. With the vastly different hormonal conditions that exist during the immature and adult periods, a unifying scheme for the control of spermatogenesis has not been possible. Indeed, it has been suggested that insect spermatogenesis may be a sequential process of differentiation that is completely independent of hormones.

The rate of mitotic divisions of spermatogonia to form spermatocytes is increased by high levels of 20-hydroxyecdysone, but high titers of JH abolish this increase. The spermatocytes then begin meiotic divisions that are arrested at prophase until the end of the larval period is reached. The postwandering peak of 20-hydroxyecdysone unblocks meiosis and allows the cells to proceed to metaphase. In some insects, JH accelerates spermatogenesis.

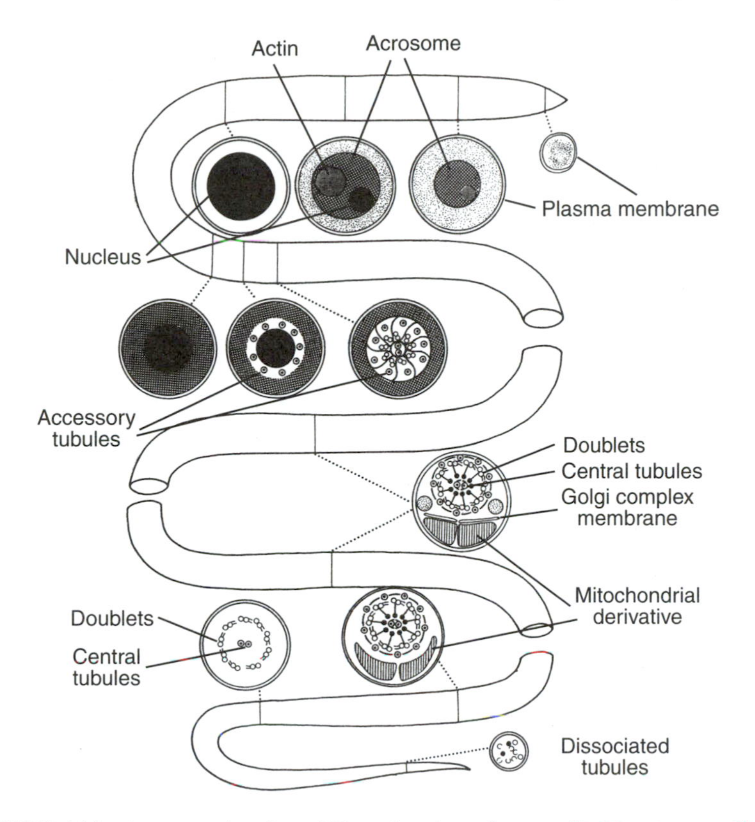

FIGURE 4.11 A cross section along different locations of a generalized insect sperm. Reprinted from Baccetti, B. (1998). *Microscopic Anatomy of Invertebrates* **11C**: 843–894. This material is used by permission of Wiley-Liss, Inc., a subsidiary of John Wiley & Sons, Inc.

The release of mature spermatozoa from the cysts in the testes displays a circadian rhythmicity that is initially inhibited by 20-hydroxyecdysone. The decline of 20-hydroxyecdysone is thus necessary in order for sperm to be released. Juvenile hormone may also control the development and secretion of the male accessory glands. The accumulation of some secretory peptides in the glands is enhanced by JH and inhibited by 20-hydroxyecdysone.

Spermatogenesis is interrupted in those lepidopterans that undergo a larval or pupal diapause but resumes once diapause has been completed. It is not any developmental activity that causes this interruption but rather the lysis of developing gametes before they become mature. The renewal of spermatogenesis occurs as a result of the increasing titers of 20-hydroxyecdysone that occur when diapause is terminated. In Lepidopterans, the differentiation of apyrene sperm from eupyrene sperm occurs with the exposure to a hemolymph-borne apyrene-spermatogenesis-inducing factor.

UNCONVENTIONAL METHODS OF INSECT REPRODUCTION

In most insects, females mate with males, fertilize the eggs just prior to oviposition and lay them outside of the body. This most common means of reproduction is termed **oviparity.** There are also some more unusual methods of producing offspring among the insects.

Parthenogenesis

Sexual reproduction involves the fusion of the two haploid nuclei from the sperm of the male parent and the egg of the female parent (Fig. 4.12). Occurring in most insects, this sexual reproduction offers an opportunity for the reassortment of genetic information that provides the phenotypic variation that is acted upon by natural selection. However, there have been reports of **parthenogenesis,** or the development of unfertilized eggs, in every insect order except the Odonata, Neuroptera, and Siphonaptera. Unlike sexual reproduction, parthenogenesis no longer provides any opportunity for the reassortment of parental genes. The populations that reproduce parthenogenetically thus have a much greater genetic stability, but this stability can be a disadvantage in a changing environment because it reduces the amount of variation upon which natural selection can operate. With a drastically changing environment, a parthenogenetic population can be wiped out because it has no phenotypic variation to fall back on. However, there are advantages to parthenogenesis that allow it to be maintained in a population. With fertilization unnecessary, parthenogenetically reproducing females do not have to expend energy to find members of the opposite sex. Allocation of resources to devote to pheromone production or swarming is not necessary. When a female is no longer required to call attention to herself to attract a male, the risk of predation from making oneself more obvious in general is also reduced.

Another potential difficulty in parthenogenesis is that when the haploid gametes no longer unite at fertilization, there must be some other way for the diploid number of chromosomes in the zygote to be restored. The mechanism of restoration can be classified as either **haploid parthenogenesis** or **diploid parthenogenesis.** In haploid parthenogenesis, meiosis by the oocyte produces a reduction in chromosome number that results in haploid gametes, and the haploid eggs may develop with or without fertilization. Eggs that are fertilized become female, while those that are not fertilized become male (Fig. 4.13). Although the cytology of oogenesis is normal, spermatogenesis is not, because the haploid males must also produce haploid sperm and therefore lack the meiotic divisions in their formation. In diploid parthenogenesis, the eggs are diploid, and the diploid number is regained from haploid gametes in several ways. In **automictic parthenogenesis,** the early stages of meiosis are normal and the chromosome complement

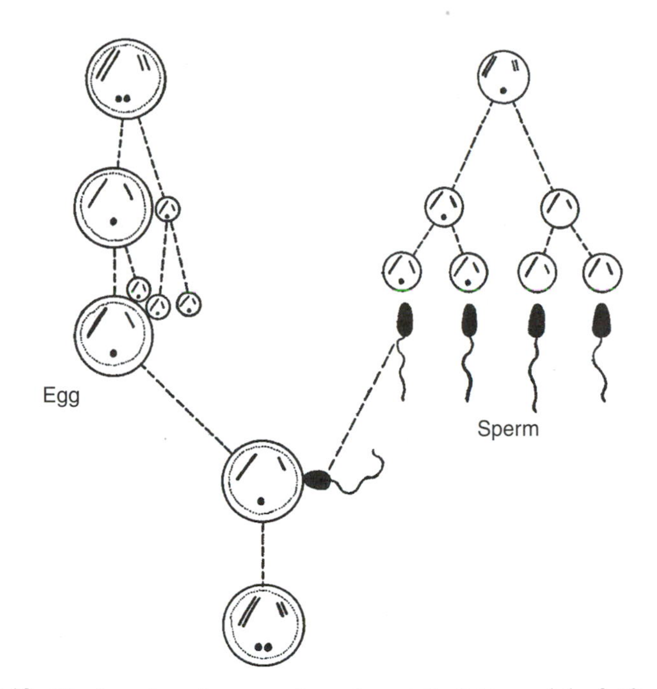

FIGURE 4.12 The formation of sperm and eggs by meiotic division and the fertilization of the egg to produce a diploid zygote by sexual reproduction. From Johansen and Butt (1941). Reprinted with permission.

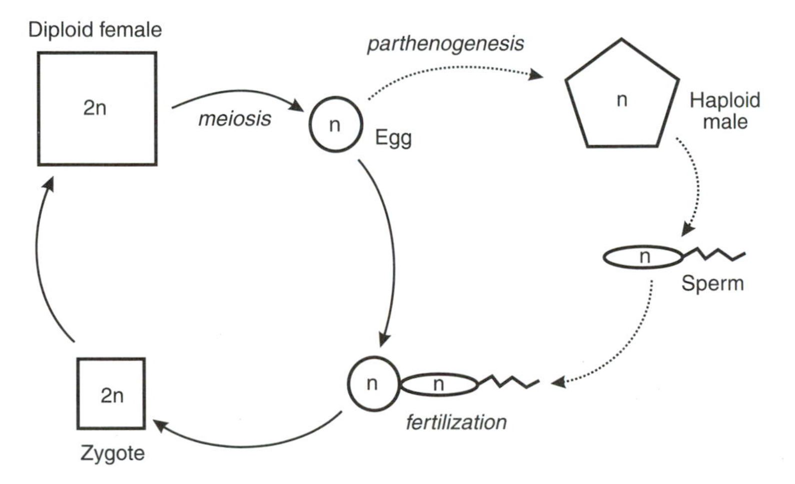

FIGURE 4.13 An example of haploid parthenogenesis. The diploid female produces haploid eggs by meiosis; if the eggs are fertilized, they become a diploid zygote that develops into a female. Unfertilized eggs develop into haploid males.

in the egg is haploid. The diploid number is restored not by the fertilization by a spermatozoan (**amphimixis**), but by fusion of the egg nucleus with a polar body so that two nuclei from the same individual fuse. Because most insects have a system of sex determination in which the female has 2 X chromosomes, this form of parthenogenesis produces all females (Fig. 4.14). In the Lepidoptera where the female is XY, reproduction by automictic parthenogenesis may produce either a male or a female (Fig. 4.15). Another form of diploid parthenogenesis is probably more common in insects. In **apomictic parthenogenesis** there is no problem in restoring the diploid number because the oocyte does not undergo a reduction division. The oocyte remains diploid and the egg develops in the normal way. With meiosis absent, offspring retain the genetic constitution of the mother. The trigger for further egg development in the absence of sperm penetration and fusion as a trigger may come from the displacement of the oocyte nucleus as it is squeezed through the oviduct during oviposition.

Viviparity

Most insects are **oviparous,** developing their eggs internally and fertilizing them just before oviposition, with the eggs hatching after being laid outside of the female. **Viviparity,** in contrast, involves the retention of eggs for varying periods after they have been fertilized, allowing embryonic development to be completed within the female. The eggs hatch while inside the female and instead of laying eggs, larvae are deposited. There are four main types of viviparity.

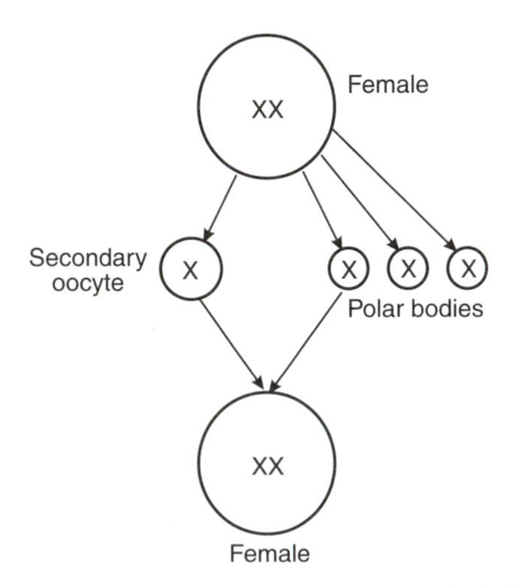

FIGURE 4.14 Automictic parthenogenetic development, in which the oocyte fuses with a polar body.

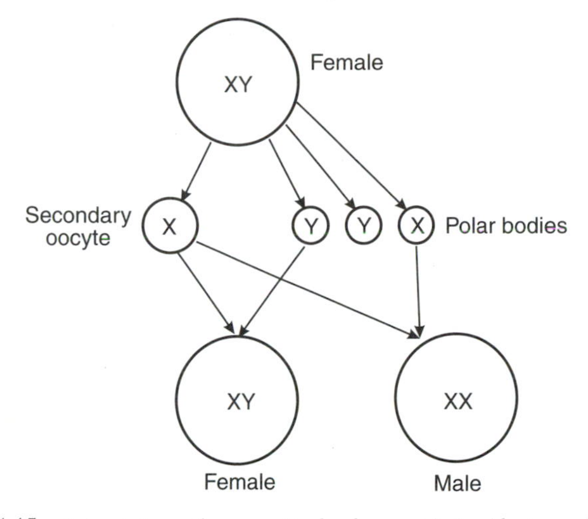

FIGURE 4.15 Automictic parthenogenetic development in Lepidoptera, where the mechanism of XY sex determination produces polar bodies that may have either an X or Y chromosome, producing either male or female progeny.

In **ovoviviparity,** the eggs contain sufficient yolk to nourish the embryo until it hatches and are retained by the female without her providing any nourishment. The female deposits the larvae soon before or immediately after they hatch. In **adenotrophic viviparity** the embryo develops within the female but when the egg hatches, the larva is retained in the modified genital chamber and feeds from secretions of the female accessory gland. The larva molts within the female and is deposited as either a mature larva or a pupa. In **hemocelous viviparity** the ovaries lie free within the fat body and disperse eggs within the hemocele. The eggs are surrounded by specialized follicle cells, the **trophamnion,** that feeds the oocyte from maternal tissues. Mature larvae either escape through a brood canal, as in Strepsiptera, or devour the tissues of the maternal larva, as in some cecidomyiid dipterans. In **pseudoplacental viviparity,** the eggs have little yolk and the developing embryo receives nourishment through a special structure formed from follicle cells, the **pseudoplacenta,** that absorbs nutrients from "milk" secreted by the female accessory glands.

Polyembryony

In **polyembryony,** found in some parasitic Hymenoptera, one sexually produced embryo is split into up to several thousand others during development once the egg is oviposited. It thus shares some of the characteristics of both sexual reproduction and parthenogenesis. Unlike parthenogenesis that produces many copies of a single genotype, the genotypes of the mother and offspring differ in polyembryony. Unlike sexual reproduction, however, all the genotypes produced are the same as each other. Polyembryony occurs after the egg is oviposited and enables the offspring, rather than the parent, to determine optimal brood size.

Paedogenesis

Some insects are able to engage in reproduction as larvae. This form of reproduction is termed **paedogenesis** and is necessarily parthenogenetic because the larvae have no external reproductive structures for mating. The best example of this is the dipteran *Miastor* that produces eggs from ovarioles dispersed throughout the larval fat body. Under inadequate nutritional conditions, the developing embryos at first absorb nutrients from the maternal fat body, but after hatching, the larvae feed on the mother's internal organs and ultimately kill her. The larvae escape through the cuticle and soon initiate another paedogenic cycle. Paedogenesis also commonly occurs in aphids when they reproduce parthenogenetically. When the development of offspring begins in the parent before the parent has reached adulthood, their development is considered to be paedogenic.

Hemocelic Insemination

There is a bizarre method of insemination in the hemipteran superfamily Cimicoidea. As ectoparasites, these insects may have difficulty finding food and surviving between blood meals, and the additional nutrients might allow them to survive longer than they would otherwise. In many members of this group, the male punctures the integument of the female and places his sperm in specialized integumental structures in the hemocele rather than introducing sperm conventionally into the genital tract. A progression toward this specialized insemination can be seen within species in this hemipteran superfamily. In the nabid, *Alloeorhynchus,* the injection of sperm occurs through the wall of the female's genital tract. A spine on the penis ruptures the genital wall and sperm move through the hemolymph and collect around the ovarioles where they fertilize the eggs. Some of the sperm that enter the hemocele are phagocytized by hemocytes and may be used as nutritional precursors by the female. In *Primicimex,* a bedbug found in bat caves, sperm are injected into the body cavity of the female through the abdominal integument. The number of sclerotized scars on the female's cuticle is a good indication of the number of times she has mated. The sperm circulate within the hemolymph and accumulate in pouches at the base of the oviducts. In other nabid species that are most highly evolved for hemocelic insemination, a cuticular pouch, the **ectospermalege,** has evolved as a special reception site. A group of mesodermal cells, apparently derived from hemocytes, forms a

mesospermalege that phagocytoses some of the sperm and also conducts others to the base of the oviducts. The value of this type of insemination is probably related to the female's absorption of some of the nutrients in the spermatozoa and seminal fluid that can be used for egg maturation.

MATING SYSTEMS

The concept of natural selection proposes that the number of descendants that an individual produces is affected by climate, predators, and competition with other species. **Sexual selection** presumes that interactions between the sexes of one species can also affect the survival of particular genes within the gene pool. The number and quality of mating encounters, and the resulting number of offspring produced, can be influenced by the ability of males to compete among themselves for females or for their capacity to attract females. The female dung beetle, *Onthophagous binodis,* produces more offspring when mated to males with horns than when mated to males without horns because the horned males assist her in raising progeny. A female choosing to mate with a horned male thus reaps the benefits of better parenting.

The evolution of insect mating systems has been influenced by the interactions among the reproductive interests of both males and females. Females tend to invest more of their resources into egg production than males do in producing sperm. They produce not only gametes, but also a large package, the egg, that represents a huge outlay of metabolic assets. For the most part, males only contribute sperm to the next generation, and because the sperm are small and plentiful, there is little risk for males to invest in a single mating. They allocate their resources to make themselves more successful at inseminating females than at investing in parental behavior. Females are thus a limiting resource for which males compete; there is considerable competition among the males of a species for the insemination of females and the subsequent transmission of their genes to ensuing generations. A strong selective pressure therefore exists for males to produce substances that give them a reproductive advantage by preventing a female from subsequently mating with other males. Lepidopteran males that produce a larger spermatophore are able to delay the subsequent mating of the female by providing a greater activation of stretch receptors in the female genital tract.

Females also exercise choice in their selection of males. Although it is to the male's advantage to appear to be fit and successfully reproduce, it is to the female's advantage to distinguish those males that are truly fit from those that only appear to be. This sexual selection by females can influence the paternity of offspring by discriminating not only between males before copulation occurs but also by choosing among sperm after copulation has taken place. For example, female moths and butterflies are able to select for the larger-sized sperm produced by certain males after insemination occurs.

Most male insects are **polygynous,** mating with more than one female during their lifetimes. Sexual selection favors males that mate with many females, because they fertilize more eggs. Monogamy in males is rare, and may be best exemplified by the honeybee drone that dies after detaching his genitalia and leaving them inserted in the female's genital tract. In contrast, most female insects, even though they produce a relatively small number of eggs, still must make a large investment in their production and the survival of offspring. It may be in the best interests of the female to mate with more than one male during her lifetime and use the genetic products of several males.

Because sperm are stored in the female's spermatheca, the process of insemination is usually temporally distinct from the fertilization of eggs. As described earlier, sperm may be primitively transferred to the female within a spermatophore, which is deposited on the ground by some apterygotes, or into the genital tract of the female as by primitive pterygotes. In many more advanced insects, direct insemination of the female occurs when the male deposits the sperm in seminal fluid into the genital tract or directly into the spermatheca. Males engage in several strategies to protect their genetic investment and prevent other males from mating. Because females can potentially store sperm from several partners in their spermathecae, the possibility exists for competition between the sperm present for use in fertilization. Rather than form a spermatophore, male accessory gland substances may form mating plugs or act chemically as pheromones to make the female physiologically refractory to subsequent males. Males of the giant water bug, *Belastoma,* accept the eggs from a female and incubate them on their backs, but mate repeatedly with a female before and during her oviposition to assure his paternity. Some odonate males also guard their mates after inseminating, rather than searching for other females to mate with, to prevent them from receiving other sperm. The penis of the male damselfly is fitted with hairs and projections that can remove the existing sperm in the female's spermatheca before adding his own, thus assuring his paternity.

Female insects can engage in several different patterns of receptivity. Females may either mate only once in their lifetimes or mate multiply. The multiple mating may occur throughout their lives or be limited to specific periods of receptivity. An example of an insect with a defined period of receptivity is the harvester ant, *Pogonomyrmex,* which copulates with several males on the day of her nuptial flight but is then unreceptive. Female *Drosophila* mate repeatedly during their lives, but only at widely spaced intervals. In contrast, *Anthidium* bees will mate with any male that the females encounter at any time during their lives.

The copulatory acts and positions in insects are believed to have originated with the indirect transfer of a spermatophore, the most primitive means of transferring gametes. In apterygotes, the male deposits the spermatophore on the ground and the female acquires it independently, but the act is more intimate in the thysanuran family, *Machilidae,* where the male uses a thread he produces leading from the spermatophore to guide the female to it. The mating position of the female when this occurs, lying above the male, is considered to be the most primitive, from which all other mating positions evolved. In male dipterans, the genitalia must rotate after the adults emerge to allow them to acquire the proper position for mating to occur.

ADDITIONAL REFERENCES

Female Reproduction

Beckemeyer, E. F., Lea, A. O. 1980. Induction of follicle separation in the mosquito by physiological amounts of ecdysterone. *Science* **209:** 819–821.

Bownes, M. 1994. The regulation of the yolk protein genes, a family of sex differentiation genes in *Drosophila melanogaster*. *Bioessays* **16:** 745–752.

Brown, M. R., Graf, R., Swiderek, K. M., Fendley, D., Stracker, T. H., Champagne, D. E., Lea, A. O. 1998. Identification of a steroidogenic neurohormone in female mosquitoes. *J. Biol. Chem.* **273:** 3967–3971.

Büning, J. 1979. The telotrophic nature of ovarioles of polyphage Coleoptera. *Zoomorphology* **93:** 51–57.

Büning, J. 1998. The ovariole: Structure, type and phylogeny. *Microscopic anatomy of invertebrates*. (F. W. Harrison, M. Locke, Eds.), pp. 897–932. Wiley-Liss, New York.

Camenzind, R., Schuepbach, P. 1981. Unorthodox oogenesis in the gall midge, *Mycophila speyeri*. *Adv. Invertebr. Reprod.* **2:** 392.

Clark, J., Lange, A. B. 2000. The neural control of spermathecal contractions in the locust, *Locusta migratoria*. *J. Insect Physiol.* **46:** 191–201.

Clark, J., Lange, A. B. 2001. Evidence of a neural loop involved in controlling spermathecal contractions in *Locusta migratoria*. *J. Insect Physiol.* **47:** 607–616.

Cummings, M. R., King, R. C. 1969. The cytology of the vitellogenic stages of oogenesis in *Drosophila melanogaster*. I. General staging characteristics. *J. Morphol.* **128:** 427–435.

Davey, K. G. 1997. Hormonal controls on reproduction in female heteroptera. *Arch. Insect Biochem. Physiol.* **35:** 443–453.

Davey, K. G., Kuster, J. E. 1981. The source of an antigonadotropin in the female of *Rhodnius prolixus Stal*. *Can. J. Zool.* **59:** 761–764.

Davey, K. G., Sevala, V. L., Gordon, D. R. B. 1993. The action of juvenile hormone and antigonadotropin on the follicle cells of *Locusta migratoria*. *Invert. Reprod. Dev.* **24:** 39–46.

Deitsch, K. W., Chen, J. -S., Raikhel, A. S., 1995. Indirect control of yolk protein genes by 20-hydroxyecdysone in the fat body of the mosquito, *Aedes aegypti*. *Insect Biochem. Mol. Biol.* **25:** 449–454.

Edwards, M. J., Severson, D. W., Hagedorn, H. H. 1998. Vitelline envelope genes of the yellow fever mosquito, *Aedes aegypti*. *Insect Biochem. Mol. Biol.* **28:** 915–925.

Engelmann, F. 1979. Insect vitellogenin: Identification biosynthesis and role in vitellogenesis. *Adv. Insect Physiol.* **14:** 49–108.

Gillott, C., Venkatesh, K. 1985. Development of secretory ability in the spermatheca of the migratory grasshopper, *Melanoplus sanguinipes*. *J. Insect Physiol.* **31:** 647–652.

Goltzene, F., Lagueux, M., Charlet, M., Hoffmann, J. A. 1978. The follicle cell epithelium of maturing ovaries of *Locusta migratoria*: A new biosynthetic tissue for ecdysone. *Hoppe-Seylers Z. Physiol. Chem.* **359:** 1427–1434.

Hagedorn, H. H., Kunkel, J. G. 1979. Vitellogenin and vitellin in insects. *Annu. Rev. Entomol.* **24:** 475–505.

Hagedorn, H. H., Maddison, D. R., Tu, Z. 1998. The evolution of vitellogenins, cyclorrhaphan yolk proteins and related molecules. *Adv. Insect Physiol.* **27:** 335–384.

Hagedorn, H. H., O'Connor, J. D., Fuchs, M. S., Sage, B., Schlaeger, D. A., Bohm, M. K. 1975. The ovary as a source of α-ecdysone in an adult mosquito. *Proc. Natl. Acad. Sci. USA* **72:** 3255–3259.

Harnish, D. G., White, B. N. 1982. Insect vitellins: Identification, purification, and characterization from eight orders. *J. Exp. Zool.* **220:** 1–10.

Huebner, E. 1984. The ultrastructure and development of the telotrophic ovary. In *Insect ultrastructure* (R. C., King, H. Akai, Eds.), vol. 2., pp. 3–48. Plenum, New York.

Koch, E. A., Smith, P. A., King, R. C. 1967. The division and differentiation of *Drosophila* cystocytes. *J. Morphol.* **121:** 55–70.

Kriger, F. L., Davey, K. G. 1984. Identified neurosecretory cells in the brain of female *Rhodnius prolixus* contain a myotropic peptide. *Can. J. Zool.* **62:** 1720–1723.

Lea, A. O. 1967. The medial neurosecretory cells and egg maturation in mosquitoes. *J. Insect Physiol.* **13:** 419–429.

Lea, A. O., Van Handel, E. 1982. A neurosecretory hormone-releasing factor from ovaries of mosquitoes fed blood. *J. Insect Physiol.* **28:** 503–508.

Li, C., Kapitskaya, M. Z., Zhu, J., Miura, K., Segraves, W., Raikhel, A. S. 2000. Conserved molecular mechanism for the stage specificity of the mosquito vitellogenic response to ecdysone. *Dev. Biol.* **224:** 96–110.

Mesnier, M. 1984. Patterns of laying behaviour and control of oviposition in insects: Further experiments on *Sphodromantis lineola* (Dictyoptera). *Int. J. Invertebr. Reprod. Dev.* **7:** 23–32.

Okelo, O. 1979. Mechanisms of sperm release from the receptaculum seminis of *Schistocerca vaga* Scudder (Orthoptera: Acrididae). *Int. J. Invert. Reprod.* **1:** 121–131.

Raikhel, A. S., Lea, A. O. 1986. Internalized proteins directed into accumulative compartments of mosquito oocytes by the specific ligand, vitellogenin. *Tissue Cell* **18:** 559–574.

Raikhel, A. S. 1984. The accumulative pathway of vitellogenin in the mosquito oocyte: A high-resolution immuno- and cytochemical study. *J. Ultrastruct. Res.* **87:** 285–302.

Raikhel, A. S., Lea, A. O. 1991. Control of follicular epithelium development and vitelline envelope formation in the mosquito; role of juvenile hormone and ecdysone. *Tissue Cell* **23:** 577–591.

Raikhel, A. S., Snigirevskaya, E. S. 1998. Vitellogenesis. In *Microscopic anatomy of invertebrates* (F. W. Harrison, M. Locke, Eds.), pp. 933–955. Wiley-Liss, New York.

Raikhel, A. S., Miura, K. 1999. Nuclear receptors in mosquito vitellogenesis. *Am. Zool.* **39:** 722–735.

Ribolla, P. E., Bijovsky, A. T., De Bianchi, A. G. 2001. Procathepsin and acid phosphatase are stored in *Musca domestica* yolk spheres. *J. Insect Physiol.* **47:** 225–232.

Robinson, D. N., Cant, K., Cooley, L. 1994. Morphogenesis of *Drosophila* ovarian ring canals. *Development* **120:** 2015–2025.

Roth, T. F., Porter, K. R. 1964. Yolk protein uptake in the oocyte of the mosquito *Aedes aegypti*. *J. Cell Biol.* **20:** 313–332.

Spradling, A. C., de Cuevas, M., Drummond-Barbosa, D., Keyes, L., Lilly, M., Peling, M., Xie, T. 1997. The *Drosophila* germarium: Stem cells, germ line cysts, and oocytes. *Cold Spring Harbor Symp. Quant. Biol.* **62:** 25–34.

Sugawara, T. 1993. Oviposition behavior of the cricket *Teleogryllus commodus*: Mechanosensory cells in the genital chamber and their role in the switch-over steps. *J. Insect Physiol.* **39:** 335–346.

Telfer, W. H. 1975. Development and physiology of the oocyte-nurse cell syncytium. *Adv. Insect Physiol.* **11:** 223–319.

Wheeler, D. 1996. The role of nourishment in oogenesis. *Ann. Rev. Entomol.* **41:** 407–431.

Male Reproduction

Alrubeai, H. F., Gorell, T. A. 191. Hormonal control of testicular protein synthesis in developing *Tenebrio molitor*. *Insect Biochem.* **11:** 337–342.

Arthur, B. I., Jr., Hauschteck-Jungen, E., Nothiger, R., Ward, P. I. 1998. A female nervous system is necessary for sperm storage in *Drosophila melanogaster*: A masculinized nervous system is as good as none. *Proc. R. Soc. London B* **265:** 1749–1753.

Baccetti, B. 1986. Evolutionary trends in sperm structure. *Comp. Biochem. Physiol.* **85A:** 29–36.

Baccetti, B. 1998. Spermatozoa. In *Microscopic Anatomy of Invertebrates* (F. W. Harrison, M. Locke, Eds.), pp. 843–894. Wiley-Liss, New York.

Butlin, R. K., Woodhatch, C. W., Hewitt, G. M. 1987. Male spermatophore investment increases female fecundity in a grasshopper. *Evolution* **41:** 221–224.

Carlson, J. G., Handel, M. A. 1988. Intercellular bridges and factors determining their patterns in the grasshopper testis. *J. Morphol.* **196:** 173–185.

Chen, P. S. 1984. The functional morphology and biochemistry of insect male accessory glands and their secretions. *Annu. Rev. Entomol.* **29:** 233–255.

Chen, P. S. 1996. The accessory gland proteins in male *Drosophila*: Structural, reproductive and evolutionary aspects. *Experientia* **52:** 503–510.

Clark, A. G., Begun, D. J., Prout, T. 1999. Female x male interactions in *Drosophila* sperm competition. *Science* **283:** 217–220.

Cook, D. 1990. Differences in courtship, mating and precopulatory behaviour between male morphs of the dung beetle *Onthophagous binodis* Thunberg (Coleoptera: Scarabaeidae). *Anim. Behav.* **40:** 428–436.

Davey, K. G. 1959. Spermatophore production in *Rhodnius prolixus*. *Q. J. Microscop. Sci.* **100:** 221–230.

Davey, K. G. 1960. The evolution of spermatophores in insects. *Proc. R. Entomol. Soc. London* **35A:** 107–113.

Dumser, J. B., Davey, K. G. 1974. Endocrinological and other factors influencing testis development in *Rhodnius prolixus*. *Can. J. Zool.* **53:** 1011–1022.

Dumser, J. B. 1980. The regulation of spermatogenesis in insects. *Annu. Rev. Entomol.* **25:** 341–369.

Dumser, J. B., Davey, K. G. 1975. The *Rhodnius* testis: Hormonal effects on cell division. *Can. J. Zool.* **53:** 1682–1689.

Friedlander, M. 1997. Control of eupyrene-apyrene sperm dimorphism in Lepidoptera. *J. Insect Physiol.* **43:** 1085–1092.

Friedlander, M., Reynolds, S. E. 1988. Meiotic metaphases are induced by 20-hydroxyecdysone during spermatogenesis of the tobacco hornworm, *Manduca sexta*. *J. Insect Physiol.* **34:** 1013–1019.

Friedlander, M., Reynolds, S. E. 1992. Intratesticular ecdysteroid titres and the arrest of sperm production during pupal diapause in the tobacco hornworm, *Manduca sexta*. *J. Insect Physiol.* **38:** 693–703.

Gage, M. J. G. 1994. Associations between body size, mating pattern, testis size and sperm lengths across butterflies. *Proc. R. Soc. Lond. B* **258:** 247–254.

Gerber, G. H. 1970. Evolution of the methods of spermatophore formation in pterygotan insects. *Can. Entomol.* **102:** 358–362.

Giebultowicz, J. M., Blackburn, M. B., Thomas-Laemont, P. A., Weyda, F., Raina, A. K. 1996. Daily rhythm in myogenic contractions of vas deferens associated with sperm release cycle in a moth. *J. Comp. Physiol. A* **178:** 629–636.

Giebultowicz, J. M., Brooks, N. L. 1998. The circadian rhythm of sperm release in the codling moth, *Cydia pomonella*. *Entomol. Exp. Appl.* **88:** 229–234.

Giebultowicz, J. M., Joy, J. E. 1992. Ontogeny of the circadian system controlling release of sperm from the insect testis. *J. Biol. Rhythms* **7:** 203–212.

Giebultowicz, J. M., Loeb, M. J., Borkovec, A. B. 1987. In vitro spermatogenesis in lepidopteran larvae: Role of the testis sheath. *Int. J. Invert. Reprod. Dev.* **11:** 211–286.

Giebultowicz, J. M., Weyda, F., Erbe, E. F., Wergin, W. P. 1997. Circadian rhythm of sperm release in the gypsy moth, *Lymantria dispar*: Ultrastructural study of transepithelial penetration of sperm bundles. *J. Insect Physiol.* **43:** 1133–1147.

Gillott, C., Gaines, S. B. 1992. Endocrine regulation of male accessory gland development and activity. *Canad. Entomol.* **124:** 871–886.

Happ, G. M. 1992. Maturation of the male reproductive system and its endocrine regulation. *Annu. Rev. Entomol.* **37:** 303–320.

Jamieson, B. G. M. 1987. *The ultrastructure and phylogeny of insect spermatozoa*. Cambridge University Press, Cambridge.

Khalifa, A. 1949. The mechanism of insemination and the mode of action of the spermatophore in *Gryllus domesticus*. *Quart. J. Microscop. Sci.* **90:** 281–292.

King, R. C., Akai, H. 1971. Spermatogenesis in *Bombyx mori*. I. The canal system joining sister spermatocytes. *J. Morphol.* **134:** 47–56.

LaMunyon, C. W., Eisner, T. 1994. Spermatophore size as determinant of paternity in an arctiid moth (*Utetheisa ornatrix*). *Proc. Natl. Acad. Sci. USA* **91:** 7081–7084.

Leloup, A. M. 1981. About the endocrine control of spermatogenesis in insects. *Ann. Endocrinol. (Paris)* **42:** 63–64.

Linley, J. R., Simmons, K. R. 1981. Sperm motility and spermathecal filling in lower Diptera. *Int. J. Invertebr. Reprod.* **4:** 137–146.

Lung, O., Kuo, L., Wolfner, M. F. 2001. *Drosophila* males transfer antibacterial proteins from their accessory gland and ejaculatory duct to their mates. *J. Insect Physiol.* **47:** 617–622.

Mann, T. 1984. Insecta. In *Spermatophores*, pp. 89–134. Springer Verlag, Amsterdam.

Mojica, J. M., Bruck, D. L. 1996. Sperm bundle coiling: Transporting long sperm bundles in *Drosophila dunni dunni*. *J. Insect Physiol.* **42:** 303–307.

Morrow, E. H., Gage, M. J. 2000. The evolution of sperm length in moths. *Proc. R. Soc. London B* **267:** 307–313.

Osanai, M., Baccetti, B. 1993. Two-step acquisition of motility by insect spermatozoa. *Experientia* **49:** 593–595.

Otronen, M. 1997. Sperm numbers, their storage and usage in the fly *Dryomyza anilis*. *Proc. R. Soc. London B* **264:** 777–782.

Phillips, D. M. 1970. Insect sperm: Their structure and morphogenesis. *J. Cell. Biol.* **44:** 243–277.

Pitnick, S., Markow, T. A. 1994. Large-male advantages associated with costs of sperm production in *Drosophila hydei*, a species with giant sperm. *Proc. Natl. Acad. Sci. USA* **91:** 9277–9281.

Quicke, D. L. J., Ingram, S. N., Baillie, H. S., Gaitens, P. V. 1992. Sperm structure and ultrastructure in the Hymenoptera (Insecta). *Zool. Scripta* **21:** 381–402.

Shimizu, T., Yagi, S., Agui, N. 1989. The relationship of testicular and hemolymph ecdysteroid titer to spermiogenesis in the common armyworm, *Leucania separata*. *Entomol. Exp. Appl.* **50:** 195–198.

Waage, J. K. 1986. Evidence for widespread sperm displacement ability among Zygoptera (Odonata) and the means for predicting its presence. *Biol. J. Linn. Soc.* **28:** 285–300.

Walker, W. 1980. Sperm utilization strategies in non-social insects. *Am. Nat.* **115:** 780–799.

Wandall, A. 1986. Ultrastructural organization of spermatocysts in the testes of *Aedes aegypti* (Diptera: Culicidae). *J. Med. Entomol.* **23:** 374–379.

Wolfner, M. F., Harada, H. A., Bertram, M. J., Stelick, T. J., Kraus, K. W., Kalb, J. M., Lung, Y. O., Neubaum, D. M., Park, M., Tram, U. 1997. New genes for male accessory gland proteins in *Drosophila melanogaster*. *Insect Biochem. Mol. Biol.* **27:** 825–834.

General Insect Reproduction

Alexander, R. P. 1964. The evolution of mating behaviour in arthropods. In *Insect reproduction* (K. C. Highnam, Ed.), vol. 2, pp. 78–94. Royal Entomological Society, London.

Bonhag, P. F., Wick, J. R. 1953. The functional anatomy of the male and female reproductive systems of the milkweed bug, *Oncopeltus fasciatus* (Dallas) (Heteroptera: Lygaeidae). *J. Morphol.* **93:** 177–230.

Carson, H. L., Chang, L. S., Lyttle, T. W. 1982. Decay of female sexual behavior under parthenogenesis. *Science* **218:** 68–70.

Chapman, T., Liddle, L. F., Kalb, J. M., Wolfner, M. F., Partridge, L. 1995. Cost of mating in *Drosophila melanogaster* females is mediated by male accessory gland products. *Nature* **373:** 241–244.

Cruz, Y. P. 1986. Development of the polyembryonic parasite *Copidosomopsis tanytmemus* (Hymenoptera: Encyrtidae). *Ann. Entomol. Soc. Am.* **79:** 121–127.

Daly, M. 1978. The cost of mating. *Am. Nat.* **112:** 771–774.

De Cuevas, M., Lily, M. A., Spradling, A. C. 1997. Germline cyst formation in *Drosophila*. *Annu. Rev. Gen.* **31:** 405–428.

Engelmann, F. 1970. *The physiology of insect reproduction*. Pergamon, New York.

Grbic, M., Nagy, L. M., Strand, M. R. 1998. Development of polyembryonic insects: A major departure from typical insect embryogenesis. *Dev. Genes Evol.* **208:** 69–81.

Greenspan, R. J., Ferveur, J. F. 2000. Courtship in *Drosophila*. *Annu. Rev. Entomol.* **34:** 205–232.

Handley, H. L., Estridge, B. H., Bradley, J. T. 1998. Vitellin processing and protein synthesis during cricket embryogenesis. *Insect Biochem. Mol. Biol.* **28:** 875–885.

He, Y., Tanaka, T., Miyata, T. 1995. Eupyrene and apyrene sperm and their numerical fluctuations inside the female reproductive tract of the armyworm, *Pseudaletia separata*. *J. Insect Physiol.* **41:** 689–694.

Hinton, H. E. 1964. Sperm transfer in insects and the evolution of haemocoelic insemination. *Symp. R. Entomol. Soc. London* **2:** 95–107.

Hinton, H. E. 1969. Respiratory systems of insect egg shells. *Annu. Rev. Entomol.* **14:** 343–368.

Ibrahim, I. A., Gad, A. M. 1975. The occurrence of paedogenesis in *Eristalis* larvae (Diptera: Syrphidae). *J. Med. Entomol.* **12:** 268.

Keller, L., Reeve, H. K. 1995. Why do females mate with multiple males? The sexually selected sperm hypothesis. *Adv. Study Behav.* **24:** 291–299.

Khalifa, A. 1950. Spermatophore production and egg-laying behaviour in *Rhodnius prolixus* Stal. (Hemiptera: Reduviidae). *Parasitology* **40:** 283–289.

Klowden, M. J. 1997. Endocrine aspects of mosquito reproduction. *Arch. Insect Biochem. Physiol.* **35:** 491–512.

Lasko, P. F., Ashburner, M. 1990. Posterior localization of vas protein correlates with, but is not sufficient for, pole cell development. *Genes Dev.* **4:** 905–921.

Leopold, R. A., Degrugillier, M. E. 1973. Sperm penetration of housefly eggs: Evidence for involvement of a female accessory secretion. *Science* **181:** 555–557.

Mahowald, A. P. 1972. Oogenesis. In *Developmental systems: Insects* (S. J. Counce, C. H. Waddington, Eds.), vol. 1, pp. 1–47. Academic Press, New York.

Meola, R., Lea, A. O. 1972. Humoral inhibition of egg development in mosquitoes. *J. Med. Entomol.* **9:** 99–103.

Mogie, M. 1986. Automixis: Its distribution and status. *Biol. J. Linn. Soc.* **28:** 321–329.

Otronen, M., Siva-Jothy, M. T. 1991. The effect of postcopulatory male behavior on ejaculate distribution within the female sperm storage organs of the fly, *Dryomyza anilis* (Diptera: Dryomyzidae). *Behav. Ecol. Sociobiol.* **29:** 33–37.

Parker, G. A. 1970. Sperm competition and its evolutionary consequences in the insects. *Biol. Rev.* **45:** 525–567.

Proctor, H. C. 1998. Indirect sperm transfer in arthropods: Behavioral and evolutionary trends. *Annu. Rev. Entomol.* **43:** 153–174.

Ridley, M. 1988. Mating frequency and fecundity in insects. *Biol. Rev.* **63:** 509–550.

Ridley, M. 1990. The control and frequency of mating in insects. *Funct. Ecol.* **4:** 75–84.

Ringo, J. 1996. Sexual receptivity in insects. *Annu. Rev. Entomol.* **41:** 473–494.

Sakai, M., Taoda, Y. 1992. Mating termination in the male cricket. *Acta Biol. Hung.* **43:** 431–440.

Sappington, T. W., Raikhel, A. S. 1998. Molecular characteristics of insect vitellogenins and vitellogenin receptors. *Insect Biochem. Mol. Biol.* **28:** 277–300.

Schaller, F. 1971. Indirect sperm transfer by soil arthropods. *Annu. Rev. Entomol.* **16:** 407–446.

Schoeters, E., Billen, J. 2000. The importance of the spermathecal duct in bumblebees. *J. Insect Physiol.* **46:** 1303–1312.

Schuepbach, P. M., Camenzind, R. 1983. Germ cell lineage and follicle formation in paedogenic development of *Mycophila speyeri* (Diptera, Cecidomyiidae). *Int. J. Insect Morphol. Embryol.* **12:** 211–224.

Schwalm, F. E. 1988. *Insect morphogenesis*. Karger, Basel.

Scudder, G. G. E. 1971. Comparative morphology of insect genitalia. *Annu. Rev. Entomol.* **16:** 379–406.

Simmons, L. W., Teale, R. J., Maier, M., Standish, R. J., Bailey, W. J., Withers, P. C. 1992. Some costs of reproduction for male bushcrickets, *Requena verticalis* (Orthoptera: Tettigoniidae): Allocating resources to mate attraction and nuptial feeding. *Behav. Ecol. Sociobiol.* **31:** 57–62.

Snell, L. C., Killian, K. A. 2000. The role of cercal sensory feedback during spermatophore transfer in the cricket, *Acheta domesticus*. *J. Insect Physiol.* **46:** 1017–1032.

Soller, M., Bownes, M., Kubli, E. 1997. Mating and sex peptide stimulate the accumulation of yolk in oocytes of *Drosophila melanogaster*. *Eur. J. Biochem.* **243:** 732–738.

Stanley-Samuelson, D. W., Loher, W. 1986. Prostaglandins in insect reproduction. *Ann. Entomol. Soc. Am.* **79:** 841–853.

Strand, M. R., Goodman, W. G., Baehrecke, E. H. 1991. The juvenile hormone titer of *Trichoplusia ni* and its potential role in embryogenesis of the polyembryonic wasp *Copidosoma floridanum*. *Insect Biochem.* **21:** 205–214.

Strand, M. R., Grbic, M. 1997. The development and evolution of polyembryonic insects. *Curr. Top. Dev. Biol.* **35:** 121–159.

Sugawara, T. 1979. Stretch reception in the bursa copulatrix of the butterfly, *Pieris rapae* crucivora, and its role in behaviour. *J. Comp. Physiol.* **130:** 191–199.

Thornhill, R., Alcock, J. 1983. *The evolution of insect mating systems*. Harvard University Press, Cambridge, MA.

Vahed, K. 1998. The function of nuptial feeding in insects: A review of empirical studies. *Biol. Rev.* **73:** 43–78.

Wanjama, J. K., Holliday, N. J. 1987. Paedogenesis in the wheat aphid *Schizaphis graminum*. *Entomol. Exp. Appl.* **45:** 297–298.

Williamson, A., Lehmann, R. 1996. Germ cell development in *Drosophila*. *Annu. Rev. Cell Dev. Biol.* **12:** 365–391.

CHAPTER 5

Behavioral Systems

THE CONTROL OF INSECT BEHAVIOR

Behavior can be most simply defined as what living organisms do. Behavioral physiology examines how they do it and the physiological reasons for why the behavior may be expressed at a particular time. Like morphological structures, behavior is a phenotype whose development and expression are influenced by both genetic and environmental factors. Although usually simple to observe and requiring little in the way of instrumentation to measure, behavior represents the end result of a very complex series of nervous impulses and muscular contractions that are initiated in a specific order that ultimately result in the expression of what we see as a spatial displacement of the insect. What may be interpreted as a single behavior may actually consist of a chain of individual behaviors that constitute it.

Behavior is largely a function of the expression of the particular genes that have been selected for when an animal has been successful at living in a particular environment. There have been a number of human genes that are linked to behavior through various heritable disorders, but in general the genetic basis of behavior applied to humans has been controversial, either because it is believed that humans are so complex that they have probably evolved to a degree where they are somehow different than other animals or that our unique consciousness allows us to override the determinism of our genes. It is virtually impossible to separate the influence of genes from that of human culture. The expression of insect behavior is significantly less affected by these factors and the evidence for a genetic basis for behavior is much more obvious. As we will see, the role of genes in the repertoire of available behaviors is relatively clear in insects and other invertebrates.

Ways of Looking at Behavior

A common pitfall in the study of animal behavior is **anthropomorphism,** the attribution of human characteristics to nonhuman animals. The true mechanisms that underlie behavior are obscured when insects are assumed to have human qualities such as motivation, anger, hunger, and lust. For example, a female mosquito does not approach a host for a meal of blood because it is hungry and wants to reproduce. Its nervous system is simply responding to the environmental stimuli that are translated by sensory receptors and that activate a series of genetically programmed behaviors that have been shaped by natural selection. It is not necessary to postulate that the insect has any purpose or goal; it is only responding to the stimuli it receives with stereotyped behaviors that involve the orderly activation of muscle groups.

If the physiological basis of behavior is most simply described as a temporal series of muscle contractions, then the units of behavior are composed of the systems of sensory receptors, nerves, and muscles that control the physical displacement. The nervous system appears to contain a number of innate prewired endogenous motor programs known as **fixed action patterns.** Generally, fixed action patterns have several common characteristics: they are fairly stereotyped, and are found in all individuals of the species that displays them; they are initiated in their complete forms by certain releasers; they occur in the absence of positive feedback once they are initiated; and they involve the coordination of several different muscle groups.

An example of this stereotypical behavior is the avoidance behavior of some noctuid moths to the ultrasonic cries of bats. The moths have a tympanum on either side of the abdomen that is innervated by two sensory receptors of differing sensitivity (Fig. 5.1). When the moth is far from the bat's attempts to echolocate, only the more sensitive of the receptors is triggered, initiating a directional response away from the area of the bat. When the moth is closer, the less sensitive receptor is triggered, initiating an evasive dive by the moth instead. With only four sensory receptors, nervous connections, and the necessary muscles, the moth can execute behaviors based on fixed action patterns that are both economical in design and also essential for its survival. It is not necessary to consider the moth's conscious awareness of the bat or its fear of being captured, and indeed such considerations only prevent the true mechanism of behavior from being understood.

The simplest innate behavior and the most primitive of the fixed action patterns is the **reflex.** In a reflex arc, the dendrites of a sensory neuron synapse with an associative neuron that then synapses with a motor neuron that innervates a muscle (Fig. 5.2). Stimulation of the receptor causes an immediate contraction of the muscle. More complex fixed action patterns include **kineses** and **taxes.** Kineses are locomotor responses to stimuli that are nondirectional. The intensity of the response may vary with the strength of the stimulus but is not related to its direction. With a photokinesis, an insect might respond to light by becoming more active and moving without any particular direction whenever the light is perceived. This would ultimately cause the insect to be displaced from areas of light and settle in dark areas where it is not stimulated to move. In contrast, taxes are directional movements toward or away from a stimulus. For example, a negative

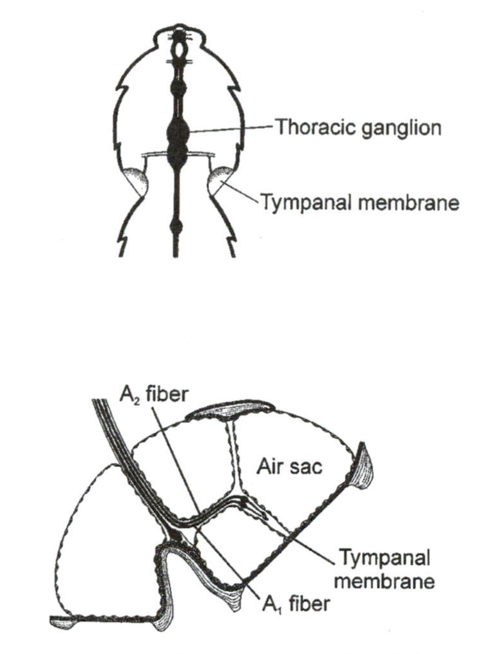

FIGURE 5.1 The moth tympanum that allows a response to the echolocation of bats. The tympanal membrane is supplied with two nerves, A_1 and A_2, which differ in their sensitivity. The more sensitive nerve is triggered when the bat's cries are far away; the less sensitive nerve is triggered when the bat's cries are close. From Roeder (1965). Reprinted with permission.

phototaxis would cause an insect to move away from the source of light. In any event, a fixed action pattern is performed in response to a specific stimulus called a **releaser** when the internal physiological state puts the insect in a condition of readiness to respond.

Genetic Basis of Insect Behavior

The best examples of how single genes can affect behavior can be seen in *Drosophila melanogaster,* where specific behaviors appear to be programmed by genes that directly regulate the development and function of the central nervous system. Mating in *Drosophila* involves a specific sequence of fixed action patterns that the male must execute before his intent to mate with a female culminates in copulation. Females are generally more passive, but are able to either accept or reject the advances of a male based on their own physiological state. The male first orients toward the female, standing about 0. 2 mm away while facing her. If she moves, he follows her and begins to tap her abdomen with a foreleg. Next, he opens one wing and vibrates it to generate a courtship song. If the female is

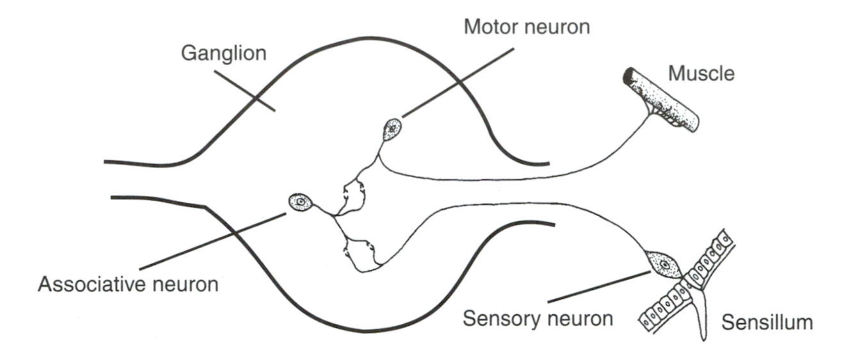

FIGURE 5.2 A reflex arc, with an associative neuron in the ganglion connecting a sensory neuron with a motor neuron.

receptive at this point, he licks the female's genitalia, grabs her wings as he mounts her, and copulates. Receptive females open the vaginal plate to allow copulation to take place and raise their wings so the male can grasp them. Unreceptive females may extrude their ovipositor to prevent the male from copulating.

Several gene mutations in *Drosophila* have been isolated that affect these specific fixed action patterns that lead to copulation. A group of mutations affect the ability of the male to initiate mating behavior. The *fruitless* (*fru*) locus contains multiple alleles that affect courtship by the male and enhance his interactions with other males. In addition, *fru* mutations affect the formation of a male-specific abdominal muscle, the Muscle of Lawrence, which is essential for proper mating behavior by the male. Some alleles of *fru* block the male from displaying a courtship song. Males with mutations in the *dissatisfaction* (*dsf*) locus actively court and attempt to copulate with both males and females but have difficulty doing so because of an inability to bend their abdomens properly. Unmated *dsf* females tend to resist the courtship of males.

Alleles at the *period* (*per*) locus affect the *Drosophila* courtship song, as well as the circadian rhythms of the male. Males with a normal *per* gene generate a song that makes the females more receptive to their advances, but males bearing mutant *per* alleles produce songs that are less effective in generating the necessary female receptivity. This male song has a species-specificity; *D. stimulans* males produce a song that differs from that of *D. melanogaster* in the intervals between song pulses. The *per* genes in the two species are the same with the exception of differences in their center regions. In an elegant example of how a single gene can be responsible for specific fixed-action patterns, a hybrid *per* gene containing the center region of the *stimulans* gene was constructed and inserted into the *melanogaster* genome. The transformed male *melanogaster* sang the *stimulans* song instead of its own. Other genes such as *dissonance, croaker,* and *cacophony* are also known to alter the male song pattern.

There are several other genes that affect copulation, the last step in mating. Normal copulation may last between 10 and 20 min in typical *Drosophila,* but males with the mutations *coitus interuptus* or *fickle* terminate copulation prematurely. The male mutants *stuck* and *lingerer* copulate normally, but have difficulty withdrawing their genitalia from the female. These males may even die with their genitalia still attached to the female. The fixed action patterns involved in mating have been linked to specific neurons that control them within the central nervous system of *Drosophila.*

The foraging behavior of larval *Drosophila* has a genetic basis. Some *Drosophila* larvae are characterized as "sitters" that do not move very far during feeding. Others are "rovers" that move in wide circles as they feed, covering as much as 5 cm every 10 min. After molting to the adult stage, rover flies also show an increased tendency to walk farther from a food source than do sitters. Although the behaviors were once thought to be attributable to the *foraging* gene, they have recently been mapped at *dg2,* which codes for a cyclic GMP-dependent protein kinase (PKG) that acts as a second messenger inside cells. The reduced levels of PKG in sitters may affect signaling pathways that influence the excitability of nerve cells. The single *dg2* gene appears to have multiple effects on behaviors that are expressed in the very different larval and adult stages.

Learning is a change in the behavior of an organism as a result of experience, and its ability to learn accounts for a large part of its behavior. The capacity of *Drosophila* to learn has been assessed by pairing an odor with an electric shock and then measuring the avoidance of the insect to the odor alone. The learning mutants that have been identified in *Drosophila* can be divided into structural brain mutants and conditioning mutants. Defects in brain structure account for a variety of the learning and behavioral alterations, with changes in the mushroom bodies of the brain having perhaps the most prominent role in odor conditioning. The mushroom bodies are groups of about 2500 neurons found in the dorsal regions of the brain hemispheres that relay sensory information. Their ablation abolishes the conditioned odor avoidance behavior. A deficit in odor learning is characteristic of several other mushroom body structural mutants. The mushroom body miniature gene (*mbm*) affects the gross anatomy of the mushroom bodies, and mutant *mbm* flies show a significant odor conditioning impairment. Conditioning mutants include *DCO,* which codes for the catalytic subunit of protein kinase A in the mushroom bodies and lessens learning ability when the mutant allele is present.

The *dunce* (*dnc*) mutants also affect *Drosophila* learning. These mutants have poor memory and show an inability to learn. The *dnc* gene is largely expressed in the mushroom bodies of the brain and encodes the enzyme cAMP phosphodiesterase, and its various alleles differ in the degree of enzyme activity that is present. There is up to an eightfold increase in cAMP in some *dnc* mutants that may be responsible for the learning deficit. The *rutabaga* (*rut*) gene encodes an adenylate cyclase that is also necessary for normal learning to occur in *Drosophila.* Associated with

Drosophila learning are capacities for both short-term memory, reflected by transient changes in the transduction cascades within neurons, and long-term memory, involving alterations in gene expression that produce structural changes in the neurons that result in the restructuring of their synapses.

The nervous system of *Drosophila* shows a considerable degree of structural plasticity. The exposure to sensory stimuli during development is capable of altering the size and number of fibers in the mushroom bodies. Rearing flies for 4 days in constant light increases the volume of the mushroom bodies by as much as 15% compared to other flies reared in total darkness. The structure of the insect nervous system is thus capable of changing in response to environmental stimuli during development and setting the stage for the expression of different behaviors as a result of experience.

Many insects, particularly social insects, are guided to their nests by landmarks. **Tinbergen,** a pioneer in the field of insect behavior, circled the entrance of a solitary wasp nest with pine cones, and then moved them to a nearby location while the wasp was foraging (Fig. 5.3). When the wasp returned, it searched for the nest entrance at the center of the pine cones rather than the true entrance. This suggests that the insects are capable of recognizing and remembering environmental patterns. Evidence indicates that insects recognize these patterns by **retinotopic matching,** in which retinal coordinates of a particular area are stored when the insect is at a certain position, and the pattern is then recognized when the insect matches the stored image as it returns to the same position. They are able to recognize the pattern when it falls on the same area of the retina that it was viewed with during learning. The insect must adopt a standard viewing position during both learning and recall in order to place the image in the same place on the optical receptor. The physical storage location of these patterns may be in the mushroom bodies or optic lobes of the brain.

Hormonal Regulation of Behavior

The genes that are associated with particular behaviors may produce their effects by establishing particular neural pathways that are responsible for the manner in which stimuli are processed. The genes also code for the production of hormones that trigger the expression of the fixed action patterns that are already present in the nervous system. These effects of hormones may be either immediate as **releasers,** or occur significantly after their release as **modifiers.** In either case they interact with the nervous system and the insect's physiological state to activate appropriate behaviors.

When hormones act as releasers, they interact with the nervous system to directly trigger a specific behavior (Fig. 5.4). Given that the fixed action patterns of insects are genetically programmed, chemical messengers such as hormones, along with exogenous sensory input, are capable of releasing the behaviors that the

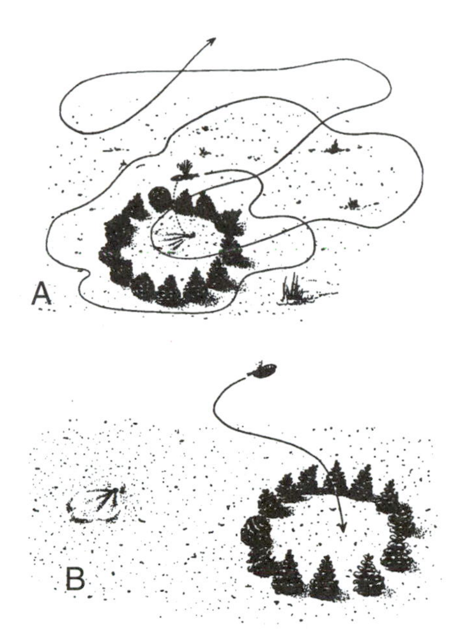

FIGURE 5.3 (A) The path followed by a wasp while on an orientation flight, after its nest is circled by pine cones. (B) The circle of pine cones is moved while the wasp is away foraging. When it returns, it attempts to enter the nest at the center of the cones instead of the real entrance. The wasp associates the nest opening with the cones after its orientation flight. Reprinted from Matthews, R.W. and J.R. Matthews (1987). *Insect Behavior.* This material is used by permission from Elsevier Science.

genes have encoded. In the case of releasers, a fixed action pattern is immediately activated by the hormone's release. For example, larvae periodically display stereotyped behaviors that allow them to ecdyse from the old cuticle at the end of a molt. These behaviors are only executed when the molt occurs. The fixed action patterns involved in this complex series of behaviors are displayed after eclosion hormone and ecdysis triggering hormone are released (see *Physiology of Behaviors Accompanying Metamorphosis,* below). In some mosquitoes, a releaser hormone that is produced during oogenesis inhibits the host-seeking behavior of the female until after her eggs have been laid. Another releaser hormone triggers an increased sensitivity to oviposition site stimuli to enable her to find a site to lay them. Allowing a hormone that circulates throughout the body to regulate behaviors assures that a sustained, coordinated response will be mounted.

Modifier effects are subtler, altering the responsive state of the central nervous system so that a given stimulus provokes a new behavioral response in the presence of the modifier hormone (Fig. 5.5). Eclosion hormone acting on adult behavior is also an example of a modifier effect. A newly emerged male adult moth readily responds to the sex pheromone produced by the female, but if the moth

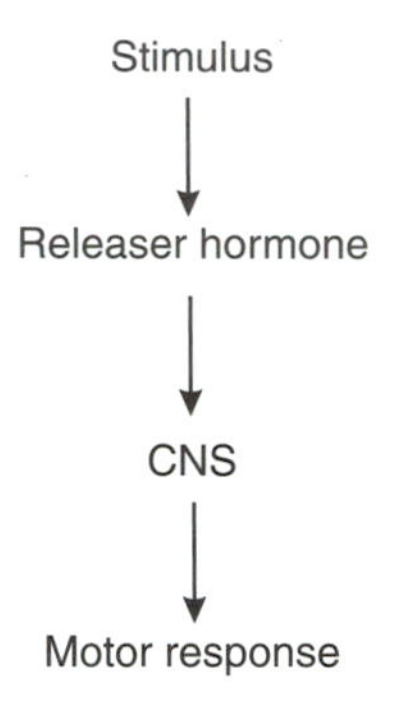

FIGURE 5.4 A releaser hormone acts on the central nervous system to produce an immediate response after a stimulus is received.

prematurely ecloses when the pupal integument is artificially peeled off the pharate adult, the resulting adult fails to respond until after eclosion hormone is released or experimentally injected. The hormone appears to remove a block in the central nervous system that allows the male to respond to the female sex pheromone. Eclosion hormone, acting as a modifier hormone, changes the way that the central nervous system is able to respond to a stimulus. A subset of modifier hormones are **organization hormones** that affect the nervous system during a critical period of development to create permanent changes in behavior.

Because fixed action patterns reside in the central nervous system, substances that affect nervous transmission can sometimes alter their expression. When the parasitoid wasp, *Ampulex compressa,* finds a cockroach prey, it stings it in the thorax and the head near the subesophageal ganglion before laying an egg on it. The venom that is released in the sting causes the cockroach to express its repertoire of grooming behavior, after which it finally becomes paralyzed. In this case it is an injected allomone, and not a hormone, that causes the fixed action patterns associated with grooming to be played out before the cockroach is no longer able to behave at all.

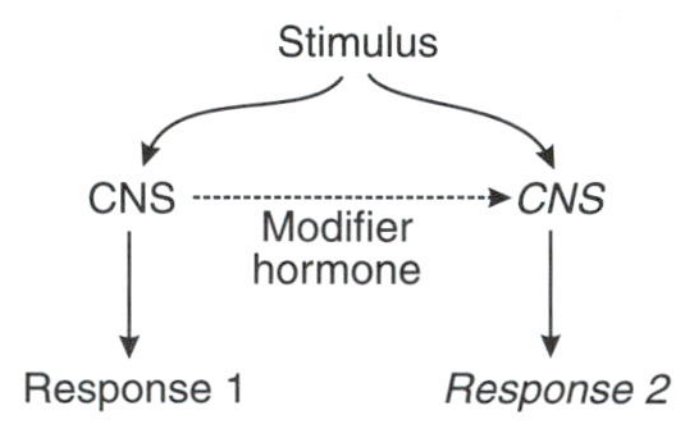

FIGURE 5.5 A modifier hormone changes the responsiveness of the central nervous system so that a given stimulus may produce a new response.

Magnetic Sensitivity

Monarch butterflies, *Danaus plexippus,* migrate from breeding sites in the north-eastern United States and Canada to overwintering sites in Mexico each autumn. The butterflies are the progeny of other monarchs that migrated northward the previous spring, yet they are precisely able to locate the overwintering sites from which their predecessors originated. This orientation behavior is largely due to an internal magnetic compass that is used to maintain the proper direction during the southward movement. Like other migratory animals including birds and fish, the monarchs have an internal compass that allows them to orient to the Earth's magnetic fields.

One way that other insects might adopt a standard viewing pattern for the retinotopic matching discussed previously is if they can orient using magnetic fields. Magnetic cues can provide bees with a framework of coordinates used to position themselves for pattern matching, especially in the absence of celestial cues on cloudy days. Even terrestrial ants have been shown to use geomagnetism for orientation behavior, employing magnetic iron oxides that are deposited in both the head and the abdomen.

Physiology of Social Behavior

Insect behavior reaches its highest level of complexity in the social insects. Their social behavior is considered to be one of the major reasons for the dominance of eusocial, or truly social, insects on the planet. The societies in which they live exist because of their capacity for communication that allows colony activities to be coordinated. In addition, social insects have a system of division of labor in which the responsibilities for the completion of different behavioral tasks are assigned to castes that are all genetically similar but physiologically distinct. Members of the worker castes give up their ability to reproduce in favor of increased ability to defend the colony and to forage for resources. The reproductive castes have the monopoly on reproduction, and caste development is basically a question of who is charged with reproduction in the colony.

In the hymenopterous social insects, an individual's sex is determined by haplodiploidy. Fertilized eggs produce diploid females that may be either workers or queens and unfertilized eggs produce haploid males. Fertilization is a cooperative venture between the queen and workers; the queen decides whether an egg is fertilized, but the workers construct the cells into which either fertilized or unfertilized eggs are expected to be laid. It is only during the reproductive season that the larger drone cells are constructed for the production of males. The queens and all workers are all diploid females and arise from fertilized eggs, but they characteristically undergo different developmental pathways. Whether a female becomes a queen or worker is not determined genetically, but is specified by the endocrine system. The passage into a particular pathway is a consequence of the endocrine milieu that is determined by the feeding program of the larva.

When worker bees feed developing diploid larvae large amounts of royal jelly, a mixture of their glandular secretions, the larvae develop into queens, but if a diet more diluted with nectar and pollen is fed to the same larvae, they will become workers. The diets affect the synthesis of JH by the larval CA, and larvae destined to be queens show significantly higher rates of JH synthesis. The information about the quality of the food may be transmitted to the endocrine system by the stomatogastric nervous system. High levels of JH act in at least two ways during development: they affect the proliferation of ovarian cells during the last larval instar and also inhibit the programmed cell death that takes place in the developing gonads, resulting in increased reproductive cells in the queens. Later in the last instar, the higher titers of JH also stimulate the prothoracic glands to produce increases in ecdysteroids in queens, which direct the caste-specific protein synthesis in ovaries. The smaller amount of JH that is present in workers still gives them the potential to reproduce, but that potential is totally controlled by an existing queen in the colony. Her production of queen pheromone, largely (*E*)-9-oxo-2-decenoic acid, inhibits whatever limited ovarian development is possible in workers. The queen pheromone acts specifically on the endocrine system of the workers to suppress JH synthesis. When the queen is removed and her queen pheromone no longer circulates, a small number of eggs is able to mature in colony workers. These eggs develop into males because the workers are unable to mate and fertilize the eggs.

Polyphenisms are discrete alternative phenotypes that arise not from an organism's genetic information but from environmental cues that are received during development. A **sequential polyphenism** occurs in holometabolous insects that undergo specific differing forms of larvae, pupae, and adults, during their development. An **alternative polyphenism** occurs in the castes found in many social insects, where workers, soldiers, and reproductive castes can all arise from the same genotype. In both cases, environmental factors affect hormone titers that in turn trigger developmental switches that alter the pattern of gene expression (Fig. 5.6).

A model of the control of the soldier-worker polyphenism in the ant, *Pheidole bicarinata,* is shown in Figure 5.7. Juvenile hormone levels determine whether development leads to the soldier or worker caste, and these levels of JH are in turn regulated by the amount of protein in the diet. A diet low in protein causes reduced JH, and when these low levels of JH are present during a JH-sensitive period, development is channeled toward the worker phenotype. In contrast, a diet high in protein causes JH levels to be elevated, and the elevated levels during the JH-sensitive period reprogram the imaginal disks in the head so that the wide-headed soldier phenotype results. Continued production of JH in the soldiers inhibits their pupation and prolongs the larval period, causing their body proportions in general to grow larger, in addition to the increase in head size.

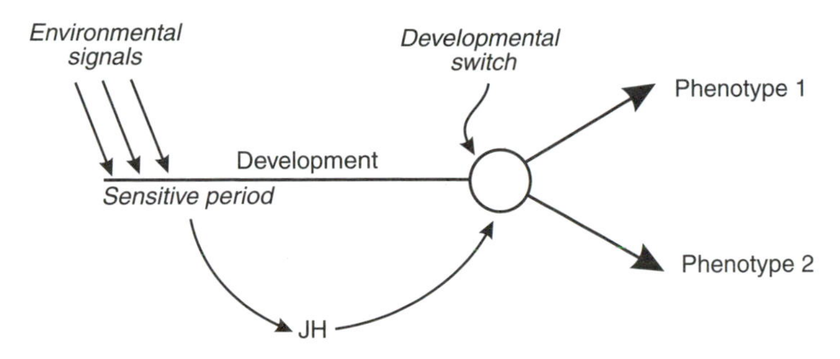

FIGURE 5.6 The control of polyphenisms by juvenile hormone. Environmental signals during a sensitive period of development trigger the release of JH later in development. The JH can switch between alternative phenotypes. Adapted from Nijhout (1999). Reprinted with permission.

Physiology of Temporal Polyethisms

During adult life, individual honeybees show a stereotyped age-related change in the tasks they perform that is known as **temporal** or **age polyethism.** The younger honeybees act as "nurse bees" that care for and feed larvae and the queen. Older bees take the roles of hive maintenance and food storage. The oldest bees, generally about 3 weeks of age until their deaths 3 weeks later, are the foragers that

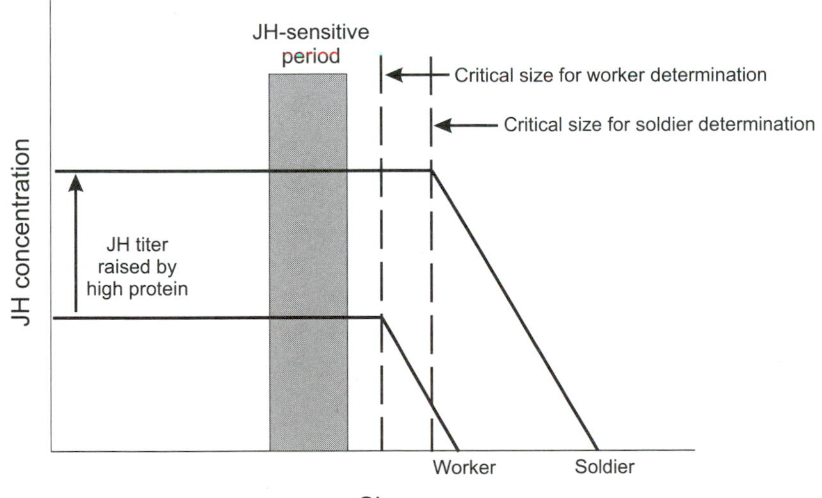

FIGURE 5.7 A model for soldier–worker polyphenism in the ant. Ingested protein raises the levels of JH, and increased JH levels during a sensitive period reprogram the imaginal discs so that development proceeds toward a soldier rather than a worker phenotype. Adapted from Nijhout (1999). Reprinted with permission.

search for nectar and pollen and defend the hive against intruders. These behavioral changes occur in an individual bee as it ages and result from a complex interaction between the endocrine environment, social environment, and the structural changes that occur during development.

The hypopharyngeal glands of the younger workers normally produce the larval food, but as these glands degenerate later in life, the abilities of the adults change. Without their functional glands they are no longer able to behave as nurse bees. There are other changes in behavior with age that result from the restructuring of the neurons of the mushroom bodies in the protocerebrum of the brain. The older foraging bees must develop a more sophisticated behavioral repertoire to learn the location of a food source in relation to the hive and to communicate that location to other members of the colony by performing an elaborate dance language. The mushroom bodies of the protocerebrum are generally larger in social insects, suggesting they may be involved in the regulation of these more complex behaviors. They are considered to be fundamental for information processing and memory and the association of visual and olfactory stimuli. In foraging honeybees, the volume of the mushroom body neuropil is increased significantly in the transition from nurse bee and is also correlated with increased JH titers as the bees age, although JH alone is insufficient for these changes.

The behavioral progression by honeybees is affected by the social environment of the worker. Other workers, generally the older bees, produce the pheromone 10-hydroxy-2-decenoic acid (10-HDA) from their mandibular glands. The exchange of food within the colony by workers circulates the 10-HDA that inhibits the behavioral transition to foraging in younger bees. The rate at which workers make the transition to foragers thus depends on the demographic information received about the proportion of older bees in the colony. This mechanism maintains the proper proportion of the various behavioral types in the colony without regard to its overall size.

Physiology of Behaviors Accompanying Metamorphosis

The immatures and adults of hemimetabolous insects look similar and have similar lifestyles. Their metamorphosis from larva to adult is not very drastic, with many larval structures persisting into the adult stage. In contrast, holometabolous insects have much more extreme differences between immatures and the adults, with an intervening pupal stage that allows the more pronounced transition to occur. When holometabolous insects undergo metamorphosis, the changes that take place involve more than just the modifications of the exterior structures of the integument. In addition to the morphological and ecological differences between holometabolous immature and adult insects, there are also significant behavioral differences. The structures that appear in the adult for the first time, such as antennae, wings, compound eyes, and new legs, require musculature and nervous innervations that did not exist previously. The type of food these stages ingest, the sensory receptors required to identify it, and the behavioral processes by which the food is located, also differ. Changes in behavioral repertoires that accommodate the changes in lifestyle result from a reorganization of the nervous system that occurs at the same time the outer integument is being reorganized.

During the metamorphosis of the moth, *Manduca sexta,* the reorganization of the nervous system and consequent behavioral alternations are linked to endocrine changes. Hormones affect the remodeling of the nervous system, causing the programmed death of many larval neurons, the remodeling of others, and the proliferation of new adult neurons from nests of neuroblasts. Behavior is also affected by the new sensilla that arise and the adult muscles that replace the degenerating larval muscles. Juvenile hormones and ecdysteroids control not only the molt, but also the development or degeneration of all these other systems that affect behavior.

Whether a specific neuron dies or proliferates during metamorphosis depends on its response to the peaks of ecdysteroids to which it is exposed. As discussed in Chapter 1, steroid hormones bind to nuclear receptors that affect gene transcription. The receptor that binds ecdysteroids is a heterodimer consisting of the ecdysone receptor (EcR) and ultraspiracle (USP). EcR exists in at least three isoforms, EcR-A, EcR-B1, and EcR-B2, and the expression of one of these isoforms by a neuron correlates with its response to ecdysteroids. Neurons that bear the EcR-A receptor undergo maturation when exposed to ecdysteroid, while those characterized by an EcR-B1 receptor isoform are usually associated with regression and synapse loss. The fate of the specific neurons and the behaviors they encode are thus determined by the hormone receptors that are displayed during the course of their development (Fig. 5.8).

The pupal stage of most insects is behaviorally passive and can be an easy mark for predators. However, one effective defense system has been incorporated into the behavioral repertoire of pupal *Manduca.* The pupae bear sharp-edged cuticular pits at the anterior margins of some of the abdominal segments that are lined with sensory hairs. These "gin-traps" are pupal-specific structures, absent in larvae and adults. Deflection of the sensillar hairs within the pits provokes an immediate contraction of the intersegmental longitudinal muscles, which draws the gin-trap under the cuticle of the next anterior segment and crushes anything that was inserted, such as the appendages of a predator (Fig. 5.9). Last instar larvae do not have gin-traps, but they do have the sensory neurons that will later be present in the pupal structure. During the transition to the pupa, the sensory neurons become associated with the pupal hairs and their axons mature in both length and the number of branches. This increased arborization of the sensory neurons occurs as the result of ecdysteroids acting in the absence of JH during the commitment peak of the last larval instar. This maturation of neurons itself is insufficient for the later operation of the gin-trap, however, because although the gin-trap and all its associated structures are fully formed in the pharate pupa, the

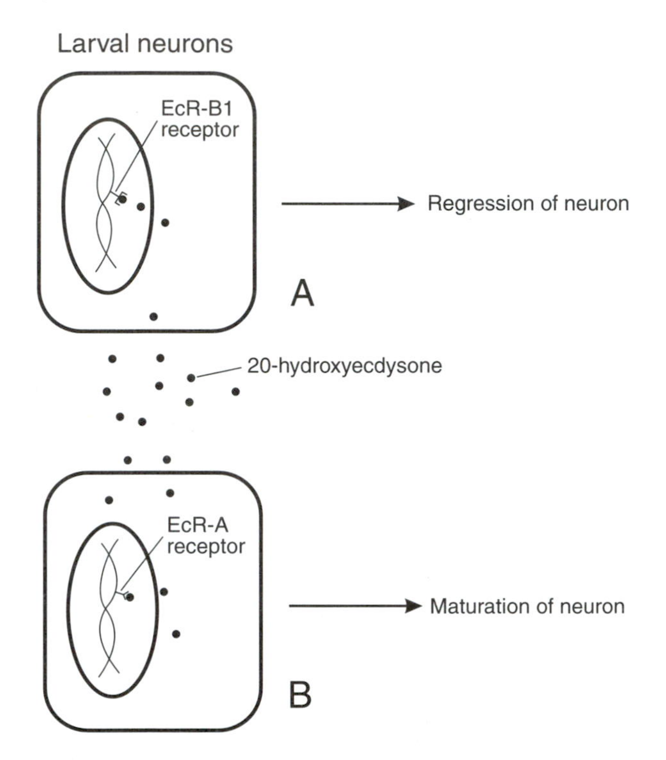

FIGURE 5.8 The fate of a neuron during metamorphosis depends upon the nuclear receptors that are present. (A) If the neuron develops an EcR-B1 receptor, the ecdysteroid that binds to it causes a regression of the neuron at metamorphosis. (B) If a neuron develops an EcR-A receptor during its development, it undergoes a maturation at metamorphosis.

mechanism of muscle contraction does not occur until after ecdysis. If the pharate pupa is "peeled" to remove the larval cuticle artificially before normal eclosion, the gin-trap reflex is not operative. It requires subsequent exposure to eclosion hormone in order to activate the reflex circuit so the gin-trap reflex becomes operational.

A more passive approach to predator avoidance is to hide. Just prior to their metamorphosis to the pupal stage, many holometabolous insects cease feeding and begin a period of sustained crawling until they find a suitable substrate for pupation. This stereotyped **wandering behavior** ultimately moves them to a place where they are less likely to be disturbed during the pupal period. The nervous system of a wandering larva shows characteristically sustained, intense bursts of activity in motor neurons. This wandering behavior is also induced by the commitment peak of 20-hydroxyecdysone in the absence of JH during the end of the last instar larval period. If the prothoracic glands, the source of ecdysteroids, are

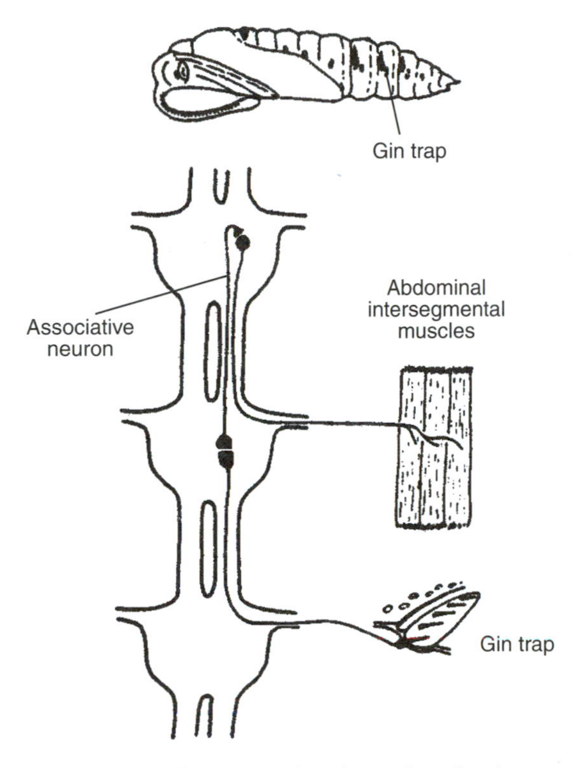

FIGURE 5.9 The mechanism of the gin trap. When the sensilla within the gin trap are stimulated, the reflex arc that is activated causes the contraction of abdominal intersegmental muscles. Reprinted from *Brain Research*, Volume 279. Levine, R.B. and J.W. Truman. Peptide activation of a simple neural circuit, pp. 335–338. 1983. Copyright 1983, with permission from Elsevier Science.

removed before the commitment peak, the larva does not begin wandering, but if ecdysteroids are then injected, the behavior is induced. Similarly, precocious wandering behavior can be induced if ecdysteroids are injected before the normal endogenous release occurs but after the decline of JH later in the instar.

There are several stereotyped behaviors associated with ecdysis itself, which are expressed only prior to and during the molt, including longitudinal peristaltic contractions of the body and the swallowing of air or water to cause the old cuticle to rupture. After the insect reaches the adult stage, however, because ecdysis no longer occurs, these behaviors no longer have a need to be expressed. There is a wave of programmed cell death in *Manduca* in which about half of the neurons of the central nervous system die within the first few days of adult life. Among these dying cells are about 50 neurons that regulate the ecdysis motor program, eliminating those cells that are responsible for outmoded or unused behaviors.

Because engaging in the molting process is wrought with danger as the immobile, defenseless insect sheds its exoskeleton, the molt is usually restricted to a specific time of the day when each species is best able to avoid predators and unfavorable environmental conditions. The timing of the molt is determined by an endogenous circadian clock that allows it to take place only during a narrow window of time. If the physiological events that are required for the molt are not completed by the onset of the window, the insect will wait for the next window to begin its molt. The coordination of the various behaviors involved in eclosion is controlled by the brain through the release of the neuropeptide, **eclosion hormone** (EH). The brain controls its circadian release and once in circulation the hormone initiates the execution of the behaviors. Truman and Riddiford first elucidated the mechanism controlling this behavior in an elegant series of experiments in the 1970s that have continued through the present.

The two species of silkmoths, *Hyalophora cecropia* and *Antheraea pernyi,* have different eclosion gates. *Hyalophora* ecloses in the morning hours while *Antheraea* ecloses just before the scotophase (Fig. 5.10A). When their brains are removed surgically, the rhythmicity of eclosion is abolished in both species (Fig. 5.10B). If the brains are removed and reimplanted in the abdomens, the rhythmicity is restored, even though there are no nervous connections between the brain and the central nervous system (Fig. 5.10C). Finally, if the brains are removed and implanted into individuals of the opposite species, their eclosion behavior is not only restored, but its restoration is based on the identity of the brain that is present and not the identity of the body and remainder of the central nervous system (Fig. 5.11). EH activates the fixed action patterns that reside in the ventral ganglia that are responsible for the eclosion behaviors. When isolated ganglia are exposed to EH, their neurons show the same characteristic bursts of activity as when they are intact and undergoing eclosion. Thus, the brain releases EH that then acts on the ventral ganglia to release the genetically encoded fixed action patterns that allow the insect to escape from its old cuticle. The hormone is not species-specific, only the tim-

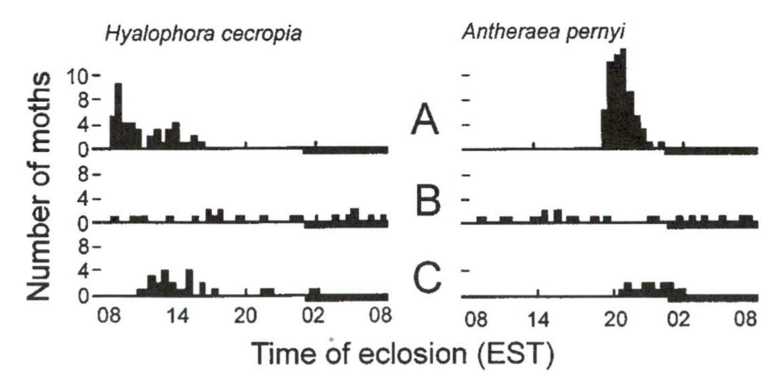

FIGURE 5.10 (A) Normal adult eclosion times for *H. cecropia* and *A. pernyi*. (B) Adult eclosion after the pupal brains were removed. (C) Adult eclosion after the brains were removed and loose brains reimplanted. The horizontal black bar indicates the scotophase. From Truman (1973). Reprinted with permission.

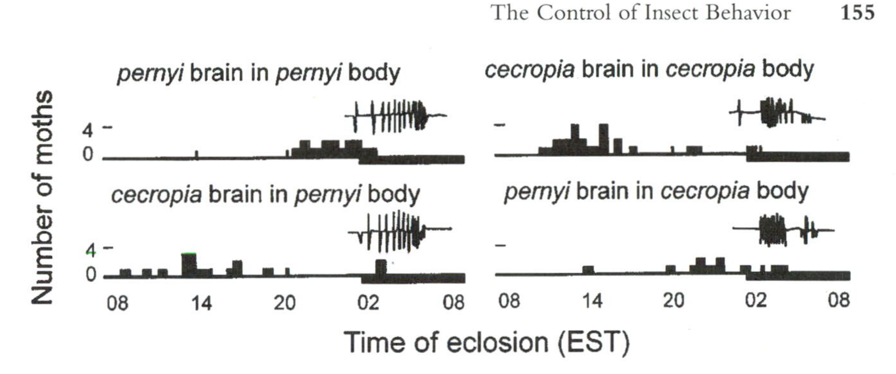

FIGURE 5.11 Effect of interchanging the brains on eclosion behavior. (Top) The brain implanted into the body of the same species. (Bottom) The brains from *H. cecropia* implanted into a *pernyi* body resulted in the *cecropia* emergence pattern, while the brains from *pernyi* implanted into a *cecropia* body resulted in the *pernyi* emergence pattern. (Inset) The traces from a lever attached to the abdominal tips lasting for approximately 2 h. The horizontal black bar indicates the scotophase. From Truman (1973). Reprinted with permission.

ing of its release. *Hyalophora* releases its EH early in the day; *Antheraea* releases its hormone later. Either brain can activate the behavior, but only at the time it is programmed to release the hormone. Eclosion activity is also determined by the responsiveness of the ganglia to EH. The ventral ganglia are responsive to EH only after ecdysteroid levels decline once the molting peak has appeared. EH initiates not only adult ecdysis, but also the ecdysial behaviors at each larval–larval molt and the larval–pupal molt.

Eclosion hormone, a 62-amino-acid polypeptide, was originally thought to be central in the triggering of these behaviors, but recent studies have suggested an even greater degree of complexity involving another hormone in eclosion behavior. A 26-amino-acid **ecdysis-triggering hormone** (ETH) is released from the epitracheal glands that are located on the large tracheal trunks near the spiracles. It appears that ETH may be the hormone that acts directly on the nervous system, with EH causing the release of ETH. When both of the hormones are released together, they activate a neural network that enhances the expression of the ecdysial behaviors. The precise relationship between the two hormones and eclosion behavior is still not yet well understood.

Physiology of Sexual Receptivity in Mosquitoes

Many insects do not mate until a few days after adult emergence. Recently emerged adult female *Aedes aegypti* mosquitoes that are approached by a male may

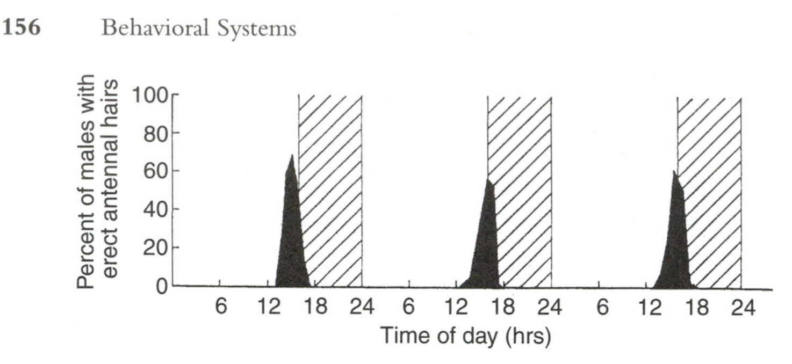

FIGURE 5.12 The rhythmicity of antennal hair erection in anopheline mosquitoes. Vertical bars indicate the scotophase. From Nijhout (1977). Reprinted with permission.

physically couple with them but they do not retain any semen. Insemination with the retention of sperm does not occur unless the females are several days old. It is the release of juvenile hormone within 1–2 days after emergence that causes the maturation that allows the female to mate with successful insemination. Juvenile hormone is acting as a modifier hormone in this case, changing the way the insect responds to a given stimulus.

Male mosquitoes are attracted to females by their species-specific wing-beat frequency. The antennae of males have many long hairs that pick up the vibrations from female wings and activate those behaviors that are associated with mating. The antennal hairs of some mosquito species are fully erect at all times, allowing the males

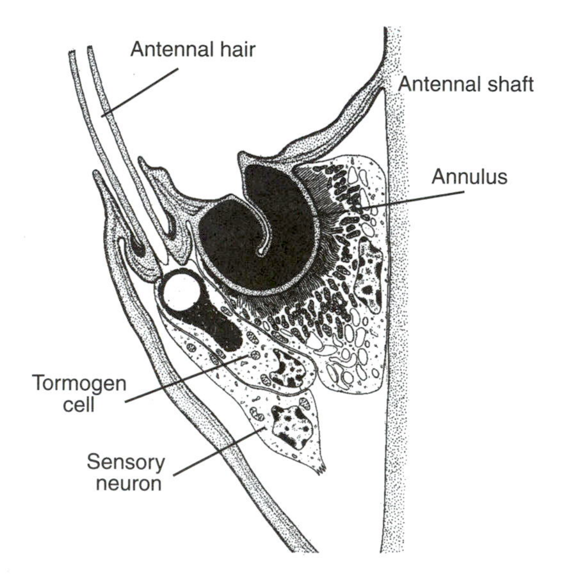

FIGURE 5.13 The mechanism of antennal hair erection in anopheline mosquitoes. The annulus swells as a result of the change in pH, causing the hair to move outward. Reprinted with permission from Nijhout, H. F., and H. G. Sheffield. Antennal hair erection in male mosquitoes: A new mechanical effector in insects. *Science* **206:** 595–596. Copyright 1979. American Association for the Advancement of Science.

to receive the vibrational stimuli from females at any time of day. In other species, the hairs lie parallel to the antennal shaft during most of the day and only become erect at certain times when mating occurs (Fig. 5.12). The antennal hairs must be unfolded in order for the males to respond to female stimuli. In those species where mating occurs only at certain times of the day, usually at dusk, males with erect antennal hairs form swarms above markers and the females fly into them. The hairs allow the males to distinguish the wing-beat of those females in flight. The antennal hairs become erect as the result of the hydration of an annulus at their base. As the annulus swells, the attached hair is moved more perpendicular to the shaft and the male is thus capable of perceiving vibrational stimuli and responding to it (Fig. 5.13).

ADDITIONAL REFERENCES

Acosta-Avalos, D., Wajnberg, E., Oliveira, P. S., Leal, I., Farina, M., Esquivel, D. M. 1999. Isolation of magnetic nanoparticles from *Pachycondyla marginata* ants. *J. Exp. Biol.* **202:** 2687–2692.

Anton, S., Gadenne, C. 1999. Effect of juvenile hormone on the central nervous processing of sex pheromone in an insect. *Proc. Natl. Acad. Sci. USA* **96:** 5764–5767.

Arthur, B. I., Jr., Jallon, J. M., Caflisch, B., Choffat, Y., Nothiger, R. 1998. Sexual behaviour in *Drosophila* is irreversibly programmed during a critical period. *Curr. Biol.* **8:** 1187–1190.

Baker, J. D., McNabb, S. L., Truman, J. W. 1999. The hormonal coordination of behavior and physiology at adult ecdysis in *Drosophila melanogaster*. *J. Exp. Biol.* **202:** 3037–3048.

Bate, C. M., 1973. The mechanism of the pupal gin trap II. The closure movement. *J. Exp. Biol.* **59:** 109–119.

Bloch, G., Hefetz, A., Hartfelder, K. 2000. Ecdysteroid titer, ovary status, and dominance in adult worker and queen bumble bees (*Bombus terrestris*). *J. Insect Physiol.* **46:** 1033–1040.

Boulay, R., Soroker, V., Godzinska, E. J., Hefetz, A., Lenoir, A. 2000. Octopamine reverses the isolation-induced increase in trophallaxis in the carpenter ant *Camponotus fellah*. *J. Exp. Biol.* **203:** 513–520.

Brembs, B., Heisenberg, M. 2000. The operant and the classical in conditioned orientation of *Drosophila melanogaster* at the flight simulator. *Learn. Mem.* **7:** 104–115.

Collett, T. S. 1993. Route following and the retrieval of memories in insects. *Comp. Biochem. Physiol.* **104A:** 709–716.

Collett, T. S. 1994. Invertebrate vision: bees learn how to look. *Curr. Biol.* **4:** 717–719.

Collett, T. S., Collett, M., Wehner, R. 2001. The guidance of desert ants by extended landmarks. *J. Exp. Biol.* **204:** 1635–1639.

Comer, C. M., Robertson, R. M. 2001. Identified nerve cells and insect behavior. *Prog. Neurobiol.* **63:** 409–439.

Davis, R. L. 1996. Physiology and biochemistry of *Drosophila* learning mutants. *Physiol. Rev.* **76:** 299–317.

de Belle, J. S., Hilliker, A. J., Sokolowski, M. B. 1989. Genetic localization of *foraging* (*for*): a major gene for larval behavior in *Drosophila melanogaster*. *Genetics* **123:** 157–163.

de Belle, J. S., Heisenberg, M. 1994. Associative odor learning in *Drosophila* abolished by chemical ablation of mushroom bodies. *Science* **263:** 692–695.

de Belle, J. S., Heisenberg, M. 1996. Expression of *Drosophila* mushroom body mutations in alternative genetic backgrounds: a case study of the mushroom body miniature gene (*mbm*). *Proc. Natl. Acad. Sci. USA* **93:** 9875–9880.

de Belle, J. S., Sokolowski, M. B., Hiller, A. J. 1993. Genetic analysis of the foraging microregion of *Drosophila melanogaster*. *Genome* **36:** 94–104.

Dill, M., Wolf, R., Heisenberg, M. 1995. Behavioral analysis of *Drosophila* landmark learning in the flight simulator. *Learn. Mem.* **2:** 152–160.

Dubnau, J., Tully, T. 1998. Gene discovery in *Drosophila*: New insights for learning and memory. *Annu. Rev. Neurosci.* **21:** 407–444.

Elekonich, M. M., Robinson, G. E. 2000. Organizational and activational effects of hormones on insect behavior. *J. Insect Physiol.* **46:** 1509–1515.

Emlen, D. J., Nijhout, H. F. 1999. Hormonal control of male horn length dimorphism in the dung beetle *Onthophagus taurus* (Coleoptera: Scarabaeidae). *J. Insect Physiol.* **45:** 45–53.

Emlen, D. J., Nijhout, H. F. 2000. The development and evolution of exaggerated morphologies in insects. *Annu. Rev. Entomol.* **45:** 661–708.

Ernst, R., Heisenberg, M. 1999. The memory template in *Drosophila* pattern vision at the flight simulator. *Vision Res.* **39:** 3920–3933.

Etheredge, J. A., Perez, S. M., Taylor, O. R., Jander, R. 1999. Monarch butterflies (*Danaus plexippus* L.) use a magnetic compass for navigation. *Proc. Natl. Acad. Sci. USA* **96:** 13,845–13,846.

Evans, J. D., Wheeler, D. E. 1999. Differential gene expression between developing queens and workers in the honey bee, *Apis mellifera*. *Proc. Natl. Acad. Sci. USA* **96:** 5575–5580.

Evans, J. D., Wheeler, D. E. 2001. Gene expression and the evolution of insect polyphenisms. *BioEssays* **23:** 62–68.

Ewer, J., Gammie, S. C., Truman, J. W. 1997. Control of insect ecdysis by a positive-feedback endocrine system: roles of eclosion hormone and ecdysis triggering hormone. *J. Exp. Biol.* **200:** 869–881.

Ewer, J., Wang, C. M., Klukas, K. A., Mesce, K. A., Truman, J. W., Fahrbach, S. E. 1998. Programmed cell death of identified peptidergic neurons involved in ecdysis behavior in the moth, *Manduca sexta*. *J. Neurobiol.* **37:** 265–280.

Fahrbach, S. E., Giray, T., Farris, S. M., Robinson, G. E. 1997. Expansion of the neuropil of the mushroom bodies in male honeybees is coincident with initiation of flight. *Neurosci. Lett.* **236:** 135–138.

Fahrbach, S. E., Robinson, G. E. 1996. Juvenile hormone, behavioral maturation, and brain structure in the honey bee. *Dev. Neurosci.* **18:** 102–114.

Farris, S. M., Robinson, G. E., Fahrbach, S. E. 2001. Experience- and age-related outgrowth of intrinsic neurons in the mushroom bodies of the adult worker honeybee. *J. Neurosci.* **21:** 6395–6404.

Gadenne, C., Anton, S. 2000. Central processing of sex pheromone stimuli is differentially regulated by juvenile hormone in a male moth. *J. Insect Physiol.* **46:** 1195–1206.

Gammie, S. C., Truman, J. W. 1999. Eclosion hormone provides a link between ecdysis-triggering hormone and crustacean cardioactive peptide in the neuroendocrine cascade that controls ecdysis behavior. *J. Exp. Biol.* **202:** 343–352.

Giebultowicz, J. M., Truman, J. W. 1984. Sexual differentiation in the terminal ganglion of the moth *Manduca sexta*: Role of sex-specific neuronal death. *J. Comp. Neurol.* **226:** 87–95.

Giebultowicz, J. M., Zdarek, J., Chroscikowska, U. 1980. Cocoon spinning behaviour in *Ephestia kuehniella*: Correlation with endocrine events. *J. Insect Physiol.* **26:** 459–464.

Giray, T., Robinson, G. E. 1996. Common endocrine and genetic mechanisms of behavioral development in male and worker honeybees and the evolution of division of labor. *Proc. Natl. Acad. Sci. USA* **93:** 11,718–11,722.

Greenspan, R. J. 1995. Understanding the genetic construction of behavior. *Sci. Am.* **272:** 72–78.

Greenspan, R. J., Ferveur, J. F. 2000. Courtship in *Drosophila*. *Annu. Rev. Entomol.* **34:** 205–232.

Gronenberg, W., Heeren, S., Hölldobler, B. 1996. Age-dependent and task-related morphological changes in the brain and the mushroom bodies of the ant *Camponotus floridanus*. *J. Exp. Biol.* **199:** 2011–2019.

Grotewiel, M. S., Beck, C. D., Wu, K. H., Zhu, X. R., Davis, R. L. 1998. Integrin-mediated short-term memory in *Drosophila*. *Nature* **391:** 455–460.

Hall, J. C. 1994. The mating of a fly. *Science* **264:** 1702–1714.

Hammer, M., Menzel, R. 1995. Learning and memory in the honeybee. *J. Neurosci.* **15:** 1617–1630.

Hartfelder, K. 2000. Insect juvenile hormone: from "status quo" to high society. *Braz. J. Med. Biol. Res.* **33:** 157–177.

Hartfelder, K., Cnaani, J., Hefetz, A. 2000. Caste-specific differences in ecdysteroid titers in early larval stages of the bumblebee *Bombus terrestris*. *J. Insect Physiol.* **46:** 1433–1439.

Hartfelder, K., Engels, W. 1998. Social insect polymorphism: hormonal regulation of plasticity in development and reproduction in the honeybee. *Curr. Top. Dev. Biol.* **40:** 45–77.

Hegstrom, C. D., Riddiford, L. M., Truman, J. W. 1998. Steroid and neuronal regulation of ecdysone receptor expression during metamorphosis of muscle in the moth, *Manduca sexta*. *J. Neurosci.* **18:** 1786–1794.

Hegstrom, C. D., Truman, J. W. 1996. Steroid control of muscle remodeling during metamorphosis in *Manduca sexta*. *J. Neurobiol.* **29:** 535–550.

Heisenberg, M. 1995. Pattern recognition in insects. *Curr. Opin. Neurobiol.* **5:** 475–481.

Heisenberg, M. 1997. Genetic approaches to neuroethology. *BioEssays* **19:** 1065–1073.

Huang, Z.Y., Plettner, E., Robinson, G. E. 1998. Effects of social environment and worker mandibular glands on endocrine-mediated behavioral development in honeybees. *J. Comp. Physiol. A* **183:** 143–152.

Huang, Z. -Y., Robinson, G. E. 1992. Honeybee colony integration: worker-worker interactions mediate hormonally regulated plasticity in the division of labor. *Proc. Natl. Acad. Sci. USA* **89:** 11,726–11,729.

Huang, Z. -Y., Robinson, G. E. 1999. Social control of division of labor in honey bee colonies. In *Information processing in social insects.* (C. Detrain, J. L. Deneubourg, J. M. Pasteels, Eds.), pp. 165–186. Birkhäuser Verlag, Basel.

Jassim, O., Huang, Z. Y., Robinson, G. E. 2000. Juvenile hormone profiles of worker honeybees, *Apis mellifera*, during normal and accelerated behavioural development. *J. Insect Physiol.* **46:** 243–249.

Jiang, C., Baehrecke, E. H., Thummel, C. S. 1997. Steroid regulated programmed cell death during *Drosophila* metamorphosis. *Development* **124:** 4673–4683.

Judd, S. P. D., Collett, T. S. 1998. Multiple stored views and landmark guidance in ants. *Nature* **392:** 710–714.

Kennedy, J. S. 1992. The new anthropomorphism. Cambridge University Press, New York.

Klowden, M. J. 1990. The endogenous regulation of mosquito reproductive behaviour. *Experientia* **46:** 660–670.

Klowden, M. J., Blackmer, J. L. 1987. Humoral control of pre-oviposition behaviour in the mosquito, *Aedes aegypti*. *J. Insect Physiol.* **33:** 689–692.

Klowden, M. J., Lea, A. O. 1979. Humoral inhibition of host-seeking in *Aedes aegypti* during oocyte maturation. *J. Insect Physiol.* **25:** 231–235.

Lea, A. O. 1968. Mating without insemination in virgin *Aedes aegypti*. *J. Insect Physiol.* **14:** 305–308.

Lemon, W. C., Levine, R. B. 1997. Multisegmental motor activity in the segmentally restricted gin trap behavior in *Manduca sexta* pupae. *J. Comp. Physiol. A* **180:** 611–619.

Levine, R. B. 1986. Reorganization of the insect nervous system during metamorphosis. *Trends Neurosci.* **9:** 315–319.

Levine, R. B., Truman, J. W. 1983. Peptide activation of a simple neural circuit. *Brain Res.* **279:** 335–338.

Levine, R. B., Weeks, J. C. 1990. Hormonally mediated changes in simple reflex circuits during metamorphosis in *Manduca*. *J. Neurobiol.* **21:** 1022–1036.

Martin, J. R., Ernst, R., Heisenberg, M. 1998. Mushroom bodies suppress locomotor activity in *Drosophila melanogaster*. *Learn. Mem.* **5:** 179–191.

Menzel, R., Brandt, R., Gumbert, A., Komischke, B., Kunze, J. 2000. Two spatial memories for honeybee navigation. *Proc. R. Soc. London B* **267:** 961–968.

Menzel, R., Geiger, K., Chittka, L., Joerges, J., Kunze, J., Müller, U. 1996. The knowledge base of bee navigation. *J. Exp. Biol.* **199:** 141–146.

Menzel, R., Müller, U. 1996. Learning and memory in honeybees: From behavior to neural substrates. *Annu. Rev. Neurosci.* **19:** 379–404.

Mesce, K. A., Truman, J. W. 1988. Metamorphosis of the ecdysis motor pattern in the hawkmoth, *Manduca sexta*. *J. Comp. Physiol.* A **163:** 287–299.

Moller, R. 2001. Do insects use templates or parameters for landmark navigation? *J. Theor. Biol.* **210:** 33–45.

Morton, D. B. 1997. Eclosion hormone action on the nervous system. Intracellular messengers and sites of action. *Ann. NY Acad. Sci.* **814:** 40–52.

Nijhout, H. F. 1977. Control of antennal hair erection in male mosquitoes. *Biol. Bull.* **153:** 591–603.

Nijhout, H. F. 1999. Control mechanisms of polyphenic development in insects. *BioScience* **49:** 181–192.

Nijhout, H. F., Emlen, D. J. 1998. Competition among body parts in the development and evolution of insect morphology. *Proc. Natl. Acad. Sci. USA* **95:** 3685–3689.

Nijhout, H. F., Sheffield, H. G., 1979. Antennal hair erection in male mosquitoes: A new mechanical effector in insects. *Science* **206:** 595–596.

Orgad, S., Rosenfeld, G., Smolikove, S., Polak, T., Segal, D. 1997. Behavioral analysis of *Drosophila* mutants displaying abnormal male courtship. *Invert. Neurosci.* **3:** 175–183.

Osborne, K. A., DeBelle, J. S., Sokolowski, M. B. 2001. Foraging behaviour in *Drosophila* larvae: Mushroom body ablation. *Chem. Senses* **26:** 223–230.

Osborne, K. A., Robichon, A., Burgess, E., Butland, S., Shaw, R. A., Coulthard, A., Pereira, H. S., Greenspan, R. J., Sokolowski, M. B. 1997. Natural behavior polymorphism due to a cGMP-dependent protein kinase of *Drosophila*. *Science* **277:** 834–836.

Page, R. E., Robinson, G. E. 1991. The genetics of division of labour in honey bee colonies. *Adv. Insect Physiol.* **35:** 117–169.

Pereira, H. S., Sokolowski, M. B. 1993. Mutations in the larval foraging gene affect adult locomotory behavior after feeding in *Drosophila melanogaster*. *Proc. Natl. Acad. Sci. USA* **90:** 5044–5046.

Rachinsky, A., Hartfelder, K. 1990. Corpora allata activity, a prime regulating element for caste-specific juvenile hormone titre in honey bee larvae (*Apis mellifera carnica*). *J. Insect Physiol.* **36:** 189–194.

Rachinsky, A., Strambi, C., Strambi, A., Hartfelder, K. 1990. Caste and metamorphosis: Hemolymph titers of juvenile hormone and ecdysteroids in last instar honeybee larvae. *Gen. Comp. Endocrinol.* **79:** 31–38.

Robinson, G. E. 1992. Regulation of division of labor in insect societies. *Annu. Rev. Entomol.* **37:** 637–665.

Robinson, G. E. 1998. From society to genes with the honey bee. *Am. Sci.* **86:** 456–462.

Robinson G. E., Fahrbach, S. E., Winston, M. L. 1997. Insect societies and the molecular biology of social behavior. *BioEssays* **19:** 1099–1108.

Robinson, G. E., Vargo, E. L. 1997. Juvenile hormone in adult eusocial Hymenoptera: Gonadotropin and behavioral pacemaker. *Arch. Insect Biochem. Physiol.* **35:** 559–583.

Roeder, K. D. 1965. Moths and ultrasound. *Sci. Am.* **212:** 94–102.

Schmidt Capella, I. C., Hartfelder, K. 1998. Juvenile hormone effect on DNA synthesis and apoptosis in caste-specific differentiation of the larval honey bee (*apis mellifera* l.) ovary. J. *Insect Physiol.* **44:** 385–391.

Shaw, P. J., Cirelli, C., Greenspan, R. J., Tononi, G. 2000. Correlates of sleep and waking in *Drosophila melanogaster*. *Science* **287:** 1834–1837.

Sokolowski, M. B., Pereira, H. S., Hughes, K. 1997. Evolution of foraging behavior in *Drosophila* by density-dependent selection. *Proc. Natl. Acad. Sci. USA* **94:** 7373–7377.

Srinivasan, M. V., S. Zhang, M. Altwein, J. Tautz. 2000. Honeybee navigation: nature and calibration of the "odometer." *Science* **287:** 851–853.

Srinivasan, M. V., Zhang, S. W. 2000. Visual navigation in flying insects. *Int. Rev. Neurobiol.* **44:** 67–92.

Sullivan, J. P., Fahrbach, S. E., Robinson, G. E. 2000. Juvenile hormone paces behavioral development in the adult worker honey bee. *Horm. Behav.* **37:** 1–14.

Taghert, P. H., Hewes, R. S., Park, J. H., O'Brien, M. A., Han, M., Peck, M. E. 2001. Multiple amidated neuropeptides are required for normal circadian locomotor rhythms in *Drosophila*. *J. Neurosci.* **21:** 6673-6686.

Tallamy, D. W. 2001. Evolution of exclusive paternal care in arthropods. *Annu. Rev. Entomol.* **46:** 139–165.

Tata, J. R. 1993. Gene expression during metamorphosis: An ideal model for post-embryonic development. *BioEssays* **15:** 239–248.

Truman, J. W. 1973. How moths turn on: A study of the action of hormones on the nervous system. *Am. Sci.* 61: 700–706.

Truman, J. W. 1978. Hormonal control of invertebrate behavior. *Horm. Behav.* **10:** 214–234.

Truman, J. W. 1990. Metamorphosis of the central nervous system of *Drosophila*. *J. Neurobiol.* **21:** 1072–1084.

Truman, J. W. 1992. Developmental neuroethology of insect metamorphosis. *J. Neurobiol.* **23:** 1404–1422.

Truman, J. W. 1992. The eclosion hormone system of insects. *Prog. Brain Res.* **92:** 361–374.

Truman, J. W. 1996. Metamorphosis of the insect nervous system. In *Metamorphosis: Postembryonic reprogramming of gene expression in amphibian and insect cells* (L. I. Gilbert, J. R. Tata, B. G. Atkinson, Eds.), pp. 283–320. Academic Press, San Diego, CA.

Truman, J. W. 1996. Steroid receptors and nervous system metamorphosis in insects. *Dev. Neurosci.* **18:** 87–101.

Truman, J. W., Reiss, S. E. 1976. Dendritic reorganization of an identified motoneuron during metamorphosis of the tobacco hornworm moth. *Science* **192:** 477–479.

Truman, J. W., Riddiford, L. M. 1970. Neuroendocrine control of ecdysis in silkmoths. *Science* **167:** 1624–1626.

Truman, J. W., Riddiford, L. M. 1974. Hormonal mechanisms underlying insect behaviour. *Adv. Insect Physiol.* **10:** 297–352.

Truman, J. W., Rountree, D. B., Reiss, S. E., Schwartz, L. M. 1983. Ecdysteroids regulate the release and action of eclosion hormone in the tobacco hornworm *Manduca sexta* (L.). *J. Insect Physiol.* **29:** 895–900.

Truman, J. W., Schwartz, L. M. 1982. Programmed death in the nervous system of a moth. *Trends NeuroSci.* **5:** 270–273.

Truman, J. W., Sokolove, P. G. 1972. Silk moth eclosion: Hormonal triggering of a centrally programmed pattern of behavior. *Science* **175:** 1491–1493.

Truman, J. W., Taghert, P. H., Copenhaver, P. F., Tublitz, N. J., Schwartz, L. M. 1981. Eclosion hormone may control all ecdyses in insects. *Nature* **291:** 70–71.

Truman, J. W., Thorn, R. S., Robinow, S. 1992. Programmed neuronal death in insect development. *J. Neurobiol.* **23:** 1295–1311.

Tully, T. 1996. Discovery of genes involved with learning and memory: An experimental synthesis of Hirschian and Benzerian perspectives. *Proc. Natl. Acad. Sci. USA* **93:** 13,460–13,467.

Waldrop, B., Levine, R. B. 1989. Development of the gin trap reflex in *Manduca sexta*: A comparison of larval and pupal motor responses. *J. Comp. Physiol. A* **165:** 743–753.

Wehner, R., Räber, F. 1979. Visual spatial memory in desert ants, *Cataglyphis fortis* (Hymenoptera: Formicidae). *Experientia* **35:** 1569–1571.

Wheeler, D. A., Kyriacou, C. P., Greenacre, M. L., Yu, Q., Rutila, J. E., Rosbash, M., Hall, J. C. 1991. Molecular transfer of a species-specific behavior from *Drosophila simulans* to *Drosophila melanogaster*. *Science* **251:** 1082–1085.

Wheeler, D. E., Nijhout, H. F. 1981. Soldier determination in ants: New role for juvenile hormone. *Science* **213:** 361–363.

Wheeler, D. E., Nijhout, H. F. 1983. Soldier determination in *Pheidole bicarinata*: Effect of methoprene on caste and size within castes. *J. Insect Physiol.* **29:** 847–854.

Withers, G. S., Fahrbach, S. E., Robinson, G. E. 1995. Effects of experience and juvenile hormone on the organization of the mushroom bodies of honeybees. *J. Neurobiol.* **26:** 130–144.

Yamamoto, D., Jallon, J. M., Komatsu, A. 1997. Genetic dissection of sexual behavior in *Drosophila melanogaster*. *Annu. Rev. Entomol.* **42:** 551–585.

Yamamoto, D., Nakano, Y. 1998. Genes for sexual behavior. *Biochem. Biophys. Res. Commun.* **246:** 1–6.

Zars, T. 2000. Behavioral functions of the insect mushroom bodies. *Curr. Opin. Neurobiol.* **10:** 790–795.

Zars, T., Fischer, M., Schulz, R., Heisenberg, M. 2000. Localization of a short-term memory in *Drosophila*. *Science* **288:** 672–675.

CHAPTER 6

Metabolic Systems

ENERGY PRODUCTION FROM FOOD

A feature that distinguishes living cells is their ability to transform the chemical energy in the environment into electrical, mechanical, osmotic, and other forms of chemical energy. These energy transformations are required in order to maintain essential life processes such as nervous transmission, muscle contraction, and the synthesis of structural components. The sources of transformed energy include the ingested food that contains complex carbohydrates, fats, and proteins that are broken down in the alimentary tract to the simpler components and absorbed through the wall of the midgut into the hemolymph. The circulatory system then transports these components to all the cells of the body, which break them down even further and capture the chemical energy they contain. Each cell may use the components immediately or they may be used to synthesize reserves for later use. These processes of food breakdown, utilization, and storage are strikingly similar in insects and vertebrates, as well as in most other living things. However, some distinctive metabolic processes that are not found in vertebrates have been recognized in insects, and these differences will be discussed in this chapter.

It must be emphasized that the metabolic systems of only a small number of insect species have been examined, most often cockroaches, blow flies, fruit flies, or caterpillars. The evidence for the existence of complete metabolic pathways in insects has often been based on the presence of certain key enzymes, reaction end products, or intermediates, which may also exist in vertebrate systems where the cycles were more completely identified. The determination of metabolic pathways in insects is complicated by the presence of symbiotic microorganisms that may provide some of the steps missing in the insect, particularly in those systems that require trace components such as vitamins. This symbiotic contribution makes it difficult to establish whether the metabolic pathways are actually present in the insect system.

THE INSECT ALIMENTARY TRACT

The ability of insects as a group to feed on practically every type of organic matter imaginable has been a major factor in their success, enabling them to expand into diverse ecological niches. This diversity in the food that can be ingested is reflected in the diversity of external mouthpart structures that serve as the gateway to the digestive tract. Accompanying the mouthpart diversity is a structural diversity of the insect digestive tract, with an enormous degree of specialization that varies with the particular type of diet. Because of this wide variation in insect feeding and the structures that are associated with feeding, a "typical" insect digestive tract does not exist and cannot be described. One useful generalization might be a division of feeding behavior into continuous versus discontinuous feeders. Predatory and carnivorous species, such as mosquitoes, go long periods between meals and their digestive tracts are modified for food storage. Many phytophagous insects, such as caterpillars, feed continuously and their guts are designed more for the absorption and processing of food.

The digestive tract consists of a tube of epithelial cells running from the mouth to the anus. It is divided into three major regions based on embryonic origins and physiological functions, the **foregut, midgut,** and **hindgut** (Fig. 6.1). The **stomodeum** and **proctodeum** both arise as invaginations of the embryonic ectoderm and produce the foregut and hindgut, respectively (Fig. 6.2). The midgut forms from endodermal tissues and connects with the foregut and hindgut during embryogenesis. They are supported in the body cavity by **extrinsic visceral muscles** whose contractions dilate the lumen of the gut (Fig. 6.3). Because they

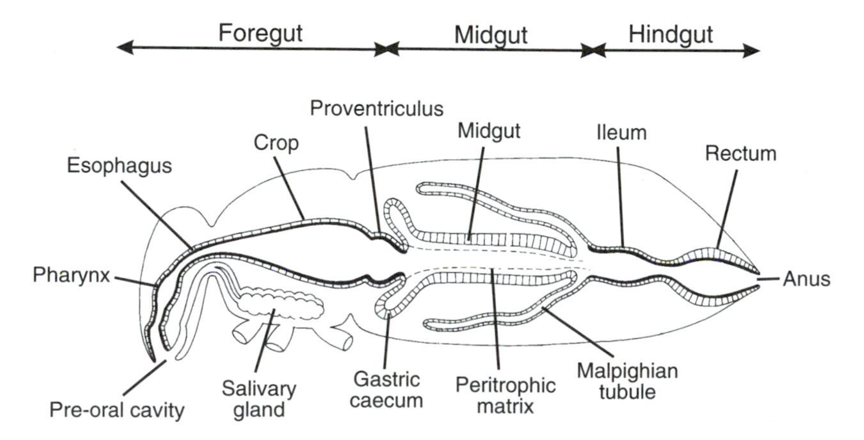

FIGURE 6.1 The three major divisions of the insect digestive tract, foregut, midgut, and hindgut, and their components. Reprinted with permission from Dow, J.A.T. (1986). *Advances in Insect Physiology* **19:** 187–328. Copyright Academic Press.

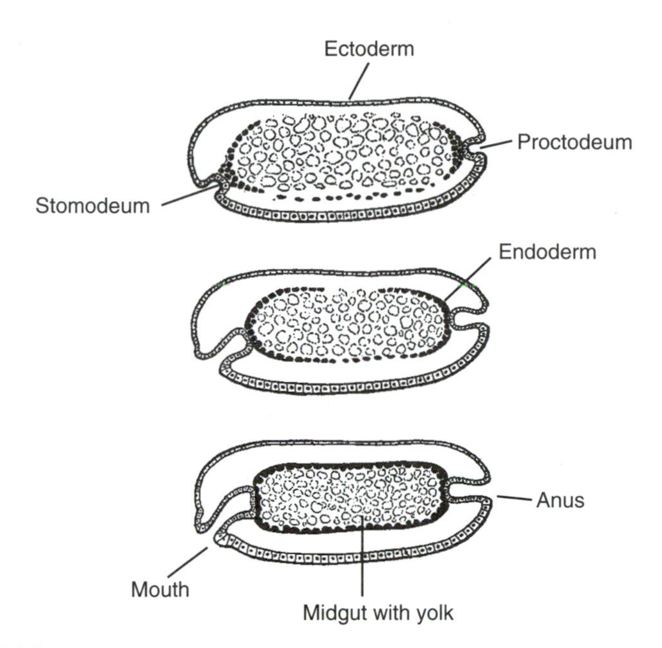

FIGURE 6.2 Embryonic derivation of the digestive tract. The hindgut and foregut are derived from embryonic ectoderm, while the midgut forms from endodermal tissues and connects with the blind ends of the stomodeal and proctodeal invaginations. From Snodgrass (1935). Reprinted with permission.

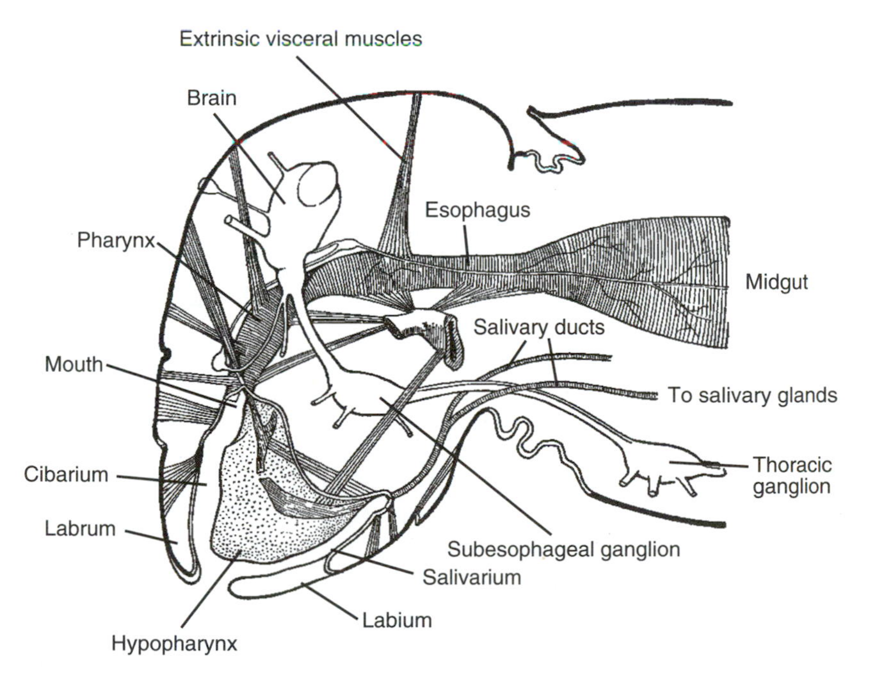

FIGURE 6.3 A section of the head showing the suspension of the digestive tract by extrinsic visceral muscles. From Snodgrass (1935). Reprinted with permission.

are derived from ectodermal cells, these two regions are lined with cuticle and molt along with the rest of the epidermis, a feature that has important implications for digestion and absorption. A number of **intrinsic visceral muscles** are also present, consisting of circular and longitudinal muscles that permit the gut to contract and undergo peristalsis. An important feature of the insect digestive tract is its spatial compartmentalization, a theme that will be apparent throughout subsequent sections.

ANTERIOR STRUCTURES AND THE FOREGUT

Primitive insect ancestors had a pair of walking appendages on each segment (Fig. 6.4). Along the evolutionary path leading to insects, the appendages on the abdomen were either lost or modified into external genitalia or sensory cerci, and the appendages borne on the segments that were destined to be the head capsule were modified to manipulate food and became closely associated with the mouth. The present-day mouthparts of insects and all their variations thus originated as external appendages that ultimately surrounded the true mouth and created a **preoral cavity.** In mandibulate insects, the cavity is divided by the **hypopharynx** into an anterior **cibarium** and a posterior **salivarium** (Fig. 6.3). A pair of salivary glands, modified from the epidermal cells of the labium segment, empties their secretions into the salivarium through the cuticle-lined **salivary duct.** In silk-producing lepidopterans, the labial glands produce the silk, and saliva is produced by the mandibular glands instead. In many mosquitoes, the cibarium is armed with cuticular spines that lyse red blood cells before they pass into the midgut.

Saliva lubricates the mouthparts as they move against each other and acts as a solvent for food. It may contain digestive enzymes that predigest the food before it is internalized and further acted upon by gut enzymes. These enzymes commonly consist of an amylase that breaks down starch to simple sugars and an invertase that converts the disaccharide sucrose in the food to glucose and fructose. Carnivorous insects may produce salivary proteases or chitinases. Many predatory insects inject saliva into their hosts and drink up the digested tissues. The saliva may also contain toxins that act on the nervous system of the host to paralyze it. In blood-feeding insects, the saliva may contain anticoagulants and other pharmacological substances that enhance blood-feeding ability. The saliva of some plant-feeding hemipterans hardens into a **stylet sheath** that forms around the mouthparts and prevents the leakage of plant liquids when the insects feed.

The true **mouth** lies just beyond the preoral cavity and is the opening of the digestive tract. The **pharynx** is the first region of the foregut and may be modified with dilator muscles in sucking insects that expand its lumen and create a partial vacuum to take up fluids (Fig. 6.3). The **esophagus** comprises the region of the foregut just beyond the pharynx and is usually a simple tube that leads to the

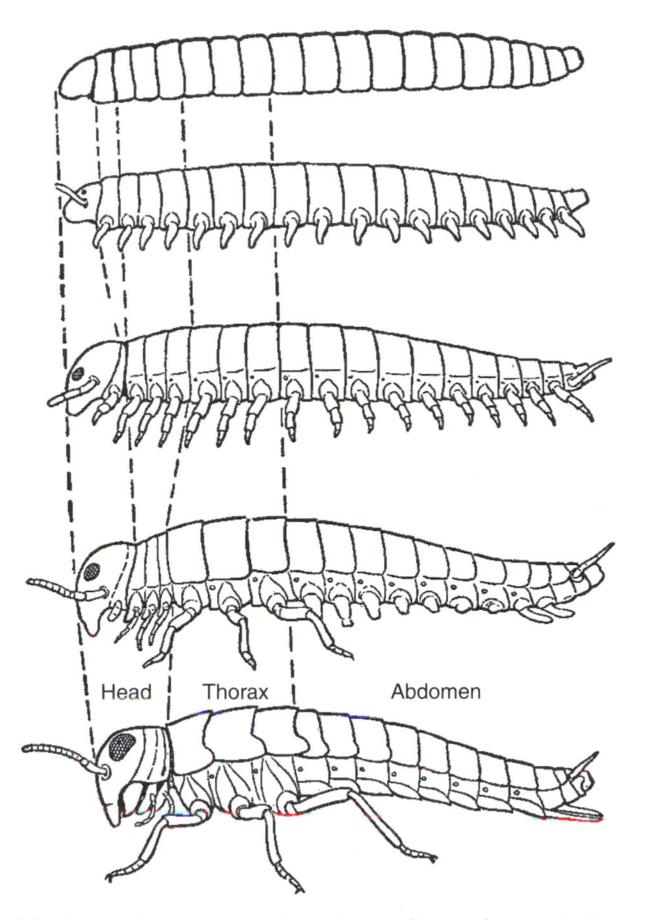

FIGURE 6.4 A probable sequence in the evolution of insects from a primitive annelid ancestor. The appendages on all body segments were reduced and segments were consolidated into functional groupings. From Snodgrass (1935). Reprinted with permission.

midgut, but it may be modified into a distensible **crop** that is used to store food. In adult Diptera and Lepidoptera, the crop is a diverticulum that is separated from the rest of the esophagus by a short duct fitted with a valve (Fig. 6.5). Pharyngeal receptors can determine whether ingested food enters the diverticulum or the crop by activating the valves that shunt the food. Sugar meals are stored in the diverticulum and passed slowly to the midgut, while protein is sent directly to the midgut. The cuticular lining of the crop limits its absorptive capacity and it therefore functions mainly as a food reservoir, but it is permeable to some ingested fats. Although absorption may be limited, digestion may still occur here as the food is acted upon by salivary enzymes. A large foregut is often characteristic of insect

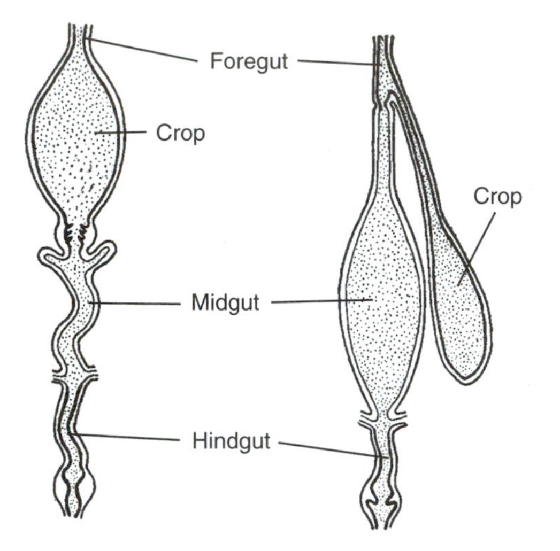

FIGURE 6.5 (Left) The generalized insect digestive tract. (Right) The evolution of the crop into a diverticulum. From Wiggleworth (1965). Reprinted with permission.

predators that may find it more difficult to locate their animal prey and gorge themselves on infrequent large meals, storing the meal for digestion away from the host.

The posterior part of the foregut is variously modified into a muscular **proventriculus.** At its simplest, as in some beetles, the proventiculus is a muscular sphincter that regulates the passage of food into the midgut. In cockroaches, it is present as a gizzard lined with teeth that grinds the food prior to its entry into the midgut (Fig. 6.6). Fleas have a proventriculus that is lined with backwardly pointing spines that break up the ingested blood cells before passing them into the midgut. In bees, the proventriculus projects into the crop, and armed with short spines, the structure retains nectar in the crop while passing grains of pollen into the midgut.

MIDGUT

The midgut contains at least four cell types in a single epithelial layer that include **columnar cells, regenerative cells, goblet cells,** and **endocrine cells.** All cells of the midgut are derived from endodermal tissue and lack the cuticular lining that is present in the foregut and that prevents absorption from occurring there. The columnar cells are most numerous, and their borders that face the lumen contain abundant microvilli and numerous folds, increasing the surface area

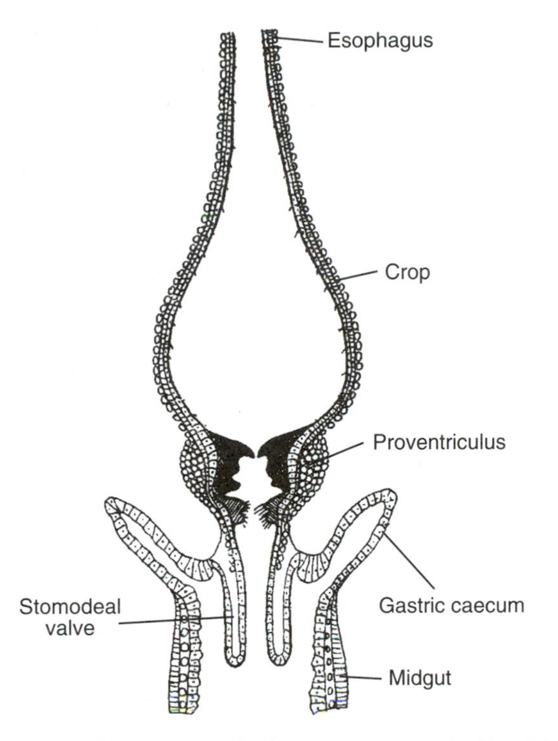

FIGURE 6.6 The location of the proventriculus between the crop and midgut. The proventriculus is modified here as a gizzard for grinding food. From Wiggleworth (1965). Reprinted with permission.

for absorption and secretion (Fig. 6.7). Most nutrients in the gut lumen are absorbed through these columnar cells. They contain extensive networks of endoplasmic reticulum that are necessary for the production of the digestive enzymes. The columnar cells have limited life spans, and new ones are continually being formed from the regenerative cells present in groups called **nidi** (Fig. 6.8). The **goblet cells** that are scattered throughout the midgut epithelium transport potassium from the hemolymph into the lumen (Fig. 6.9). This movement of ions may be important for the flow of water in the gut that is necessary for nutrients to be absorbed.

In addition to these conventional gut cells, there are a number of endocrine cells that are dispersed throughout the midgut, usually occurring as single cells but are sometimes in small groups. They may represent a way for the insect to integrate its digestive and endocrine systems, assessing food in the midgut and transmitting the information to other cells by endocrine pathways. Antibodies against a large number of mammalian hormones react to the contents of these cells, including members of the insulin family, glucagons, somatostatin, β-endorphins, members of the

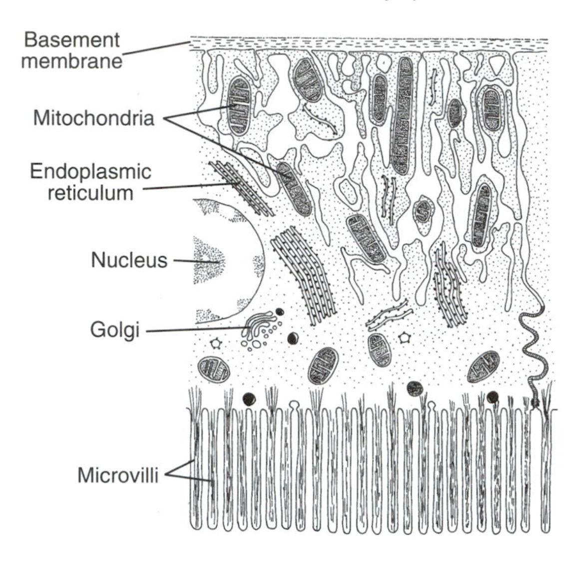

FIGURE 6.7 A typical columnar midgut cell. From Berridge (1969). Reprinted with permission.

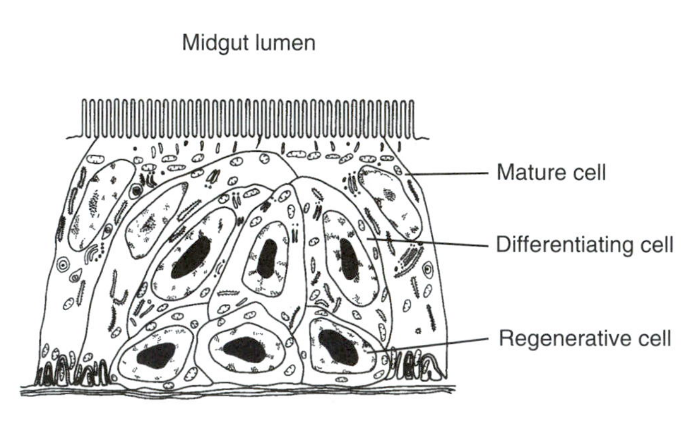

FIGURE 6.8 A group of nidi, the regenerative cells of the midgut. Reprinted from Chapman, R.F. *Comprehensive Insect Physiology, Biochemistry and Pharmacology* 1985, Volume 4, pp. 165–211. With permission from Elsevier Science.

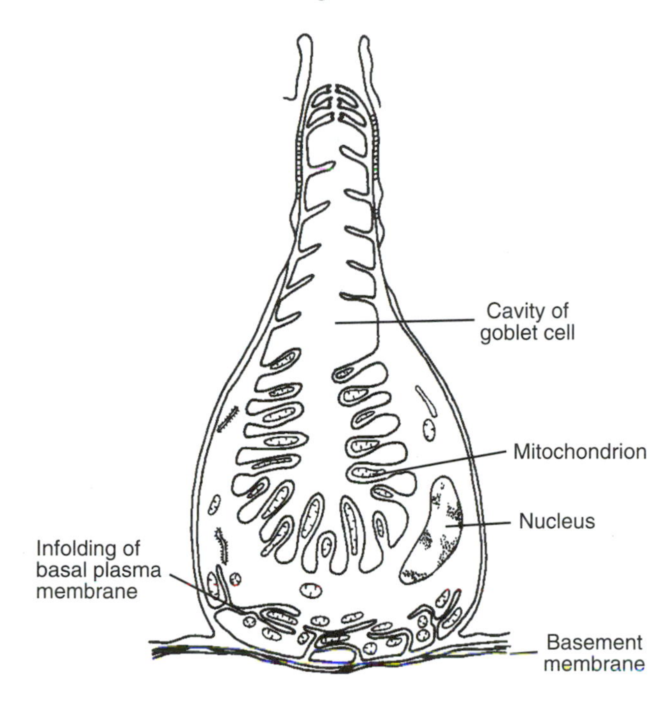

FIGURE 6.9 A midgut goblet cell. Reprinted from Chapman, R.F. *Comprehensive Insect Physiology, Biochemistry and Pharmacology* 1985, Volume 4, pp. 165–211. With permission from Elsevier Science.

tachykinin family, which are myotropins that may stimulate muscle contraction or act as cardioaccelerators, FMRFamide-immunoreactive peptides that may function in digestion, and allatostatin-like peptides that may regulate CA activity or have some other presently unknown functions. The physiological roles of these gut neuropeptides are not well understood.

The anterior region of the midgut may contain diverticula called **gastric caecae.** The caecae increase the surface area of the midgut for secretion and absorption, and create a countercurrent flow within the gut as a result of their differential absorption of water (Fig. 6.10). As water is secreted into the lumen by the posterior midgut, it moves forward to be resorbed in the caecal region, allowing the products of digestion to pass through the gut while solid undigested food moves backward. The caecae may be associated with mechanisms of food detoxification that allow insects to ingest plant materials that contain potentially toxic secondary compounds.

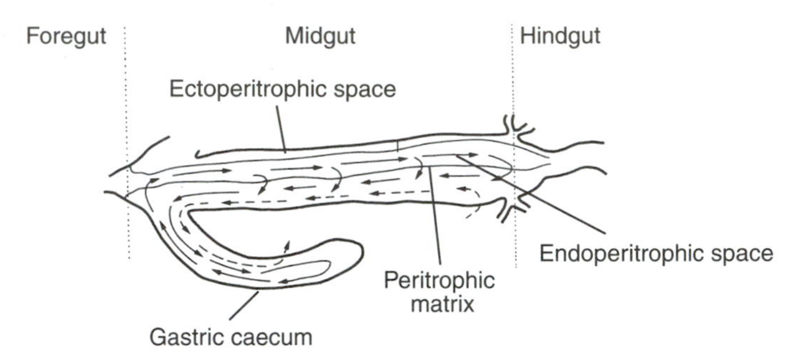

FIGURE 6.10 The compartmentalization of the midgut by the peritrophic matrix into a ectoperitrophic and endoperitrophic spaces. Arrows show the direction of water flow. Reprinted from *Journal of Insect Physiology*, Volume 27. Terra, W., C. Ferreira. The physiological role of the peritrophic membrane and trehalase: Digestive enzymes in the midgut and excreta of starved larvae of *Rhynchosciara*, pp. 325–331. Copyright 1981, with permission from Elsevier Science.

Because it lacks a cuticular lining, the cells that make up the midgut are susceptible to abrasion by food. Depending on the insect, either certain anterior midgut cells or all the cells of the midgut produce a **peritrophic matrix (PM)** that consists of a network of chitin microfibrils within a matrix of carbohydrate and protein. When secreted by all midgut cells it condenses to form the matrix that surrounds the bolus of food. This Type I PM is commonly produced in Orthoptera, Odonata, Coleoptera, and Hymenoptera. The Type II PM is formed by a ring of cells at the anterior region of the midgut, forming a tube that moves backward to enclose the gut contents. A Type II PM is found in Diptera and Dermaptera.

A PM is also found in fluid-feeding insects that have no food particles to abrade the gut cells, but here it may have a more important function in the creation of compartments for digestion. The PM contains pores and is permeable to some digestive enzymes and the products of digestion. This selective permeability creates a compartmentalization into an **endoperitrophic space** surrounded by the PM and an **ectoperitrophic** space that lies between the midgut wall and the PM (Fig. 6.10). These compartments may also aid the caecae in maintaining the countercurrent water flows that result in the movement of solutes in the gut and a compartmentalization of digestive enzymes and products, with large enzymes restricted to the ectoperitrophic space. Fluid carrying the products of digestion moves forward through the ectoperitrophic space as a result of the absorption that occurs in the anterior midgut and gastric caecae. The PM may also serve as a barrier to parasites such as the microbial control agent, *Bacillus thuringiensis,* and its impermeability can protect the insect against toxins in food to prevent them from crossing the PM and directly affecting the midgut cells.

An important consideration in understanding the adaptive features of the insect digestive tract and its morphological and physiological variation is the particular insect's phylogenetic position and the diet of its ancestors, rather than what it may currently eat. The true generalists are typified by the cockroaches, which eat a wide variety of foods and have a gut that is divided evenly among storage, digestion, and osmoregulation. Ingested food mixed with salivary secretions undergoes preliminary digestion in the crop and moves backward into the midgut for further digestion and absorption. Protein digestion and final digestion of carbohydrates occur in the midgut, which has a pH that is slightly acidic to neutral. The anterior region of the hindgut is adapted as a fermentation chamber and is alkaline in pH, conditions that favor the survival of populations of microorganisms. The rectum removes water from the feces and allows a dry fecal pellet to be deposited. Solid food requires about 20 h to pass through the entire alimentary canal.

Plant feeding is considered to be a more advanced specialization. Insects that feed exclusively on plants generally ingest suboptimal levels of nutrients and are forced to process tremendous volumes to gain sufficient amounts of necessary components. The food, together with saliva, travels through the short foregut to the midgut for absorption. No digestion appears to occur in the foregut of Lepidoptera larvae, with initial digestion taking place within the endoperitrophic space of the midgut. The circulation of digesting food between the endo- and ectoperitrophic spaces is driven by fluid fluxes produced by the columnar cells in the anterior and posterior midgut and the gastric caecae. Liquid plant feeding insects, primarily those in the Homoptera and Hemiptera, ingest plant juices that contain very dilute nutrients along with large amounts of water. The length of the gut is generally much longer than the body length to provide for the processing of large amounts of dilute fluids. The gut pH is usually high, with the alkaline conditions serving to protect the insects from the toxic secondary substances that are ingested with the meal. The cibarium of these insects may be equipped with a muscular pump to suck up the liquids.

To accommodate the large amount of water in the diet, cicadas and cercopid Homoptera have a gut modification known as a **filter chamber** (see Chapter 8, Excretory Systems), in which the anterior midgut is expanded to wrap around the posterior midgut and the proximal ends of the Malpighian tubules, and the entire structure is enclosed in a cellular sheath. Water entering the anterior midgut can thus pass directly to the Malpighian tubules for excretion so it is not absorbed and the hemolymph does not become diluted.

Digestion of Proteins

Proteins are broken down into their amino acid constituents by proteolytic enzymes and absorbed passively through the midgut wall when they are present in high concentrations. In lepidopteran larvae, low concentrations of amino acids are actively transported across the gut wall with different areas of the midgut responsible for taking up specific amino acids. The peptidases that initially digest the proteins may be **endopeptidases,** which cleave internal peptide bonds, or **exopeptidases,** which remove terminal amino acids from the protein chain. The most common endopeptidases are classified as serine proteases, including trypsin and chymotrypsin, which have serine at the active site. Trypsin cleaves protein chains on the carboxyl side of basic amino acids. Trypsin-like activity has been identified in most insect species. Chymotrypsin cleaves protein chains on the carboxyl side of aromatic amino acids. **Carboxypeptidases** are exopeptidases that remove the terminal amino acids from the carboxyl end of the chain. **Aminopeptidases** remove single amino acids from the N-terminal of the peptide chain. They require metal ions for activity and generally have an alkaline pH optimum. (Fig. 6.11). Both types of enzymes may be present in the same insects.

In mosquitoes, the presence of initial digestion products in the blood induces the release of proteolytic enzymes by a **secretagogue** stimulus associated with the presence of the protein in the midgut. There are two forms of trypsin induced by blood ingestion: an early form in small amounts appears within 2 h of the blood meal, followed by a late form in larger amounts that appears after 12 h. The late form is responsible for most of the proteolytic activity in the gut. The transcription of the late trypsin gene is dependent both on the activity of the early trypsin as well as the quantity and quality of blood present.

Digestion of Carbohydrates

Carbohydrates are ingested in the food as polysaccharides and disaccharides and are broken down to monosaccharides for absorption through the gut wall. The absorption of monosaccharides generally occurs passively through the midgut wall, but there is some evidence that this diffusion is facilitated by an active glu-

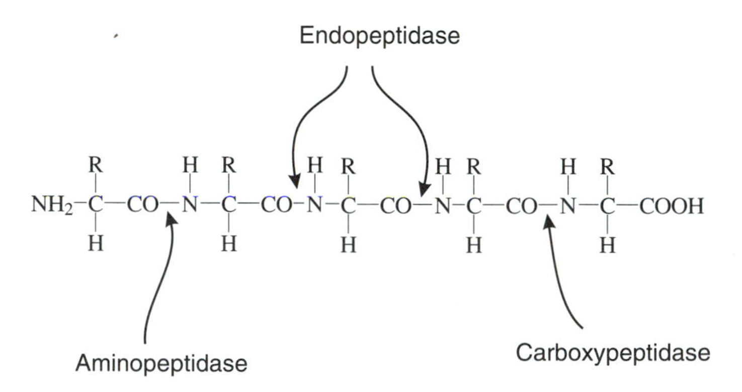

FIGURE 6.11 The action of aminopeptidases, endopeptidases, and carboxypeptidases on a polypeptide chain.

cose transporter. The sugars that pass into the hemolymph are then taken up by the fat body cells that surround the gut. **Glycosidases** secreted in the midgut hydrolyze the glycosidic bonds between the sugar residues, and their specificity depends on the type of bond and whether the linkage is α or β. An α-glucosidase hydrolyzes the common α-glucosides of sucrose, maltose, trehalose, and melezitose; β-glucosidase breaks down the β-glucosides of cellobiose and gentiobiose; α-galactosidase acts on the α-galactosides melibiose and raffinose; β-fructosidase hydrolyses the β-fructosides sucrose, gentianose, and raffinose. Amylases act on the α-glucosidic linkages in starch and glycogen (Fig. 6.12). One of the most common carbohydrases in insects is **trehalase,** that hydrolyzes trehalose into 2 glucose molecules.

Although cellulose is common in the diet of phytophagous insects, the innate ability to digest it is rare. There are three classes of enzymes involved in the hydrolysis of celluloses: endo-β-1,4-glucanases randomly cleave the β-1,4-glucosidic bonds in the cellulose chain; exo-β-1,4-glucanases cleave cellobioise residues from one end of the chain; and β-1,4-glucosidases break down cellobiose to glucose. These enzymes are usually produced by endosymbiotic microorganisms that live in the gut and not the insects themselves. In termites and wood roaches, a population of protozoa is established in the hindgut that digests cellulose. Bacterial symbionts are responsible for cellulose digestion in many cockroaches and beetles. Some fungus-growing termites and ants cultivate the fungi in gardens and ingest the fungal enzymes that are responsible for their ability to digest cellulose. Cellulose digestion that is independent of symbionts has been identified only in cerambycid beetle larvae, in silverfish and firebrats, and in a few species of higher termites and the Australian wood-eating cockroach, *Panesthia cribrata.*

Digestion of Lipids

Lipids are absorbed both as fatty acids or diacylglycerols in the anterior midgut and gastric caecae. In the cockroach, *Periplaneta,* some absorption of lipids also occurs in the cuticle-lined crop. Phytophagous insects ingest mostly the plant constituents monogalactosyl diglycerides, found mainly in chloroplasts, and digalactosyl diglycerides (Fig. 6.13). Triacylglycerols are present in plant storage tissues such as seeds. The major types of lipids in animals are triacylglycerols, phospholipids, and cholesterol (Fig. 6.14).

Although lipids are insoluble in water, they must diffuse through the aqueous hemolymph in order to be absorbed. In vertebrates, bile produced in the liver solubilizes the lipids in the gut, but no such emulsifiers have been identified in insects. Instead, dietary lipids may first be incorporated into polar fractions that increase their solubilization for absorption.

Lipolytic digestive enzymes have not been well-studied in insects and the process by which triacylglycerols in the diet are broken down and transferred to

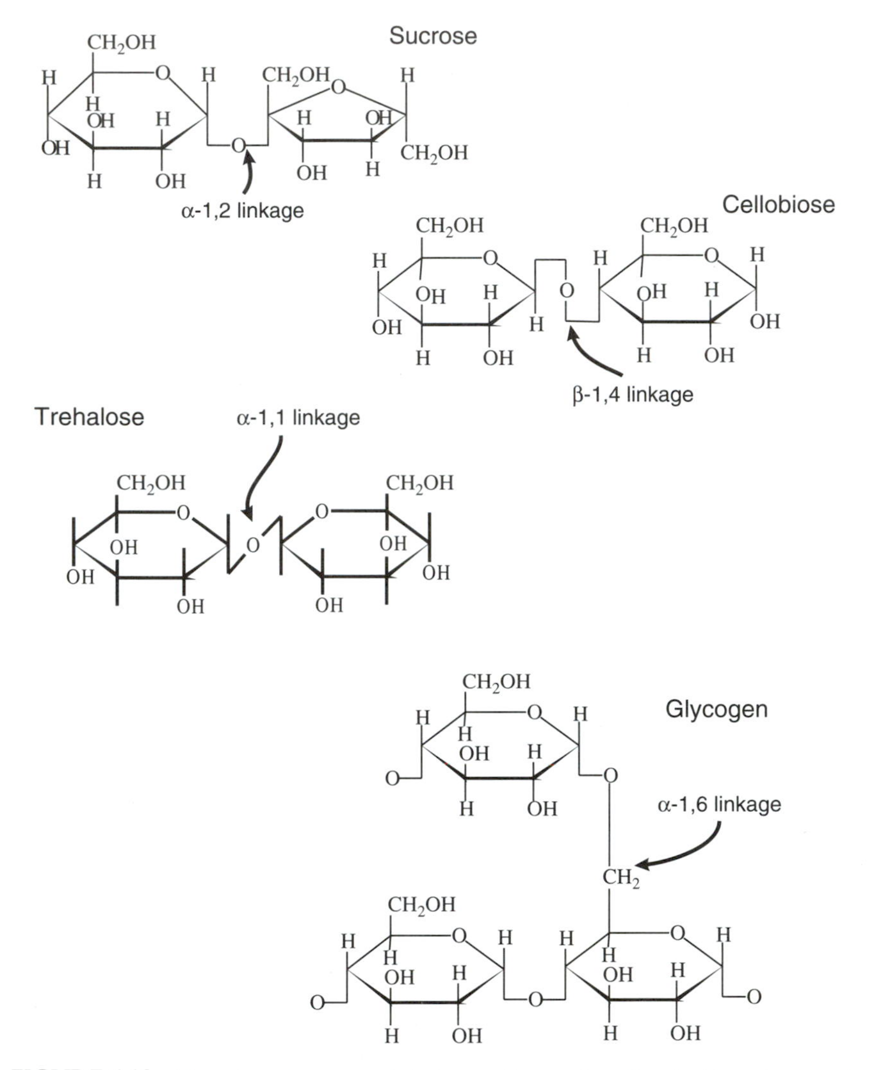

FIGURE 6.12 The linkages between carbohydrate residues to form disaccharides and polysaccharides.

the hemolymph through the gut is largely unknown. In general, **phospholipases** remove the fatty acid portion from phosphatides. They break down the cell membranes of ingested food, allowing the cell contents to be acted upon by other enzymes. **Esterases** act on molecules that are dissolved in water, hydrolyzing car-

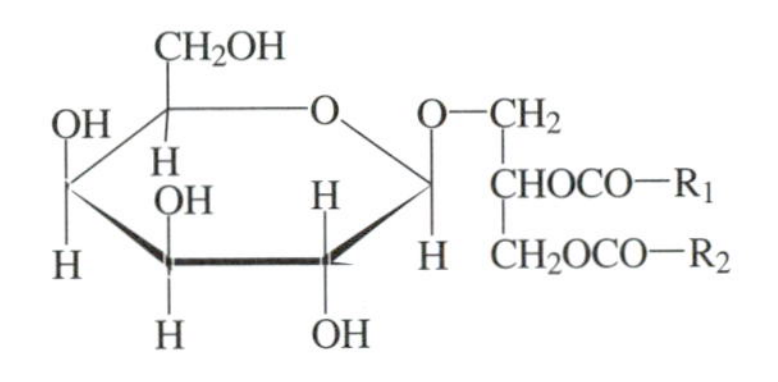

Monogalactosyl diglyceride

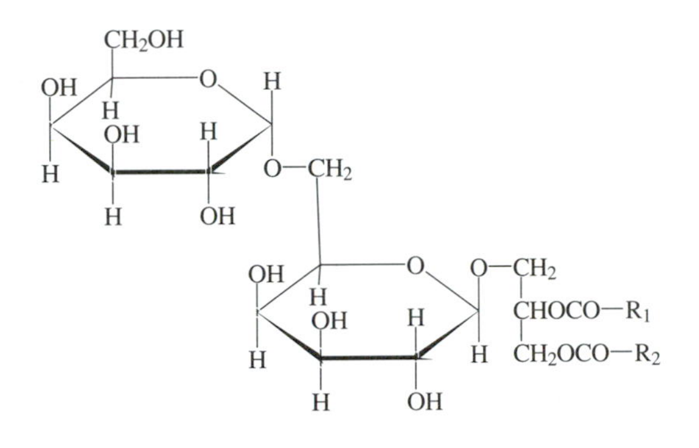

Digalactosyl diglyceride

FIGURE 6.13 The major diglycerides found in plants.

boxyl esters into alcohol and carboxylate. They may break down cholesterol and are important in the resistance to insecticides and plant secondary substances.

The major lipid components that appear in the hemolymph are diacylglycerols. These are resynthesized by midgut cells from digested components before they are released into the hemolymph. Insoluble in the aqueous hemolymph by themselves, the diacylglycerols are bound to a **lipophorin** that allows them to be transported throughout the body. Lipophorins also transport the cholesterol and phospholipids that are present in the hemolymph. This lipid transport will be described further in the section on the metabolism of lipids.

HINDGUT

The insect hindgut, along with the Malpighian tubules, is primarily concerned with osmoregulation and will be discussed in Chapter 8, Excretory Systems. The Malpighian tubules produce a primary isoosmotic urine that is rich in potassium and low in sodium, and contains various ions, amino acids, and waste materials.

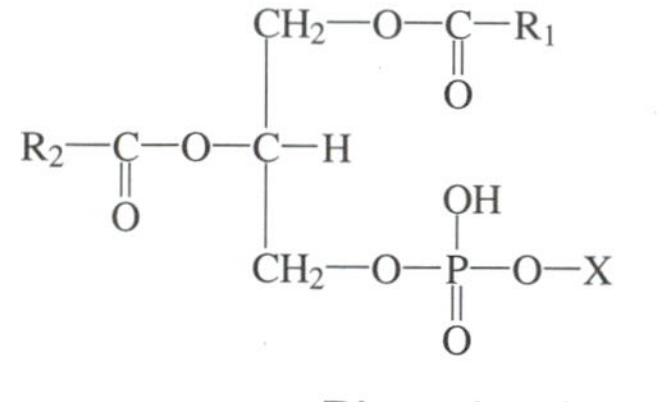

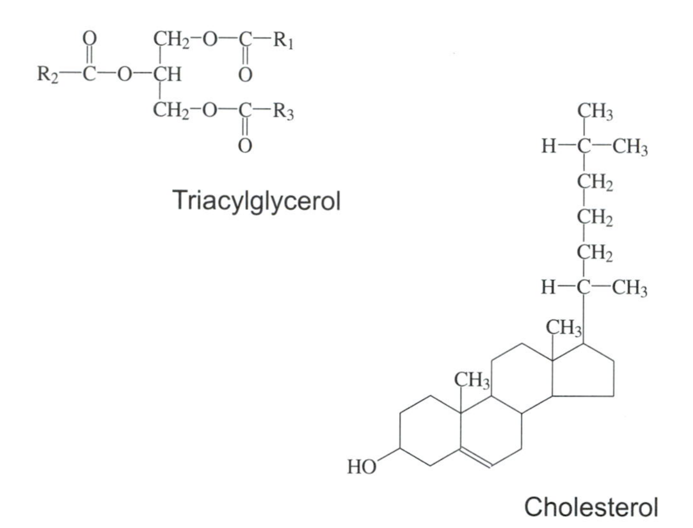

FIGURE 6.14 The major types of animal lipids. X = alcohol in phosphoglyceride.

The hindgut is capable of a selective resorption of the amino acids, water, and ions and produces a hyper- or hypoosmotic urine that is deposited to the outside. Undigested food and waste products from digestion also pass through the hindgut, which can recover a number of important substances.

METABOLIC PROCESSES IN INSECTS

Metabolism is differentiated into the two processes of catabolism and anabolism. **Catabolism** involves the enzymatic degradation of large nutrient molecules from an organism's reserves or from the environment. **Anabolism** is the enzymatic synthesis of larger cell components from smaller precursors. Whenever molecules are degraded or synthesized, there is a change in the energy states between the starting substrates and ending products, and living systems can capture this change in energy or store it in molecules that conserve the energy as phosphate bonds. By coupling these degradative, energy-releasing steps with equivalent energy-conserving steps, the difference in energy can be transferred efficiently without generating too much heat that is generally wasted energy.

Adenosine triphosphate, or **ATP,** is the universal energy currency of cells, serving as a transfer medium for energy and its temporary storage. The energy derived from the stepwise oxidation of food is harnessed and parceled out by ATP to perform cellular work. The ATP molecule consists of adenine, ribose, and a triphosphate unit (Fig. 6.15). The phosphate bonds are what give ATP its ability to store energy, which is liberated when it is hydrolyzed to adenosine diphosphate (ADP). The turnover of cellular ATP is very high, as it is typically utilized shortly after it is formed. There are other high-energy nucleotides that play an equally important, but less major role in the transfer of cellular energy, including **guanosine triphosphate** (**GTP**) and **uridine triphosphate** (**UTP**).

The pathways by which ingested and stored carbohydrate, fat, and protein are catabolized involve the transfer of energy in small amounts during their degradation

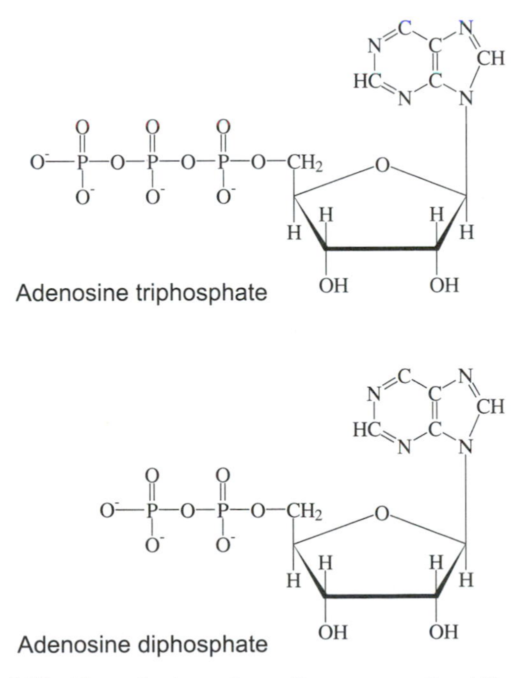

FIGURE 6.15 The molecules used most for energy transfer within the cell.

or synthesis. The energy locked up in organic molecules is released primarily by **oxidation,** the removal of electrons from the molecules and the transfer of those electrons to other molecules within the cell that undergo **reduction** when they acquire the electrons. This generation of energy occurs in three general stages in all animals. In the first stage, the large food molecules are converted into smaller ones within the alimentary tract. We have already seen that proteins are degraded to amino acids, fats to fatty acids and glycerol, and polysaccharides to monosaccharides (Fig. 6.16). No energy is produced during this stage, but, in fact, some may be consumed in the process of synthesizing enzymes needed to break the large molecules down. In the second stage, these simpler molecules are further broken down to two carbon molecules that are acceptable for entry into the citric acid cycle. These molecules consist primarily of acetic acid that is combined with Co-enzyme A to form acetyl Co-enzyme A. A small amount of energy is generated anaerobically in this stage with the production of a few molecules of ATP and carbon dioxide. In the third stage, the molecules enter the citric acid cycle where they are oxidized to carbon dioxide and their electrons are ultimately transferred to oxygen to form water. It is during this

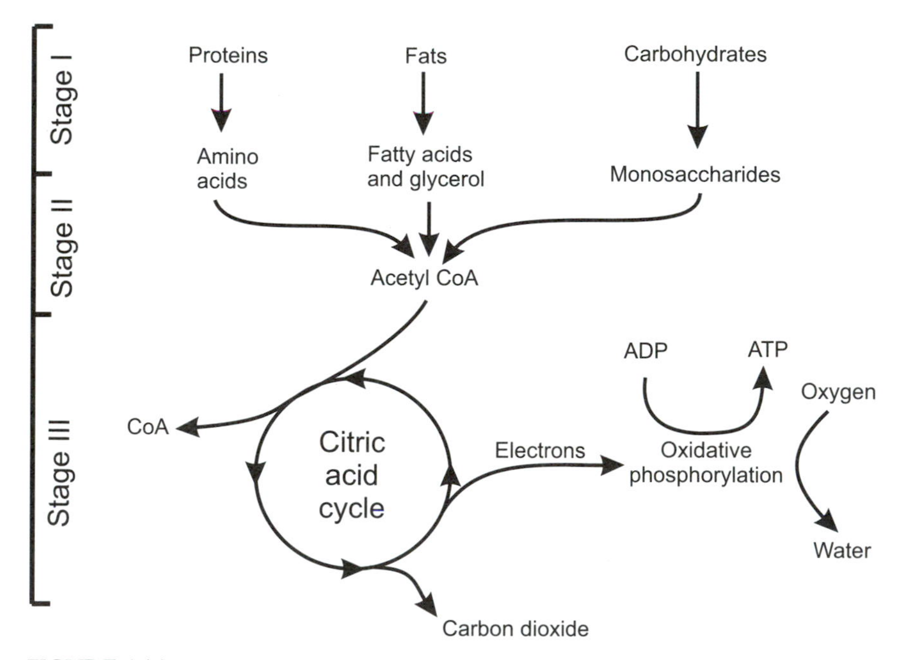

FIGURE 6.16 Stages in the oxidation of food. In Stage I, proteins, fats, and carbohydrates are broken down into their constituents. In Stage II, these building blocks are reduced to two carbon molecules for entry in the citric acid cycle. In Stage III, the two carbon molecules enter the citric acid cycle, with carbon dioxide and water produced along with the bulk of the energy transfer to ATP.

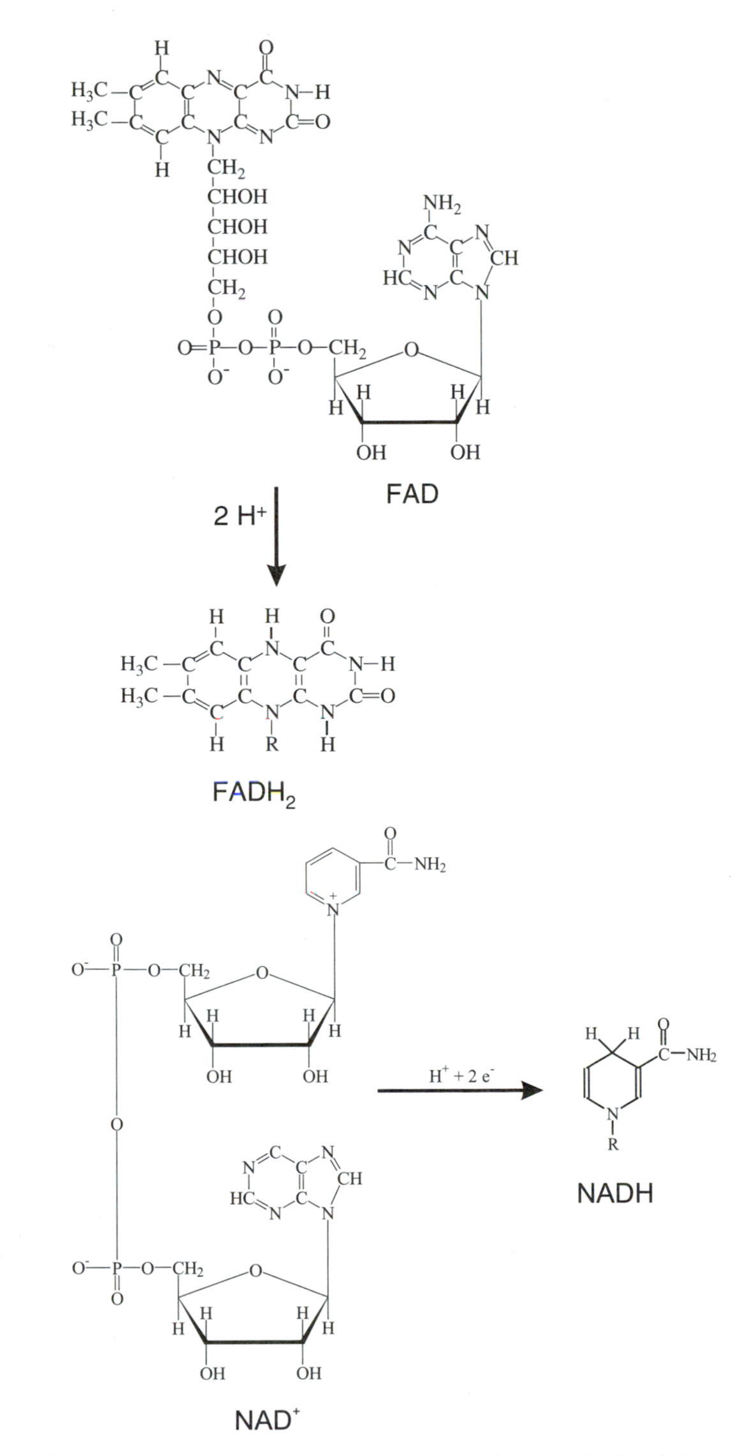

FIGURE 6.17 The major electron acceptor molecules during food oxidation.

process of oxidative phosphorylation that much of the energy is coupled to the generation of large amounts of ATP.

The difference in energy from the electrons being removed from food is eventually transferred to ATP by the oxidation reactions. Electrons are shuttled within the cell and are ultimately passed to oxygen, resulting in the formation of water and the synthesis of ATP. The major electron acceptor molecules during food oxidation are **nicotinamide adenine dinucleotide** (**NAD$^+$**) and **flavin adenine dinucleotide** (**FAD**) (Fig. 6.17). Successive decarboxylations of the food substrate also account for the carbon dioxide that is released into the environment as cells respire. During the degradation of proteins, nitrogen is additionally removed from the molecules and released as ammonia. Because ammonia is so toxic, most terrestrial animals incorporate the nitrogen into more complex but less toxic molecules such as urea and uric acid. The net result of these biochemical transformations is the production of useful energy from the degradation of organic molecules. Energy that is not required immediately is stored as trehalose, glycogen, or fat.

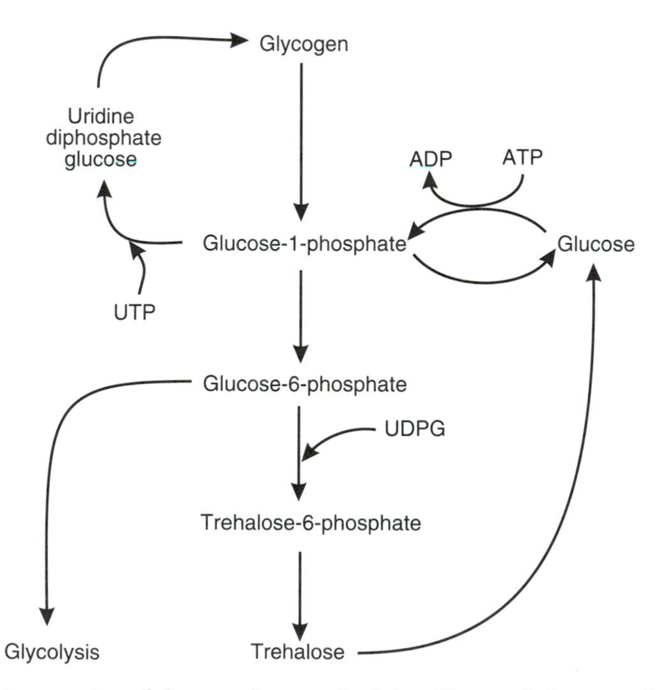

FIGURE 6.18 Interconversions of glycogen, glucose, and trehalose. Glycogen is the storage form of carbohydrate and is converted to glucose and trehalose when energy is required.

Metabolism of Carbohydrates

The most important carbohydrate reserves in insects are glycogen and trehalose, synthesized when the intake of carbohydrate is greater than what is immediately required. Both can easily be converted to glucose when the reserves need to be mobilized (Fig. 6.18). Glycogen is a polymer of many glucose residues existing as a branched chain storage form (Fig. 6.12). Supplies of glycogen are stored in the flight muscles, fat body, and around the digestive tract. Because it is able to be stored within cells, glycogen can provide an immediate source of energy for rapidly respiring muscles, such as the flight muscles. Glycogen stored in the fat body can be immediately converted to trehalose for release into the hemolymph.

Glycogen reserves can be mobilized in order to increase the levels of the disaccharide trehalose and monosaccharide glucose in the hemolymph. The mobilization is hormonally controlled by a **hyperglycemic hormone,** which is released from the corpus cardiacum. The release of the hormone is triggered by the declining concentrations of hemolymph sugars as they are consumed during metabolism or during starvation. Through a second messenger system, the hyperglycemic hormone activates the normally inactive enzyme, phosphorylase kinase, that results in glycogen-1-phosphate being released from glycogen and the ultimate formation of trehalose. In some insects, an **adipokinetic hormone** plays a similar role in mobilizing glycogen reserves. This hormone will be discussed in more detail in the section on lipid metabolism.

Trehalose, a disaccharide of glucose, is the major hemolymph sugar in insects that serves as a circulating energy source, as glucose does in the blood of vertebrates (Fig. 6.12). However, trehalose concentrations in insect hemolymph are considerably higher, ranging from 0. 5 to 5. 0 g/100 ml, compared to the levels of less than 0. 1 g/100 ml of glucose in the blood of vertebrates. As a disaccharide, it is a larger molecule than glucose and diffuses more slowly, so trehalose can be maintained at a higher concentration in the hemolymph before it diffuses across membranes and into cells. Trehalose also accounts for reduced osmotic effects compared to the same concentrations of the monosaccharide glucose. The higher concentration of hemolymph trehalose that can be maintained facilitates its distribution by diffusion to all the cells of the insect. Another reason for the utilization of trehalose rather than glucose in insects relates to the ease with which glucose in the diet is absorbed through the gut wall. If high levels of glucose were maintained in the hemolymph, they would interfere with the uptake of glucose from the gut because the uptake occurs primarily by passive diffusion. By maintaining the low concentration of glucose in the hemolymph, the uptake of glucose from the gut is facilitated. Once in the hemolymph, the glucose is then converted to trehalose by the fat body. Other cells can readily hydrolyze the trehalose to glucose that is oxidized to provide energy.

The first step in the generation of metabolic energy from carbohydrate is **glycolysis,** the sequence of reactions that break glucose down to pyruvate with the accompanying generation of ATP. The process of glycolysis occurs in nearly all cells and serves as a prelude to the more complete breakdown of the molecules into carbon dioxide and water in the citric acid cycle. The basic steps of glycolysis and the enzymes involved are no different in insects than in other organisms. The process occurs in the cell cytoplasm, generating two molecules of ATP and two molecules of pyruvate for every molecule of glucose. The subsequent fate of pyruvate can vary depending on the organism and the circumstances. In yeast and a few other microorganisms, pyruvate is broken down into ethanol and CO_2 (Fig. 6.19). During intense metabolic activity when oxygen is limited, as in the skeletal muscles of insects and vertebrates during exertion, pyruvate is converted to lactic acid. This regenerates NAD^+ and buys the organism some time that allows glycolysis to continue. Pyruvate can also be transaminated to α-ketoglutarate by glutamate in those insects that use proline as fuel for flight. This pathway will be described in more detail under the metabolism of proteins. Because each molecule of glucose gives rise to two molecules of pyruvate, a total of four molecules of ATP results from glycolysis, but after subtracting the two molecules of ATP that are used, a net production of two molecules of ATP and two molecules of NADH result.

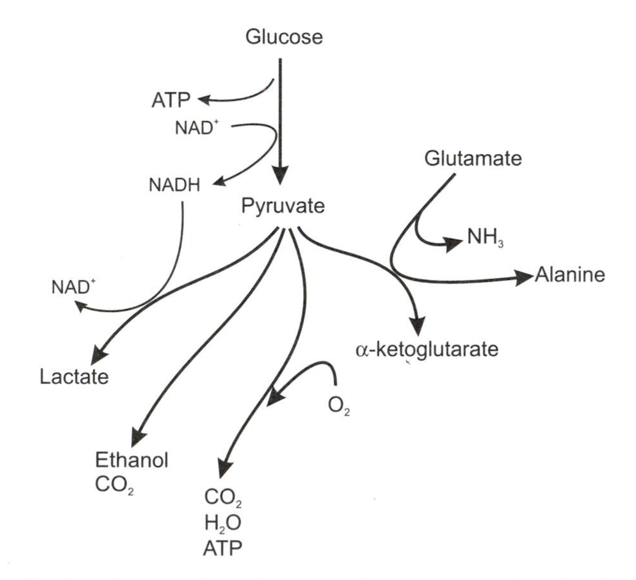

FIGURE 6.19 The fate of pyruvate in respiring tissues. In yeast, pyruvate is broken down into carbon dioxide and ethanol. In insect and vertebrate skeletal muscles when oxygen is limited, the pyrvate is broken down to lactate. In the presence of sufficient oxygen, the pyruvate is converted to acetyl CoA and enters the citric acid cycle where it is released as carbon dioxide and water, and large amounts of ATP are produced. In some insects, the pyruvate may be transaminated by glutamate to α-ketoglutarate in a pathway that utilizes proline for flight energy.

Much more energy can be extracted from glucose when oxygen is plentiful if the pyruvate is converted to acetyl CoA that is able to enter the oxidative pathway of the citric acid cycle. When sufficient oxygen is present, the pyruvate that is first converted to acetyl CoA enters the mitochondria to participate in a series of oxidative reactions called the **citric acid cycle.** In addition to the oxidation of pyruvate, the citric acid cycle provides a mechanism for the interconversion of many biochemical intermediates. Here the acetyl CoA is oxidized to carbon dioxide and water and the electrons are passed through an electron transport chain with their final acceptor of oxygen. The energy that is released during this transfer is used to phosphorylate ADP to ATP (Fig. 6.20).

A unique pathway of carbohydrate metabolism is found in some insects that rely on carbohydrate substrates to produce energy for flight. Glycolysis takes place in the cytoplasm and electron transport occurs in the mitochondria, but because of the impermeability of the mitochondrial membrane, pyridine nucleotides such as NAD^+ are prevented from passing between the mitochondrion and the cytoplasm. Thus, the NADH that is formed with the oxidation of glyceraldehyde-3-phosphate during glycolysis within the cytoplasm cannot become oxidized by the electron transport chain because it is unable to enter the mitochondrion, and without the regeneration of NAD^+, glycolysis cannot continue. The operation of the **glycerol 3-phosphate shuttle** in insects allows the reducing equivalents from the cytoplasmic pool of NADH to cross the permeability barrier of the mitochondrial membrane and be oxidized by the cytochromes, ultimately transferring their electrons to oxygen (Figs. 6.20 and 6.21). The shuttle thus serves to reoxidize the NADH that is produced during the glycolysis occurring in flight muscles by carrying the electrons from NADH across the mitochondrial membrane rather than the NADH itself. One carrier is glycerol 3-phosphate, formed when NADH transfers its electrons to dihydroxyacetone phosphate. The glycerol 3-phosphate is able to cross the outer mitochondrial membrane, where it is reoxidized to dihydroxyacetone phosphate by the electron acceptor FAD that transfers its electrons to the respiratory chain. The dihydroxyacetone phosphate diffuses back into the cytoplasm to complete the shuttle. Other insect skeletal muscles lack this pathway, and there the NAD^+ is regenerated by the reduction of pyruvate to lactic acid. The lactic acid is a metabolic dead end; it must be converted back to pyruvate before it can be metabolized, but the reduction of pyruvate at least generates the NAD^+ that allows glycolysis to continue. The glycerol-3-phosphate shuttle prevents the wasteful formation of lactic acid in rapidly respiring insect tissues when oxygen supplies might be limited.

Chitin is a major component of the insect procuticle, often comprising half the dry weight of the cuticle. It is produced by most cells that have an epidermal origin. It is a polymer of *N*-acetyl-D-glucosamine residues (some glucosamine residues may also be present; see Fig. 2.12) joined in 1-4 β linkages (Fig. 6.22). The synthesis of chitin begins with glucose and involves a phosphorylation, amination, acetylation, and a conjugation with uridine diphosphate (Fig. 6.23). The mature

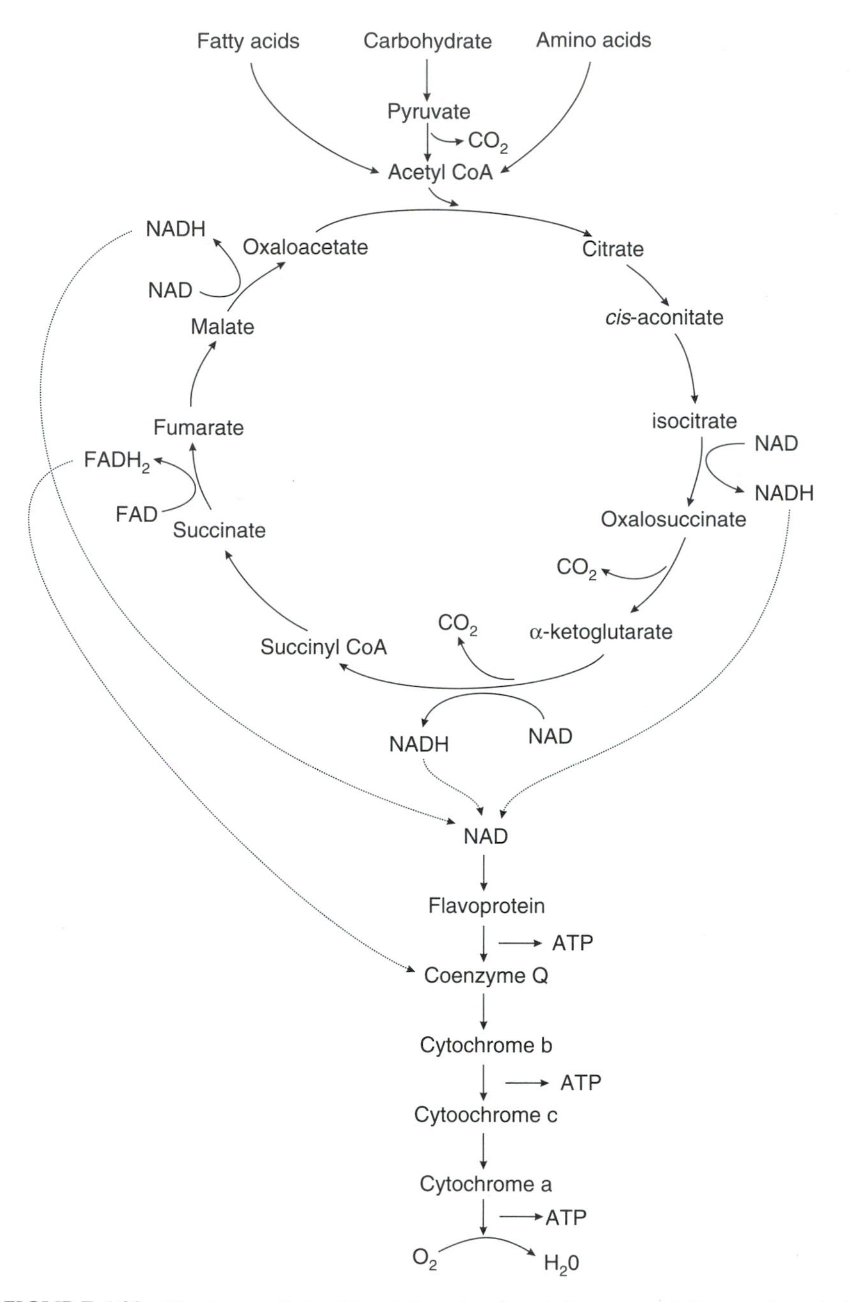

FIGURE 6.20 The intermediates of the citric acid cycle and the passage of electrons through the electron transport chain, where ATP is generated.

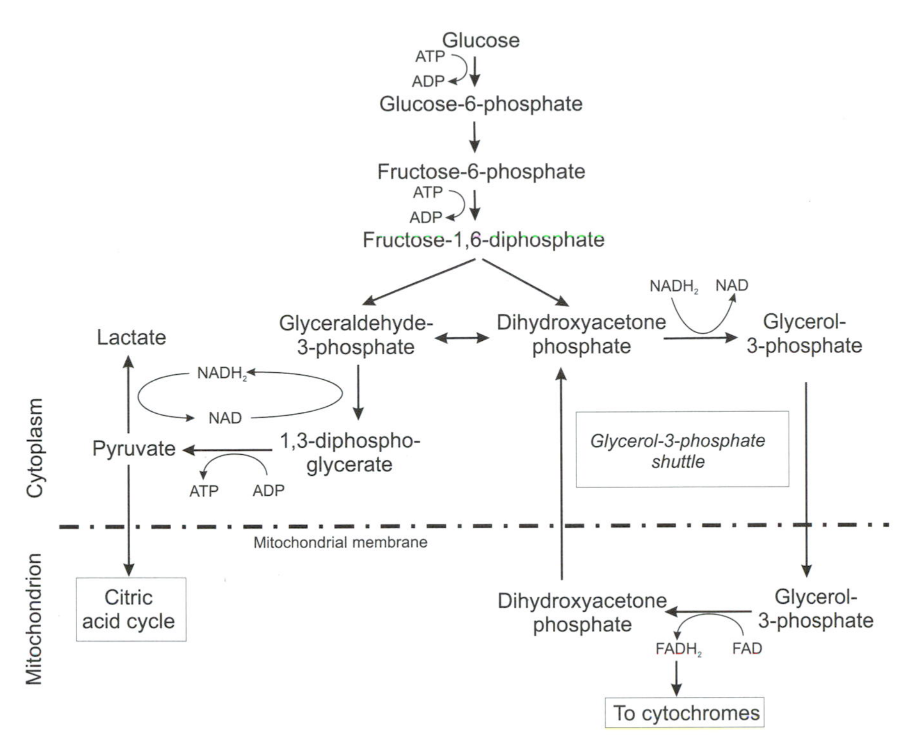

FIGURE 6.21 The glycerol-3-phosphate shuttle that operates in insect flight muscles.

helical polymer of chitin may consist of as many as 1500 *N*-acetyl-D-glucosamine residues, with an occasional glucosamine also present. It can be degraded during the molting process by molting fluid and its raw materials recovered for use in the new cuticle that is synthesized. During the molt, the enzyme **chitinase** breaks the polymer down to chitobiose and *N*-acetylglucosamine.

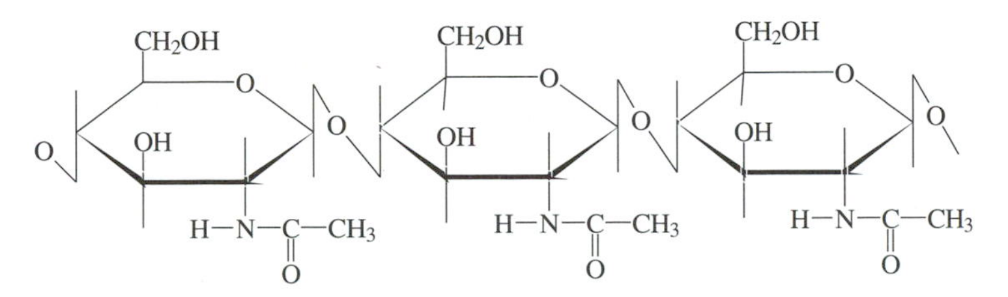

FIGURE 6.22 Residues of *N*-acetyl-D-glucosamine, forming a polymer of chitin.

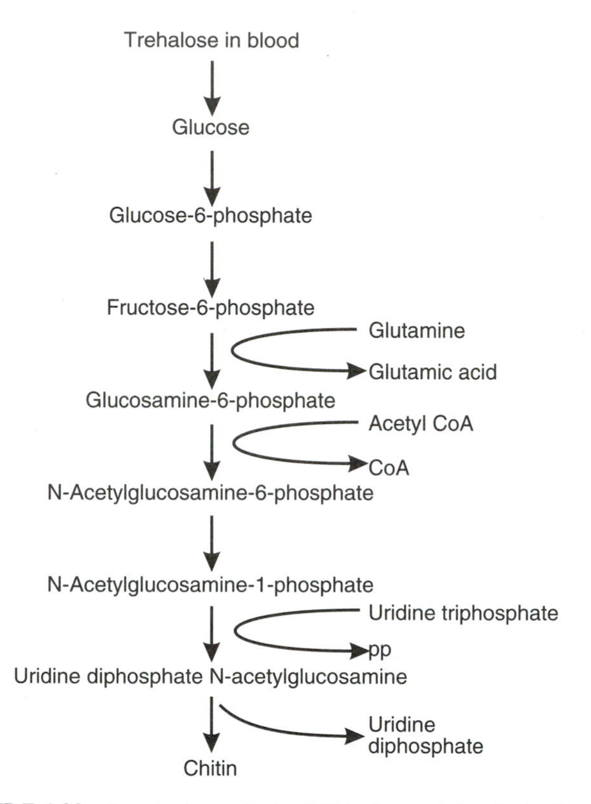

FIGURE 6.23 Steps in the synthesis of chitin from trehalose in the blood.

Metabolism of Proteins

The most important functions of amino acids, the building blocks of the proteins that are derived from the insect diet, include the synthesis of structural proteins of the integument and the synthesis of hormones and enzymes that participate in metabolic reactions. The same 20 amino acids are involved in building the proteins of all living things from bacteria to vertebrates, and insects also employ those 20 amino acids in constructing proteins. Of these 20 amino acids, 9 of them cannot be synthesized by the interconversion of other amino acids. These 9 **essential amino acids** must be ingested as dietary components, while the remaining 11 others are **nonessential** and may be derived by biochemical conversions (Table 6.1; Fig. 6.24). Insects need the same 9 essential amino acids that are required by vertebrates. Because the central carbon atom allows a tetrahedral array of different groups that bond to it, two mirror image isomers that are designated D and L are possible. However, only the L isomers of amino acids are found in proteins.

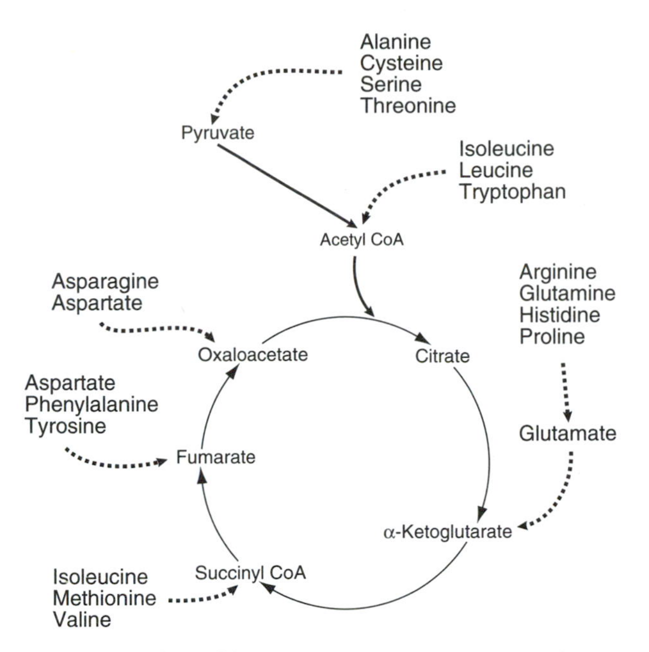

FIGURE 6.24 Interconversion of some amino acids.

TABLE 6.1 Basic Amino Acids

Essential	Nonessential
Histidine	Alanine
Isoleucine	Arginine
Leucine	Asparagine
Lysine	Aspartate
Methionine	Cysteine
Phenylalanine	Glutamate
Threonine	Glutamine
Tryptophan	Glycine
Valine	Proline
	Serine
	Tyrosine

Amino acids that are present in the hemolymph make a large osmotic contribution, sometimes accounting for more than 30% of its total osmotic activity. Amino acids also figure prominently in many biochemical pathways. Some insects use the amino acid proline as a metabolic substrate for flight energy in addition to the use of fats and carbohydrates. Tyrosine is necessary for cuticular sclerotization, and glutamate is involved in neurotransmission. In insects, among the most important requirements for protein synthesis are for the proteins that will be deposited in the newly formed cuticle and that participate in cuticular sclerotization. The cuticular proteins are largely synthesized by epidermal cells, and the mechanisms involved are further discussed in Chapter 2. Another important protein consists of the vitellogenins that supply the protein requirements for the egg. The control of vitellogenin synthesis by cells of the fat body and ovarian follicle is discussed in Chapter 4. The production of both cuticular and vitellogenic proteins occurs in response to levels of juvenile hormone and 20-hydroxyecdysone.

In contrast to fats and carbohydrates that can easily be stored, the amino acids that are ingested in excess of the immediate needs of most animals are generally either used as metabolic fuel or are excreted. However, there is a particular need for proteins to be stored by many holometabolous insects. The protein that is acquired during the larval stage must often be carried over to the pupal and adult stages that require protein but may be unable to acquire it. This protein is essential for metamorphosis, reproduction, and general body maintenance. Indeed, the consumption of the cast skin by some insects after molting is a way to recover some of the nitrogen that is lost during the molt.

There is a special family of proteins, the **storage hexamerins,** that have been described in many other arthropods besides insects. The hexamerins present in crustaceans and celicerate arthropods are considered to be hemocyanins that are used to carry oxygen in the blood. However, as the hexamerins evolved in insects, they apparently lost this respiratory function. Instead, they act primarily as storage proteins that provide amino acids that are required for protein synthesis in the developmental phases that do not feed. The hexamerins are synthesized by the fat body and are released into the hemolymph during the larval stage, but just prior to metamorphosis, they are recaptured by the fat body and stored in cytoplasmic granules that can be utilized during adult development. They comprise a family of proteins that are each composed of six similar subunits of between 70,000 and 90,000 molecular weight and with a total molecular weight of approximately 500,000.

The **lipophorins** are lipoproteins found in the hemolymph that serve as a vehicle for the transport of lipids that are otherwise not soluble in the aqueous blood. They have been found in all life stages of all the insect species that have been examined. They are loaded with the dietary lipid that is absorbed through the midgut wall and carry it to developing tissues or to the fat body where it may be stored. Juvenile hormones may also be transported by lipophorins. These will be discussed in more detail in the section on lipid metabolism later in this chapter.

The degradation of dietary amino acids generally yields acetyl CoA, pyruvate, or other citric acid cycle intermediates. A common reaction in amino acid metabolism is **transamination,** in which the amino groups are transferred from amino acids to keto acids for conversion into ammonia (Fig. 6.25). The amino acids glutamate, aspartate, and alanine and their corresponding keto acids are mostly involved, with glutamate serving as one of the key intermediates. Keto acids can also be formed by an oxidative deamination. Amino acid degradation can thus commonly form metabolic intermediates that can be converted to glucose or that can enter the citric acid cycle.

The amino acids proline and glutamate appear in high concentrations in the hemolymph of many insects and can serve as substrates for the citric acid cycle. During the first few seconds of the flight of the blow fly, proline concentrations decline and alanine concentrations increase, along with an accumulation of pyruvate because its production occurs faster than the rate at which it is oxidized by the mitochondria. The proline is converted to glutamate by the enzyme proline dehydrogenase, which is activated by high levels of pyruvate, and the transamination of the glutamate with pyruvate produces the alanine and additionally α-ketoglutarate. The α-ketoglutarate is further metabolized to oxaloacetate through the citric acid cycle, and its condensation with acetyl CoA yields citrate. This allows the complete oxidation of pyruvate through the citric acid cycle, supplying citric acid cycle intermediates to prime the cycle and speed it up. Proline is first converted to glutamate that is then used for the transamination of pyruvate, so that the proline is able to enter the citric acid cycle as α-ketoglutavate.

Proline may also play a direct role as the predominant substrate for flight metabolism. In the tsetse and the Colorado potato beetle, proline is utilized to a

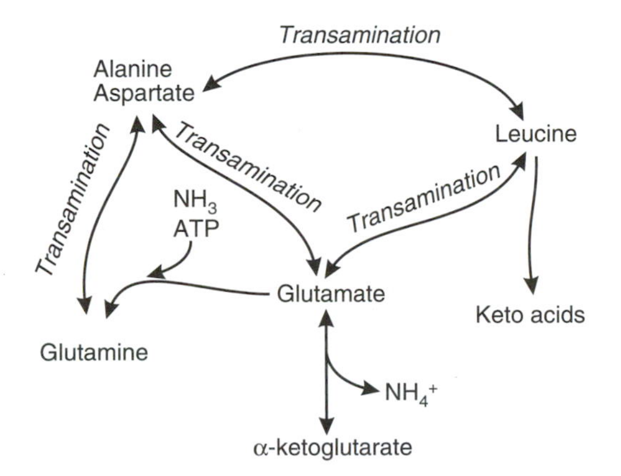

FIGURE 6.25 Transamination, the transfer of amino groups from amino acids to keto acids.

much greater degree than is pyruvate during flight. Again as the levels of proline decline during flight, levels of alanine increase. This results from the proline first being converted to glutamate, which is then transaminated with pyruvate to form α-ketoglutarate and alanine. The alanine is taken up by fat body cells and reconverted to proline, while the α-ketoglutarate enters the citric acid cycle (Fig. 6.26). The conversion of 1 mol of proline into 1 mol of alanine by this mechanism yields 14 mol of ATP, comparing favorably to the 15 mol of ATP produced by 1 mol of pyruvate passing through the citric acid cycle.

Metabolism of Lipids

Lipids are a heterogeneous group of compounds that are defined by their insolubility in water and high solubility in nonpolar organic solvents. Fatty acids are lipids that contain a long hydrocarbon chain and a terminal carboxylate group. They vary in their chain length and the degree of unsaturation, or double bonds, within the chain, but those in biological systems usually contain an even number of carbon atoms and are unbranched. The physical characteristics of fatty acids and the lipids that are derived from them are largely based on the degree of saturation

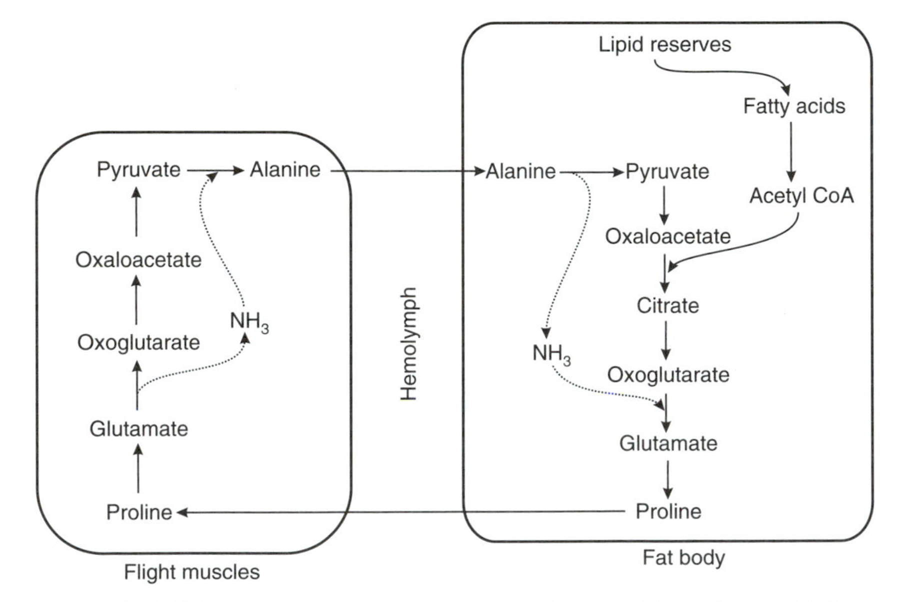

FIGURE 6.26 Utilization of fat body proline for flight. Adapted from Chapman (1998).

and the length of their chains. Those with shorter chains and more unsaturation tend to be more fluid in biological systems.

Fatty acids are significant molecules in biological systems. They are building blocks for the synthesis of the phospholipids that are important constituents of the cell membrane. They also act as hormones and as sources of metabolic energy that can be mobilized to meet the energy requirements of the insect. They make up a large part of the cuticular lipids that protect insects from desiccating. Many of the sex pheromones that are synthesized by insects are derived from fatty acids as well as some defensive secretions such as quinones, phenols, and carboxylic acids.

The predominant lipids in insects are in the form of uncharged esters of glycerol known as **triacylglycerols** (Fig. 6.27), and most of the triacylglycerol is located in the fat body where it may make up over 50% of its wet weight in the cockroach, *Periplaneta americana*. Triacylglycerols are a very concentrated store of energy, with most of the energy coming from the fatty acid component of the molecule. The complete oxidation of carbohydrates and proteins yields about 4 kcal/g, but in contrast, the oxidation of fatty acids yields over twice this amount, about 9 kcal/g. This is undoubtedly the reason that fatty acids have evolved as the

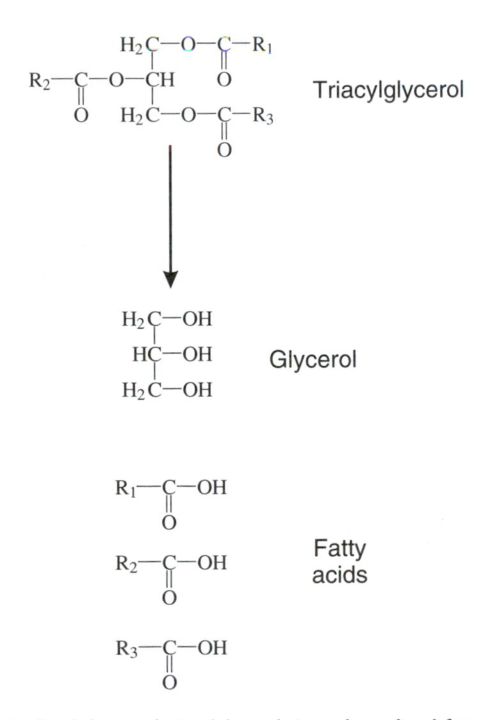

FIGURE 6.27 The breakdown of triacylglycerols into glycerol and fatty acids.

major energy reservoir in animals. As very small animals, insects in particular benefit from the large amount of energy that can be stored in a relatively compact form. Although they are stored in the fat body as triacylglycerides, they are transported in the hemolymph to target cells as diacylglycerides.

Many of the biochemical pathways involving lipid metabolism in insects are similar to those in vertebrates. Like mammals, most insects are unable to synthesize fatty acids that contain two or more double bonds and these are therefore required in the diet. Linoleic acid, with two double bonds, is a component of the lipid bilayer of the cell membrane and is an essential fatty acid in vertebrates. Although it was long assumed that insects also must meet their requirement for linoleic acid by its ingestion, a survey of about 35 insect species demonstrated that at least 15 species including several cockroaches, crickets, aphids, and termites were indeed able to synthesize linoleic acid from actetate. Of those insects apparently capable of synthesizing all their requirements for fatty acids, the extent to which microorganisms contribute to this ability is not always clear.

In contrast to vertebrates, however, all insects require a dietary source of sterols because they are unable to synthesize them from precursors, as do most other animals and plants. Sterols are important components of cell membranes, cuticular surface waxes, and precursors for the synthesis of ecdysteroids, the hormones involved in molting. In the cells of most other animals, cholesterol is synthesized from the two-carbon acetate in several sequential steps, but insects lack this ability and must ingest it in the diet. Because most plants do not contain any cholesterol, phytophagous insects must obtain their cholesterol by converting the predominant phytosterols, sitosterol, campesterol, and stigmasterol to cholesterol by the dealkylation of the C-24 alkyl group. In those phytophagous insects that are unable to make this conversion, makisterone A, or 24-methyl 20-hydroxyecdysone, is used as the molting hormone. Makisterone A is thus a 28-carbon ecdysteroid, compared to the 27-carbon skeletons of the other ecdysteroids.

Ingested triacylglycerides are hydrolyzed to diacylglycerides and fatty acids by digestive enzymes, and as they enter the hemolymph they are bound to lipophorins that carry them to target cells. Intracellular lipases break down the acylglycerides into glycerol and fatty acids. Glycerol is phosphorylated to glycerol 3-phosphate and joins the glycolytic pathway. The fatty acids undergo β-oxidation in the mitochondria, which involves the sequential removal of acetyl CoA and two pairs of hydrogen atoms from the fatty acyl CoA and the ensuing generation of $FADH_2$ and NADH. The acetyl CoA is then able to enter the citric acid cycle (Fig. 6.28). Before they can cross the inner mitochondrial membrane, however, they must first form an ester with the compound **carnitine,** which facilitates this transport. The production of acetyl CoA from fatty acid oxidation is also a key intermediate for the interconversion of fats, proteins, and carbohydrates.

The capacity of many insects to store polysaccharides tends to be somewhat limited, and the carbohydrate that is ingested above the immediate caloric require-

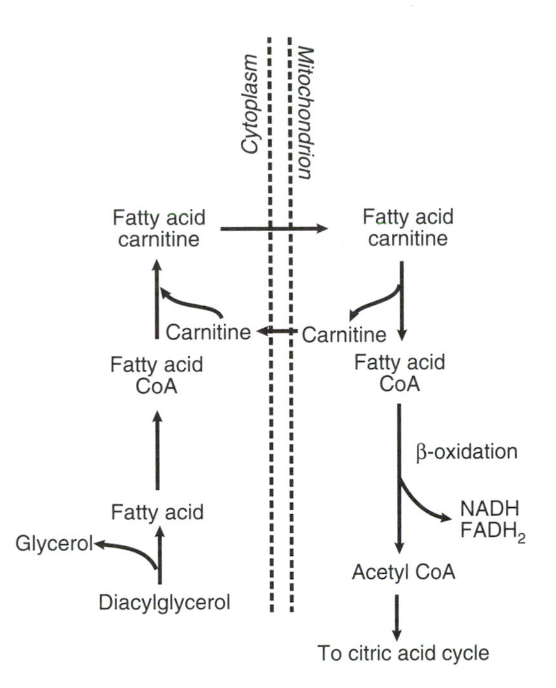

FIGURE 6.28 The entry of fatty acids into the mitochondrion, using carnitine for transport.

ments is often converted to fatty acids that are in turn stored as triacylglycerols in the fat body. The synthesis of fatty acids does not simply occur by a reversal of the same enzymatic steps within the mitochondria that are involved in degradation. Instead, the synthesis of fatty acids occurs in the cytoplasm and is catalyzed by a group of enzymes known as the **fatty acid synthetase complex.** Although the enzyme complex has been isolated from several insects, the details of the synthesis are far from complete, with most of the pathway assumed to be similar to that in vertebrates. There have been differences reported in the mechanisms of fatty acid biosynthesis in the few insects that have been studied, and these differences may be responsible for the variation in the types of fatty acids that are synthesized in each. A single molecule of acetyl CoA serves as a primer, with growth of the chain proceeding by the successive additions of acetyl residues at the carboxyl end. Each acetyl residue is derived from two carbon atoms of malonyl CoA, with the third carbon atom lost as CO_2. The acyl intermediates that form in the process are thioesters of a low-molecular-weight protein, **acyl carrier protein,** or **ACP** (Fig. 6.29). A major difference between the fatty acid synthesis in insects and vertebrates is that insects cannot elongate unsaturated fatty acids or introduce additional double bonds into a fatty acid.

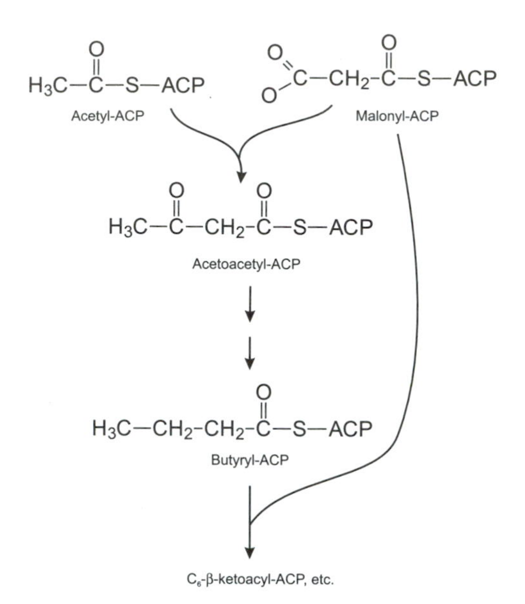

FIGURE 6.29 The synthesis of fatty acids from acetyl-ACP.

The site of most lipid synthesis and storage is the fat body, and lipid must be transported from this site to other target cells throughout the insect. However, the hemolymph is mostly water, making it difficult for the lipid molecules to dissolve and be carried. A mechanism for the transport of lipids consists of the **lipophorins,** hemolymph lipoproteins that serve as lipid shuttles in the aqueous hemolymph to load and unload a variety of lipid molecules at target sites. The high-density lipophorins have a molecular weight of about 600,000 Da and contain single molecules of apolipophorin I (apoLp-I) and apolipophorin II (apoLp-II). A third apolipophorin particle, apolipophorin III (apoLp-III) circulates in the hemolymph and is taken up by the HDLp when it acquires diacylglycerols from the fat body. The apoLp-III increases the capacity of the lipophorin to bind the diacylglycerols, and as the ratio of lipid to protein increases with this binding, the molecule becomes less dense and changes the HDLp into a low-density lipophorin (LDLp) (Fig. 6.30). This low-density lipophorin transfers the diacylglycerols to cells in the flight muscles, ovaries, epidermal cells, and oenocytes. These target cells unload the diacylglycerols, hydrolyze them, and oxidize the liberated fatty acids for energy. The apoLp-III is released into the hemolymph and the high-density lipoprotein shuttle molecule is then ready to be loaded once again.

Because lipids are such a compact form of energy storage, they are the storage molecules of choice in migratory insects and those that otherwise engage in pro-

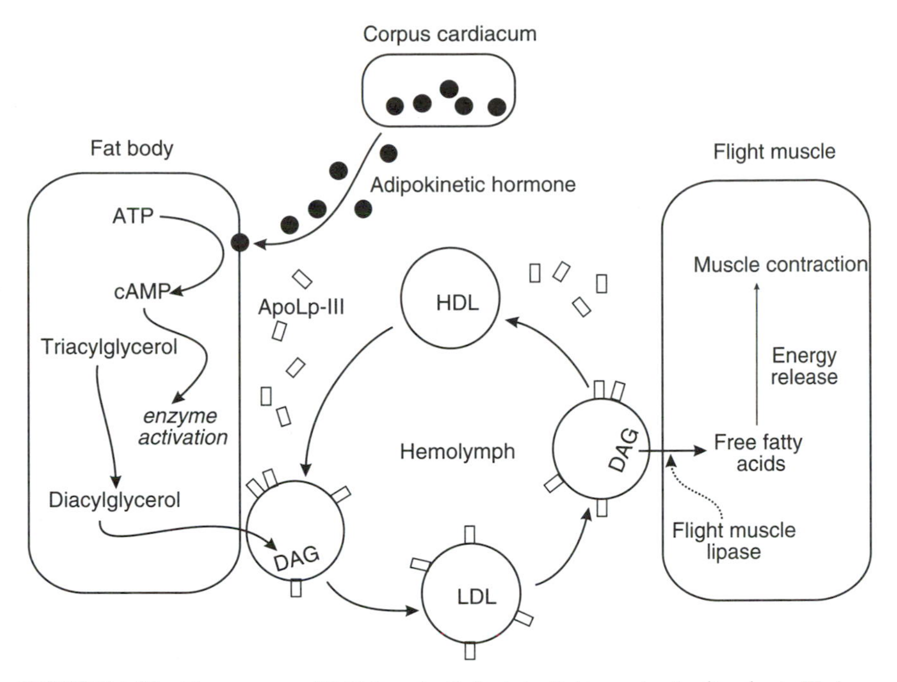

FIGURE 6.30 The transport of lipid from the fat body to flight muscles. Apolipophorin III circulates in the hemolymph and is taken up by the high density lipophorin (HDL) when it acquires the diacylglycerols (DAG) from the fat body. This transforms the HDL into a low-density lipophorin (LDL), which transfers the DAG to the flight muscles. From Ryan (1990). Reprinted with permission.

longed flight. Migratory locusts can fly over 200 km during flights of 10 h or more. Although their initial energy substrate is carbohydrate, triacylglycerides stored in the fat body begin to be mobilized after 15–30 min. The neuropeptide, **adipokinetic hormone** (**AKH**) is synthesized by the intrinsic neurosecretory cells in the glandular lobe of the corpus cardiacum. When released into the hemolymph during flight, AKH activates a fat body lipase that hydrolyzes the stored triacylglycerides to diacylglercides that are then transported to target tissues by lipophorins. The release of AKH is inhibited by blood trehalose and stimulated by octopamine. Over 30 different AKHs have been identified from a variety of insect orders, with some insects containing multiple forms of the hormone; the locust, *Locusta migratoria,* utilizes at least three different AKHs. The hormones consist of 8–10 amino acids with a tryptophan residue at position 8. AKH is also involved in the mobilization of carbohydrate reserves in some insects by its activation of the enzyme glycogen phosphorylase.

ADDITIONAL REFERENCES

Digestion

Adang, M. J., Spence, K. D. 1981. Surface morphology of peritrophic membrane formation in the cabbage looper, *Trichoplusia ni*. *Cell Tissue Res.* **218:** 141–147.

Azuma, M., Takeda, S., Yamamoto, H., Endo, Y., Eguchi, M. 1991. Goblet cell alkaline phosphatase in silkworm midgut epithelium: Its entity and role as an ATPase. *J. Exp. Zool.* **258:** 294–302.

Barbehenn, R.V., Martin, M. M. 1992. The protective role of the peritrophic membrane in the tannin-tolerant larvae of *Orgyia leucostigma* (Lepidoptera). *J. Insect Physiol.* **38:** 973–980.

Barbehenn, R.V., Martin, M. M. 1995. Peritrophic envelope permeability in herbivorous insects. *J. Insect Physiol.* **41:** 303–311.

Barillas-Mury, C. V., Noriega, F. G., Wells, M. A. 1995. Early trypsin activity is part of the signal transduction system that activates transcription of the late trypsin gene in the midgut of the mosquito, *Aedes aegypti*. *Insect Biochem. Mol. Biol.* **25:** 241–246.

Bernays, E. A. 1981. A specialized region of the gastric cacea in the locust, *Schistocerca gregaria*. *Physiol. Entomol.* **6:** 1–6.

Brown, M. R., Lea, A. O. 1990. Neuroendocrine and midgut endocrine systems in the adult mosquito. In *Advances in disease vector research* (K. F. Harris, Ed.), vol. 6, pp. 29–58. Springer-Verlag, New York.

Champagne, D. E., Smartt, C. T., Ribeiro, J. M., James, A. A. 1995. The salivary gland-specific apyrase of the mosquito *Aedes aegypti* is a member of the 5'-nucleotidase family. *Proc. Natl. Acad. Sci. USA* **92:** 694–698.

Chapman, R. F. 1985. Structure of the digestive system. In *Comprehensive insect physiology, biochemistry, and pharmacology* (G. A. Kerkut, L. I. Gilbert, Eds.), vol. 4, pp. 165–211. Pergamon, Oxford.

Chapman, R. F. 1988. The relationship between diet and the size of the midgut caecae in grasshoppers (Insecta: Orthoptera: Acridoidea). *Zool. J. Linn. Soc.* **94:** 319–338.

Charlab, R., Valenzuela, J. G., Rowton, E. D., Ribeiro, J. M. 1999. Toward an understanding of the biochemical and pharmacological complexity of the saliva of a hematophagous sand fly *Lutzomyia longipalpis*. *Proc. Natl. Acad. Sci. USA* **96:** 15,155–15,160.

Cherqui, A., Tjallingii, W. F. 2000. Salivary proteins of aphids, a pilot study on identification, separation and immunolocalisation. *J. Insect Physiol.* **46:** 1177–1186.

Cook, B. J., Holman, G. M. 1979. The pharmacology of insect visceral muscle. *Comp. Biochem. Physiol. C* **64:** 183–190.

Dow, J. A. T. 1986. Insect midgut function. *Adv. Insect Physiol.* **19:** 187–328.

Endo, Y. 1984. Ontogeny of endocrine cells in the gut of the insect *Periplaneta americana*. *Cell Tissue Res.* **238:** 421–423.

Endo, Y., Ferreira, C., Torres, B. B., Terra, W. R. 1998. Substrate specificities of midgut beta-glycosidases from insects of different orders. *Comp. Biochem. Physiol. B.* **119:** 219–225.

Endo, Y., Nishiitsutsuji-Uwo, J. 1981. Gut endocrine cells in insects: The ultrastructure of the gut endocrine cells of the lepidopterous species. *Biomed. Res.* **2:** 270–280.

Freyvogel, T. A., Staubli, W. 1965. The formation of the peritrophic membrane in Culicidae. *Acta Trop.* **22:** 118–147.

Fuse, M., Orchard, I. 1998. The muscular contractions of the midgut of the cockroach, *Diploptera punctata*: Effects of the insect neuropeptides proctolin and leucomyosuppressin. *Regul. Pept.* **77:** 163–168.

Garayoa, M., Villaro, A. C., Sesma, P. 1994. Myoendocrine-like cells in invertebrates: Occurrence of noncardiac striated secretory-like myocytes in the gut of the ant *Formica polyctena*. *Gen. Comp. Endocrinol.* **95:** 133–142.

Gelperin, A. 1965. Control of crop emptying in the blowfly. *J. Insect Physiol.* **12:** 331–345.

Greenberg, B., Kowalski, J., Karpus, J. 1968. Micro-potentiometric pH determinations of muscoid maggot digestive tracts. Ann. Entomol. *Soc. Am.* **61:** 365–368.

Hansen Bay, C. M. 1978. Control of salivation in the blowfly *Calliphora*. *J. Exp. Biol.* **75:** 189–201.

Harrison, J. F. 2001. Insect acid-base physiology. *Annu. Rev. Entomol.* **46:** 221–250.

Henry, S. M. 1962. The significance of microorganisms in the nutrition of insects. *Trans. NY Acad. Sci.* **24:** 676–683.

House, H. L. 1974. Digestion. In *The physiology of insecta* (M. Rockstein, Ed.), vol. 5, pp. 63–117. Academic Press, New York.

Jahan, N., Docherty, P. T., Billingsley, P. F., Hurd, H. 1999. Blood digestion in the mosquito, *Anopheles stephensi*: The effects of *Plasmodium yoelii nigeriensis* on midgut enzyme activities. *Parasitology* **119:** 535–541.

Jeffs, L. B., Phillips, J. E. 1996. Pharmacological study of the second messengers that control rectal ion and fluid transport in the desert locust (*Schistocerca gregaria*). *Arch. Biochem. Physiol.* **31:** 169–184.

Jones, D. 1998. The neglected saliva: Medically important toxins in the saliva of human lice. *Parasitology* **116:** S73–81.

Kingan, T. G., Zitnan, D., Jaffe, H., Beckage, N. E. 1997. Identification of neuropeptides in the midgut of parasitized insects: FLRFamides as candidate paracrines. *Mol. Cell. Endocrinol.* **133:** 19–32.

Kramer, K. J., Muthukrishnan, S. 1997. Insect chitinases: Molecular biology and potential use as biopesticides. *Insect Biochem. Mol. Biol.* **27:** 887–900.

Lehane, M. J. 1997. Peritrophic matrix structure and function. *Annu. Rev. Entomol.* **42:** 525–550.

Lehane, M. J., Billingsley, P. F. 1996. *The biology of the insect midgut*. Chapman and Hall, London.

Lorenz, M. W., Kellner, R., Volkl, W., Hoffmann, K. H., Woodring, J. 2001. A comparative study on hypertrehalosaemic hormones in the hymenoptera: Sequence determination, physiological actions and biological significance. *J. Insect Physiol.* **47:** 563–571.

Marinotti, O., James, A. A., Ribeiro, J. M. C. 1990. Diet and salivation in female *Aedes aegypti* mosquitoes. *J. Insect Physiol.* **36:** 545–548.

Martin, M. M. 1991. The evolution of cellulose digestion in insects. *Phil. Trans. R. Soc. London B* **333:** 281–288.

Mira, A. 2000. Exuviae eating: A nitrogen meal? *J. Insect Physiol.* **46:** 605–610.

Moskalyk, L. A., Oo, M. M., Jacobs-Lorena, M. 1996. Peritrophic matrix proteins of *Anopheles gambiae* and *Aedes aegypti*. *Insect Mol. Biol.* **5:** 261–268.

Nassel, D. R. 1999. Tachykinin-related peptides in invertebrates: A review. *Peptides* **20:** 141–158.

Noriega, F. G., Colonna, A. E., Wells, M. A. 1999. Increase in the size of the amino acid pool is sufficient to activate translation of early trypsin mRNA in *Aedes aegypti* midgut. *Insect Biochem. Mol. Biol.* **29:** 243–247.

Noriega, F. G., Edgar, K. A., Goodman, W. G., Shah, D. K., Wells, M. A. 2001. Neuroendocrine factors affecting the steady-state levels of early trypsin mRNA in *Aedes aegypti*. *J. Insect Physiol.* **47:** 515–522.

Noriega, F. G., Pennington, J. E., Barillas-Mury, C., Wang, X. Y., Wells, M. A. 1996. *Aedes aegypti* midgut early trypsin is post-transcriptionally regulated by blood feeding. *Insect. Mol. Biol.* **5:** 25–29.

Noriega, F. G., Shah, D. K., Wells, M. A. 1997. Juvenile hormone controls early trypsin gene transcription in the midgut of *Aedes aegypti*. *Insect Mol. Biol.* **6:** 63–66.

Pabla, N., Lange, A. B. 1999. The distribution and myotropic activity of locustatachykinin-like peptides in locust midgut. *Peptides* **20:** 1159–1167.

Phillips, J. E., Hanrahan, J., Chamberlin, A., Thompson, B. 1986. Mechanisms and control of reabsorption in insect hindgut. *Adv. Insect Physiol.* **19:** 329–422.

Reichwald, K., Unnithan, G. C., Davis, N. T., Agricola, H., Feyereisen, R. 1994. Expression of the allatostatin gene in endocrine cells of the cockroach midgut. *Proc. Natl. Acad. Sci. USA* **91:** 11,894–11,898.

Ribeiro, J. M. 1987. Role of saliva in blood-feeding by arthropods. *Annu. Rev. Entomol.* **32:** 463–478.

Ribeiro, J. M. C. 1989. Vector saliva and its role in parasite transmission. *Exp. Parasitol.* **69:** 104–106.

Richards, A. G., Richards, P. A. 1977. The peritrophic membranes of insects. *Annu. Rev. Entomol.* **22:** 219–240.

Shen, Z., Jacobs-Lorena, M. 1998. A type I peritrophic matrix protein from the malaria vector *Anopheles gambiae* binds to chitin. Cloning, expression, and characterization. *J. Biol. Chem.* **273:** 17,665–17,670.

Siviter, R. J., Coast, G. M., Winther, A. M., Nachman, R. J., Taylor, C. A., Shirras, A. D., Coates, D., Isaac, R. E., Nassel, D. R. 2000. Expression and functional characterization of a *Drosophila* neuropeptide precursor with homology to mammalian preprotachykinin A. *J. Biol. Chem.* **275:** 23,273–23,280.

Slaytor, M. 1992. Cellulose digestion in termites and cockroaches: What role do symbionts play? *Comp. Biochem. Physiol. B* **103:** 775–784.

Smith, A. F., Tsuchida, K., Hanneman, E., Suzuki, T. C., Wells, M. A. 1992. Isolation, characterization, and cDNA sequence of two fatty acid- binding proteins from the midgut of *Manduca sexta* larvae. *J. Biol. Chem.* **267:** 380–384.

Stanley-Samuelson, D., Jurenka, R. A., Cripps, C., Bloomquist, G. J., De Renobales, M. 1988. Fatty acids in insects: Composition, metabolism, and biological significance. *Arch. Insect Biochem. Physiol.* **9:** 1–33.

Stanley-Samuelson, D. W., Dadd, R. H. 1983. Long-chain polyunsaturated fatty acids: Patterns of occurence in insects. *Insect Biochem.* **13:** 549–558.

Stanley-Samuelson, D. W., Pedibhotla, V. K. 1996. What can we learn from prostaglandins and related eicosanoids in insects. *Insect Biochem. Mol. Biol.* **26:** 223–234.

Stark, K. R., James, A. A. 1996. Salivary gland anticoagulants in culicine and anopheline mosquitoes (Diptera:Culicidae). *J. Med. Entomol.* **33:** 645–650.

Tellam, R. L., Eisemann, C. 2000. Chitin is only a minor component of the peritrophic matrix from larvae of *Lucilia cuprina. Insect Biochem. Mol. Biol.* **30:** 1189–1201.

Tellam, R. L., Eisemann, C., Casu, R., Pearson, R. 2000. The intrinsic peritrophic matrix protein peritrophin-95 from larvae of *Lucilia cuprina* is synthesised in the cardia and regurgitated or excreted as a highly immunogenic protein. *Insect Biochem. Mol. Biol.* **30:** 9–17.

Tellam, R. L., Wijffels, G., Willadsen, P. 1999. Peritrophic matrix proteins. *Insect Biochem. Mol. Biol.* **29:** 87–101.

Terra, W., Ferreira, C. 1981. The physiological role of the peritrophic membrane and trehalase: Digestive enzymes in the midgut and excreta of starved larvae of *rhynchosciara. J. Insect Physiol.* **2:** 325–331.

Terra, W. R. 1990. Evolution of digestive systems of insects. *Annu. Rev. Entomol.* **35:** 181–200.

Terra, W. R., Ferreira, C. 1994. Insect digestive enzymes: Properties, compartmentalization and function. *Comp. Biochem. Physiol. B* **103:** 775–784.

Treherne, J. E. 1967. Gut absorption. *Annu. Rev. Entomol.* **12:** 43–58.

Turunen, S. 1979. Digestion and absorption of lipids in insects. *Comp. Biochem. Physiol. A* **63:** 455–460.

Turunen, S. 1985. Absorption. In *Comprehensive insect physiology, biochemistry, and pharmacology* (G. A. Kerkut, L. I. Gilbert, Eds.), vol. 4, pp. 241–277. Pergamon Press, Oxford.

Turunen, S. 1990. Plant leaf lipids as fatty acid sources in two species of Lepidoptera. *J. Insect Physiol.* **36:** 665–672.

Turunen, S. 1993. Metabolic pathways in the midgut epithelium of *Pieris brassicae* during carbohydrate and lipid assimilation. *Insect Biochem. Mol. Biol.* **23:** 681–689.

Turunen, S., Crailsheim, K. 1996. Lipid and sugar absorption. In *Biology of the insect midgut* (M. J. Lehane, P. F. Billingsley, Eds.), pp. 293–320. Chapman and Hall, London.

Veenstra, J. A., Lambrou, G. 1995. Isolation of a novel RFamide peptide from the midgut of the American cockroach, *Periplaneta americana. Biochem. Biophys. Res. Commun.* **213:** 519–524.

Wijffels, G., Hughes, S., Gough, J., Allen, J., Don, A., Marshall, K., Kay, B., Kemp, D. 1999. Peritrophins of adult dipteran ectoparasites and their evaluation as vaccine antigens. *Int. J. Parasitol.* **29:** 1363–1377.

Wolfersberger, M. G. 1996. Localization of amino acid absorption systems in the larval midgut of the tobacco hornworm *Manduca sexta. J. Insect Physiol.* **42:** 975–982.

Yamauchi, Y., Hoeffer, C., Yamamoto, A., Takeda, H., Ishihara, R., Maekawa, H., Sato, R., Su-Il, S. Sumida, M., Wells, M. A., Tsuchida, K. 2000. cDNA and deduced amino acid sequences of apolipophorin-IIIs from *Bombyx mori* and *Bombyx mandarina*. *Arch. Insect Biochem. Physiol.* **43:** 16–21.

Metabolism

Andersen, S. O., Hojrup, P., Roepstorff, P. 1995. Insect cuticular proteins. *Insect Biochem. Mol. Biol.* **25:** 153–176.

Arrese, E., Gazard, J., Flowers, M., Soulages, J., Wells, M. 2001. Diacylglycerol transport in the insect fat body. Evidence of involvement of lipid droplets and the cytosolic fraction. *J. Lipid Res.* **42:** 225–234.

Arrese, E. L., Canavoso, L. E., Jouni, Z. E., Pennington, J. E., Tsuchida, K., Wells, M. A. 2001. Lipid storage and mobilization in insects: Current status and future directions. *Insect Biochem. Mol. Biol.* **31:** 7–17.

Arrese, E. L., Rojas-Rivas, B. I., Wells, M. A. 1996. The use of decapitated insects to study lipid mobilization in adult *Manduca sexta*: Effects of adipokinetic hormone and trehalose on fat body lipase activity. *Insect Biochem. Mol. Biol.* **26:** 775–782.

Arrese, E. L., Wells, M. A. 1997. Adipokinetic hormone-induced lipolysis in the fat body of an insect, *Manduca sexta*: Synthesis of SN-1,2-diacylglycerols. *J. Lipid Res.* **38:** 68–76.

Atella, G. C., Gondim, K. C., Masuda, H. 1992. Transfer of phospholipids from fat body to lipophorin in *Rhodnius prolixus*. *Arch. Insect Biochem. Physiol.* **19:** 133–144.

Atella, G. C., Arruda, M. A., Masuda, H., Gondim, K. C. 2000. Fatty acid incorporation by *Rhodnius prolixus* midgut. *Arch. Insect Biochem. Physiol.* **43:** 99–107.

Becker, A., Schloder, P., Steele, J. E., Wegener, G. 1996. The regulation of trehalose metabolism in insects. *Experientia* **52:** 433–439.

Blacklock, B. J., Ryan, R. O. 1994. Hemolymph lipid transport. *Insect Biochem. Mol. Biol.* **24:** 855–873.

Blomquist, G. J., Borgeson, C. E., Vundla, M. 1991. Polyunsaturated fatty acids and eicosanoids in insects. *Insect Biochem.* **21:** 99–106.

Blomquist, G. J., Dwyer, L. A., Chu, A. J., Ryan, R. O., de Renobales, M. 1982. Biosynthesis of linoleic acid in a termite, cockroach and cricket. *Insect Biochem.* **12:** 349–353.

Burmester, T., Massey, Jr., H. C., Zakharkin, S. O., Benes, H. 1998. The evolution of hexamerins and the phylogeny of insects. *J. Mol. Evol.* **47:** 93–108.

Bursell, E. 1977. Synthesis of proline by fat body of the tsetse fly (*Glossina morsitans*): Metabolic pathways. *Insect Biochem.* **7:** 427–434.

Canavoso, L. E., Jouni, Z. E., Karnas, K. J., Pennington, J. E., Wells, M. A. 2001. Fat metabolism in insects. *Annu. Rev. Nutr.* **21:** 23–46.

Canavoso, L. E., Wells, M. A. 2001. Role of lipid transfer particle in delivery of diacylglycerol from midgut to lipophorin in larval *Manduca sexta*. *Insect. Biochem. Mol. Biol.* **31:** 783–790.

Candy, D. J. 1985. Intermediary metabolism. In *Comprehensive insect physiology biochemistry and pharmacology* (G. A. Kerkut, L. E. Gilbert, Eds.), vol. 10, pp. 1–41. Pergamon Press, New York.

Candy, D. J., Hall, L. J., Spencer, I. M. 1976. The metabolism of glycerol in the locust *Schistocerca gregaria* during flight. *J. Insect Physiol.* **22:** 583–587.

Chapman, R. F. 1998. *The insects: Structure and function*. Cambridge University Press, Cambridge.

Chen, A. C. 1987. Chitin metabolism. *Arch. Insect Biochem. Physiol.* **6:** 267–277.

Cleveland, L. R. 1924. The physiology and symbiotic relationships between the intestinal protozoa of termites and their host, with special reference to *Reticulitermes flavipes* Kollar. *Biol. Bull.* **46:** 117–227.

Cleveland, L. R., Burke Jr., A. W., Karlson, P. 1960. Ecdysone induced modifications in the sexual cycles of the protozoa of *Cryptocercus*. *J. Protozool.* **7:** 229–239.

de Renobales, M., Cripps, C., Stanley-Samuelson, D. W., Jurenka, R. A., Blomquist, G. J. 1987. Biosynthesis of linoleic acid in insects. *Trends Biochem. Sci.* **12:** 364–366.

Friedman, S. 1978. Treholose regulation, one aspect of metabolic homeostasis. *Annu. Rev. Entomol.* **23:** 389–407.

Gilmour, D. 1961. *The biochemistry of insects.* Academic Press, New York.

Golodne, D. M., Van Heusden, M. C., Gondim, K. C., Masuda, H., Atella, G. C. 2001. Purification and characterization of a lipid transfer particle in *Rhodnius prolixus*: Phospholipid transfer. *Insect Biochem. Mol. Biol.* **31:** 563–571.

Haunerland, N. H. 1996. Insect storage proteins: Gene families and receptors. *Insect Biochem. Mol. Biol.* **26:** 755–765.

Howard, R. W., Stanley, D. W. 1999. The tie that binds: Eicosanoids in invertebrate biology. *Ann. Entomol. Soc. Am.* **92:** 880–890.

Kanost, M. R., Kawooya, J. K., Law, J. H., Ryan, R. O., van Heusden, M. C., Ziegler, R. 1990. Insect hemolymph proteins. *Adv. Insect Physiol.* **22:** 229–396.

Kawooya, J. K., Law, J. H. 1988. Role of lipophorin in lipid transport to the insect egg. *J. Biol. Chem.* **263:** 8748–8753.

Law, J. H., Ribeiro, J. M. C., Wells, M. A. 1992. Biochemical insights derived from insect diversity. *Annu. Rev. Biochem.* **61:** 87–111.

Locke, M., Nichol, H. 1992. Iron economy in insects: Transport, metabolism and storage. *Annu. Rev. Entomol.* **37:** 195–215.

Oudejans, R. C., Harthoorn, L. F., Diederen, J. H., van der Horst, D. J. 1999. Adipokinetic hormones. Coupling between biosynthesis and release. *Ann. NY Acad. Sci.* **897:** 291–299.

Oudejans, R. C., Vroemen, S. F., Jansen, R. F., van der Horst, D. J. 1996. Locust adipokinetic hormones: carrier-independent transport and differential inactivation at physiological concentrations during rest and flight. *Proc. Natl. Acad. Sci. USA* **93:** 8654–8659.

Ryan, R. O. 1990. Dynamics of lipophorin metabolism. *J. Lipid Res.* **31:** 1725–1739.

Ryan, R. O., Schmidt, J. O., Law, J. H. 1984. Chemical and immunological properties of lipophorins from seven insect orders. *Arch. Insect Biochem. Physiol.* **1:** 375–383.

Ryan, R. O., van der Horst, D. J. 2000. Lipid transport biochemistry and its role in energy production. *Annu. Rev. Entomol.* **45:** 233–260.

Shapiro, J. P., Keim, P. S., Law, J. H. 1984. Structural studies on lipophorin: An insect lipoprotein. *J. Biol. Chem.* **259:** 3680–3685.

Shapiro, J. P., Law, J. H., Wells, M. A. 1988. Lipid transport in insects. *Annu. Rev. Entomol.* **33:** 297–318.

Siegert, K. J. W. M. 1994. Adipokinetic hormone and developmental changes of the response of a fat body glycogen phosphorylase in *Manduca sexta*. *J. Insect Physiol.* **40:** 759–764.

Stanley-Samuelson, D., Jurenka, R. A., Cripps, C., Bloomquist, G. J., de Renobales, M. 1988. Fatty acids in insects: Composition, metabolism, and biological significance. *Arch. Insect Biochem. Physiol.* **9:** 1–33.

Steele, J. E. 1982. Glycogen phosphorylase in insects. *Insect Biochem.* **12:** 131–147.

Steele, J. E. 1985. Control of metabolic processes. In *Comprehensive insect physiology, biochemistry and pharmacology* (G. A. Kerkut, L. I. Gilbert, Eds.), vol. 8, pp. 99–145. Pergamon Press, Oxford.

Stevenson, E., Wyatt, G. R. 1964. Glycogen phosphorylase and its activation in silkmoth fat body. *Arch. Biochem. Biophys.* **108:** 420–429.

Soulages, J. L., Arrese, E. L. 2000. Dynamics and hydration of the alpha-helices of apolipophorin III. *J. Biol. Chem.* **275:** 17,501–17,509.

Soulages, J. L., Pennington, J., Bendavid, O., Wells, M. A. 1998. Role of glycosylation in the lipid-binding activity of the exchangeable apolipoprotein, apolipophorin-III. Biochem. *Biophys. Res. Commun.* **243:** 372–376.

Soulages, J. L., Salamon, Z., Wells, M. A., Tollin, G. 1995. Low concentrations of diacylglycerol promote the binding of apolipophorin iii to a phospholipid bilayer: A surface plasmon resonance spectroscopy study. *Proc. Natl. Acad. Sci. USA* **92:** 5650–5654.

Soulages, J. L., Van Antwerpen, R., Wells, M. A. 1996. Role of diacylglycerol and apolipophorin-iii in regulation of physiochemical properties of the lipophorin surface: Metabolic implications. *Biochemistry* **35:** 5191–5198.

Soulages, J. L., Wells, M. A. 1994. Lipophorin: The structure of an insect lipoprotein and its role in lipid transport in insects. *Adv. Protein Chem.* **45:** 371–415.

Svoboda, J. A. 1999. Variability of metabolism and function of sterols in insects. *Crit. Rev. Biochem. Mol. Biol.* **34:** 49–57.

Svoboda, J. A., Feldlaufer, M. F. 1991. Neutral sterol metabolism in insects. *Lipids* **26:** 614–618.

Telfer, W. H., Kunkel, J. G. 1991. The function and evolution of insect storage hexamers. *Annu. Rev. Entomol.* **36:** 205–228.

Tsuchida, K., Soulages, J. L., Moribayashi, A., Suzuki, K., Maekawa, H., Wells, M. A. 1997. Purification and properties of a lipid transfer particle from *Bombyx mori*: Comparison to the lipid transfer particle from *Manduca sexta*. *Biochim. Biophys. Acta* **1337:** 57–65.

Tsuchida, K., Wells, M. A. 1990. Isolation and characterization of a lipoprotein receptor from the fat body of an insect, *Manduca sexta*. *J. Biol. Chem.* **265:** 5761–5767.

Weeda, E., 1981. Hormonal regulation of proline synthesis and glucose release in the fat body of the Colorado potato beetle *Leptinotarsa decemlineata*. *J. Insect Physiol.* **27:** 411–417.

Weeda, E., Koomanschap, A. B., De Kort, C. A. D., Beenakkers, A. M. Th. 1980. Proline synthesis in fat body of *Leptinotarsa decemlineata* Say. *Insect Biochem.* **10:** 631–636.

Wyatt, G. R. 1967. The biochemistry of sugars and polysaccharides in insects. *Adv. Insect Physiol.* **4:** 287–360.

CHAPTER 7

Circulatory Systems

REQUIREMENTS OF A CIRCULATORY SYSTEM

All cells need to exchange materials with their environments. To function properly, cells must take up nutrients, discard metabolic wastes, and receive chemical messages from hormones. These substances can pass through the cell membrane only if they are dissolved in water, and for this reason, every living cell is required to be surrounded by an aqueous medium. In single-celled animals, this exchange of materials occurs by simple diffusion or active transport through the cell membrane, because the large ratio of surface area to volume in these small organisms provides an area sufficient for the exchange of materials by diffusion. In larger multicellular animals, including humans and other vertebrates, diffusion alone could never accomplish this exchange. Nutrients that are absorbed through the digestive tract would never reach cells in fingers or toes if they had to travel through the body cavity by diffusion. In vertebrates, a closed system of internal transport consisting of the heart, blood vessels, and the enclosed blood is necessary to bathe cells and transport materials to them throughout the body.

Although insects are also multicellular animals, they are small enough to allow diffusion to serve as a mechanism of metabolic exchange. The extensive tracheal system in insects carries oxygen to the cells, and except for some chironomid fly larvae that use hemoglobin in their hemolymph, the circulatory system has a limited function in oxygen transport. As a medium that is used primarily for the transport of chemical agents, its role is far less demanding than for vertebrate blood, which must also transport oxygen. The most important roles of insect hemolymph are to serve as a medium that bathes cells and transfers substances to and from them, as a reservoir of water and metabolic substances, as a medium for cellular and humoral defense, and, in soft-bodied insects, to provide the necessary hydrostatic pressure for molting and maintenance of body shape.

STRUCTURE OF THE INSECT CIRCULATORY SYSTEM

The body cavity of insects consists of a series of sinuses that are collectively referred to as the **hemocele.** The only closed portion of the circulatory system is the **dorsal vessel,** a tube that extends the length of the body from the posterior end of the abdomen into the head (Figs. 7.1A and 7.1B). Within the head, it passes under the brain just above the digestive tract and then opens anteriorly. The dorsal vessel is not a uniform tube, but consists of two segments, the **heart** and **aorta,** which are formed during embryogenesis from the cardioblasts of the embryonic mesoderm.

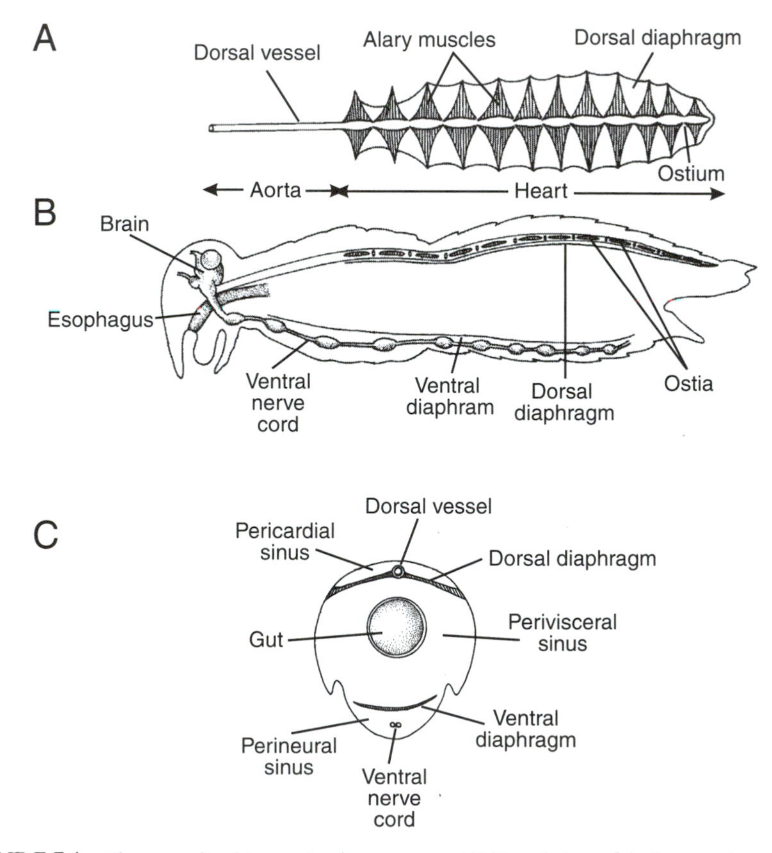

FIGURE 7.1 The generalized insect circulatory system. (A) Dorsal view of the heart and aorta. (B) Longitudinal section of the insect body, showing the dorsal and ventral diaphragms. (C) Cross section through the abdomen, showing the sinuses created by the diaphragms. From Romoser and Stoffolano (1998). Reprinted with permission.

The segment of the dorsal vessel that is known as the heart makes up the posterior portion that is found mainly within the abdomen. The heart is comprised of a series of segments or chambers that contain paired lateral openings called **ostia** (Fig. 7.2). Incurrent ostia are simple valves that allow hemolymph to enter the lumen of the heart from the pericardial sinus at diastole but prevent its outflow during systole directly into the body cavity. Excurrent ostia have no valves and sometimes open caudally from the posterior of the heart in some Diptera or laterally in the more primitive Thysanura, Orthoptera, and Plecoptera. In a few insects, internal valves are present that prevent the backward flow of hemolymph between the chambers within the heart as it pumps the hemolymph forward to the head. These are generally rare, however, although the appearance of the ostia in many insects may often suggest the presence of valved chambers.

The heart is supported in the hemocele by connective tissue strands and from 2 to 12 pairs of fanned-shaped **alary muscles** that along with connective tissue form the dorsal diaphragm that lies just below the heart. The diaphragm extends through the abdomen and creates a **pericardial sinus** in the compartment that surrounds the dorsal vessel (Fig. 7.1C). Large **pericardial cells** are usually located on the dorsal diaphragm and function as phagocytic organs to filter out large particles from the hemolymph. In some insects, segmental blood vessels may also extend laterally from the heart (Fig. 7.3).

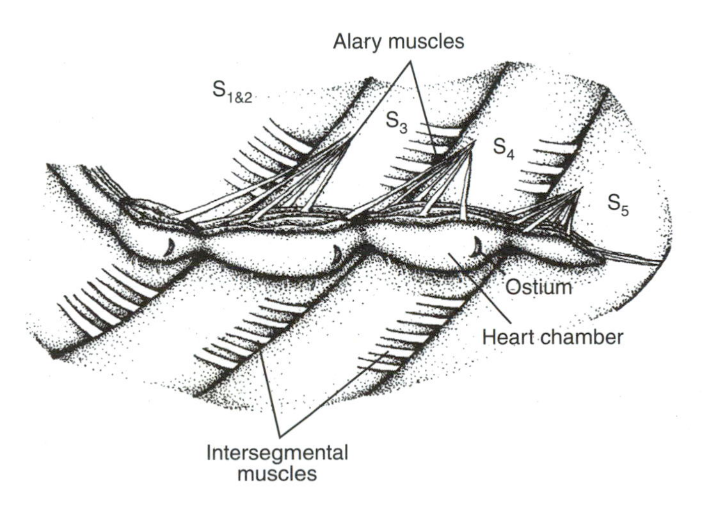

FIGURE 7.2 The ostia and alary muscles of the heart. From Cook and Meola (1983). Reprinted with permission.

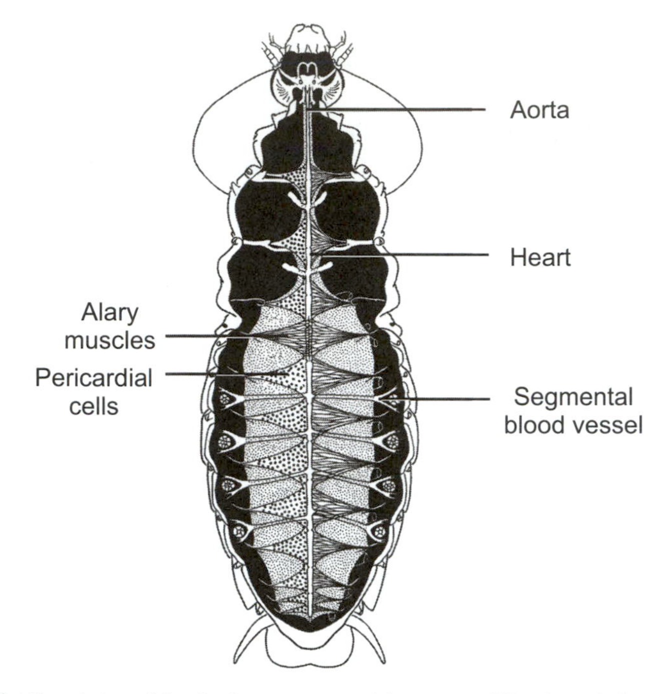

FIGURE 7.3 Dorsal view of the circulatory system and the segmental blood vessels that extend laterally from the heart. From Nutting, W.L. (1951). A comparative anatomical study of the heart and accessory structures of the orthopteroid insects. *Journal of Morphology* **89:** 501–598. Reprinted by permission of Wiley-Liss, Inc., a subsidiary of John Wiley & Sons, Inc.

The aorta, the anterior portion of the dorsal vessel, extends from the thorax into the head. It is a simple unbranched tube that is thinner than the heart and lacks ostia. It is often attached to the brain and pharynx by connective tissue, passing underneath the brain and opening behind the pharynx. In most adult Lepidoptera and Hymenoptera, the aorta loops through the throracic flight muscles before it enters the head. In some higher Diptera, a cephalic pulsatile organ is also present in the posterior region of the head that facilitates the distribution of hemolymph into either the head or the thorax.

A more advanced feature that further compartmentalizes the hemocele is a **ventral diaphragm,** a feature generally absent in the most primitive insect orders. This diaphragm is located just above the ventral nerve cord in the abdomen and divides the hemocele into a **perineural sinus** (Fig. 7.1C), with the remaining portion of the hemocele that surrounds the gut termed the **perivisceral sinus**. Along with the dorsal diaphragm, undulations of the ventral diaphragm can control the distribution of hemolymph within the compartments of the hemocele. Higher dipterans lack a ventral diaphragm but have large tracheal air sacs that partition the hemocele so that a smaller volume of hemolymph can function more economically when a reduced weight is required for flight.

Accessory Pulsatile Organs

One problem with an open circulatory system is that without a directional flow of hemolymph, outlying dead-end structures such as legs, antennae, and wings have difficulty getting circulating fluid. The dorsal vessel pumps hemolymph only to the head where it then must diffuse backward passively through the hemocele, and diffusion alone is insufficient to provide the distant cells of these appendages with nutrients. To deal with this problem, special **accessory pulsatile organs** at the base of each of these structures channel the hemolymph into them. A large number of these accessory pulsatile organs may be present and are completely separate from the functioning of the dorsal vessel.

Accessory pulsatile organs that are present at the bases of the legs pump hemolymph into a ventral sinus and out of a dorsal sinus created by an internal septum that divides them internally to maintain a directional flow (Fig. 7.4). In locusts, the mesothoracic legs contain a diaphragm between the trochanter and femur that is moved by muscles, and in coordination with changes in the size of tracheal air sacs during respiration, hemolymph is pumped into the leg sinus that is created and divided into channels by a septum (Fig. 7.5). The pulsatile organ at the base of the wings in some insects is a diverticulum of the dorsal vessel that creates a sac when its volume is increased that draws blood out of the wings.

Antennal pulsatile organs consist of a noncontractile sac connected to a long blood vessel that reaches to the tip of the antenna. More advanced insects retain a muscle that compresses the sacs and pumps hemolymph through the antenna (Fig. 7.6). In Diplura, extensions of the aorta reach into the antennae and cerci

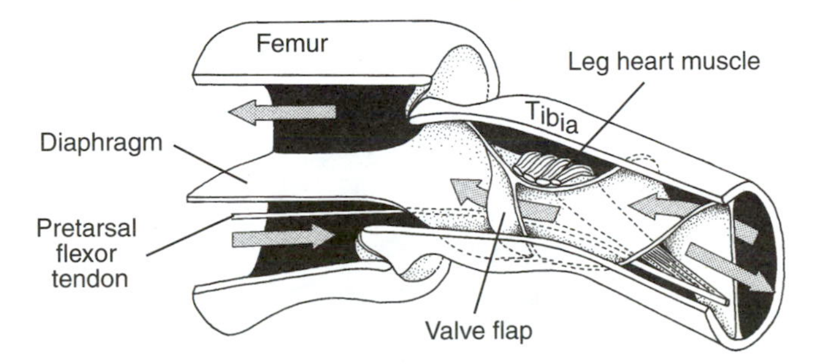

FIGURE 7.4 An accessory pulsatile organ in the insect leg. Arrows show the direction of hemolymph flow. From Hantschk (1991). Reprinted with permission.

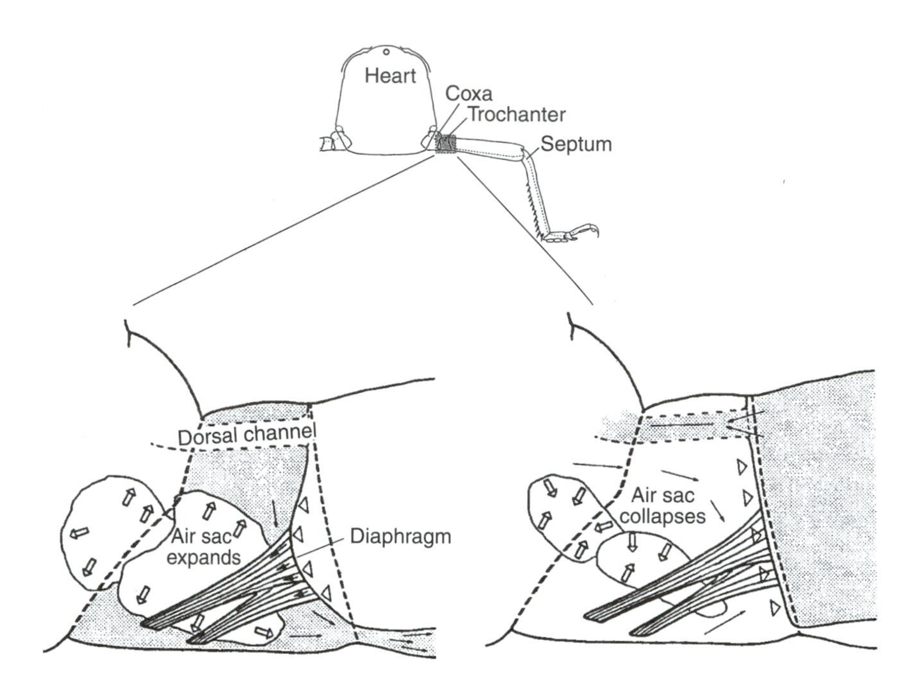

FIGURE 7.5 A tracheal air sac that regulates the movement of hemolymph into the leg. From Hustert (1999). Reprinted with permission.

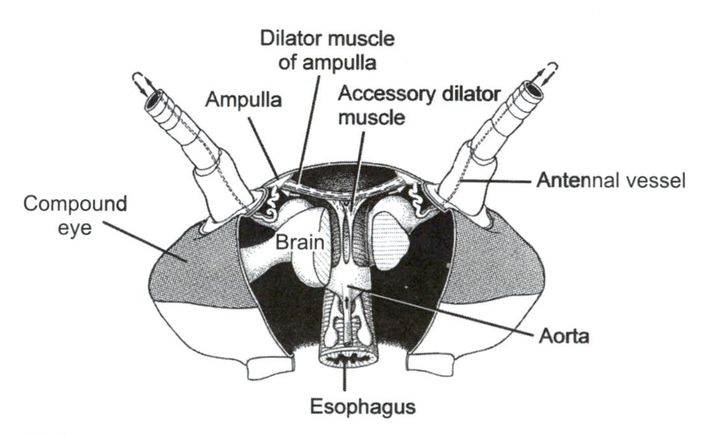

FIGURE 7.6 An ampulla and its dilator muscles that pumps hemolymph into the antennae. From Pass (1985). Reprinted with permission.

(Fig. 7.7). The contractions of these accessory pulsatile organs are usually myogenic and originate in the muscles themselves, but they may be modulated by the nervous system.

Heartbeat and Its Regulation

The contraction of the chambers of the heart results from the contractions of the muscles that lie within its wall. A wave of peristaltic contractions moves forward along the entire length of the heart, pushing hemolymph forward that had entered the heart though the ostia during relaxation (Fig. 7.8). The rate of heart contraction varies between species and between stages of development within a species, making it impossible to generalize the rate throughout the insect orders. Environmental factors such as temperature affect heartbeat, with higher temperatures causing increased rates. At extreme high and low temperatures, the heartbeat may stop entirely. It is not uncommon, especially in pupae, for the heart to stop beating for several seconds and even to reverse its contraction so that peristaltic waves move from the front to the back. These backward contractions may result when the pressure in the anterior of the insect is high or when excurrent ostia become blocked by the histolyzed tissues in the pupa. The heart of the adult blow fly, *Calliphora vomitoria,* has been observed to beat forward as fast as 376 beats per minute and rearward at about 175 beats per minute. The contraction of accessory pulsatile organs is generally slower. Although the aorta may show rhythmic pulsations from hemolymph being pumped through it, it does not itself beat.

Like the accessory pulsatile organs, the contractions of the heart are regulated myogenically but can be influenced by nervous and endocrine stimuli. In many

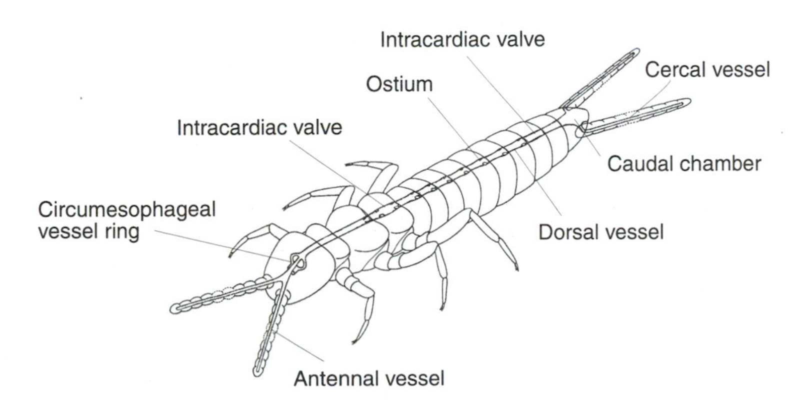

FIGURE 7.7 The extension of the circulatory system into the antennae of diplurans. From Gereben-Krenn and Pass (1999). Reprinted with permission.

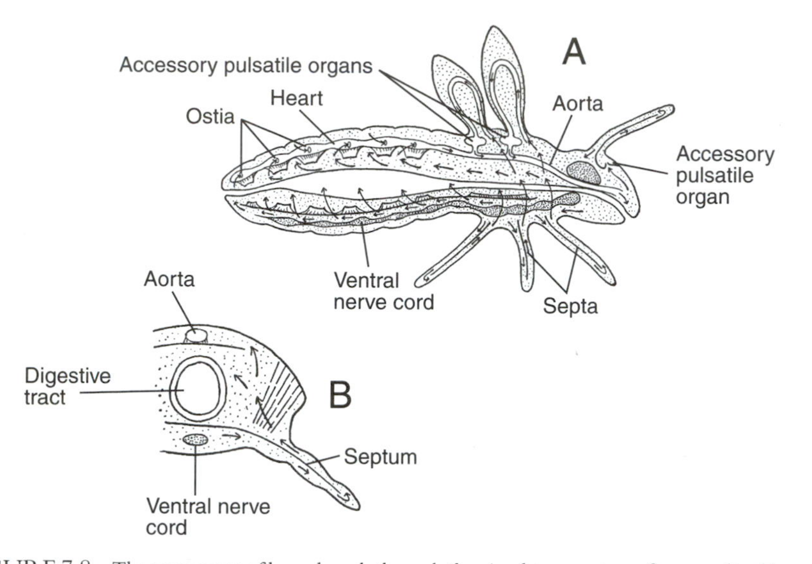

FIGURE 7.8 The movement of hemolymph through the circulatory system of a generalized insect. (A) A sagittal section showing the hemolymph moving forward through the dorsal vessel and backward through the hemocele. Dorsal and ventral diaphragms and septa in appendages create a directional flow. (B) The movement of hemolymph through an appendage. From Wigglesworth (1974). Reprinted with permission.

insects, the dorsal vessel has no known innervation. In others, the aorta is innervated by a median nerve from the hypocerebral ganglion of the stomatogastric nervous system. Segmental nerves arising from the ventral nerve cord may also innervate the alary muscles and the heart. The alary muscles are well innervated and may contribute to the heartbeat, especially during diastole.

Endocrine factors have a large influence on the rate of the heartbeat. **Cardioacceleratory peptides (CAPs)**, which are produced by the corpus cardiacum and ventral nerve cords in some insects, modulate heartbeat. In *Manduca sexta,* cardioacceleratory peptides stimulate the heart to contract immediately after adult emergence and during flight. This increase in hemolymph pumping after adult emergence may aid in wing inflation. During flight, the increased rate may facilitate hemolymph transfer between the abdomen and thorax so the flight muscles do not overheat with activity. CAP axons also innervate the hindgut and may be responsible for hindgut myotropic activity. CAP causes an increase in hindgut contractions that are associated with the gut purging in *Manduca* larvae that occurs during their wandering behavior just prior to pupation.

Neurosecretory cells may be contained within the aorta or fused with its walls. Neurosecretory axons arise from cell bodies in the brain and often terminate on the aorta wall where they may form a neurohemal organ. The heart may also be secretory itself. The hearts of the lepidopteran *Calpodes* and the Hemipteran

Rhodnius synthesize peptides of unknown function that may be released into the hemolymph.

Hemolymph Composition

The hemolymph is the major extracellular fluid in insects. It makes up from 15 to 75% of the volume of the insect, varying significantly with species and individual physiological state. The hemolymph is the major transport medium for the exchange of materials between cells, such as hormones, waste materials, and nutrients. Through its regulation of ionic and chemical composition, it maintains the proper internal environment for cells as an extracellular extension of intracellular fluids. In this role, it contributes to the ability of the insect to live at both high and low temperatures. It serves as a major compartment and storage reserve for water. The hemolymph is far from being a static reservoir for the storage of metabolites, however. It is a dynamic tissue that changes with the changing physiological state of the insect. Finally, it maintains the hydrostatic pressure required to maintain body shape in soft-bodied insects and to facilitate the splitting of the cast skin at ecdysis. The volume of the hemolymph often increases toward the end of each stadium and contributes to the rupture of the old cuticle.

The hemolymph consists of the liquid **plasma** and the cellular **hemocytes.** Plasma composition is variable. It is usually clear but may be colored green or yellow in some insects, reflecting the pigments that are present. Its pH can be variable; it is generally slightly acid but may be alkaline in some species. A variety of soluble components contribute to its total osmotic pressure. In most other animals, the major inorganic components of the body fluid are sodium and chloride but in insect blood the composition can be quite different. In primitive apterygotes, sodium and chloride do indeed appear to be the most important osmotic effectors. Primitive exopterygotes, including Ephemeroptera, Odonata, and Dictyoptera, also largely use sodium and chloride, but with contributions from magnesium, potassium, and calcium. In endopterygotes, such as Diptera, Mecoptera, and Neuroptera, sodium is again an important cation, but chloride is replaced by higher concentrations of amino acids and other organic components (Fig. 7.9). In the endopterygotes Hymenoptera and Lepidoptera, amino acids and other organic molecules play a major role along with potassium but with a reduced involvement of sodium. These ionic differences were once attributed to either phytophagous or carnivorous diets, because plant-feeding insects contained higher levels of potassium and insects feeding on other diets had higher levels of sodium. However, this generalization has numerous contradictions and the relationship between the ionic composition of the hemolymph and basic diet of the insect is not entirely clear.

There are much higher concentrations of free amino acids in insect plasma than in vertebrate blood. As discussed above, these become even more important in

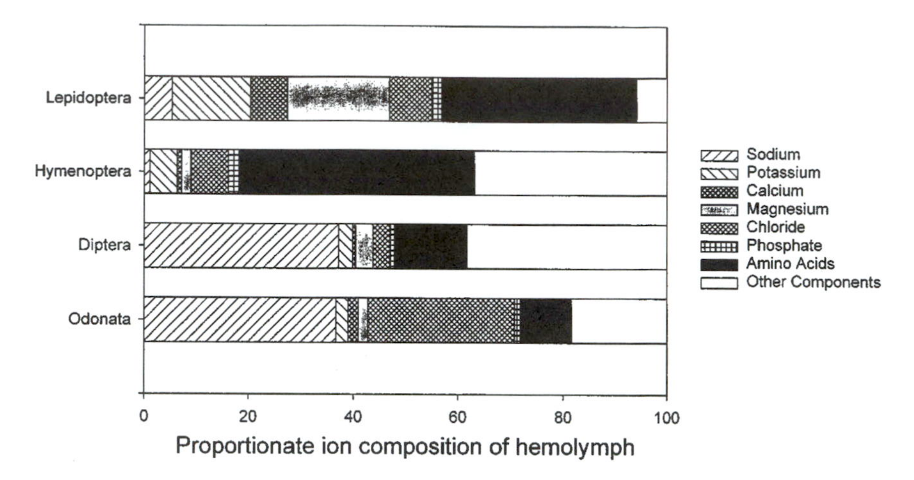

FIGURE 7.9 Components of the hemolymph in four insect orders. Adapted from Sutcliffe (1963). Reprinted with permission.

more evolutionarily advanced groups. Exopterygotes are characterized by lower levels of amino acids than in endopterygotes and show uniform concentrations of the amino acids that are present. In contrast, endopterygotes contain some amino acids in much higher concentrations than others. For example, glutamic acid and proline are often found in significantly higher concentrations in the amino acid pool depending on the physiological state of the particular insect. In the adult tsetse fly, proline, which is used as a substrate for flight, declines during activity and alanine concentrations increase. The concentrations of methionine, glutamic acid, and aspartic acid are correlated with the activity of the silk glands during the development of the silkworm moth. The amino acids leucine and isoleucine tend to be present in lower concentrations in most insects.

Other organic components of the hemolymph include carbohydrates, various citric acid cycle intermediates, uric acid, and soluble proteins. As discussed in Chapter 6, trehalose, an α-1,1 disaccharide of two glucose residues, serves as the major circulating energy source in most insects (Fig. 6.12). It is present at concentrations of 5–50 times higher than glucose, the circulating energy source in vertebrates. This higher level in insects may be one of the compromises necessary to make up for the inefficiency of the circulatory system in distributing these materials by diffusion. The blood sugar in insects must exist in a higher concentration to ensure that it reaches remote areas of the hemocele in sufficient concentration. If glucose were to serve this function as it does in vertebrates, the higher concentration required in the blood would interfere with the uptake of glucose by diffusion through the digestive tract. By using the disaccharide trehalose in the blood, which is usually rare in the diet, high levels of blood sugar can be maintained without interfering with the uptake of glucose.

Soluble proteins in the hemolymph include the vitellogenins, which are yolk proteins produced by the female fat body and that are taken up by the oocytes.

Several enzymes, including esterases, chitinases, and proteases, appear in the hemolymph depending on the developmental stage of the insect.

The hemolymph may be a physical deterrent to predation. Autohemorrhaging, or **reflexive bleeding,** occurs in some insects when they are attacked. Hemolymph that is fortified with defensive terpenoids such as cantharadin may be released outside the body through intersegmental membranes to discourage ants and other insect predators. The loss of hemolymph may be substantial; in chrysomelid beetles, as much as 13% of the wet weight of the larvae can be lost through autohemorrhage. The hemolymph clots immediately and can bind ants in the coagulum that is formed.

Hemocytes

The cellular hemocytes are suspended in the plasma, but may often remain attached to other body tissues rather than circulate with the blood. They originate from embryonic mesodermal tissue and differentiate during embryogenesis into several distinct types. They have a variety of functions, including phagocytosis of foreign particulate matter, encapsulation of multicellular parasites, and coagulation and wound healing after injury and also play an important role in metabolism.

The descriptions of the types of hemocytes present in insects have been varied, perhaps reflecting the procedures used to study them and the differences that may exist between species. For example, hemocytes of different morphologies may be reported when they are observed either through transparent wing veins or in fixed blood smears. Hemocytes are generally classified according to their size, shape, nuclear characteristics, and cytoplasmic inclusions (Fig. 7.10) and can be placed into one of three general categories: **prohemocytes** that give rise to other hemocyte types, **plasmatocytes** that are phagocytic, and **granular hemocytes** that are involved in intermediary metabolism.

Within these general categories, an additional classification of insect hemocytes is also recognized. **Prohemocytes** are the small, round cells that contain a large nucleus and do not undergo any phagocytosis. These cells are believed to be stem cells that postembryonically give rise to other types. **Plasmatocytes** are larger, more ameboid, pleiomorphic cells with a nucleus surrounded by large amounts of cytoplasm. Plasmatocytes are usually among the most abundant of the hemocytes and are frequently engaged in phagocytosis. **Granulocytes** are compact cells with a small nucleus surrounded by a large cytoplasm with abundant granules and may differentiate into the remaining granulocyte types. **Adipohemocytes** are round cells containing a small nucleus surrounded by a large amount of cytoplasm that contains a number of lipid vacuoles. **Spherule cells** are nonmotile and also have large inclusions that may obscure the observation of the small nucleus. **Oenocytoids** are ovoid and variable in size with a small nucleus and a large complex cytoplasm. They are also nonmotile. **Cystocytes** are fragile cells that rapidly degenerate upon fixation. Fixed cells are ovoid with a small nucleus and granular inclusions within the cytoplasm. A possible differentiation pathway is shown in Fig. 7.11.

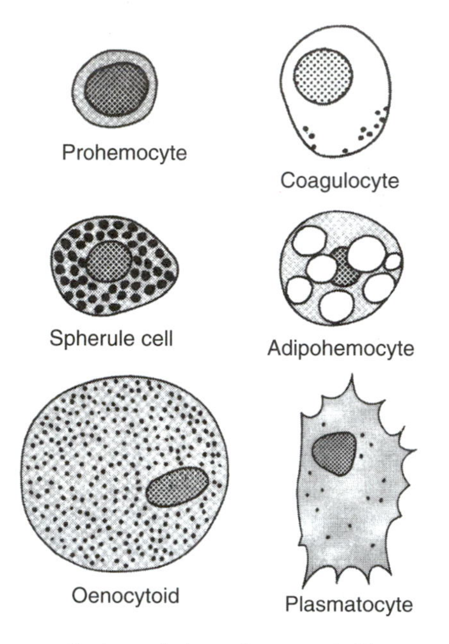

FIGURE 7.10 The generalized morphology of some typical hemocytes that are found in insect hemolymph. From Bursell (1970). Reprinted with permission.

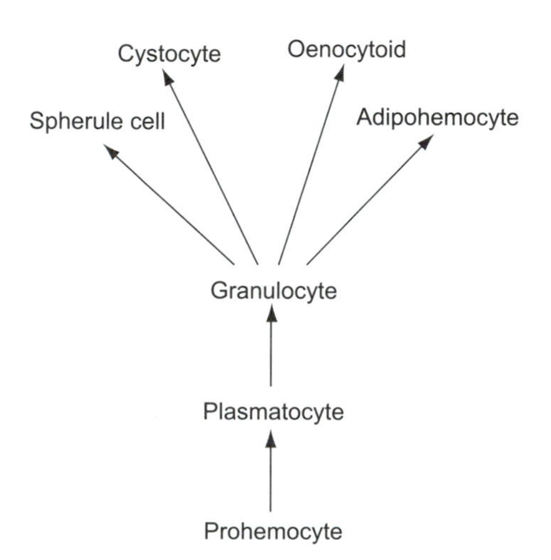

FIGURE 7.11 A possible scheme of hemocyte differentiation. Adapted from Gupta (1985). Reprinted with permission.

Like all other insect cells, hemocytes require a source of oxygen for respiration. However, unlike other insect cells, they are suspended in the hemolymph and thus have difficulty deriving their oxygen directly from the tracheal system. In the larvae of *Calpodes,* tufts of special aerating trachea fill a compartment, the **tokus,** at the tip of the abdomen near the heart (Fig. 7.12). Hemocytes migrate to the tokus, where they become aerated, and are then circulated through the dorsal vessel. This modification of the tracheal system in the tokus thus serves as a lung for the oxygenation of hemocytes.

IMMUNE MECHANISMS IN INSECTS

Most insects live in close proximity to microorganisms that could easily establish infections if lines of defense were not established. The primary barriers to infection are the cuticle and digestive tract that physically exclude potential parasites. Once inside the hemocele after these primary defenses are breached, parasites face additional cellular and humoral immune mechanisms. Insects display a very basic

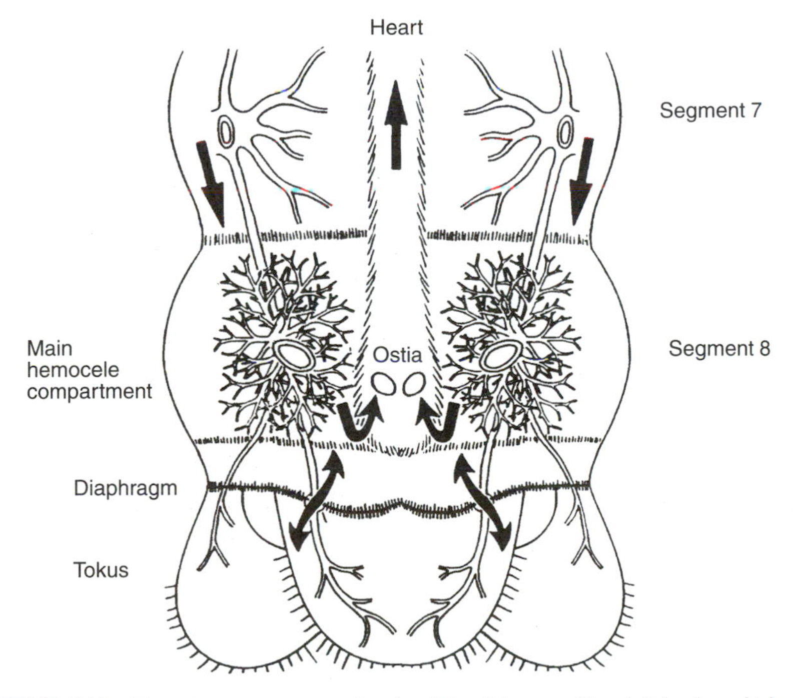

FIGURE 7.12 The tokus, a structure at the tip of the abdomen of larval Calpodes, which oxygenates hemocytes. Reprinted from Locke, M. *Journal of Insect Physiology* **44:** 1–20. With permission from Elsevier Science.

form of immunity, consisting of the ability to distinguish between self and non-self. Cellular and humoral immune mechanisms both identify foreign tissue and act to eliminate it. Insects do not produce the immunoglobulins that mediate immune reactions in vertebrates, but there is an array of fairly specific humoral substances that combat infection.

Cell-Mediated Immunity

Hemocytes are able to recognize some foreign bodies in the hemolymph and phagocytose them. Plasmatocytes and granulocytes are the major cell types involved in this phagocytosis. The process involves first the recognition of foreignness, based on surface receptors on the hemocyte membrane. This recognition is followed by the formation of pseudopodia and the ingestion of the foreign particles within a membrane-bound phagosome. After the phagosome moves to the cell interior, it fuses with a lysosome where the foreign tissue is digested and destroyed by the agents that are released.

The response of hemocytes to larger foreign bodies that cannot be phagocytosed, either those living or nonliving, is to isolate them from the other insect cells by the process of encapsulation. Through encapsulation, multiple layers of hemocytes wall off the foreign object and prevent it from contacting host cells and sources of oxygen and nourishment. The plasmatocytes and granulocytes are also involved in this defensive response which begins with the recognition of foreignness of an invading body of cells. Once recognized, the invader triggers the release of chemotactic aggregation factors from granulocytes that bring more plasmatocytes to the site. Initially surrounding the foreign body, the plasmatocytes flatten out and die. As more plasmatocytes attach to the site, they also flatten but remain alive, walling off the body with a layer of cells as deep as 50 or more. The recruitment of hemocytes ends when the capsule that is formed becomes coated with glycoaminoglycans similar to that of the basement membrane that covers all tissue surfaces in the hemocele. Some plasmatocytes contain the enzyme polyphenoloxidase and deposit the melanin they synthesize, further walling off the invader. A humoral encapsulation is also possible in which melanin is deposited on the foreign surface in the absence of participation by hemocytes. This mechanism has been reported to occur in only a few dipterans that have relatively small populations of hemocytes and is a fast and efficient way to deal with invading bacteria and fungi.

Nodule formation may also trap large numbers of invading bacteria. Nodules consist of aggregations of hemocytes that produce extracellular material that forms a matrix to catch large numbers of bacteria that are too large to be phagocytized. Larger nodules may eventually also be encapsulated.

Nephrocytes are mesodermal cells that are found throughout the hemocele that are able to sequester high-molecular-weight colloids but not bacteria. Included in this category of cells are the **pericardial cells,** the best known of the nephrocytes. They are primarily located on each side of the heart, attached to the dorsal vessel and alary muscles by connective threads, but are sometimes numerous around the fat body as well. These cells absorb chemicals by pinocytosis and return the degraded substances to the hemolymph. They contain numerous granular inclusions. In addition to this detoxification function, there is evidence that the pericardial cells also synthesize and secrete hemolymph proteins.

Humoral Immunity

A humoral mechanism exists in addition to the cell-mediated immunity from hemocytes. The fat body may also synthesize inducible antibacterial peptides that are released into the hemolymph at the time of a microbial infection. It has long been observed that the injection of pathogenic bacteria that have been heat-killed leads to a reduced insect mortality when the live bacteria are injected at various times afterward. The injection of peptidoglycans or lipopolysaccharide from bacterial cell walls into *Manduca* or *Bombyx* larvae induce the same synthesis of various antibacterial hemolymph proteins by the fat body as do whole bacteria, suggesting that the receptors for this induction may exist on the fat body cells.

Once wounded, the injured area around the cuticular site can acquire a dark pigmentation due to the action of hemolymph phenoloxidases. The proteins in the surrounding cuticle thus become sclerotized, which can more effectively block the invasion of microorganisms. **Melanin** also forms from the quinones that are produced and surrounds objects that may have been encapsulated.

Lectins are multimeric proteins consisting of 30- to 40-kDa subunits that are capable of agglutinating vertebrate red blood cells. Their production in insects is induced by injury, and they circulate within the insect plasma of many insects and bind to the carbohydrates in the cell walls of microorganisms. With their multiple binding sites, they can cause an aggregation of lectin-linked cells. The binding of the lectins to the surface of bacteria may be an initial signal for recognition and mobilization by hemocytes.

The first of the antibacterial peptides to be characterized were the **cecropins,** isolated from the pupae of *Hyalophora cecropia,* but theses peptides also have since been isolated from dipterans. The cecropins are low-molecular-weight peptides (<5 kDa with between 35 and 39 amino acids), most of which are basic at the *N*-terminal and give the molecule a net positive charge. They are active against both Gram-positive and Gram-negative bacteria as membrane-active antibiotics that create channels in their lipid bilayer. Cecropins may also be active in insect vectors against the parasites that cause malaria and might possibly contribute to the refractoriness to infection by some mosquitoes.

The **defensins** are small peptides of between 4 and 5 kDa, homologs of which have been identified in molluscs and scorpions and are widely distributed in mammals as well as in insects. There are over 40 varieties of insect defensins, also referred to as **sapecins,** which are active against Gram-positive bacteria. As with cecropins, the insect defensins disrupt the permeability of the bacterial cytoplasmic membrane. Some mosquitoes also produce defensins that can interfere with the potential development of the protozoan malaria parasite. In *Drosophila,* the fat body synthesizes the defensin-like **drosomysin** in response to infection. Drosomysin consists of 44 amino acids with 8 cysteine residues associated with 4 disulfide bridges. It has more of a structural similarity to plant defensins than insect defensins and is similarly active against filamentous fungi but is ineffective against bacteria and yeasts.

Hyalophora cecropia and *Manduca sexta* pupae produce **hemolin,** a 47-kDa protein, in response to infection. Hemolin is among the first proteins to bind to the surface of invading bacteria and forms a protein complex that may initiate the immune response. Its amino acid sequence contains four internal repeats that are characteristic of immunogloblin-like domains and place it in the immunoglobulin superfamily. Hemolin inhibits the aggregation of hemocytes, suggesting that it may affect their adhesive properties during a defensive response and convert them to an activated state.

Other inducible antibacterial peptides can be characterized as belonging to proline-rich or glycine-rich families. The proline-rich peptides are active mostly against Gram-negative bacteria and are relatively small, containing 15–34 residues and a high proportion of proline and arginine. They include the apidaecins from Hymenoptera, the drosocins from *Drosophila,* and the hemipteran metalnikowins. The glycine-rich family of peptides has larger members ranging from 9 to 30 kDa, and contain a high proportion of glycine residues. These are active primarily against Gram-negative bacteria. Included in this family are the various attacins, sarcotoxins, coleoptericins, hemiptericins, diptericins, pyrrhocoricins, gloverins, and hymenoptaecins, named after the insect from which they were isolated.

By its synthesis of a battery of antimicrobial substances that are similar to those produced by vertebrates and plants, the fat body is thus able to respond quickly to infections by microorganisms. These substances have even been proposed as candidates for dealing with the problem of drug resistance in human infections. They rapidly kill a broad array of pathogens, do not affect mammalian cells, and have a mode of action that minimizes the development of resistance.

CONTRIBUTION OF THE CIRCULATORY SYSTEM TO WITHSTANDING TEMPERATURE VARIATIONS

As small terrestrial animals that thrive in temperate climates, insects have successfully adapted to the cold temperatures of winter. Some insects have evolved the ability to migrate to warmer climates when they recognize the signals of seasonal photoperiod. Others remain in the cold, having evolved physiological adaptations to subzero temperatures. Although low temperatures can affect the physical properties of biologically important molecules and the rate of biological processes, the exposure to freezing temperatures itself is not necessarily critical. However, when the low temperatures induce the formation of intracellular ice crystals, physical damage can occur to cell membranes that cannot be repaired and is generally lethal for the organism. In addition to postural adjustments to better absorb solar radiation on cool days, insects use the heat generated by flight muscles and distributed by the circulatory system to raise body temperatures.

Cold Hardiness

Cold hardiness describes the ability of insects to survive exposure to low temperatures. Insects can tolerate winter temperatures if they first undergo a physiological preparedness that may take several weeks to develop. The acclimatization process occurs when insects are first exposed to low temperatures, which trigger the accumulation of the cryoprotectants. Their production typically occurs in the early autumn and declines by early spring. The biosynthesis of all known carbohydrate cryoprotectants begins with glycogen (Fig. 7.13).

There are two physiological strategies that have been recognized to occur in species that overwinter successfully. Insects that are categorized as being **freeze-tolerant** are able to survive the formation of extracellular ice crystals. They synthesize

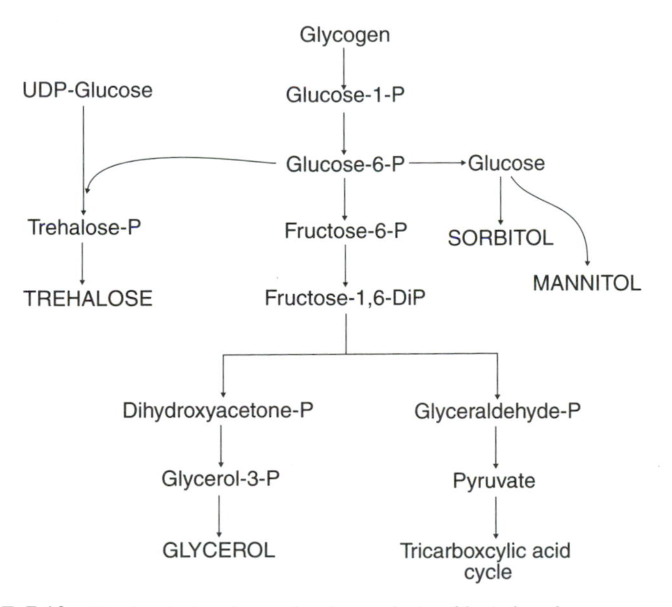

FIGURE 7.13 Biochemical pathways for the synthesis of hemolymph cryoprotectants. Adapted from Storey (1988). Reprinted with permission.

ice-nucleating proteins that raise the supercooling point of body fluids and serve as catalysts for the extracellular nucleation of ice. The extracellular freezing that occurs at the raised temperatures (generally above –8°C) gives the cells time to adjust to the osmotic changes that result from the formation of ice crystals and reduces the likelihood of intracellular freezing.

Freeze-intolerant species produce hemolymph cryoprotectants that allow the insect to supercool and remain in a liquid state without the formation of ice crystals. These species can often supercool to as low as –35°C. Gall-forming larval dipterans in northern Canada can supercool to as low as –60°C. The cryoprotectants that are produced include glycerol, sorbitol, trehalose, and mannitol, often in concentrations approaching 25% of the insect's total body weight. In addition to preventing the formation of ice, these components may stabilize enzymes and cell membranes.

Thermoregulation

Insect activity is largely restricted to the warmer parts of the year, but a few insects, such as gryloblattids and antarctic collembolans, are active during the winter months because their enzyme systems have been tuned to optimal activity at those lower temperatures. Most other species are completely at the mercy of environmental temperatures and are inactive when the temperatures are low. However, some insects are able to regulate their internal temperatures by capturing metabolic heat and using the circulatory system to distribute that heat so they may be active when ambient temperatures are too low to otherwise allow this activity to occur.

Temperature maintenance may be based on either external or internal sources. **Ectothermy,** based on external heat acquisition, is the more primitive means of thermoregulation in nonflying insects, occurring when the insects position themselves in an area of preferred temperature. By behaviorally regulating their exposure to the sun, they actively control the temperature of their bodies. In contrast, **endothermic** insects use the heat generated from flight muscles to raise their body temperatures. Almost inconsequential in smaller insects, the generation of heat by the flight muscles of larger insects can be substantially higher than ambient levels. Although insect flight muscles are the most metabolically active tissue known, they have a mechanical efficiency of no more than 20%, and the remaining 80% of energy expended during flight is degraded as heat that can be used to regulate body temperature.

The flight muscles of larger insects must operate at high enough temperatures to produce the sufficient lift for flight to occur, but the reliance on the flight muscles to generate heat to raise body temperature presents a bit of a paradox. Insects can use flight muscles as a source of heat, but no flight is possible at low ambient temperatures to provide this heat. The problem has been solved by a **preflight**

warm-up in which antagonistic flight muscles contract simultaneously to generate the initial heat that allows them to reach their temperature optima. The behavior is sometimes known as **shivering** because the contractions may cause the wings to quiver slightly. Bumblebees require a thoracic temperature above 30°C before the muscles can generate sufficient lift, and by shivering, can raise their temperatures to this point even when ambient temperatures are as low as 3°C.

The heat is largely concentrated in the thorax, maintaining the flight muscles at the higher temperatures even though the circulatory system is pumping cooler thoracic hemolymph into the warmer thorax. The temperature differential is maintained by the operation of a **countercurrent heat exchange** in the circulatory system. In the bumblebee, a narrow petiole connects the thorax and the abdomen, and through it pass the dorsal vessel, the digestive tract, tracheae, and the ventral diaphragm, all closely oppressed (Fig. 7.14). Blood is pumped forward

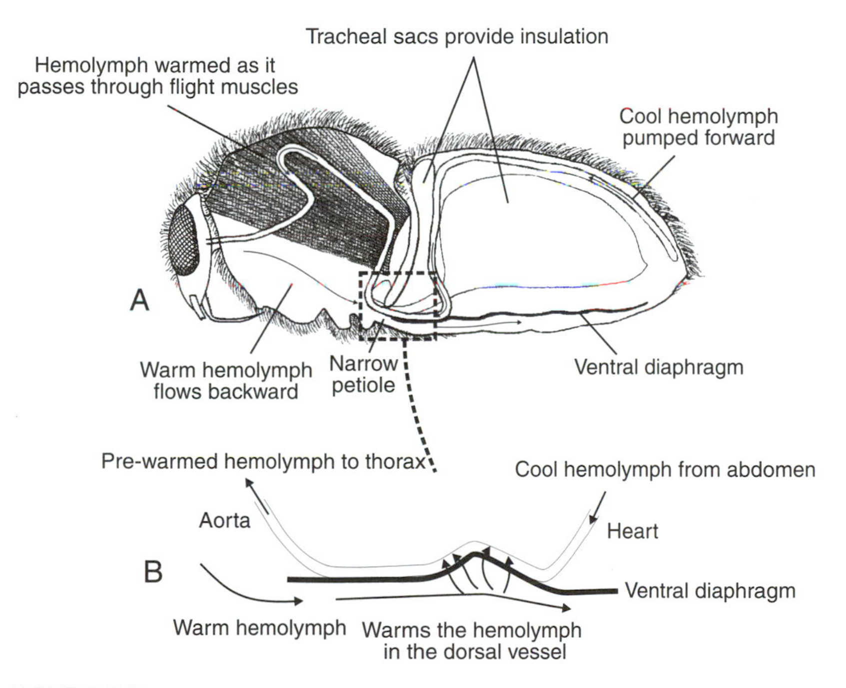

FIGURE 7.14 (A) Longitudinal section of the bumblebee, showing the path of the dorsal vessel through the narrow petiole and the flight muscles. (B) The countercurrent heat exchange occurs with the passage of cool hemolymph from the abdomen within the heart at the same time that warm hemolymph flows backward in the hemocele. The warm hemolymph from the thorax prewarms the hemolymph from the abdomen before its heat can be dissipated when it flows into the abdomen. Reprinted from Heinrich, B. (1976). *Journal of Experimental Biology* **64:** 561–585. With kind permission from the Company of Biologists Ltd.

within the dorsal vessel from the cool abdomen through the warm flight muscles in the thorax and into the head. The warm blood then moves posteriorly within the hemocele through the petiole and back into the abdomen, with the ventral diaphragm regulating its rearward passage. To maintain the heat in the thorax, contractions of the ventral diaphragm and dorsal vessel are timed so that hemolymph both moving forward and backward pass through the petiole simultaneously. As they pass each other, some of the heat acquired by blood in the thorax is passed to the cooler blood from the abdomen, warming it before it enters the thorax. The countercurrent exchange thus keeps most of the heat in the thorax. The winter moth, *Eupsilia morrisoni,* can be airborne on days when the temperatures are as low as 0°C because it can maintain its thorax at temperatures between 30 and 35°C. These moths are especially efficient in maintaining elevated thorax temperatures because in addition to the heat exchanger in the anterior abdomen, an additional countercurrent heat exchanger is present in the thorax (Fig. 7.15).

On warm days when there is no need to maintain a warmer thorax, the countercurrent exchange in bumblebees is reduced so that hemolymph passing forward through the petiole alternates with hemolymph passing backward (Fig. 7.16). This brings the warmer hemolymph into the abdomen where it can be dissipated. The bee can also use this heated abdomen to incubate the brood. At higher ambient temperatures, honeybees are able to reduce their metabolic heat production by regulating their wingbeat frequency and reducing it when necessary so as to lessen the accumulation of internal heat.

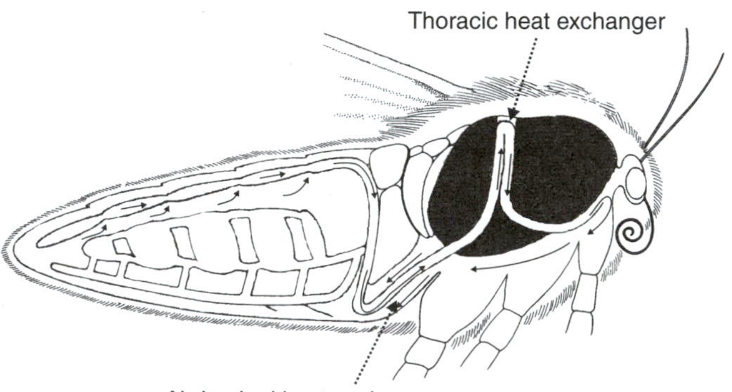

FIGURE 7.15 A second heat exchanger in the thorax of winter moths. From Heinrich (1987). Reprinted with permission.

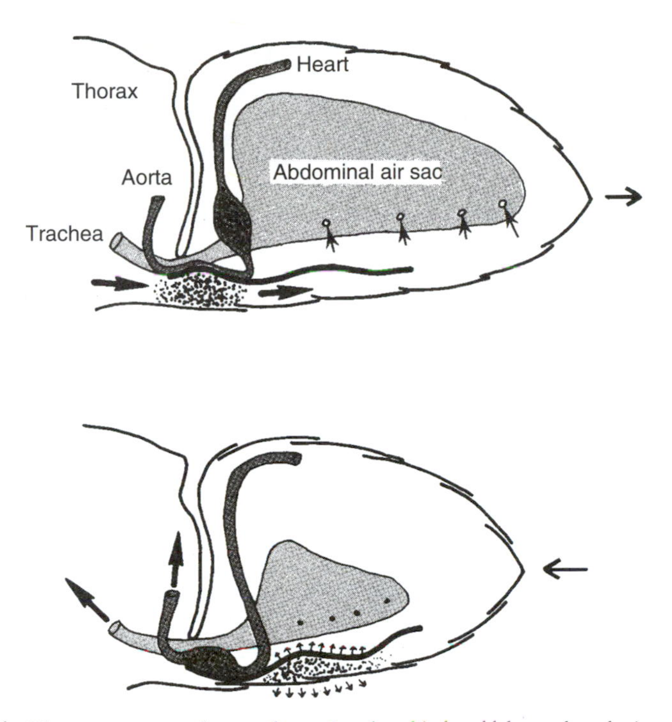

FIGURE 7.16 The countercurrent heat exchange is reduced in bumblebees when the insect must rid itself of heat on a warm day. The passage of warm hemolymph backward from the thorax alternates with cool hemolymph passing forward to the thorax. Hemolymph movements are coordinated by the inflation of abdominal air sacs and the extension and contraction of the abdomen. Reprinted from Heinrich, B. (1976). *Journal of Experimental Biology* **64:** 561–585. With kind permission from the Company of Biologists Ltd.

ADDITIONAL REFERENCES

Circulatory System

Arnold, J. W. 1972. A comparative study of the haemocytes (blood cells) of cockroaches (Insecta: Dictyoptera: Blattaria) with a view of their significance in taxonomy. *Can. Entomol.* **104:** 309–348.

Arnold, J. W. 1974. The hemocytes of insects. In *The physiology of insecta* (M. Rockstein, Ed), vol. 5, pp. 201–254. Academic Press, New York.

Arnold, J. W., Salkeld, E. H. 1967. Morphology of the haemocytes of the giant cockroach *Blaberus giganteus* with histochemical tests. *Can. Entomol.* **99:** 1138–1146.

Braunig, P. 1999. Structure of identified neurons innervating the lateral cardiac nerve cords in the migratory locust, *Locusta migratoria migratoriodes* (Reiche & Fairmaire) (Orthoptera, Acrididae). *Int. J. Insect Morphol. Embryol.* **28:** 81–89.

Cook, B. J., Meola, S. 1983. Heart structure and beat in the stable fly *Stomoxys calcitrans*. *Physiol. Entomol.* 8: 139–149.

Florkin, M., Jeuniaux, C. 1974. Hemolymph: Composition. In *The physiology of insecta* (M. Rockstein, Ed.), vol. 5, pp. 255–307. Academic Press, New York.

Gereben-Krenn, B. -A., Pass, G. 1999. Circulatory organs of Diplura (Hexapoda): The basic design in Hexapoda? *Int. J. Insect Morphol. Embryol.* **28:** 71–79.

Hantschk, A. M. 1991. Functional morphology of accessory circulatory organs in the legs of Hemiptera. *Int. J. Insect Morphol. Embryol.* **20:** 259–273.

Hertel, W., Pass, G., Penzlin, H. 1985. Electrophysiological investigation of the antennal heart of *Periplaneta americana* and its reactions to proctolin. *J. Insect Physiol.* **31:** 563–572.

Hustert, R. 1999. Accessory hemolymph pump in the mesothoracic legs of locusts (*Schistocerca gregaria* Forskal) (Orthoptera, Acrididae). *Int. J. Insect Morphol. Embryol.* **28:** 91–96.

Jones, J. C. 1974. Factors affecting heart rates in insects. In *The physiology of insecta* (M. Rockstein, Ed.), vol. 5, pp. 119–168. Academic Press, New York.

Kaufman, W. R., Davey, K. G. 1971. The pulsatile organ in the tibia of *Triatoma phyllosoma pallidipennis. Can. Entomol.* **103:** 487–496.

Locke, M. 1998. Caterpillars have evolved lungs for hemocyte gas exchange. *J. Insect Physiol.* **44:** 1–20.

Markou, T., Theophilidis, G. 2000. The pacemaker activity generating the intrinsic myogenic contraction of the dorsal vessel of *Tenebrio molitor* (Coleoptera). *J. Exp. Biol.* **203:** 3471.

Matus, S., Pass, G. 1999. Antennal circulatory organ of *Apis mellifera* L. (Hymenoptera: Apidae) and other Hymenoptera: Functional morphology and phylogenetic aspects. *Int. J. Insect Morphol. Embryol.* **28:** 97–109.

Miller, T. A. 1997. Control of circulation in insects. *Gen. Pharmacol.* **29:** 23–38.

Normann, T. C. 1975. Neurosecretory cells in insect brain and production of hypoglycaemic hormone. *Nature* **254:** 259–261.

Nutting, W. L. 1951. A comparative anatomical study of the heart and accessory structures of the Orthopteroid insects. *J. Morphol.* **89:** 501–598.

Pass, G. 1985. Gross and fine structure of the antennal circulatory organ in cockroaches (Blattodea, Insecta). *J. Morphol.* **185:** 255–268.

Pass, G. 1991. Antennal circulatory organs in Onychophora, Myriapoda and Hexapoda: Functional morphology and evolutionary implications. *Zoomorphology* **110:** 145–164.

Pass, G. 2000. Accessory pulsatile organs: Evolutionary innovations in insects. *Annu. Rev. Entomol.* **45:** 495–518.

Richards, A. G. 1963. The ventral diaphragm of insects. *J. Morphol.* **113:** 17–47.

Richter, M., Hertel, W., 1997. Contributions to physiology of the antenna-heart in *Periplaneta americana* (L.) (Blattodea: Blattidae). *J. Insect Physiol.* **43:** 1015–1021.

Sutcliffe, D. W. 1963. The chemical composition of haemolymph in insects and some other arthropods, in relation to their phylogeny. *Comp. Biochem. Physiol.* **9:** 121–135.

Tublitz, N. 1989. Insect cardioactive peptides: Neurohormonal regulation of cardiac activity by two cardioacceleratory peptides during flight in the tobacco hawkmoth, *Manduca sexta. J. Exp. Biol.* **142:** 31–48.

Tublitz, N. J., Cheung, C. C., Edwards, K. K., Sylwester, A. W., Reynolds, S. E. 1992. Insect cardioactive peptides in *Manduca sexta*: A comparison of the biochemical and molecular characteristics of cardioactive peptides in larvae and adults. *J. Exp. Biol.* **165:** 265–272.

Wasserthal, L. T. 1999. Functional morphology of the heart and of a new cephalic pulsatile organ in the blowfly *Calliphora vicina* (Diptera: Calliphoridae) and their roles in hemolymph transport and tracheal ventilation. *Int. J. Insect Morphol. Embryol.* **28:** 111–129.

Wigglesworth, V. B. 1959. Insect blood cells. *Annu. Rev. Entomol.* **4:** 1–16.

Immune Mechanisms

Armstrong, P. B., Melchior, R., Quigley, J. P. 1996. Humoral immunity in long-lived arthropods. *J. Insect Physiol.* **42:** 53–64.

Axen, A., Carlsson, A., Engstrom, A., Bennich, H. 1997. Gloverin, an antibacterial protein from the immune hemolymph of *Hyalophora* pupae. *Eur. J. Biochem.* **247:** 614–619.

Bettencourt, R., Lanz-Mendoza, H., Lindquist, K. R., Faye, I. 1997. Cell adhesion properties of hemolin, an insect immune protein in the Ig superfamily. *Eur. J. Biochem.* **250:** 630–637.

Bettencourt, R., Assefaw-Redda, Y., Faye, I. 2000. The insect immune protein hemolin is expressed during oogenesis and embryogenesis. *Mech. Dev.* **95:** 301–304.

Boman, H. 1995. Peptide antibiotics and their role in innate immunity. *Annu. Rev. Immunol.* **13:** 61–92.

Bulet, P., Cociancich, S., Dimarcq, J. L., Lambert, J., Reichhart, J. M., Hoffmann, D., Hetru, C., Hoffmann, J. A. 1991. Insect immunity. Isolation from a coleopteran insect of a novel inducible antibacterial peptide and of new members of the insect defensin family. *J. Biol. Chem.* **266:** 24,520–24,525.

Bulet, P., Hetru, C., Dimarcq, J. L., Hoffmann, D. 1999. Antimicrobial peptides in insects: Structure and function. *Dev. Comp. Immunol.* **23:** 329–344.

Carlsson, A., Nystrom, T., De Cock, H., Bennich, H. 1998. Attacin—an insect immune protein—binds LPS and triggers the specific inhibition of bacterial outer-membrane protein synthesis. *Microbiology* **144:** 2179–2188.

Chernysh, S., Cociancich, S., Briand, J. -P., Hetru, C., Bulet, P. 1996. The inducible antibacterial peptides of the hemipteran insect *Palomena prasina*: Identification of a unique family of proline-rich peptides and of a novel insect defensin. *J. Insect Physiol.* **42:** 81–89.

Clark, K. D., Pech, L. L., Strand, M. R. 1997. Isolation and identification of a plasmatocyte-spreading peptide from the hemolymph of the lepidopteran insect *Pseudoplusia includens*. *J. Biol. Chem.* **272:** 23,440–23,447.

Dunn, P. E. 1990. Humoral immunity in insects. *BioScience* **40:** 738–744.

Ekengren, S., Hultmark, D. 1999. *Drosophila* cecropin as an antifungal agent. Insect *Biochem. Mol. Biol.* **29:** 965–972.

Engstrom, Y. 1999. Induction and regulation of antimicrobial peptides in *Drosophila*. *Dev. Comp. Immunol.* **23:** 345–358.

Elrod-Erickson, M., Mishra, S., Schneider, D. 2000. Interactions between the cellular and humoral immune responses in *Drosophila*. *Curr. Biol.* **10:** 781–784.

Fehlbaum, P., Bulet, P., Chernysh, S., Briand, J. P., Roussel, J. P., Letellier, L., Hetru, C., Hoffmann, J. A. 1996. Structure-activity analysis of thanatin, a 21-residue inducible insect defense peptide with sequence homology to frog skin antimicrobial peptides. *Proc. Natl. Acad. Sci. USA* **93:** 1221–1225.

Fife, H., Palli, S. R., Locke, M. 1987. A function for pericardial cells in an insect. *Insect Biochem.* **17:** 829–840.

Finnerty, C. M., Granados, R. R. 1997. The plasma protein scolexin from *Manduca sexta* is induced by baculovirus infection and other immune challenges. *Insect Biochem. Mol. Biol.* **27:** 1–7.

Finnerty, C. M., Karplus, P. A., Granados, R. R. 1999. The insect immune protein scolexin is a novel serine proteinase homolog. *Protein Sci.* **8:** 242–248.

Ganz, T., Lehrer, R. I. 1994. *Defensins. Curr. Opin. Immunol.* **6:** 584–589.

Gao, Y., Hernandez, V. P., Fallon, A. M. 1999. Immunity proteins from mosquito cell lines include three defensin A isoforms from *Aedes aegypti* and a defensin D from *Aedes albopictus*. *Insect Mol. Biol.* **18:** 311–318.

Gillespie, J. P., Kanost, M. R., Trenczek, T. 1997. Biological mediators of insect immunity. *Annu. Rev. Entomol.* **42:** 611–643.

Hoffmann, J. A. 1995. Innate immunity of insects. *Curr. Opin. Immunol.* **7:** 4–10.

Hoffmann, J. A., Kafatos, F. C., Janeway, C. A., Ezekowitz, R. A. 1999. Phylogenetic perspectives in innate immunity. *Science* **284:** 1313–1318.

Ip, Y. T., Levine, M. 1994. Molecular genetics of *Drosophila* immunity. *Curr. Opin. Genet. Dev.* **4:** 672–677.

Karp, R. D. 1990. Cell-mediated immunity in invertebrates. *BioScience* 40: 732–737.

Kanost, M. R., Kawooya, J. K., Law, J. H., Ryan, R. O., Van Heusden, M. C., Ziegler, R. 1990. Insect hemolymph proteins. *Adv. Insect Physiol.* **22:** 229–396.

Kanost, M. R., Jiang, H. 1997. Serpins from an insect, *Manduca sexta*. *Adv. Exp. Med. Biol.* **425:** 155–161.

Kanost, M. R. 1999. Serine proteinase inhibitors in arthropod immunity. *Dev. Comp. Immunol.* **23:** 291–301.

Lackie, A. M. 1988. Immune mechanisms in insects. *Parasitol. Today* **4:** 98–105.

Levashina, E. A., Ohresser, S., Bulet, P., Reichhart, J. M., Hetru, C., Hoffmann, J. A. 1995. Metchnikowin, a novel immune-inducible proline-rich peptide from *Drosophila* with antibacterial and antifungal properties. *Eur. J. Biochem.* **233:** 694–700.

Levashina, E. A., Ohresser, S., Lemaitre, B., Imler, J. L. 1998. Two distinct pathways can control expression of the gene encoding the *Drosophila* antimicrobial peptide metchnikowin. *J. Mol. Biol.* **278:** 515–527.

Lowenberger, C. A., Kamal, S., Chiles, J., Paskewitz, S., Bulet, P., Hoffmann, J. A., Christensen, B. M. 1999. Mosquito-*Plasmodium* interactions in response to immune activation of the vector. *Exp. Parasitol.* **91:** 59–69.

Lowenberger, C. A., Smartt, C. T., Bulet, P., Ferdig, M. T., Severson, D. W., Hoffmann, J. A., Christensen, B. M. 1999. Insect immunity: Molecular cloning, expression, and characterization of cDNAs and genomic DNA encoding three isoforms of insect defensin in *Aedes aegypti*. *Insect. Mol. Biol.* **8:** 107–118.

Lowenberger, C. 2001. Innate immune response of *Aedes aegypti*. *Insect Biochem. Mol. Biol.* **31:** 219–229.

Mackintosh, J. A., Gooley, A. A., Karuso, P. H., Beattie, A. J., Jardine, D. R., Veal, D. A. 1998. A gloverin-like antibacterial protein is synthesized in *Helicoverpa armigera* following bacterial challenge. *Dev. Comp. Immunol.* **22:** 387–399.

Meister, M., Lemaitre, B., Hoffmann, J. A. 1997. Antimicrobial peptide defense in *Drosophila*. *BioEssays* **19:** 1019–1026.

Paskewitz, S. M., Riehle, M. 1998. A factor preventing melanization of sephadex CM C-25 beads in *Plasmodium*-susceptible and refractory *Anopheles gambiae*. *Exp. Parasitol.* **90:** 34–41.

Ramos-Onsins, S., Aguade, M. 1998. Molecular evolution of the cecropin multigene family in *Drosophila*. Functional genes vs. pseudogenes. *Genetics* **150:** 157–171.

Sun, D., Eccleston, E. D., Fallon, A. M. 1998. Peptide sequence of an antibiotic cecropin from the vector mosquito, *Aedes albopictus*. *Biochem. Biophys. Res. Commun.* **249:** 410–415.

Uttenweiler-Joseph, S., Moniatte, M., Lagueux, M., Van Dorsselaer, A., Hoffmann, J. A., Bulet, P. 1998. Differential display of peptides induced during the immune response of *Drosophila*: A matrix-assisted laser desorption ionization time-of- flight mass spectrometry study. *Proc. Natl. Acad. Sci. USA* **95:** 11,342–11,347.

Vilmos, P., Kurucz, E. 1998. Insect immunity: Evolutionary roots of the mammalian innate immune system. *Immunol. Lett.* **62:** 59–66.

Vizioli, J., Bulet, P., Charlet, M., Lowenberger, C., Blass, C., Muller, H., Dimopoulos G., Hoffmann, J., Kafatos, F. C., Richman, A. 2000. Cloning and analysis of a cecropin gene from the malaria vector mosquito, *Anopheles gambiae*. *Insect. Mol. Biol.* **9:** 75–84.

Vizioli, J., Richman, A. M., Uttenweiler-Joseph, S., Blass, C., Bulet, P. 2001. The defensin peptide of the malaria vector mosquito *Anopheles gambiae*: Antimicrobial activities and expression in adult mosquitoes. *Insect Biochem. Mol. Biol.* **31:** 241–248.

Wilson, R., Chen, C., Ratcliffe, N. A. 1999. Innate immunity in insects: The role of multiple, endogenous serum lectins in the recognition of foreign invaders in the cockroach, *Blaberus discoidalis*. *J. Immunol.* **162:** 1590–1596.

Yamakawa, M., Tanaka, H. 1999. Immune proteins and their gene expression in the silkworm, *Bombyx mori*. *Dev. Comp. Immunol.* **23:** 281–289.

Yu, X. Q., Kanost, M. R. 1999. Developmental expression of *Manduca sexta* hemolin. *Arch. Insect Biochem. Physiol.* **42:** 198–212.

Zhao, L., Kanost, M. R. 1996. In search of a function for hemolin, a hemolymph protein from the immunoglobulin superfamily. *J. Insect Physiol.* **42:** 73–79.

Thermoregulation

Bale, J. S. 1987. Insect cold hardiness: Freezing and supercooling—An ecophysiological perspective. *J. Insect Physiol.* **33:** 899–908.

Bale, J. S. 1991. Insects at low temperature: A predictable relationship? *Funct. Ecol.* **5:** 291–298.

Bale, J. S. 1993. Classes of insect cold hardiness. *Funct. Ecol.* **7:** 751–753.

Baust, J. G., Rojas, R. R. 1985. Insect cold hardiness: Facts and fancy. *J. Insect Physiol.* **31:** 755–759.

Block, W. 1990. Cold tolerance of insects and other arthropods. *Phil. Trans. R. Soc. London B* **326:** 613–633.

Heinrich, B. 1971. Temperature regulation of the sphinx moth, *Manduca sexta*. I. Flight energetics and body temperature during free and tethered flight. *J. Exp. Biol.* **54:** 141–152.

Heinrich, B. 1971. Temperature regulation of the sphinx moth, *Manduca sexta*. II. Regulation of heat loss by control of blood circulation. *J. Exp. Biol.* **54:** 153–166.

Heinrich, B. 1974. Thermoregulation in endothermic insects. *Science* **185:** 747–756.

Heinrich, B. 1976. Heat exchange in relation to blood flow between thorax and abdomen in bumblebees. *J. Exp. Biol.* **64:** 561–585.

Heinrich, B. 1979. *Bumblebee economics*. Harvard University Press, Cambridge.

Heinrich, B. 1987. Thermoregulation in winter moths. *Sci. Am.* **256:** 104–111.

Heinrich, B. 1993. *The hot-blooded insects*. Harvard University Press, Cambridge.

Heinrich, B., Bartholomew, G. A. 1972. Temperature control in flying moths. *Sci. Am.* **226:** 71–77.

Heinrich, B., Kammer, A. 1973. Activation of the fibrillar muscles in the bumblebee during warm-up, stabilization of thoracic temperature and flight. *J. Exp. Biol.* **58:** 677–688.

Kammer, A. E., Heinrich, B. 1978. Insect flight metabolism. *Adv. Insect Physiol.* **13:** 133–228.

Lee, R. E., Jr. 1989. Insect cold-hardiness: To freeze or not to freeze. *BioScience* **39:** 308–313.

Storey, K. B., Storey, J. M. 1988. Freeze tolerance in animals. *Physiol. Rev.* **68:** 27–84.

CHAPTER 8

Excretory Systems

MAJOR EXCRETORY PRODUCTS IN INSECTS

The ability of insects to occupy terrestrial niches is largely due to two important adaptations: their impermeable exoskeleton and an excretory system of considerable sophistication. As small terrestrial animals with a high ratio of surface area to body volume, insects lose water readily and this loss must not be compounded by efforts to get rid of wastes. The major function of excretory systems is to maintain the internal environment of an organism by separating and eliminating metabolic wastes and other toxic substances. Because these wastes are often dissolved in water, excretory processes are also closely associated with osmoregulation and the maintenance of water balance.

Insects ingest a wide variety of foods from which they derive energy. When these foods are broken down and the nutrients extracted, there are usually products left over that must be eliminated or sequestered so they do not accumulate and eventually poison other metabolic pathways. When carbohydrates or lipids are ingested, the energy captured from their oxidation is accompanied by the end products of carbon dioxide and water, neither of which causes any great problems with accumulation or toxicity. Carbon dioxide easily diffuses through soft insect cuticles or is eliminated through the tracheal system. Excess water is generally not a problem and can be easily eliminated when it accumulates.

In contrast, the metabolism of proteins and nucleic acids produces nitrogen in addition to carbon dioxide and water. Amino acids cannot be stored to as great a degree as carbohydrates and lipids, and when more protein is consumed than is needed for maintenance, the excess nitrogen must be eliminated quickly. The nitrogen itself is not toxic, but in biological systems it readily forms ammonia. Some of this ammonia can be recycled into amino acid synthesis by the formation of glutamate from α-ketoglutarate and glutamine from glutamate (Fig. 8.1), but the excess ammonia that remains is highly toxic unless it is diluted with water.

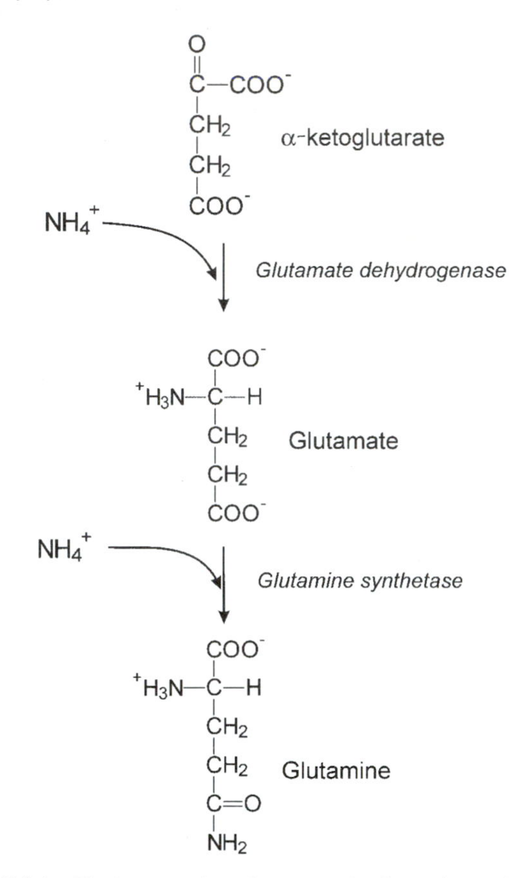

FIGURE 8.1 The incorporation of ammonia for the synthesis of amino acids.

High levels of ammonia can interrupt nervous transmission by substituting for necessary potassium and can also alter carbohydrate and lipid metabolism.

Organisms must therefore have excretory systems that are designed to avoid the toxic accumulation of ammonia. Because ammonia is very soluble in water, it cannot be sequestered from critical biological reactions. It easily crosses biological membranes and the only way to deal with it is to maintain its concentrations below levels that are toxic. Generally, for each gram of ammonia that is retained within an organism, about 400 ml of water is necessary to reduce it to this safe level. Aquatic organisms, such as odonate larvae, have little trouble finding these large quantities of water for dilution and many even excrete the nitrogen from protein metabolism directly as ammonia. However, terrestrial organisms are unable to carry a water supply that is large enough to dilute the ammonia, and there are needs for both water conservation and the incorporation of the nitrogen into a less toxic molecule than ammonia. Most terrestrial organisms have taken the pathway of the incorporation of the nitrogen into either **urea** or **uric acid** (Fig. 8.2). As a consequence of their lower toxicity, these molecules can be concentrated in body fluids to a much greater extent than can ammonia and require less water for dilution. Urea is more soluble than ammonia but far less toxic and requires about 10 times less water to be diluted to nontoxic concentrations. For a relatively large terrestrial animal that has an easy access to water or that can carry sufficiently large water reserves, urea is an adequate molecule into which the nitrogen may be incorporated. However, in insects, the need for water conservation may have been the driving force for the incorporation of their nitrogen wastes into uric acid, which has been described as the ideal excretory product for small terrestrial animals. Uric acid does not dissolve well in water and therefore fails to reach toxic levels in body fluids, so it requires about 50 times less water to dilute than does ammonia. It is only slightly soluble in biological fluids below pH 7, but becomes much more soluble above pH 9. 5, and its insolubility in water allows it to be excreted in a dry form without having a significant effect on water balance. Because hydrogen atoms are derived from water, their loss when incorporated in an excretory molecule affects water balance. Uric acid has the lowest ratio of H:N (1:1) for any of the excretory products, compared to the ratio of 2:1 for urea and 3:1 for ammonia.

Insects pay a high price for the benefits they derive from employing uric acid as a way to excrete nitrogen and still maintain a positive water balance. The synthesis of uric acid from protein results in the loss of several carbon atoms that could be used for other biosyntheses and requires a substantial amount of energy to build a larger, less toxic molecule. Eight ATP are required to first make the intermediary metabolite, **inosine monophosphate (IMP)**, synthesized by the

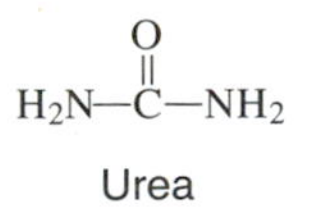

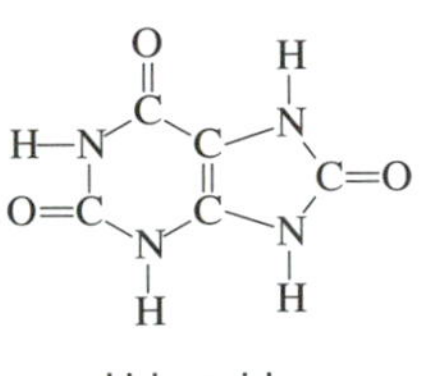

FIGURE 8.2 Excretory molecules that incorporate nitrogen.

successive addition of amino groups from glycine, glutamine, and aspartate (Fig. 8.3). The IMP is then converted to uric acid through hypoxanthine and xanthine, the entry points for the catabolism of purines. It has been estimated that when the tsetse fly, *Glossina,* which feeds exclusively on nitrogen-rich blood, ingests a 100-mg meal, it would use the energy from 47 mg of the blood as overhead to dispose of the excess nitrogen through uric acid synthesis and excretion. In contrast, if the insect could excrete the nitrogen as ammonia, only 15 mg of the meal would be

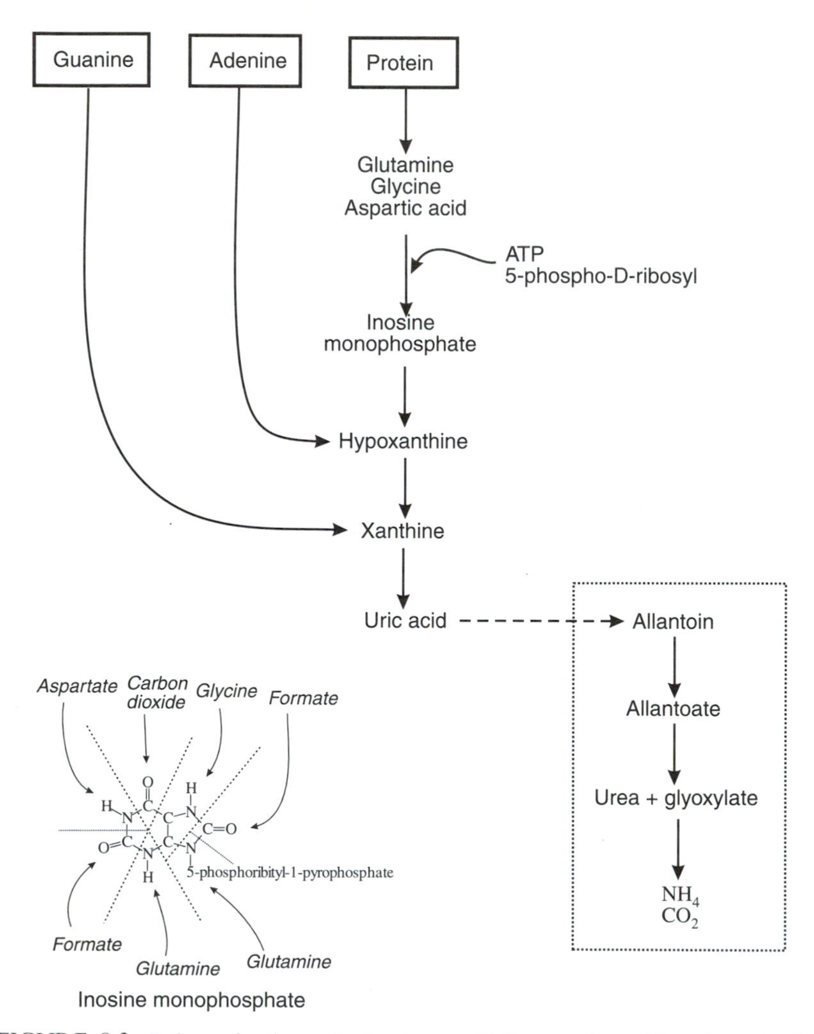

FIGURE 8.3 Pathway for the synthesis of uric acid from nucleic acids and protein. Inosine monophosphate requires ATP for its synthesis as well as the several other constituents shown. The steps toward the synthesis of ammonia and carbon dioxide, shown in the box, do not take place in insects.

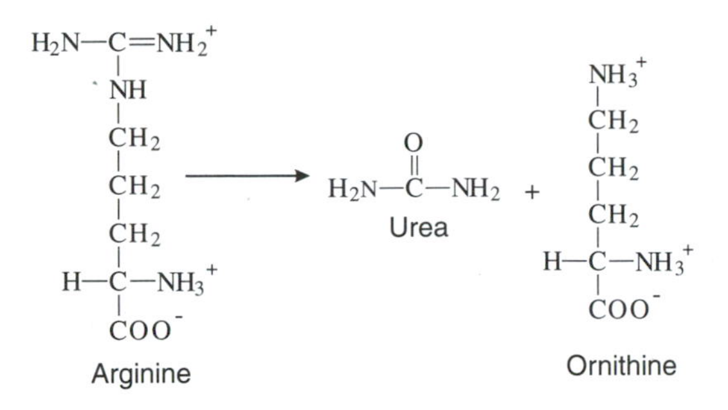

FIGURE 8.4 One way to account for the release of urea by some insects. Dietary arginine can be broken down to urea and ornithine, but the complete pathway of the ornithine cycle has yet to be identified in insects.

used for processing. Other terrestrial organisms can convert uric acid to urea and finally ammonia and carbon dioxide, but insects lack the necessary enzymes.

Not all insects excrete uric acid, and not all excrete one product exclusively. The type of excretory product produced is often a function of diet, developmental stage, and ecological niche. Allantoin is excreted by the hemipteran *Dysdercus,* and certain lepidopterans and larval dipterans excrete allantoic acid. Urea is also a minor component of some insects, but the complete pathway of urea synthesis in insects is not known. Several enzymes of the ornithine cycle that operates in vertebrates have been identified, but evidence for the complete cycle in insects is lacking. Arginine is converted to ornithine and urea by the enzyme arginase, which has been identified in insects, and the amount of urea excreted may correlate with levels of arginine in the diet. Thus, it appears that urea excretion in insects may result from the breakdown of dietary arginine (Fig. 8.4). The ammonia that is excreted by some terrestrial insects, including blowfly larvae and some cockroaches and locusts occurs as a result of the deamination of amino acids and not by the breakdown of urea (Fig. 8.5).

MALPIGHIAN TUBULES

As might be expected given the importance of osmoregulation in a small terrestrial animal, an excretory system of considerable sophistication has evolved in

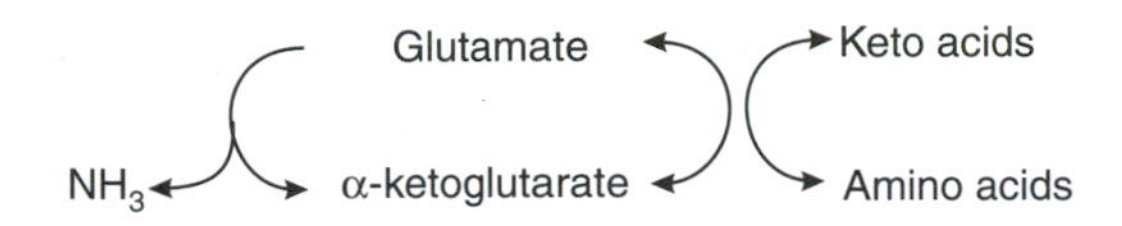

FIGURE 8.5 Release of ammonia by the deamination of amino acids.

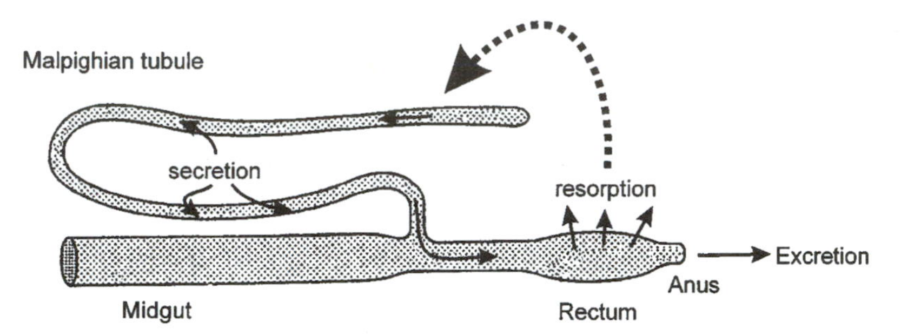

FIGURE 8.6 The overall mechanism of insect excretion. Fluid is taken up by the Malpighian tubules and moves to the hindgut, where the rectum resorbs some of the water, salts, and amino acids, and the remainder is excreted. Reprinted with permission from Hevert, F. (1984). *Environmental Physiology and Biochemistry of Insects,* pp. 184–205. Copyright Springer-Verlag GmbH & Co. KG.

insects. Analogous to the kidneys of vertebrates, the **Malpighian tubules** are the primary excretory organs of insects but operate in a different manner than kidneys that base their filtration on hydrostatic pressure. The Malpighian tubules are but one major part of the system that regulates salt and water balance in insects, with the other being the rectum. The process of excretion is a two-step process, with much of the fluid that is taken up by the tubules resorbed by the hindgut before it passes out of the body (Fig. 8.6).

Their name is derived from Malpighi, who first made reference to them in 1669 in his work with the silkworm. They arise during embryogenesis as evaginations of the gut, usually originating at the junction of the midgut and hindgut as long tubes containing a hollow center, or lumen, that is closed at the distal end (Fig. 8.7). The open end of the lumen empties into the digestive tract. Their walls

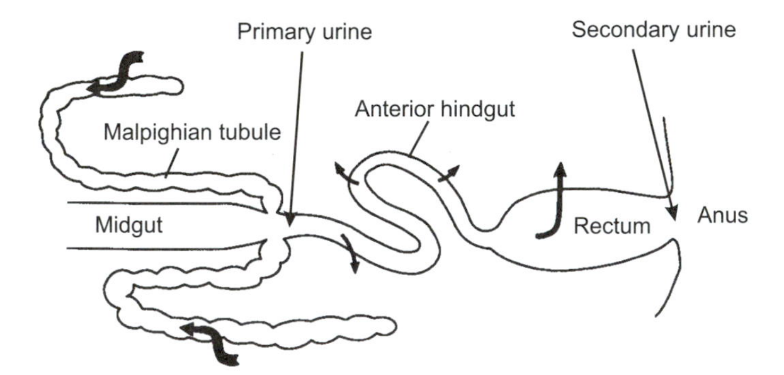

FIGURE 8.7 Formation of primary and secondary urine in the excretory system. Reprinted with permission from Hadley, N.F. (1994). *Water Relations of Terrestrial Arthropods.* Copyright Academic Press.

consist of a single cell layer of epithelial cells that often contain large polyploid nuclei and are differentiated by structure and function along the length of the tubule. For example, each of the 2-mm-long tubules of *Drosophila* contain about 150 cells that can be differentiated into six regions and six different cell types. On the hemolymph side, the cells are covered by a basement membrane and a network of plasma membrane infoldings with trachea investing the tubules on the outside of the basement membrane. The cellular membrane that faces the tubule lumen contains a brush border with abundant microvilli that are rich in endoplasmic reticulum and mitochondria. These microvilli increase the surface area available for salt and water transport between the cell and the tubule lumen. The intermediate interior region of the cell contains numerous Golgi, rough endoplasmic reticulum, and vacuoles.

The tubules lie free in the body cavity and are surrounded by longitudinal muscles that allow them to engage in writhing movements that increase the contact between the tubules and the hemolymph. In some insects the distal ends of the tubule are physically associated with the hindgut, arranged into a **cryptonephridial system** that increases the resorption of water from the hindgut. This system will be described in more detail in a subsequent section. The total number of tubules varies in different species from 2 to over 250, with a few species, including aphids and collembolans, which have none. There appears to be a direct relationship between the surface area of the tubules and the total body mass of the insect. There are four tubules in *Drosophila*, each of a length approximately twice that of the body. In the cockroach, *Periplaneta*, there are 60 tubules with a total surface area of 132 mm^2.

The Malpighian tubules initiate the excretory process. Their net uptake of fluid, ions, and waste materials is termed the **primary urine** and is transported into the tubule lumen (Fig. 8.7). This primary urine travels through the lumen from the distal to proximal regions of the tubule and then to the alimentary canal where it mixes with the end products of digestion from the midgut. Generally isosmotic with the hemolymph, in most insects it contains a high concentration of K^+ and lower concentrations of Na^+ than are present in the hemolymph. In addition to excretory products such as uric acid, the primary urine may contain other ions, sugars, and amino acids that are resorbed later in the excretory process. In *Drosophila* larvae, the main segment is the only region reported to secrete K^+ fluid, with the lower tubule region resorbing water and K+ and actively transporting Ca^{2+} into the lumen. The rapid secretion by the main segment in conjunction with the resorption that occurs in the lower tubule region and the hindgut allows metabolic wastes to be cleared from the hemolymph without the significant loss of water. From there the primary urine passes backward through the hindgut and rectum, where it is modified by continued resorption by the rectal glands to produce a **secondary urine** that is expelled through the anus.

MECHANISM OF MALPIGHIAN TUBULE SECRETION

Early ideas regarding the mechanism of primary urine formation were based on the experimental techniques developed by **Ramsay.** Preparations of Malphigian tubules were bathed in insect Ringer solution, with the open end of the tubule that normally discharges into the gut placed into a layer of oil above the Ringers with a silk ligature (Fig. 8.8). Fluid flowing through the tubule gathered in an aqueous droplet that then could be measured and collected within the oil. By analyzing the primary urine formed in these droplets, it was discovered that it was isosmotic with the hemolymph, but with potassium concentrations up to 20 times higher. From these initial experiments, the mechanism involving the active transport of potassium into the lumen against an electrochemical gradient was established. Because the primary urine is isosmotic with the hemolymph, the transport of water appeared to follow the potassium and create the flow of urine within the tubule.

The driving force for the movement of ions results from a vacuolar-proton-adenosine triphosphatase (V-ATPase), also known as a proton pump, which regulates the concentration of protons. This transduction of ATP hydrolysis into a current of protons, in the absence of any other ions, energizes the cell membranes of the tubules. The translocation of protons into the tubule lumen also plays a key role in the regulation of pH and the acidification of cell compartments. Protons transported into the lumen return to the cell by way of a cation-H^+ exchanger that promotes the movement of sodium or potassium from the cell into the lumen and that drives the process of fluid and waste excretion. Thus, in addition to the proton pump located on the cell membrane, there are antiport transport processes that carry ions including K^+, Na^+, and H^+ in opposite directions. Na/K ATPases actively transport these ions from the hemolymph, and ion channels allow chloride ions to be passively transported. Although phytophagous insects primarily

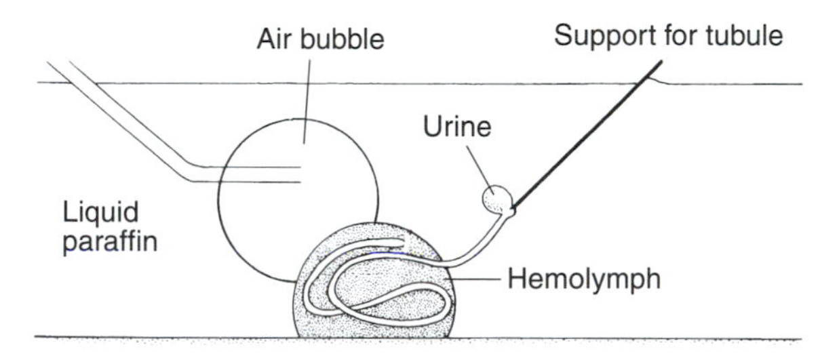

FIGURE 8.8 The bioassay of Malpighian tubule function devised by Ramsay. The excised tubule is immersed in paraffin within a drop of hemolymph, and a bubble of air to allow the tubule cells to respire. As wastes are processed by the tubule, a drop of urine forms at its end. Because the urine is insoluble in paraffin, its diameter can be measured as an estimate of Malpighian tubule activity. Reprinted from Ramsay, J. A. (1954). *Journal of Experimental Biology* **31:** 104–113. With kind permission from the Company of Biologists Ltd.

transport potassium, hematophagous insects such as the mosquito transport sodium and chloride, while the bloodsucking hemipteran *Rhodnius* transports both sodium and potassium. The V-ATPase accomplishes the secretion of all these ions by driving the secretion of sodium or potassium through the antiport ion exchangers, setting up a gradient for the passive movement of small hemolymph solutes into the tubule lumen. Water follows the major ions and the solutes in the hemolymph diffuse via a **paracellular route** between the cells. Active transport across cells, following a **transcellular route,** occurs for some toxins and metabolites, including uric acid (Fig. 8.9). Low-molecular-weight substances move through the tubule cell by passive diffusion, but several larger, toxic compounds, such as plant alkaloids, are removed from the hemolymph by an active transport.

HINDGUT AND RECTUM

If insects used the primary urine from the Malpighian tubules as their major excretory product, they would soon undergo a depletion of their potassium and water. However, superimposed on the Malpighian tubule excretion is a second system involving the rectum of the hindgut that recovers most of the ions and water,

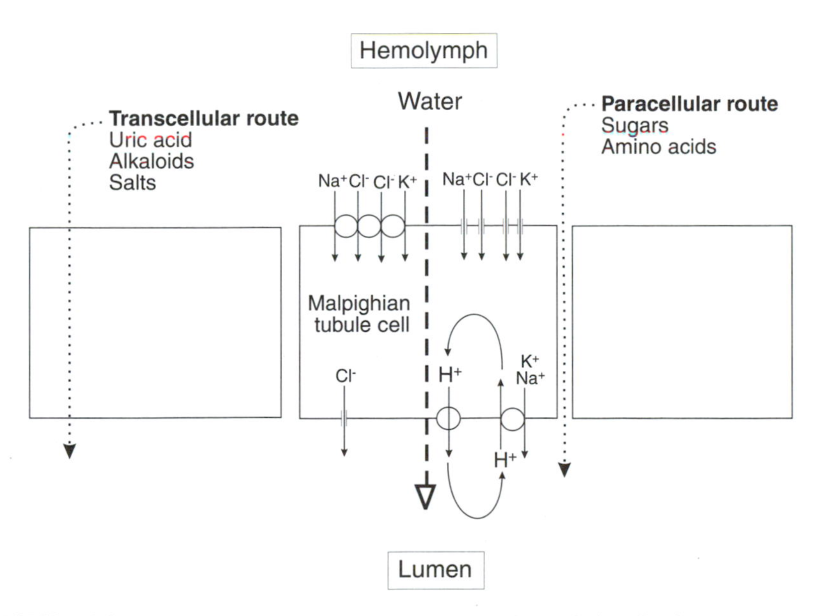

FIGURE 8.9 Transport of substances through the Malpighian tubule cells. The major ion movements result from the action of the V-ATPase that moves protons and energizes the cell membrane. Na/K ATPases actively transport ions from the hemolymph into the lumen, with ion channels allowing some passive transport.

adjusting the excretory product so that it achieves the necessary osmoregulatory balance for the insect and excretes the secondary urine.

The **rectum** consists of the enlarged posterior-most section of the hindgut, often containing specialized structures called **papillae** or **rectal pads** that are enlarged epithelial cells. Unlike those of the Malpighian tubules, the cells of the rectum lack the infoldings of the basal membrane, but do contain a brush border underneath the cuticle that faces the rectal lumen. Derived from the invagination of epidermal cells, the rectal cells have a cuticular lining on the lumen side that is shed at each molt. Some insects show little morphological differences that distinguish these rectal pads. Development of the rectum ranges from no modifications in some insects such as *Drosophila,* to the specialized rectal pads that are present in many orthopterans and coleopterans. Fluid feeding insects that do not resorb the fluid in their primary urine may not have rectal pads present. The extent to which the recovery of water and ions occurs is determined by the extent to which water must be resorbed and is a function of environmental and physiological conditions. Insects living under more xeric conditions, such as flour beetles, must resorb more of the water in the hindgut than those living under moister conditions. The rate of this rectal absorption is not constant. *Manduca sexta* larvae are able to maintain a steady state and regulate the water content of their bodies based on the water content of the food they ingest by varying the rate of rectal resorption.

Because the pH of the rectum is often more acidic than the rest of the hindgut and the solubility of uric acid considerably reduced under acidic conditions, the acidity in the hindgut precipitates uric acid and allows its excretion to occur in the absence of much water. With its cuticular lining, the rectal epithelium is able to act as a molecular sieve that restricts the passage of larger molecules through the cells. Toxic wastes are retained in the rectal lumen by this lining and are excreted with little loss of water.

The rectum works in an opposite manner to that of the Malpighian tubules. It transports water and ions from the material within the gut lumen into the hemolymph. An electrogenic Cl^- pump on the lumen side of the cell membrane that is not coupled to any other ions drives the resorption process, moving Cl^- from the gut lumen into the hemolymph. A Na/K-ATPase mediates the transport of sodium into the hemolymph and generates a positive electrical potential on the hemolymph side of the membrane and also drives fluid transport. Chloride exits the cell passively into the hemolymph through cAMP-stimulated channels, and potassium also follows passively by an electrical coupling with potassium channels (Fig. 8.10).

CRYPTONEPHRIDIAL SYSTEM

In some insects, including tenebrionid beetles that live under extremely dry conditions, the ends of the Malpighian tubules do not lie free in the hemocele. Instead,

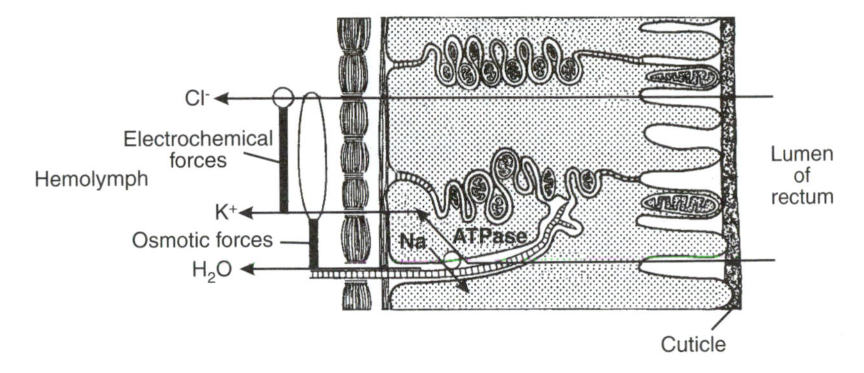

FIGURE 8.10 A rectal cell and its ion transport. Reprinted with permission from Hevert, F. (1984). *Environmental Physiology and Biochemistry of Insects,* pp. 184–205. Copyright Springer-Verlag GmbH & Co. KG.

the terminal segments of the tubules are closely associated with the wall of the rectum in what is called a **cryptonephridial complex** (Fig. 8.11). The hidden nature of the tubules, surrounded by a perinephric membrane and creating a special chamber, the perinephric space, gives the complex its name. The complex is found in most lepidopteran larvae and many coleopterans. The cryptonephridial complex

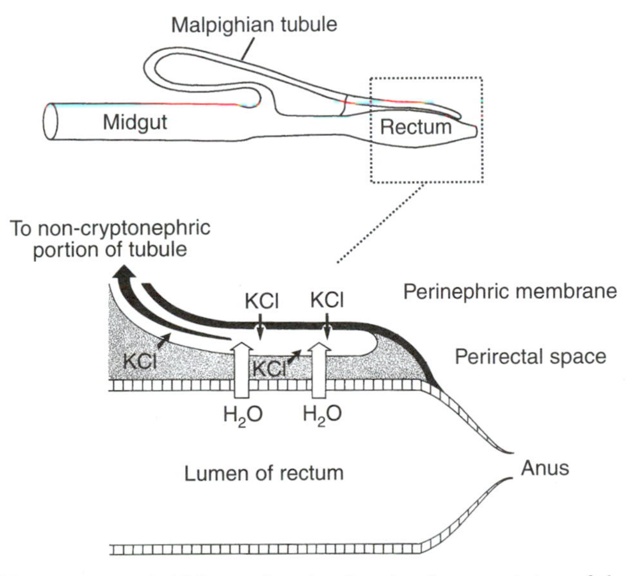

FIGURE 8.11 A cryptonephridial complex, showing the close association of the Malpighian tubules with the rectum to serve as a more efficient mechanism of water recovery. Reprinted from Bradley, T.J. *Comprehensive Insect Physiology, Biochemistry and Pharmacology* 1985, Volume 4, pp. 421–465. With permission from Elsevier Science.

performs two functions: it resorbs water from the hindgut very efficiently, and in some insects is able to absorb atmospheric water from the humidity in the hindgut. Both functions rely on the ability of the complex to maintain osmotic gradients through the transport of potassium by the tubules. The tubule cells take up K^+, Na^+, and H^+ ions, creating a sufficiently elevated osmotic pressure for water to be drawn from the rectal lumen into the perinephric space and then into the tubules, where water and some ions are resorbed into general circulation. In other insects, such as the Psocoptera or booklice, atmospheric moisture is absorbed through the specialized hypopharynx within the cibarium at the anterior region of the alimentary canal.

FILTER CHAMBER

In some Homopterans that feed exclusively on plant juices containing low concentrations of nutrients, the digestive tract forms an arrangement known as a **filter chamber.** In this arrangement, the anterior and posterior regions of the midgut come into close contact, sometimes also involving the proximal ends of the Malpighian tubules, and are surrounded by a sac that is comprised of thin epithelial cells (Fig. 8.12). The large amount of fluid acquired during feeding is concentrated and eliminated more quickly by transfer directly from the anterior region of the midgut to the posterior region without being absorbed into and diluting the hemolymph. Potassium is actively secreted into the tubules and posterior midgut, drawing water from the anterior midgut. Much of this fluid containing dilute concentrations of amino acids and sugars is ultimately excreted as "honeydew."

HORMONAL CONTROL OF EXCRETION AND OSMOREGULATION

Although the insect excretory system is designed to conserve water by maintaining a low rate of excretion under most conditions, it can be accelerated at times by various hormones. During some periods of the life cycle, there may be sound reasons for this acceleration. For example, as a result of ingesting close to their own weight in grass each day, locusts must rid themselves of the excess water present in the food to prevent their hemolymph from being diluted. Other fluid feeders like *Rhodnius* that ingest large volumes of blood during each stadium must process the liquid portion of the diet quickly so as not to overwhelm the circulatory system with water. Insects that must make the transition from a terrestrial phase to a flying phase, such as lepidopterans that metamorphose from a pupa to an adult, undergo an enhanced loss of water to make them better able to fly. Newly emerged *Pieris brassicae* adults thus lose about 40% of their body weight within 3

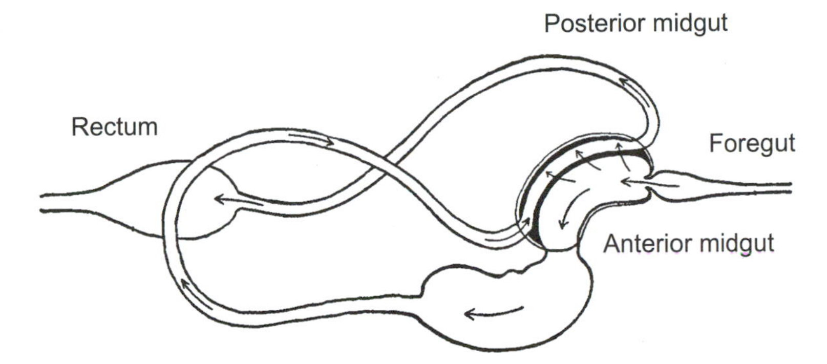

FIGURE 8.12 A filter chamber, in which the anterior and posterior midgut regions are in close contact so that water can be diverted directly to the hindgut without diluting the contents of the midgut. From Snodgrass (1935). Reprinted with permission.

h of emergence. Obviously, this rate of water loss is reserved for special times during the life cycle and could not continue indefinitely. Surprisingly, there are also many terrestrial insects that live most of their lives under xeric conditions that also employ diuretic hormones. The Malpighian tubules of the locust are stimulated by diuretic hormone during flight, accompanied by increased absorption by the rectum. In spite of the demands of the environment for an efficiency of water conservation, Namib Desert beetles also utilize a diuretic hormone. In both these cases, the increased turnover from hormone activation results in a more rapid clearance of the hemolymph of metabolic wastes without the loss of water because of the efficiency of the rectum in recovering the fluids.

There are many targets for which hormones may modulate excretory rates. The longitudinal muscles that run the length of the tubules and cause them to writhe about in the hemoceole have no nervous innervations but are activated hormonally by diuretic and myotropic hormones in the hemolymph. These increased writhing movements may increase the circulation of hemolymph over the tubule walls and facilitate the movement of primary urine within the lumen.

Another class of hormones includes the **diuretic hormones** that act on the Malpighian tubules to increase the processing of fluids through the excretory system. Their release at specific times causes an increase in the rate of primary urine production. The Ramsay assay (Fig. 8.8) using isolated tubule preparations has served as an excellent bioassay and an indication of this increased rate. By adding suspected diuretic hormones to the drop of hemolymph in which the tubules are bathed, it is possible to record increases in the production of primary urine. These increases can be dramatic: the mosquito *Aedes aegypti* produces a diuretic hormone after blood ingestion that causes the loss of over 40% of the water in the blood meal within 2 h of feeding. The diuretic hormone in *Rhodnius* can increase the secretion rates of Malpighian tubules by a factor of 1000 soon after a blood meal.

The diuretic neuropeptides that have been isolated from insects fall into one of two hormone families, the **corticotropin-releasing factor** (**CRF**)-related peptides and the insect **kinins.** The CRF-related family of hormones is so named because of their sequence similarity to this family of vertebrate peptides. Their structures are well conserved in those species that are known to have diverged about 300 million years ago and may be present in all insects. The insect kinins were first identified from the cockroach, *Leucophaea maderae,* and caused the contraction of hindgut preparations. They are named according to the genus from which they were isolated: **leucokinins** from the cockroach *L. maderae,* **culekinins** from *Culex* mosquitoes, **achetakinins** from the cricket *Acheta domesticus,* and **locustakinins** from the locust, *Locusta migratoria.*

The first CRF-related diuretic peptide was isolated from head extracts of larval *M. sexta.* The *Manduca*-diuretic hormone (*Manduca*-DH) consists of 41 amino acids and has approximately 30% sequence homology with the CRF family of vertebrate peptides. It stimulates production of cyclic AMP as a second messenger and the subsequent fluid secretion in isolated tubules. Another diuretic peptide, *Manduca*-DPII, was isolated from *Manduca* adults and is only 30 residues in length. Other CRF-related peptides have been isolated from the mealworm, *Tenebrio molitor,* the house fly *Musca domestica,* the locust *L. migratoria,* and the cockroach *Periplaneta americana.* The *Locusta*-DH is synthesized in the pars intercerebralis of the brain and transported to the storage lobe of the corpus cardiacum where it is released. Malpighian tubule activity is increased by CRF-related peptides by the ability of these hormones to open sodium channels.

The insect kinins also stimulate fluid secretion by isolated Malpighian tubules. Unlike the CRF-related peptides, the kinins have no effect on cyclic AMP but instead act to increase intracellular calcium in tubule cells. They are produced by neurosecretory cells of the pars intercerebralis and transported to the storage lobes of the corpus cardiacum via the NCC I where they are released. The kinins are small peptides of between 6 and 14 amino acids that were first characterized as myotropins before their diuretic function was discovered. In contrast to the CRF-related peptides, the insect kinins causes sodium transport to decline and potassium transport to increase. Specific kinin-binding sites have been identified on the Malpighian tubules of crickets.

The presence of CRF-related peptides and the kinins is not mutually exclusive and both may be present in insects, perhaps acting in a synergistic or modulatory fashion to precisely control Malpighian tubule function and the specific ions to be transported. In the blood-sucking bug, *Rhodnius prolixus,* there are two factors that regulate diuresis: a peptide diuretic hormone and **5-hydroxytryptamine** (**5-HT**), both of which act synergistically to stimulate fluid secretion by the tubules after a blood meal. The 5-HT inhibits the Na/K ATPase, and the resulting inhibition of the movement of sodium into the hemolymph causes it to move into the tubule lumen. 5-HT is also active in *M. sexta* and adult *A. aegypti.* In the mosquito, 5-HT also abolishes hindgut peristalsis. Another peptide, insect **cardioaccelera-**

tory peptide 2b (**CAP_{2b}**), stimulates the production of cyclic GMP in the tubules that in turn increases fluid secretion. **Nitric oxide** (**NO**), which also stimulates fluid secretion in the tubules, may do so by stimulating the enzymatic pathways for cyclic GMP production.

The absence or inactivation of diuretic hormones generally terminates the accelerated diuresis by Malpighian tubules. Another way to prevent fluid loss is to increase the resorption by the hindgut. Antidiuretic hormones act on the hindgut to increase its resorption of the primary urine produced by the Malpighian tubules. The **chloride transport-stimulating hormone** (**CTSH**) is a peptide of approximately 8 kDa and in the grasshopper *Schistocerca* is produced in the pars intercerebralis of the brain and transported to storage and glandular lobes of the corpus cardiacum via the NCC I. It activates the electrogenic uptake of chloride and opens potassium channels in rectal cells using cAMP as a second messenger. As with diuretic hormone, CTSH is released after feeding in the locust and the combination of the two hormones may increase fluid cycling to enhance the clearance of the hemolymph without altering the excretory rate.

Another antidiuretic hormone acts on the ileum of the hindgut to increase its rate of resorption. The **ion transport peptide** (**ITP**) has a molecular weight between 7.7 and 8.7 kDa and is located in the storage lobe of the CC. It is a member of a family of crustacean neuropeptides that include a crab hyperglycemic hormone and lobster molt-inhibiting hormone. It acts with cyclic AMP as a second messenger, causing increases in sodium absorption, entry of chloride via an electrogenic pump, and ammonia secretion into the lumen.

STORAGE EXCRETION

One strategy for dealing with wastes is to sequester them within the body rather than eliminate them. Because uric acid is so insoluble, it can be easily stored without it interacting with other physiological processes. Some cockroaches accumulate up to 10% of their dry weight in uric acid stored in specialized **urate cells** in the fat body, which can be utilized during periods of dietary stress. However, because the degradation of the stored uric acid occurs by microbial symbionts in specialized cells of the fat body, this may not be a good example of strict storage excretion. In this case, the net amount of uric acid in the fat body may simply be a result of the balance between synthesis by the insect and degradation by the symbionts.

In some lepidoptera, the fat body shifts from excretion of uric acid to its storage during the last larval instar. With the onset of wandering, normal uric acid excretion is terminated and uric acid begins to be accumulated within the insect. This uric acid is stored in the fat body during the pupal stage but is transported to the rectum for elimination shortly before adult eclosion. The switch from elimination to storage is the result of the exposure to 20HE in the absence of JH. The switch back to transport results from declining levels of 20HE after the molting peak. The males of some cockroach species store as much as 5% of their total live weight in uric acid maintained in the male accessory glands. In the tropical cockroach, *Xestoblatta hamata,* the male excretes this urate and offers it to the female before copulation. The status of the female's dietary nitrogen level determines how much of this secretion is fed upon, and the nitrogen is incorporated into her eggs. This nuptial gift shortens the time from mating to oviposition.

Other waste products may be utilized in other ways. Some locusts are able to feed on plants containing high levels of phenols that serve as feeding deterrents for most other species. The locust incorporates these phenols into the cuticle by using them to cross-link cuticular proteins that usually are linked by tyrosine, and saving the amino acid for other functions. The coloration of some scarab beetles results from the deposition of uric acid in the cuticle, and *Pieris* butterflies deposit pterins in their cuticles, which are end products of nitrogen metabolism.

OTHER FUNCTIONS OF THE MALPIGHIAN TUBULES

The Malpighian tubules of some insects synthesize silk. In the larvae of the lacewing, *Chrysopa,* some of the tubule cells become thickened and produce the silk used to construct the pupal cocoon. The tubules of larval antlions also produce silk that is stored in a rectal sac and used during pupation. Larval spittle bugs (Cercopidae, Homoptera) produce the spittle in which they live from their Malpighian tubules. The calcium used to fortify the puparium of the face fly, *Musca autumnalis,* is transferred from the tubules to the cuticle. Microfilariae of *Dirofilaria immitis,* the nematode responsible for dog heartworm, are ingested by adult mosquitoes and develop in the distal cells of their Malpighian tubules. Large numbers of parasites within the tubules can kill the mosquitoes.

ADDITIONAL REFERENCES

Audsley, N., McIntosh, C., Phillips, J. E. 1992. Actions of ion-transport peptide from locust corpus cardiacum on several hindgut transport processes. *J. Exp. Biol.* **173:** 275–288.

Audsley, N., Goldsworthy, G. J., Coast, G. M. 1997. Circulating levels of *Locusta* diuretic hormone: The effects of feeding. *Peptides* **18:** 59–65.

Audsley, N., Goldsworthy, G. J., Coast, G. M. 1997. Quantification of *Locusta* diuretic hormone in the central nervous system and corpora cardiaca: Influence of age and feeding status, and mechanism of release. *Regul. Pept.* **69:** 25–32.

Audsley, N., Kay, I., Hayes, T. K., Coast, G. M. 1995. Cross reactivity studies of CRF-related peptides on insect Malpighian tubules. *Comp Biochem. Physiol. A. Physiol.* **110:** 87–93.

Becker, B. F. 1993. Towards the physiological function of uric acid. *Free Rad. Biol. Med.* **14:** 615–631.

Bernays, E. A., Woodhead, S. 1982. Plant phenols utilized as nutrients by a phytophagous insect. *Science* **216:** 201–202.

Beyenbach, K. W. 1995. Mechanism and regulation of electrolyte transport in Malpighian tubules. *J. Insect Physiol.* **41:** 197–208.

Beyenbach, K. W., Petzel, D. H. 1987. Diuresis in mosquitoes: Role of a natriuretic factor. *NIPS* **2:** 171–175.

Beyenbach, K. W., Oviedo, A., Aneshansley, D. J. 1993. Malphigian tubules of *Aedes aegypti*: Five tubules, one function. *J. Insect Physiol.* **39:** 639–648.

Bradley, T. J. 1983. Functional design of microvilli in the Malpighian tubules of the insect *Rhodnius prolixus*. *J. Cell Sci.* **60:** 117–135.

Bradley, T. J. 1985. The excretory system: Structure and physiology. In *Comprehensive Insect Physiology, Biochemistry and Pharmacology* (G. A. Kerkut, L. I. Gilbert, Eds.), vol. 4, pp. 421–465. Pergamon Press, Oxford.

Bradley, T. J. 1987. Physiology of osmoregulation in mosquitoes. *Annu. Rev. Entomol.* **32:** 439–462.

Bresler, V. M., Belyaeva, E. A., Mozhayeva, M. G. 1990. A comparative study on the system of active transport of organic acids in Malpighian tubules of insects. *J. Insect Physiol.* **36:** 259–270.

Buckner, J. S. 1982. Hormonal control of uric acid storage in the fat body during last-larval instar of *Manduca sexta*. *J. Insect Physiol.* **28:** 987–993.

Buckner J. S., Caldwell, J. M., Knoper, J. A. 1985. Subcellular localization of uric acid storage in the fat body of *Manduca sexta* during the larval-pupal transformation. *J. Insect Physiol.* **31:** 741–753

Buckner J. S., Caldwell, J. M. 1980. Uric acid levels during last larval instar of *Manduca sexta* an abrupt transition from excretion to storage in fat body. *J. Insect Physiol.* **26:** 27–32.

Clark, T. M., Bradley, T. J. 1996. Stimulation of Malpighian tubules from larval *Aedes aegypti* by secretagogues. *J. Insect Physiol.* **42:** 593–602.

Coast, G. M. 1996. Neuropeptides implicated in the control of diuresis in insects. *Peptides* **17:** 327–336.

Coast, G. M. 1998. Insect diuretic peptides: Structures, evolution and actions. *Am. Zool.* **38:** 442–449.

Coast, G. M. 1998. The influence of neuropeptides on Malpighian tubule writhing and its significance for excretion. *Peptides* **19:** 469–480.

Coast, G. M. 1998. The regulation of primary urine in insects. In *Recent advances in arthropod endocrinology* (G. M. Coast, S. G. Webster, Eds.), pp. 189–209. Cambridge University Press, Cambridge.

Coast, G. M., Audsley, N., Goldsworthy, G. J. 1997. The regulation of postfeeding diuresis in the migratory locust, *Locusta migratoria*. *Ann. NY Acad. Sci.* **814:** 324–326.

Cochran, D. G. 1985. Nitrogen excretion in cockroaches. *Annu. Rev. Entomol.* **30:** 29–50.

Cochran, D. G., Mullins, D. E. 1982. Physiological processes related to nitrogen excretion in cockroaches. *J. Exp. Zool.* **222:** 227–285.

Davies, S. A., Stewart, E. J., Huesmann, G. R., Skaer, N. J., Maddrell, S. H., Tublitz, N. J., Dow, J. A. 1997. Neuropeptide stimulation of the nitric oxide signaling pathway in *Drosophila melanogaster* Malpighian tubules. *Am. J. Physiol.* **273:** R823–R827.

Dow, J. A. T., Davies, S. A., Sozen, M. A. 1998. Fluid secretion by the *Drosophila* Malpighian tubule. *Am. Zool.* **38:** 450–460.

Dow, J. A., Maddrell, S. H., Gortz, A., Skaer, N. J., Brogan, S., Kaiser, K. 1994. The Malpighian tubules of *Drosophila melanogaster*: A novel phenotype for studies of fluid secretion and its control. *J. Exp. Biol.* **197:** 421–428.

Dow, J. A., Maddrell, S. H., Davies, S. A., Skaer, N. J. V., Kaiser, K. 1994. A novel role for the nitric oxide/cyclic GMP signaling pathway: The control of fluid secretion in *Drosophila*. *Am. J. Physiol.* **266:** R1716–R1719.

Edney, E. B. 1977. Excretion and osmoregulation. In *Water balance in land arthropods*, pp. 96–171. Springer-Verlag, New York.

Ehresmann, D. D., Buckner, J. S., Graf, G. 1990. Uric acid translocation from the fat body of *Manduca sexta* during the pupal-adult transformation: Effects of 20-hydroxyecdysone. *J. Insect Physiol.* **36:** 173–180.

Florey, E. 1982. Excretion in insects: Energetics and functional principles. *J. Exp. Biol.* **99:** 417–424.
Furuya, K., Milchak, R. J., Schegg, K. M., Zhang, J., Tobe, S. S., Coast, G. M., Schooley, D. A. 2000. Cockroach diuretic hormones: Characterization of a calcitonin-like peptide in insects. *Proc. Natl. Acad. Sci. USA* **97:** 6469–6474.
Gee, J. D. 1975. Diuresis in the tsetse fly *Glossina austeni*. *J. Exp. Biol.* **63:** 381–390.
Gee, J. D. 1975. The control of diuresis in the tsetse fly *Glossina austeni*: A preliminary investigation of the diuretic hormone. *J. Exp. Biol.* **63:** 391–401.
Grimstone, A. V., Mullinger, A. M., Ramsay, J. A. 1968. Further studies on the rectal complex of the mealworm *Tenebrio molitor*, l. (Coleoptera, Tenebrionidae). *Phil. Trans. R. Soc. London B* **253:** 343–382.
Gringorten J. L., Friend, W. G. 1982. Water balance in *Rhodnius prolixus* during flight: Quantitative aspects of diuresis and its relation to changes in haemolymph and flight-muscle water. *J. Insect Physiol.* **28:** 573–577.
Hadley, N. F. 1994. *Water relations of terrestrial arthropods*. Academic Press, San Diego.
Harvey, W. R. 1992. Physiology of V-ATPases. *J. Exp. Biol.* **172:** 1–17.
Hazelton, S. R., Felgenhauer, B. E., Spring, J. H. 2001. Ultrastructural changes in the Malpighian tubules of the house cricket, *Acheta domesticus*, at the onset of diuresis: A time study. *J. Morphol.* **247:** 80–92.
Hevert, F. 1984. Water and salt relations. In *Environmental physiology and biochemistry of insects* (K. H. Hoffmann, Ed.), pp. 184–205. Springer Verlag, Berlin.
Hirayama, C., Konno, K., Shinbo, H. 1996. Utilization of ammonia as a nitrogen source in the silkworm, *Bombyx mori*. *J. Insect Physiol.* **42:** 983–988.
Hirayama, C., Konno, K., Shinbo, H. 1997. The pathway of ammonia assimilation in the silkworm, *Bombyx mori*. *J. Insect Physiol.* **43:** 959–964.
Jungreis, A. M., Ruhoy, M., Cooper, P. D. 1982. Why don't tobacco hornworms (*Manduca sexta*) become dehydrated during larval–pupal and pupal–adult development? *J. Exp. Zool.* **222:** 265–276.
Kataoka, H., Troetschler, R. G., Li, J. P., Kramer, S. J., Carney, R. L., Schooley, D. A. 1989. Isolation and identification of a diuretic hormone from the tobacco hornworm, *Manduca sexta*. *Proc. Natl. Acad. Sci. USA* **86:** 2976–2980.
Kim, I. S., Spring, J. 1992. Excretion in the house cricket (*Aceta domesticus*): Relative contribution of distal and mid-tubule to diuresis. *J. Insect Physiol.* **38:** 373–381.
Kim, I. S., Spring, J. 1993. Characteristics of *Locusta migratoria* diuretic hormone. *Arch. Insect Biochem Physiol.* **22:** 133–140.
King, D. S., Meredith, J., Wang, Y. J., Phillips, J. E. 1999. Biological actions of synthetic locust ion transport peptide (ITP). *Insect Biochem. Mol. Biol.* **29:** 11–18.
Lehmberg, E., Schooley, D. A., Ferenz, H. J., Applebaum, S. W. 1993. Characteristics of *Locusta migratoria* diuretic hormone. *Arch. Insect Biochem. Physiol.* **22:** 133–140.
Machin, J. 1981. Water compartmentalisation in insects. *J. Exp. Zool.* **215:** 327–333.
Machin, J., O'Donnell, M. J., Kestler, P. 1986. Evidence against hormonal control of integumentary water loss in *Periplaneta americana*. *J. Exp. Biol.* **121:** 339–348.
Machin, J., O'Donnell, M. J., Coutchie, P. A. 1982. Mechanisms of water vapor absorption in insects. *J. Exp. Biol.* **222:** 309–320.
Maddrell, S. H., Gardiner, B. O. 1976. Excretion of alkaloids by Malpighian tubules of insects. *J. Exp. Biol.* **64:** 267–281.
Maddrell, S. H., Herman, W. S., Mooney, R. L., Overton, J. A. 1991. 5–Hydroxytryptamine: A second diuretic hormone in *Rhodnius prolixus*. *J. Exp. Biol.* **156:** 557–566.
Maddrell, S. H., O'Donnell, M. J., Caffrey, R. 1993. The regulation of haemolymph potassium activity during initiation and maintenance of diuresis in fed *Rhodnius prolixus*. *J. Exp. Biol.* **177:** 273–285.
Maddrell, S. H. P. 1963. Excretion in the blood-sucking bug *Rhodnius prolixus* Stal. I. The control of diuresis. *J. Exp. Biol.* **40:** 247–256.
Maddrell, S. H. P. 1981. The functional design of the insect excretory system. *J. Exp. Biol.* **90:** 1–15.
Maddrell, S. H. P., Herman, W. S., Farndale, R. W., Riegel, J. A. 1993. Synergism of hormones controlling epithelial fluid transport in an insect. *J. Exp. Biol.* **174:** 65–80.

Moffett, D. F., Koch, A. 1992. Driving forces and pathways for H+ and K+ transport in insect midgut goblet cells. *J. Exp. Biol.* **172:** 403–415.

Mordue W. 1970. Evidence for the existence of diuretic and anti–diuretic hormones in locusts. *J. Endocrinol.* **46:** 119–120.

Mordue, W. 1969. Hormonal control of Malpighian tube and rectal function in the desert locust *Schistocerca gregaria. J. Insect Physiol.* **15:** 273–285.

Morgan P. J., Mordue, W. 1984. Diuretic hormone: Another peptide with widespread distribution within the insect CNS. *Physiol. Entomol.* **9:** 197–206.

Mullins, D. E., Cochran, D. G. 1972. Nitrogen excretion in cockroaches: Uric acid is not a major product. *Science* **177:** 699–700.

Mullins, D. E., Cochran, D. G. 1974. Nitrogen metabolism in the American cockroach: An examination of whole body and fat body regulation of cations in response to nitrogen balance. *J. Exp. Biol.* **61:** 557–570.

Mullins, D. E., Cochran, D. G. 1975. Nitrogen metabolism in the American cockroach—I. An examination of positive nitrogen balance with respect to uric acid stores. *Comp. Biochem. Physiol.* **50A:** 489–500.

Mullins, D. E., Cochran, D. G. 1975. Nitrogen metabolism in the American cockroach—II. An examination of negative nitrogen balance with respect to mobilization of uric acid stores. *Comp. Biochem. Physiol.* **50A:** 501–510.

Nicolson, S. W. 1976. Diuresis in the cabbage white butterfly, *Pieris brassicae*: Fluid secretion by the Malpighian tubules. *J. Insect Physiol.* **22:** 1347–1356.

Nicolson, S. W. 1980. Diuresis and its hormonal control in butterflies. *J. Insect Physiol.* **26:** 841–846.

Nicolson, S. W. 1993. The ionic basis of fluid secretion in insect Malpighian tubules: Advances in the last 10 years. *J. Insect Physiol.* **39:** 451–458.

Nicolson, S. W., Hanrahan, S. A. 1986. Diuresis in a desert beetle? Hormonal control of the Malpighian tubules of *Onymacris plana* (Coleoptera: Tenebrionidae). *J. Comp. Physiol. B* **156:** 407–413.

Nittoli, T., Coast, G. M., Sieburth, S. M. 1999. Evidence for helicity in insect diuretic peptide hormones: Computational analysis, spectroscopic studies, and biological assays. *J. Pept. Res.* **53:** 99–108.

Noble–Nesbitt, J., Al–Shukur, M. 1988. Involvement of the terminal abdominal ganglion in neuroendocrine regulation of integumentary water loss in the cockroach *Periplaneta americana. J. Exp. Biol.* **137:** 107–117.

O'Donnell, M. J., Dow, J. A., Huesmann, G. R., Tublitz, N. J., Maddrell, S. H. 1996. Separate control of anion and cation transport in Malpighian tubules of *Drosophila melanogaster. J. Exp. Biol.* **199:** 1163–1175.

O'Donnell, M. J., Maddrell, S. H., Gardiner, B. O. 1984. Passage of solutes through walls of Malpighian tubules of *Rhodnius* by paracellular and transcellular routes. *Am. J. Physiol.* **246:** R759–R769.

O'Donnell, M. J., Maddrell, S. H. P. 1995. Fluid resorption and ion transport by the lower Malpighian tubules of adult female *Drosophila. J. Exp. Biol.* **198:** 1647–1653.

O'Donnell, M. J., Spring, J. H. 2000. Modes of control of insect Malpighian tubules: Synergism, antagonism, cooperation and autonomous regulation. *J. Insect Physiol.* **46:** 107–117.

Pannabecker, T. 1995. Physiology of the Malpighian tubule. *Ann. Rev. Entomol.* **40:** 493–510.

Pannabecker, T. L., Smith, C. A., Beyenbach, K. W., Wasserman, R. H. 1995. Immunocytochemical localization of a plasma membrane calcium pump in the insect (*Lymantria dispar*) Malpighian tubule. *J. Insect Physiol.* **41:** 1105–1112.

Phillips J. 1981. Comparative physiology of insect renal function. *Am. J. Physiol.* **241:** R241–R257.

Phillips, J. E., Hanrahan, J., Chamberlin, M., Thomson, B. 1986. Mechanisms and control of reabsorption in insect hindgut. *Adv. Insect Physiol.* **19:** 329–422.

Phillips, J. E., Audsley, N. 1995. Neuropeptide control of ion and fluid transport across locust hindgut. *Am. Zool.* **35:** 503–514.

Plawner, L., Pannabecker, T. L., Laufer, S., Baustian, M. D., Beyenbach, K. W. 1991. Control of diuresis in the yellow fever mosquito *Aedes aegypti*: Evidence for similar mechanisms in the male and female. *J. Insect Physiol.* **37:** 119–128.

Proux, J. P., Picquot, M., Herault, J. –P., Fournier, B. 1988. Diuretic activity of a newly identified neuropeptide—The arginine–vasopressin–like insect diuretic hormone: Use of an improved bioassay. *J. Insect Physiol.* **34:** 919–927.

Quinlan, M. C., O'Donnell. M. J. 1998. Anti–diuresis in the blood–feeding insect *Rhodnius prolixus* Stal: Antagonistic actions of cAMP and cGMP and the role of organic acid transport. *J. Insect Physiol.* **44:** 561–568.

Quinlan, M. C., Tublitz, N. J., O'Donnell, M. J. 1997. Anti–diuresis in the blood–feeding insect *Rhodnius prolixus* Stal: The peptide CAP2b and cyclic GMP inhibit Malpighian tubule fluid secretion. *J. Exp. Biol.* **200:** 2363–2367.

Ramsay, J. A. 1954. Active transport of water by the Malpighian tubules of the stick insect *Dixippus morosus* (Orthoptera Phasmidae). *J. Exp. Biol.* **31:** 104–113.

Reagan, J. D. 1994. Expression cloning of an insect diuretic hormone receptor: A member of the calcitonin/secretin receptor family. *J. Biol. Chem.* **269:** 9–12.

Reagan, J. D. 1995. Functional expression of a diuretic hormone receptor in baculovirus–infected insect cells: Evidence suggesting that the N–terminal region of diuretic hormone is associated with receptor activation. *Insect Biochem. Mol. Biol.* **25:** 535–539.

Reagan, J. D. 1995. Molecular cloning of a putative Na^+ K^+ 2Cl– cotransporter from the Malpighian tubules of the tobacco hornworm, *Manduca sexta. Insect Biochem. Mol. Biol.* **25:** 875–880.

Reagan, J. D., Li, J. P., Carney, R. L., Kramer, S. J. 1993. Characterization of a diuretic hormone receptor from the tobacco hornworm, *Manduca sexta. Arch. Insect Biochem. Physiol.* **23:** 135–145.

Reagan, J. D., Patel, B. C., Li, J. P., Miller, W. H. 1994. Characterization of a solubilized diuretic hormone receptor from the tobacco hornworm, *Manduca sexta. Insect Biochem. Mol. Biol.* **24:** 569–572.

Reynolds, S. E., Bellward, K. 1989. Water balance in *Manduca sexta* caterpillars: Water recycling from the rectum. *J. Exp. Biol.* **141:** 33–46.

Ryerse, J. S. 1978. Ecdysterone switches off fluid secretion at pupation in insect Malpighian tubules. *Nature* **271:** 745–746.

Ryerse, J. S. 1980. The control of Malpighian tubule developmental physiology by 20–hydroxyecdysone and juvenile hormone. *J. Insect Physiol.* **26:** 449–457.

Sasaki, T., Ishikawa, H. 1995. Production of essential amino acids from glutamate by mycetocyte symbionts of the pea aphid, *Acyrthosiphon pisum. J. Insect Physiol.* **41:** 41–46.

Schal, C., Bell, W. J. 1982. Ecological correlates of paternal investment of urates in a tropical cockroach. *Science* **218:** 170–173.

Schwartz, L. M., Reynolds, S. E. 1979. Fluid transport in *Calliphora* Malpighian tubules: A diuretic hormone from the thoracic ganglion and abdominal nerves. *J. Insect Physiol.* **25:** 847–854.

Spring, J. 1990. Endocrine regulation of diuresis in insects. *J. Insect Physiol.* **36:** 13–22.

Spring, J. H., Albarwani, S. A. 1993. Excretion in the house cricket: Stimulation of rectal reabsorption by homogenates of the corpus cardiacum. *J. Exp. Biol.* **185:** 305–323.

Spring, J. H., Morgan, A. M., Hazelton, S. R. 1988. A novel target for antidiuretic hormone in insects. *Science* **241:** 1096–1098.

Strathie, L. W., Nicolson, S. W. 1993. Post–eclosion diuresis in the flightless insect, the silkmoth *Bombyx mori. Physiol. Entomol.* **18:** 435–439.

Troetschler, R. G., Kramer, S. J. 1992. Mode of action studies on a *Manduca sexta* diuretic hormone. *Arch. Insect Biochem. Physiol.* **20:** 35–47.

Van Kerkhove, E. 1994. Cellular mechanisms of salt secretion by the Malpighian tubules of insects. *Belg. J. Zool.* **124:** 73–90.

Veenstra, J. A. 1988. Effects of 5–hydroxytryptamine on the Malpighian tubules of *Aedes aegypti. J. Insect Physiol.* **34:** 299–304.

Veenstra J. A., Pattillo, J. M., Petzel, D. H. 1997. A single cDNA encodes all three *Aedes* leucokinins, which stimulate both fluid secretion by the Malpighian tubules and hindgut contractions. *J. Biol. Chem.* **272:** 10,402–10,407.

Wessing, A., Zierold, K., Hevert, F. 1992. Two types of concretions in *Drosophila* Malpighian tubules as revealed by X–ray microanalysis: A study in urine formation. *J. Insect Physiol.* **38:** 543–554.

Wharton, G. W. 1985. Water balance of insects. In *Comprehensive Insect Physiology Biochemistry Pharmacology* (G. A. Kerkut, L. I. Gilbert, Eds.), vol. 4, pp. 565–601. Pergamon Press, Oxford.

Wheeler, C. H., Coast, G. M. 1990. Assay and characterization of diuretic factors in insects. *J. Insect Physiol.* **36:** 23–34.

Wheelock, G. D., Petzel, D. H., Gillett, J. D., Beyenbach, K. W., Hagedorn, H. H. 1988. Evidence for hormonal control of diuresis after a blood meal in the mosquito *Aedes aegypti*. *Arch. Insect Biochem. Physiol.* **7:** 75–89.

Wigglesworth, V. B. 1931. The physiology of excretion in a blood–sucking insect, *Rhodnius prolixus* (Hemiptera: Reduviidae). I. Compostion of the urine. *J. Exp. Biol.* **8:** 411–427.

Wigglesworth, V. B. 1931. The physiology of excretion in a blood–sucking insect, *Rhodnius prolixus* (Hemiptera: Reduviidae). II. Anatomy and histology of the excretory system. *J. Exp. Biol.* **8:** 428–442.

Wigglesworth, V. B. 1931. The physiology of excretion in a blood–sucking insect, *Rhodnius prolixus* (Hemiptera: Reduviidae). III. The mechanism of uric acid excretion. *J. Exp. Biol.* **8:** 448–451.

Wilkens, A. S. 1995. Singling out the tip cell of the Malpighian tubules—lessons from neurogenesis. *BioEssays* **17:** 199–202.

Williams Jr., J. C., Hagedorn, H. H., Beyenbach, K. W. 1983. Dynamic changes in flow rate and composition of urine during the post–bloodmeal diuresis in *Aedes aegypti* (L.). *J. Comp. Physiol. B* **153:** 257–265.

Woods, H. A., Bernays, E. A. 2000. Water homeostasis by wild larvae of *Manduca sexta*. *Physiol. Entomol.* **25:** 82–87.

Wright, P. A. 1995. Nitrogen excretion: Three end products, many physiological roles. *J. Exp. Biol.* **198:** 273–281.

Yoder, J. A., Denlinger, D. L. 1991. Water balance in flesh fly pupae and water vapor absorption associated with diapause. *J. Exp. Biol.* **157:** 273–286.

Zeiske, W. 1992. Insect ion homeostasis. *J. Exp. Biol.* **172:** 323–334.

CHAPTER 9

Respiratory Systems

BRINGING OXYGEN TO INSECT CELLS

There are striking similarities between most of the physiological systems of vertebrates and insects, but the most conspicuous difference is in the way that oxygen is brought to the cells. The exchange of materials with the environment is a fundamental requirement for living cells. Oxygen and nutrients must be taken up and metabolic wastes must be discarded. Because all cells are bounded by a cell membrane and consist largely of water, every cell must be bathed in an aqueous medium and the materials to be transferred must be dissolved in water. However, membranes that allow the larger oxygen molecule to enter through it also allow the smaller water molecule to exit. Thus, the ability to take up oxygen through a membrane is also associated with the potential loss of water. In small single-celled animals such as protozoa that live in an aqueous medium, the exchange occurs easily over the surface of the cell. For larger multicellular terrestrial animals, where not every cell is exposed to the environment, systems have evolved that bathe each cell in body fluids that can mediate this exchange.

The hemolymph of the insect circulatory system is responsible for bathing all cells and allowing them to exchange nutrients and metabolites, but the transport of oxygen to the cells occurs through a tracheal system in which external surfaces are invaginated into the body cavity to provide an oxygen pipeline from the outside. Oxygen is more soluble in air than it is in water, so the diffusion of gases in the tracheal system is far more efficient than if it was first dissolved in the hemolymph. By providing a direct supply of oxygen to their cells, insects have eliminated the reliance on an extensive network of fluids that circulates oxygen as in vertebrates. The circulatory system and the blood within it play a very minor role in gas exchange.

Because oxygen is more soluble and diffuses much faster in air than in water, the use of air in the tracheal system as an internal transport medium instead of using water is not only more efficient, but also provides the saving of water.

Although the tracheal system provides an enormous surface area that is permeable to both water and oxygen, it is only open to the outside at the small area that the spiracles present to the environment. At the same time, the system is very effective in supplying oxygen when required. Compared to rates at rest, the rate of oxygen uptake during flight rises over 30 times in adult blowflies with no measurable oxygen debt, while in vertebrates the maximum increase during physical activity is only about 12-fold.

There are some drawbacks to using tracheae for respiration. It has been speculated that the dynamics of the tracheal system have kept insects out of the most abundant habitat on the planet, the oceans. Oxygen travels through the insect respiratory system by diffusion, and because the extreme pressures at lower depths would prevent the tracheal system from operating this way, insects were never able to colonize the harborages that are deep at the ocean floor. In contrast are the gill-bearing crustaceans, a group of arthropods that have successfully carved out an aquatic niche in part because their respiratory system operates at greater depths. The insects are thus relegated to the upper ranges of the oceans, where in the absence of hiding places, they are more likely to be preyed upon by larger aquatic organisms.

THE TRACHEAL SYSTEM

The tracheal system arises as invaginations of the epidermis and consequently is surrounded by epidermal cells that secrete their cuticles inward toward the lumen (Fig. 9.1). The tracheal cuticle typically consists of a chitinous endocuticle and the epicuticle, which is thrown into spiral folds called **taenidia** (Fig. 9.2). The taenidia provide the rigidity that prevents the tracheal tubes from collapsing due to changes in air pressure. As a consequence of being epidermal in origin, the tra-

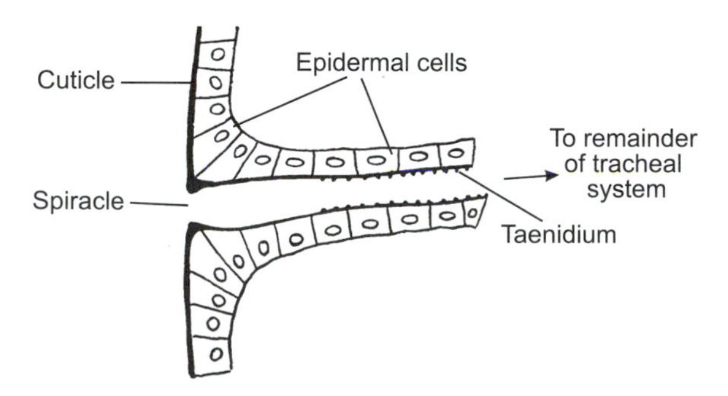

FIGURE 9.1 The development of the tracheal system by the ectodermal invaginations that occur during embryogenesis. From Snodgrass (1935). Reprinted with permission.

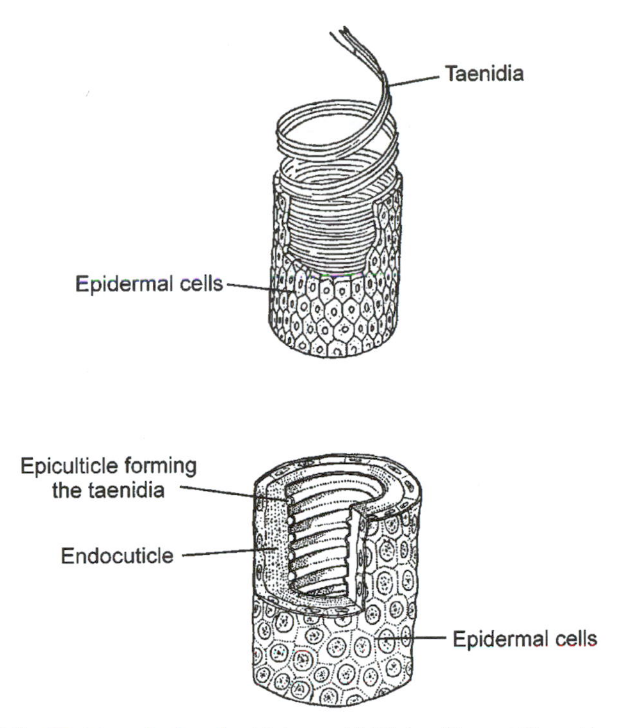

FIGURE 9.2 (Top) A tracheal trunk with its taenidial lining. (Bottom) The epidermal cells produce the taenidia as part of the epicuticular layer. From Wigglesworth (1965). Reprinted with permission.

cheal cells shed the cuticle at each molt, first digesting the old endocuticle and then shedding the thin epicuticle that remains.

The tracheal tubes open to the outside at the **spiracle,** sometimes surrounded by a circular sclerite known as the **peritreme** (Fig. 9.3). An **atrium** may be present, which is an internal cavity lined with hairs to reduce water loss and collect particulates. There is never more than one pair of spiracles per body segment, and the evolutionary trend has been toward the reduction of total numbers of spiracles. Spiracles are never present in the head capsule. The primitive number of spiracles was probably 12 pairs, but present-day insects have no more than 11, consisting of 3 thoracic and 8 abdominal pairs, a condition found in diplurans. There are fewer spiracles present in other insects.

When all spiracles are present and functional, it is known as the **holopneustic** condition (Fig. 9.4). When several pairs have become nonfunctional through evolutionary loss, the condition is referred to as **hemipneustic.** The **propneustic** and **metapneustic** conditions are found in pupal and larval dipterans that have only the first or last abdominal pair functional. Insects that have no functional spiracles are termed **apneustic,** although they still retain an internal tracheal system

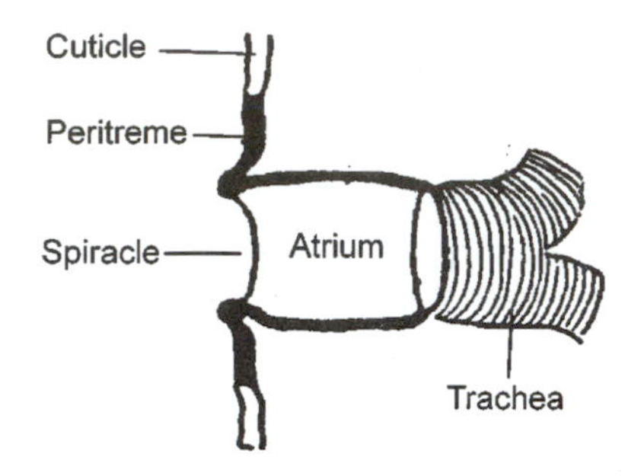

FIGURE 9.3 The opening of the tracheal system at the spiracle. The atrium is an internal cavity just beyond the spiracle that opens to the tracheal system. From Snodgrass (1935). Reprinted with permission.

and extract oxygen through a cuticle that is modified. With the exception of apterygote insects, the tracheae in most other insects are usually connected by large tracheal trunks to allow even a single spiracle to ventilate the entire system. The tracheal system is completely absent in some primitive collembolans that live in moist environments and these insects respire directly through the cuticle.

Spiracles in terrestrial pterygotes are often equipped with closing mechanisms that can restrict water loss by shutting off the tracheal system when oxygen uptake

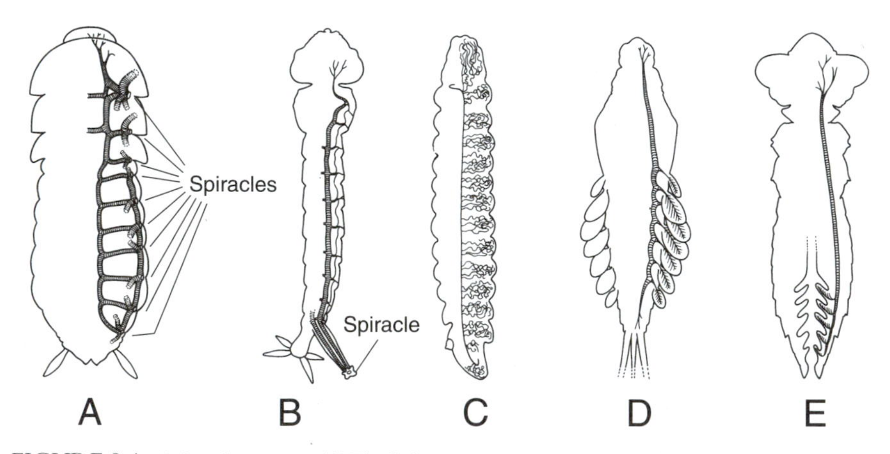

FIGURE 9.4 Spiracular systems. (A) The holopneustic arrangement in which all spiracles are open. (B) A metapneustic arrangement, with the only open spiracle borne on a siphon tube on the last abdominal segment. (C) An apneustic arrangement, with no spiraclular openings, but a cuticle heavily invested with tracheoles to permit cuticular respiration. (D) An apneustic arrangement with tracheoles concentrated in cuticular flaps. (E) An apneustic arrangment with rectal gills, in which oxygen is extracted from the tracheoles that invest the hindgut. From Gullan and Cranston (2000). Reprinted with permission.

is not required (Fig. 9.5). There may be cuticular flaps on the outside that cover the spiracle, or internal valves that close off the tracheal system from the spiracular opening. When the valves bear only one closer muscle, the antagonistic action of cuticular elasticity causes them to open when the muscle is relaxed. There may also be an additional opener muscle associated with some spiracles. The muscles that control these valves are innervated from the ganglia of the ventral nerve cord (Fig. 9.6). Some spiracles, especially when a single muscle controls their opening,

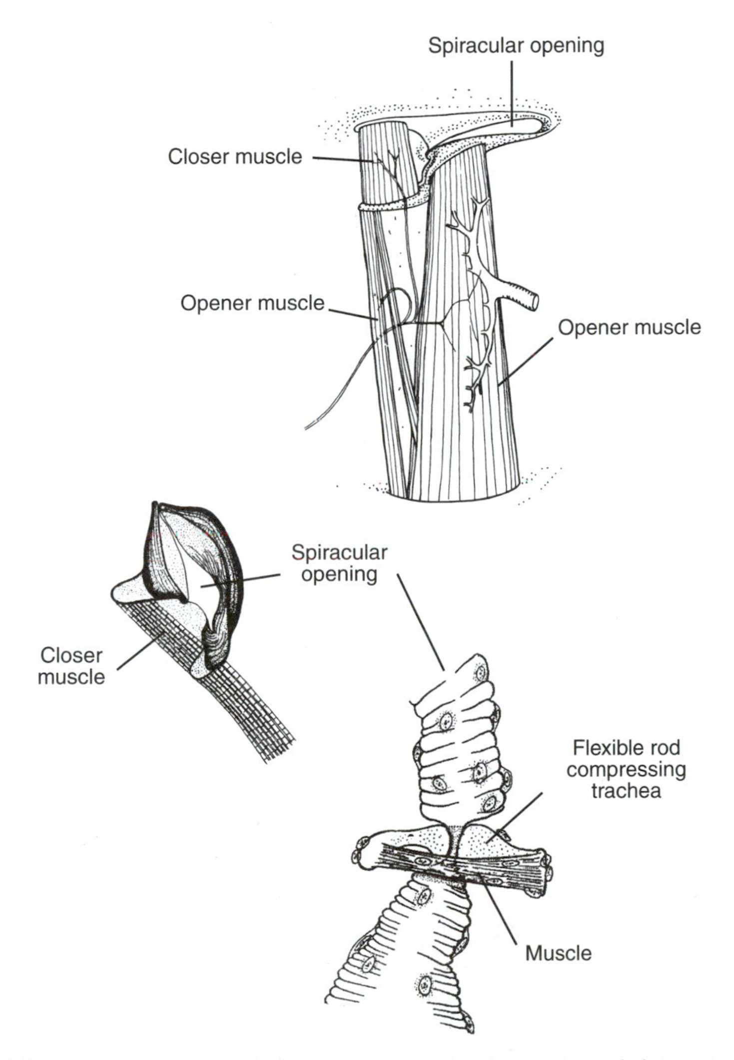

FIGURE 9.5 Some examples of the mechanisms of spiracular opening and closing. From Miller (1974), Nikam and Khole (1988), and Wigglesworth (1965). Reprinted with permission.

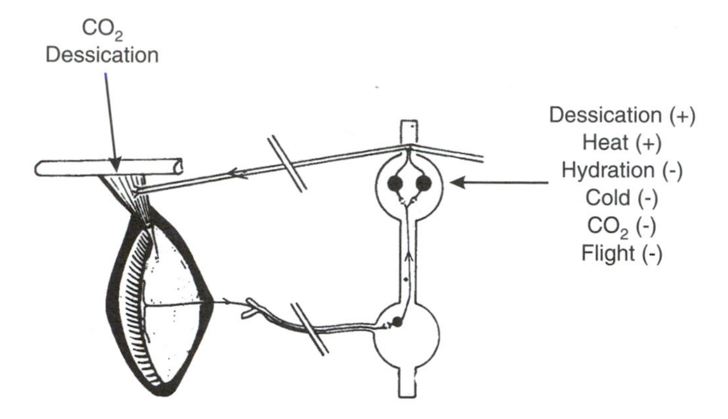

FIGURE 9.6 Nervous innervation of the spiracular closing muscle and the mechanisms that affect its action. From Miller (1964). Reprinted with permission.

may be affected by high levels of CO_2 that act directly on the muscle to cause it to relax and open the valve by the elastic action of the cuticle. When the hemolymph becomes more concentrated with water loss, the increased concentration of ions that results causes the spiracles to stay closed longer. The accumulation of CO_2 during respiration can also increase the H^+ concentrations, and the consequent change in the pH of body fluids affects spiracular opening. The ganglia also can be affected by low oxygen concentrations in the hemolymph, causing the spiracular valves to initiate fluttering. The spiracles need only remain open for a brief moment to admit air without engendering a significant water loss.

The diameter of tracheae at the spiracle may be several millimeters in large insects, but they branch and grow smaller in diameter until tapering to 1–2 μm. At this diameter, the tracheae give rise to **tracheoblasts** that produce the **tracheoles** that mediate the transfer of oxygen to cells. Tracheoles taper to approximately 0.1 μm and likewise produce a chitinous cuticle and taenidia but this cuticle is not shed during the molt. The surfaces of the tracheoles lie close to sites of oxygen uptake, but they do not penetrate cell membranes. Tissues are covered with an investment of tracheoles and based on diffusion constants, it has been determined that the cell mitochondria must be spaced at least 4–8 μm away from a tracheole in order to receive a sufficient supply of oxygen (Fig. 9.7). In insect flight muscles, the tracheoles may actually be spaced about 2–3 μm apart and invaginated into the muscles but never actually penetrate the cell membranes to become functionally intracellular. Tracheoles arising from tracheoblasts grow like the nodes on a plant but are not shed during the molt (Fig. 9.8). A ring of cement secures the connections between the old tracheole and the new trachea after growth.

A small amount of liquid may be present in the tracheole at the cell interface if the insect is not respiring actively, but it is drawn into the cell during muscular

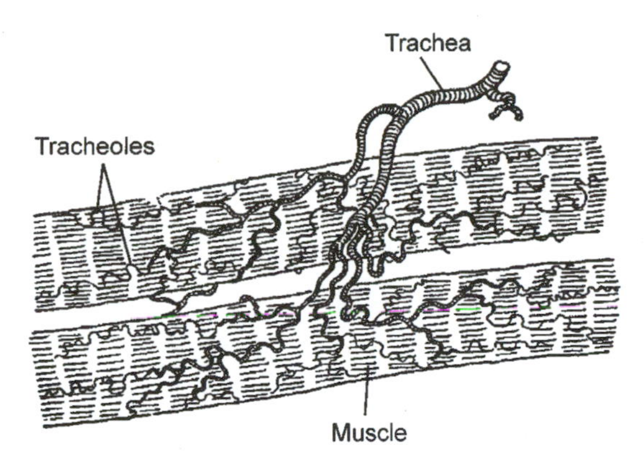

FIGURE 9.7 Trachea branching into tracheoles at the muscle. From Snodgrass (1956). Reprinted with permission.

activity. This movement of liquid appears to be the result of the balance between capillary action pulling out the fluid from cells and the osmotic uptake resulting from the buildup of metabolites within the cell.

Cutaneous respiration occurs in some aquatic insects and endoparasites. Chironomid larvae have a thin cuticle that allows oxygen to be taken up from the water. This uptake is supplemented by the respiratory pigment **hemoglobin** present in the hemolymph, which is the smallest hemoglobin known, consisting of only two subunits and a molecular weight of 31.4 kDa. Hemoglobin is also a com-

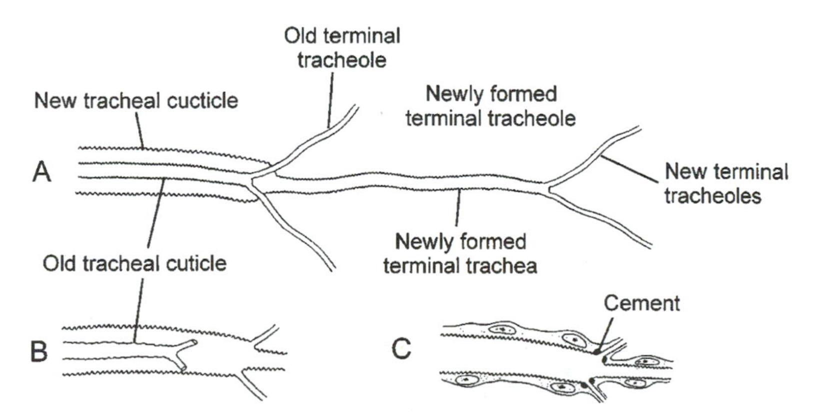

FIGURE 9.8 The mechanism of new tracheal and tracheole growth during a molt. (A) After apolysis and deposition of new tracheal cuticle. (B) The old tracheal cuticle is detached and pulled out of the tracheal tube. (C) Cement produced by the tracheole cells secures them to the old new tracheal cuticle. From Wigglesworth (1981). Reprinted with permission.

ponent of the modified fat body cells in the larvae of the dipteran *Gasterophilus* and in notonectid hemipterans. In the notonectids, the affinity of the hemoglobin for oxygen has been identified as a major determinant of the bug's ability to remain submerged for a significantly longer time. The tracheal systems of some endoparasitic hymenopterans are filled with fluid during the first instar and contain air only beginning during the second instar. The spiracles are nonfunctional during most of the larval stage until the parasite is about to leave the host.

MODIFICATIONS THAT INCREASE OXYGEN UPTAKE

The diffusion of oxygen through the tracheal system is adequate for a small insect to respire, but larger, more active insects use additional means to supplement this diffusion. Abdominal pumping in some insects ventilates the tracheal system by changing its volume. Muscular contractions of the abdomen change its shape, and contract and expand the volume of the tracheal tubes within it, allowing them to be ventilated. There are areas of the tracheal trunks that may also be dilated into sacs that have reduced taenidia, allowing the trunks to be compressed by changes in hemolymph pressure so that air can be pumped in and out like a bellows (Fig. 9.9). Many larger insects appear to be breathing as they pump their abdomens and thereby change the volume of these tracheal sacs.

When the tracheal sacs are located around the flight muscles, they can be pumped when the insect flies. In flying locusts, the peak demands of the flight muscles for oxygen are better met when the volume of the sacs is automatically increased and decreased by contractions of those muscles during flight. Each muscle is supplied with a primary tracheal trunk and air sac so that the tracheoles are well supplied with oxygen. In the locust, *Schistocerca,* abdominal pumping can ventilate the tracheal system by 40 liters/kg/h, but the additional thoracic pumping

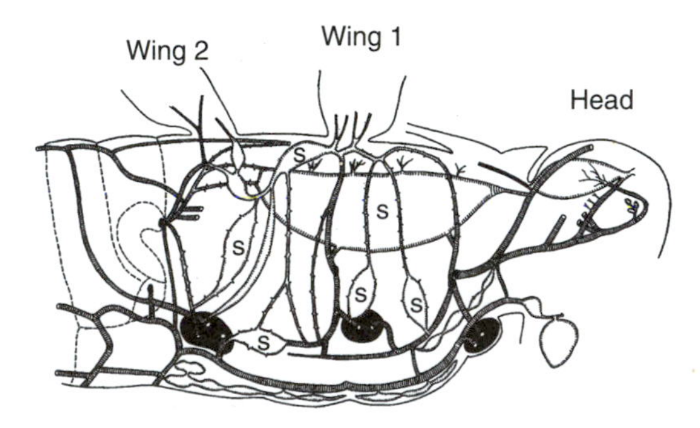

FIGURE 9.9 Tracheal air sacs (s) within the tracheal system can be increased and decreased in volume to pump air through the system. From Albrecht (1953). Reprinted with permission.

during flight can increase this rate to 250 liters/kg/h. The opening and closing of spiracular valves along the body segments is coordinated to produce a directed flow of air through the tracheal system.

The circulatory system can affect the rate at which air enters the tracheal system. The heartbeat, normally directed anteriorly by peristaltic contractions of the dorsal vessel, can reverse its direction and beat backward. During this reversal, hemolymph accumulates in the abdomen, compressing the abdominal air sacs and forcing air out of the tracheal system. At the same time, the thoracic air sacs expand, bringing air into the spiracle that supplies these tracheae. By alternating its flow of hemolymph, the blowfly *Calliphora* can modulate the ventilation of the tracheal system (Fig. 9.10).

NONRESPIRATORY FUNCTIONS OF TRACHEAL SYSTEMS

Spiracular systems may have accessory functions in addition to their role in gas exchange. In the grasshopper *Romalea,* the tracheal system is modified for the release of defensive secretions. A phenolic compound is produced by a glandular epithelium in the spiracular trunks that is expelled along with air. Similarly, in the cockroaches *Diploptera punctata* and *Blaberus discoidalis,* defensive secretions are forced out with air pressure through the spiracular openings. In the Madagascar hissing cockroach, *Gromphadorhina portentosa,* the hissing sounds are made when air is expelled through a modified spiracle.

DISCONTINUOUS GAS EXCHANGE

Diapausing moth pupae produce discrete bursts of carbon dioxide during their development, presumably as a means of restricting water loss when drinking is obviously not possible. The pattern of this release begins when the spiracles are

FIGURE 9.10 The regulation of tidal air flow in the tracheal system by the movement of hemolymph. By moving the hemolymph into the thorax or abdomen, it can compress tracheal sacs causing air to move in or out. From Wasserthal (1996). Reprinted with permission.

closed, during which time oxygen is utilized from the air already present in the tracheal system. As the oxygen is consumed, the carbon dioxide that is produced is first expressed as increased bicarbonate concentrations in the hemolymph. When the oxygen levels become reduced, the spiracular opener muscles relax and the spiracular valves flutter, allowing oxygen to enter the tracheal system. As higher levels of carbon dioxide accumulate in the tracheae, the spiracles open and allow the gas to escape. Thus, oxygen entry is fairly continuous but the loss of carbon dioxide, along with water, occurs only discontinuously during the brief open phase of the spiracles. The pupa remains within the bounds of water balance because the metabolic water generated by the hydrolysis of stored fats equals the water lost during this discontinuous respiration. Discontinuous respiration has also been identified in beetles and grasshoppers and may be a general rule for most insects when they are at rest, allowing them to exchange gases with a minimum of water loss.

AQUATIC RESPIRATION

Aquatic insects have evolved from terrestrial ancestors, and in order for them to return to the water, various adaptations have been necessary. The oxygen content of water is considerably lower than that of air because of the physical characteristics of gases in water. Therefore, to obtain a comparable degree of oxygen in the water, an aquatic insect must ventilate its gas exchange surface at a much higher rate than that of an air-breathing animal. However, the spiracles of terrestrial insects are too small to function in water and their cuticles are impermeable to gas exchange. Obviously, to enable some insects to re-exploit aquatic niches, it was necessary for them to evolve certain adaptations that allowed them to breathe in water.

Renewal of Air Supplies

The least modified adaptation is found in insects that have retained the open respiratory system of terrestrial insects. With the retention of a conventional open tracheal system, the challenge was to prevent its flooding with water. Once this problem was solved, the basic strategies for survival underwater included either the total reliance on atmospheric oxygen or the capacity to bring air underwater to satisfy the insect's short-term demands for oxygen.

The problem of keeping water out of the terrestrial tracheal system was addressed by the evolution of hydrofuge surfaces, which have a greater affinity for air than for water due to the waxes produced by glands at their base. The hydrofuge hairs that surround a spiracle will cover the spiracle when they are submerged and open from surface tension when the insect surfaces. Several metapneustic dipteran larvae, such as mosquitoes, have spiracles at the end of an abdominal siphon that allow the larvae to feed at a lower level in the water while they simultaneously respire at the surface. The hydrofuge hairs cover the spiracular opening when the insect submerges (Fig. 9.11).

Another modification is the ability to capture a small air store that opens into the spiracle and allow the insect to remain active underwater for short periods. Surfacing is still necessary to replenish the bubble, but the air store extends the time it can remain submerged. For example, Dytiscid beetles carry air that is held in place by hydrofuge surfaces beneath the elytra. The air store also provides the insect with buoyancy as it swims.

Several insects have retained their terrestrial tracheal system and spiracles while evolving the means to tap into underwater plants to obtain oxygen while submerged. Thus, all stages of the coleopteran *Donacia* can live at the bottom of bodies of water and by using a sharp posterior siphon, can penetrate the roots and extract the oxygen from them. Similarly, larval mosquitoes in the genus *Mansonia* are able to remain underwater by tapping the stems of plants with their sharply pointed spiracle.

Cutaneous Respiration

Another way to prevent water from entering the tracheal system is to completely close off the spiracles and respire through the cuticle. Larvae of the aquatic dipteran, *Chironomus,* have a thin cuticle that allows oxygen to diffuse into the

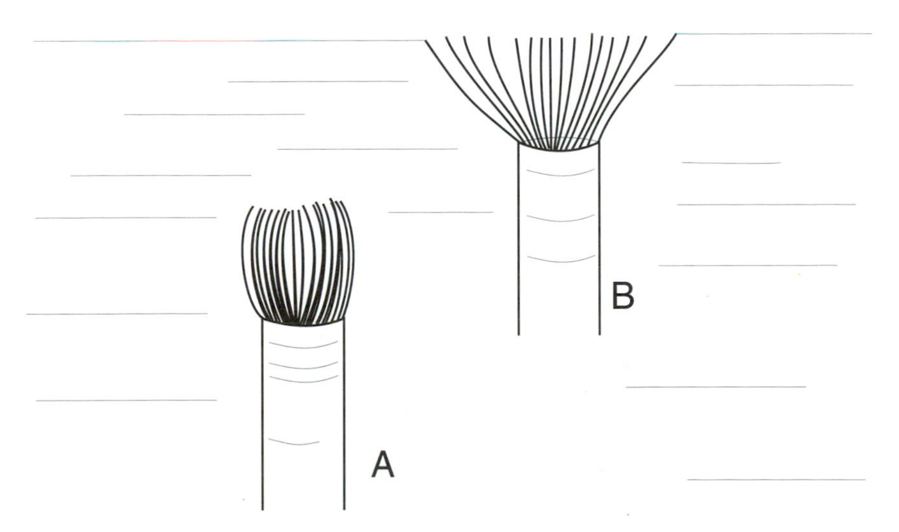

FIGURE 9.11 Hydrofuge hairs on the tip of the siphon tube of an aquatic insect. When underwater, the hairs remain over the spiracular opening and prevent water from entering. At the surface, the hairs open up and allow air to enter. Adapted from Wigglesworth (1965). Reprinted with permission.

well-developed tracheal system beneath it. In another dipteran, *Atrichopogon,* the tracheae are associated with the cuticle in eight specific areas through which respiration takes place.

Tracheal Gills

Another step toward respiration through the cuticle is the development of tracheal gills that are outgrowths of the body wall covered by relatively thin cuticles and have rich supplies of tracheae. The abdominal gills of ephemeropterans are plate-like outgrowths that undulate continuously to circulate oxygenated water over their surfaces as the insect swims (Fig. 9.12). Zygopteran odonate larvae have three caudal gills that are similarly configured to take up oxygen, and their undulations also serve as rudders to aid in swimming.

Plastron Respiration

When the density of hydrofuge hairs on the cuticle is high, the bubble of air that the insect captures when surfacing is held tightly enough when it is underwater so that it may serve as a physical gill to extract oxygen from the water. Initially, the bubble consists of gases in the same proportion as in the atmosphere, mainly nitrogen and oxygen (Fig. 9.13). As the oxygen diffuses from the bubble into the tracheal system, the oxygen tension in the bubble decreases, causing oxygen to diffuse into the gill from the water. At the same time, the decline in oxygen tension alters the nitrogen tension in the gill, causing this gas to slowly diffuse out of the bubble until the bubble itself declines in volume and can no longer act as a gill. At that point, the insect must surface and replenish the atmospheric bubble, but the gill can provide up to 13 times the quantity of oxygen initially present. This system is also known as a compressible gas gill because the volume of the gill collapses as it is used.

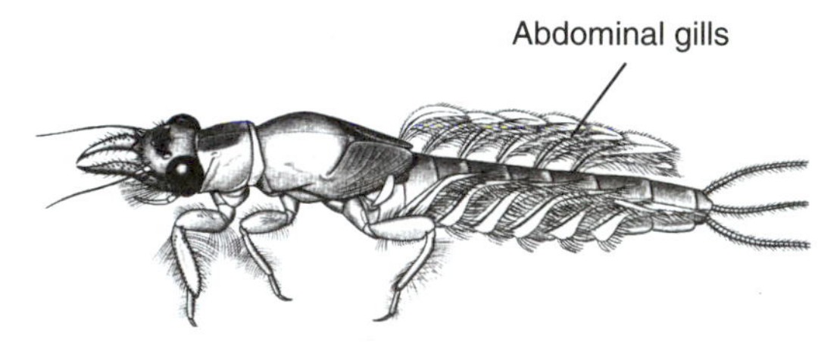

FIGURE 9.12 Abdominal gills of the mayfly. From McCafferty (1981). Reprinted with permission.

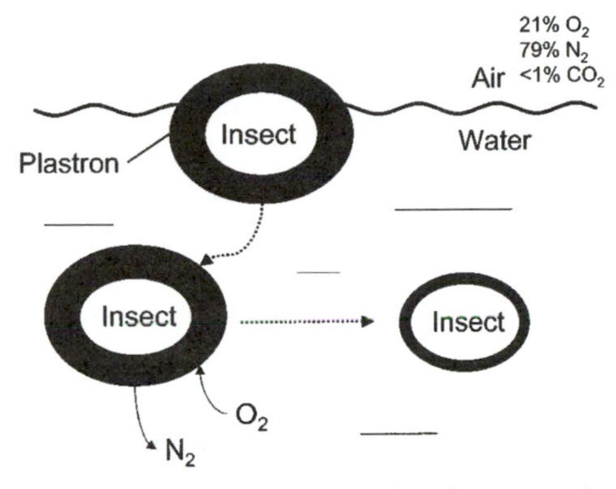

FIGURE 9.13 The mechanism of plastron respiration. When the insect surfaces, a bubble of air that is held in place by hydrofuge hairs surrounds it. When the insect submerges, the bubble remains in place. As oxygen in the bubble is used by the insect to respire, it also diffuses from the water and replenishes what is lost. In insects with a compressible gill, the nitrogen eventually diffuses into the water and the insect must resurface. In insects with an incompressible gill, the hydrofuge hairs are more dense and hold the nitrogen bubble more firmly in place, allowing the plastron to operate for long periods.

In insects with a permanent or incompressible gas gill, the bubble is held very tightly by a dense array of 10^6 to 10^8 hydrofuge hairs per cm^2. Even when the concentration of oxygen in the bubble is reduced, the volume of the gill remains the same size because of the efficient retention by the hairs. Because it remains in place over a long period, insects with incompressible gas gills can stay underwater for months without surfacing. The only requirement is that the insect remains in well-oxygenated water or the plastron will work in reverse to pass oxygen from the bubble to the water.

As described in Chapter 3, insect eggs form a relatively closed system, with the female parent packaging all the materials embryonic cells need for growth and differentiation during embryogenesis except for oxygen. The cells in the developing embryo must engage in gas exchange, obtaining oxygen without the concurrent loss of water. The mechanism of gas exchange in the egg is similar to cutaneous respiration in postembryonic stages, with gases able to pass through the intricate meshwork of the chorion via the numerous spaces in the chorionic framework (Fig. 3.1). Insect eggs are frequently glued to substrates and thus inundated with water from rain and even dew. Given that these periods of inundation may represent a significant proportion of their embryonic stages, even these terrestrial eggs must be equipped with aquatic adaptations. The chorionic meshwork, when filled with air, can also serve as a plastron that operates when the egg is submerged for these brief periods. Some insect eggs have additional chorionic horns that bear plastrons.

Spiracular Gills

Spiracular gills are found in insects that inhabit running water that is highly oxygenated but also subject to periodical drying, present in the pupal stages of several dipterans and coleopterans (Fig. 9.14). The advantage of spiracular gills is that under dry conditions respiration can still occur with the spiracular gill minimizing water loss when out of the water.

The gills are outgrowths of the spiracle, sometimes including outgrowths of the body wall in their structures. Most spiracular gills are covered with a plastron that provides a large surface area for oxygen transfer to occur by diffusion when they are immersed in water. When dry, the plastron allows the direct uptake of oxygen from areas closest to the spiracle with the distal remainder of the structure nonfunctional. Oxygen can thus be taken up without reducing the permeability of the cuticle and involving a loss of water.

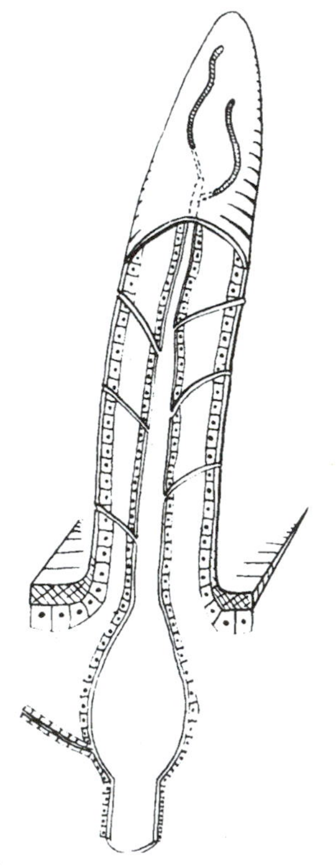

FIGURE 9.14 Spiracular gills of the blackfly pupa. Reprinted with permission from Mill, P.J. (1974). *The Physiology of Insecta* 6: 403–467. Copyright Academic Press.

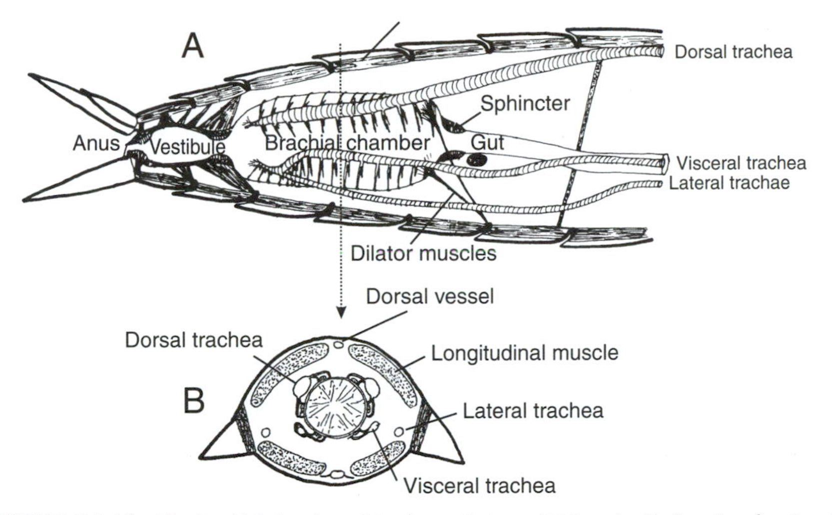

FIGURE 9.15 The brachial chamber of the dragonfly larva. (A) Longitudinal section showing the relationship of the gut and tracheae. (B) Cross section at the arrow. Reprinted from Hughes, G.M. and P.J. Mill (1966). *Journal of Experimental Biology* **44:** 317–333. With kind permission from the Company of Biologists Ltd.

Brachial Chamber

Tracheal gills are present in the modified hindgut of dragonfly larvae, creating a **brachial chamber** that extracts oxygen from the water contained there (Fig. 9.15). The wall of the chamber is lined with circular and longitudinal muscles that change its volume, causing oxygenated water to be alternately ventilated and ejected. Approximately 85% of the water is regularly renewed. The walls are richly supplied with tracheoles that take up oxygen from the water. Digested food passing through the hindgut is enclosed in a peritrophic membrane and does not foul the brachial chamber with wastes. The rapid expulsion of water from the anus also allows the larva to engage in jet propulsive swimming.

ADDITIONAL REFERENCES

Baccetti, B., Burrini, G., Gabbiani, G., Leoncini, P. 1984. Insect tracheal taenidia contain a keration-like protein. *Physiol. Entomol.* **9:** 239–245.

Beitel, G. J., Krasnow, M. A. 2000. Genetic control of epithelial tube size in the *Drosophila* tracheal system. *Development* **127:** 3271–3282.

Burkett, B. N., Schneiderman, H. A. 1974. Roles of oxygen and carbon dioxide in the control of spiracular function in cecropia pupae. *Biol. Bull.* **147:** 274–293.

Bursell, E. 1957. Spiracular control of water loss in the tsetse-fly. *Proc. R. Entomol. Soc. London A* **32:** 21–29.

Bursell, E. 1970. *An introduction to insect physiology.* Academic Press, New York.

Chappell, M. A., Rogowitz, G. L. 2000. Mass, temperature and metabolic effects on discontinuous gas exchange cycles in eucalyptus-boring beetles (Coleoptera: Cerambycidae). *J. Exp. Biol.* **203:** 3809–3820.

Chiang, C., Young, K. E., Beachy, P. A. 1995. Control of *Drosophila* tracheal branching by the novel homeodomain gene unplugged, a regulatory target for genes of the bithorax complex. *Development* **121:** 3901–3912.

Cooper, P. D. 1983. Components of evaporative water loss in the desert tenebrionid beetles *Eleodes armata* and *Cryptoglossa verrucosa*. *Physiol. Zool.* **56:** 47–55.

Davis, A. L., Chown, S. L., Scholtz, C. H. 1999. Discontinuous gas-exchange cycles in *Scarabaeus* dung beetles (Coleoptera: Scarabaeidae): Mass-scaling and temperature dependence. *Physiol. Biochem. Zool.* **72:** 555–565.

Eulenberg, K. G., Schuh, R. 1997. The tracheae defective gene encodes a bZIP protein that controls tracheal cell movement during *Drosophila* embryogenesis. *EMBO J.* **16:** 7156–7165.

Guillemin, K., Groppe, J., Ducker, K., Treisman, R., Hafen, E., Affolter, M., Krasnow, M. A. 1996. The pruned gene encodes the *Drosophila* serum response factor and regulates cytoplasmic outgrowth during terminal branching of the tracheal system. *Development* **122:** 1353–1362.

Gulinson S., Harrison, J. 1996. Control of resting ventilation rate in grasshoppers. *J. Exp. Biol.* **199:** 379–389.

Hadley, N. F. 1994. Ventilatory patterns and respiratory transpiration in adult terrestrial insects. *Physiol. Zool.* **67:** 75–189.

Hadley, N. F., Quinlan, M. C. 1993. Discontinuous carbon dioxide release in the eastern lubber grasshopper *Romalea guttata* and its effect on respiratory transpiration. *J. Exp. Biol.* **177:** 169–180.

Harrison J., Hadley, N., Quinlan, M. 1995. Acid–base status and spiracular control during discontinuous ventilation in grasshoppers. *J. Exp. Biol.* **198:** 1755–1763.

Hebets, E. A., Chapman, R. F. 2000. Surviving the flood: Plastron respiration in the non-tracheate arthropod *Phrynus marginemaculatus* (Amblypygi: Arachnida). *J. Insect Physiol.* **46:** 13–19.

Hinton, H. E. 1968. Spiracular gills. *Adv. Insect Physiol.* **5:** 65–162.

Hinton, H. E. 1969. Respiratory systems of insect egg-shells. *Annu. Rev. Entomol.* 14: 343–368.

Hinton, H. E. 1976. Plastron respiration in bugs and beetles. *J. Insect Physiol.* **22:** 1529–1550.

Hinton, H. E. 1976. Respiratory adaptations of marine insects. In *Marine Insects* (L. Cheng, Ed.), pp. 43–78. North-Holland, Amsterdam.

Hoback, W. W., Stanley, D. W. 2001. Insects in hypoxia. *J. Insect Physiol.* **47:** 533–542.

Holdgate, M. W., Seal, M. 1956. The epicuticular wax layers of the pupa of *Tenebrio molitor* L. *J. Exp. Biol.* **33:** 82–106.

Hood, W. G., Tschinkel, W. R. 1990. Dessication resistance in arboreal and terrestrial ants. *Physiol. Entomol.* **15:** 23–35.

Hoyle, G. 1960. The action of carbon dioxide gas on an insect spiracular muscle. *J. Insect Physiol.* **4:** 63–79.

Hughes, G. M., Mill, P. J. 1966. Patterns of ventilation in dragonfly larvae. *J. Exp. Biol.* **44:** 317–333.

Johnson, R. A. 2000. Water loss in desert ants: Caste variation and the effect of cuticle abrasion. *Physiol. Entomol.* **25:** 48–53.

Komai, Y. 1998. Augmented respiration in a flying insect. *J. Exp. Biol.* **201:** 2359–2366.

Krafsur, E. S., Willman, J. R., Graham, C. L., Williams, R. E. 1970. Observations on spiracular behaviour in *Aedes* mosquitoes. *Ann. Entomol. Soc. Am.* **63:** 684–691.

Krolikowski, K., Harrison, J. 1996. Haemolymph acid-base status, tracheal gas levels and the control of post-exercise ventilation rate in grasshoppers. *J. Exp. Biol.* **199:** 391–399.

Lehmann, F. O., Dickinson, M. H., Staunton, J. 2000. The scaling of carbon dioxide release and respiratory water loss in flying fruit flies (*Drosophila* spp.). *J. Exp. Biol.* **203:** 1613–1624.

Lighton, J. R. B. 1996. Discontinuous gas exchange in insects. *Annu. Rev. Entomol.* **41:** 309–324.

Lighton, J. R. B., Garrigan, D. A., Duncan, F. D., Johnson, R. A. 1993. Water-loss rate and cuticular permeability in foragers of the desert ant *Pogonomyrmex rugosus*. *Physiol. Zool.* **62:** 1232–1256.

Lighton, J. R. B., Fukushi, T., Wehner, R. 1993. Ventilation in *Cataglyphis bicolor*: Regulation of carbon dioxide release from the thoracic and abdominal spiracles. *J. Insect Physiol.* **39:** 687–699.

Locke, M. 1958. The formation of tracheae and tracheoles in *Rhodnius prolixus*. *Q. J. Microsc. Sci.* **99:** 29–46.

Locke, M. 1958. The coordination of growth in the tracheal system of insects. *Q. J. Microsc. Sci.* **99:** 373–391.

Locke, M. 1998. Caterpillars have evolved lungs for hemocyte gas exchange. *J. Insect Physiol.* **44:** 1–20.

Maddrell, S. H. P. 1998. Why are there no insects in the open sea? *J. Exp. Biol.* **201:** 2461–2464.

Meyer, E. P. 1989. Corrosion casts as a method for investigation of the insect tracheal system. *Cell Tissue Res.* **256:** 1–6.

Mill, P. J. 1985. Structure and physiology of the respiratory system. In *Comprehensive Insect Physiology, Biochemistry, and Pharmacology* (G. A. Kerkut, L. I. Gilbert, Eds.), vol. 3, pp. 517–593. Pergamon Press, Oxford.

Mill, P. J. 1974. Respiration: Aquatic insects. In *The physiology of insecta* (M. Rockstein, Ed.), vol. 6, pp. 403–467. Academic Press, New York.

Mill, P. J., Pickard, R. S. 1972. Anal valve movement and normal ventilation in aeshnid dragonfly larvae. *J. Exp. Biol.* **56:** 537–543.

Miller, P. L. 1962. Spiracle control in adult dragon-flies (Odonata). *J. Exp. Biol.* **39:** 513–535.

Miller, P. L. 1964. Possible function of haemoglobin in *Anisops*. *Nature* **201:** 1052.

Miller, P. L. 1974. Respiration-aerial gas transport. In *The physiology of insecta* (M. Rockstein, Ed.), vol. 6, pp. 345–402. Academic Press, New York.

Miller, P. L. 1982. Respiration. In *The American cockroach* (W. J. Bell, K. G. Adiyodi, Eds.), pp. 87–116. Chapman and Hall, London.

Nikam, T. B., Khole, V. V. 1988. *Insect spiracular systems*. Halsted Press, New York.

Noirot, C., Noirot-Timothee, C. 1982. The structure and development of the tracheal system. In *Insect ultrastructure* (R. C. King, H. Akai, Eds.), pp. 351–381. Plenum, New York.

Phillips, D. M. 1970. Insect sperm: Their structure and morphogenesis. *J. Cell Biol.* **44:** 243–277.

Rahn, H., Paganelli, C. V. 1968. Gas exchange in gas gills of diving insects. *Respir. Physiol.* **5:** 145–164.

Samakovlis, C., Hacohen, N., Manning, G. 1996. Development of the *Drosophila* tracheal system occurs by a series of morphologically distinct but genetically coupled branching events. *Development* **122:** 1395–1407.

Samakovlis, C., Manning, G., Steneberg, P., Hacohen, N., Cantera, R., Krasnow, M. A. 1996. Genetic control of epithelial tube fusion during *Drosophila* tracheal development. *Development* **122:** 3531–3536.

Sass, M., Kiss, A., Locke, M. 1994. The localization of surface integument peptides in tracheae and tracheoles. *J. Insect Physiol.* **40:** 561–575.

Schneiderman, H. A. 1960. Discontinuous respiration in insects: role of the spiracles. *Biol. Bull.* **119:** 494–528.

Shilo, B. Z., Gabay, L., Glazer, L., Reichman-Fried, M., Wappner, P., Wilk, R., Zelzer, E. 1997. Branching morphogenesis in the *Drosophila* tracheal system. *Cold Spring Harbor Symp. Quant. Biol.* **62:** 241–247.

Slama, K. 1988. A new look at insect respiration. *Biol. Bull.* **175:** 289–300.

Slama, K. 1999. Active regulation of insect respiration. *Ann. Entomol. Soc. Am.* **92:** 916–929.

Snyder, G. K., Sheafor, B., Scholnick, D., Farrelly, C. 1995. Gas exchange in the insect tracheal system. *J. Theor. Biol.* **172:** 199–207.

Walshe, B. M. 1950. The function of haemoglobin in *Chironomous plumosus* under natural conditions. *J. Exp. Biol.* **27:** 73–95.

Wasserthal, L. T. 1996. Interaction of circulation and tracheal ventilation in holometabolous insects. *Adv. Insect Physiol.* **26:** 297–351.

Weis-Fogh, T. 1964. Functional design of the tracheal system of flying insects as compared with the avian lung. *J. Exp. Biol.* **41:** 207–227.

Weis-Fogh, T. 1964. Diffusion in insect wing muscle, the most active tissue known. *J. Exp. Biol.* **41:** 229–256.

Weis-Fogh, T. 1967. Respiration and tracheal ventilation in locusts and other flying insects. *J. Exp. Biol.* **47:** 561–587.

Whitten, J. M. 1972. Comparative anatomy of the tracheal system. *Annu. Rev. Entomol.* **17:** 373–402.

Wigglesworth, V. B. 1945. Transpiration through the cuticle of insects. *J. Exp. Biol.* **21:** 97–113.

Wigglesworth, V. B. 1981. The natural history of insect tracheoles. *Physiol. Entomol.* **6:** 121–128.

Wigglesworth, V. B. 1983. The physiology of insect tracheoles. *Adv. Insect Physiol.* **17:** 85–148.

Wigglesworth, V. B. 1990. The direct transport of oxygen in insects by large tracheae. *Tissue Cell* **22:** 239–243.

Wigglesworth, V. B. 1990. The properties of the lining membrane of the insect tracheal system. *Tissue Cell* **22:** 231–238.

Wigglesworth, V. B., Lee, W. M. 1982. The supply of oxygen to the flight muscles of insects: a theory of tracheole physiology. *Tissue Cell* **14:** 501–518.

Woods, H. A., Bernays, E. A. 2000. Water homeostasis by wild larvae of *Manduca sexta. Physiol. Entomol.* **25:** 82–87.

Zelzer, E., Shilo, B. Z. 2000. Cell fate choices in *Drosophila* tracheal morphogenesis. *BioEssays* **22:** 219–226.

CHAPTER 10

Muscular Systems

INSECT MUSCLE

In order for ancestral insects to leave the water and establish life on land, it was necessary for them to evolve an additional means of maneuverability. Once this occurred, their efficient movement on land and in the air, as well as in water, was a major factor in their domination of terrestrial ecosystems. The investment made in the muscles that power this movement can be substantial; flight muscles alone can comprise as much as 65% of the total body mass of some insects.

It is sometimes baffling that these small creatures can move with such impressive speed and accuracy, as anyone who has tried to catch a fly will attest. In addition to their maneuverability, insects can lift many times their own weight and jump many times their body lengths. The small size of insects and their apparent strength might at first suggest that their muscles are somehow different from those of vertebrates. However, there is a surprising similarity between the muscles of the two groups; the organization and the basic structure of the muscle fibers in insects are not appreciably different from that of vertebrates, except that the muscles and muscle fibers are smaller in insects and the number of muscle fibers in each muscle is reduced. Although the total strength of an insect is obviously less, the absolute power of insect muscle, defined by the load it can raise per cross-sectional area, does not differ significantly from that of vertebrates. Their size belies their complexity; the number of individual muscles in some insect species exceeds that found in humans.

BASIC STRUCTURE OF INSECT MUSCLES

Only a relatively small number of insect muscles has been studied and none is known as well as the frog muscle that has served as the basis for much of what is known about vertebrate muscle structure. However, the muscles of the few insect species that have been examined show strong similarities to the structure of vertebrate muscles. Because the study of muscles predates modern cell theory, many of the usual terms used to describe cell structures are different when they are applied to muscle cells.

All insect muscles are **striated,** owing to the regular arrangement of their components that produce repeating patterns of light and dark bands under the light microscope. The composition of a skeletal muscle consists of many elongated **muscle fibers,** each of which is a single multinucleate cell (Fig. 10.1). Visceral muscles, in contrast, tend to have muscle fibers that contain only a single nucleus. Each fiber is surrounded by an electrically excitable **sarcolemma,** an outer plasma membrane, that encloses the **sarcoplasm,** or cytoplasm of the cell. Enclosed within the sarcoplasm of muscle fibers are many smaller **myofibrils** that are arranged longitudinally and stretch the length of the fiber. The distinctive banding pattern is visible within the myofibrils because they are in turn composed of two types of overlapping **myofilaments,** consisting of the proteins **actin** or **myosin.** The thinner myofilaments consist of actin that is complexed with the regulatory peptides **tropomyosin** and **troponin.** The thicker myofilaments are composed of myosin. A specialized endoplasmic reticulum, the **sarcoplasmic reticulum,** serves as a store for calcium and transports it from the interior of the reticulum into the sarcoplasm. Invaginations of the sarcolemma form the **transverse tubule,** or **T system,** that is closely associated with the sarcoplasmic reticulum (Fig. 10.2).

The specific areas of the myofibrils where actin and myosin overlap are optically birefringent, or anisotropic, and are therefore referred to as **A bands** (Fig. 10.1). At the center of the A band is the **H zone,** a lighter region consisting of the myosin myofilaments alone. The regions containing only actin are isotropic and thus called **I bands,** The I band is divided by a thin, dense **Z line** in which the actin filaments terminate. A functional contractile unit, the **sarcomere,** is defined as the area between the two Z lines. The lining up of the Z lines accounts for the striated pattern that is visible under the light microscope.

A single muscle can only contract; it can lengthen only when stretched by other antagonistic muscles. The contraction of skeletal muscles can occur to about 50% of their normal length, occurring when the myofilaments slide past each other, shortening the length of the sarcomere as the Z lines move closer together (Fig. 10.3). Although the individual myofilaments do not change in length, the overlap areas of the I band, consisting of only actin, and the H zone, consisting of only myosin, decrease in size during contraction. The unchanging distance between the Z line and the edge of the H zone is another indication that the actin myofilaments remain the same size during contraction.

Actin, Myosin, and Muscle Activation

The shortening of the sarcomere results from the interactions between actin and myosin in the muscle fibers. Myosin has three important biological properties: it

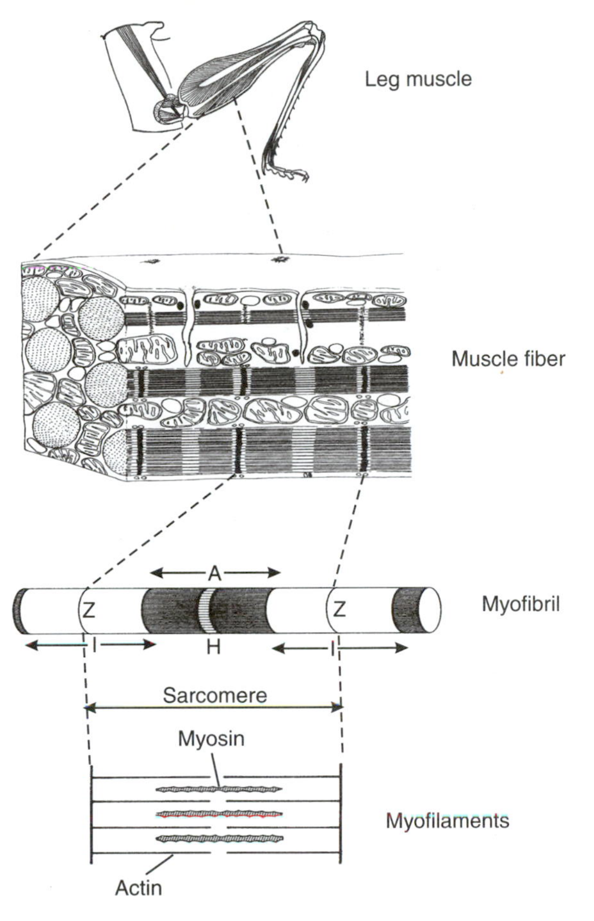

FIGURE 10.1 The structure of insect muscle. From the top: A leg containing several muscles, consisting of many muscle fibers. Each muscle fiber is a cell surrounded by an electrically excitable cell membrane. Within the cytoplasm of muscle fibers are longitudinal arrays of myofibrils that extend its length. The banding pattern visible in the myofibrils results from the degree of overlap of actin and myosin myofilaments. The dark A band results from actin and myosin overlap. The H zone within the A band represents that portion of the myosin that does not overlap. The I bands are areas of actin alone, and the Z line is the actin end plate.

assembles into filaments, it binds to actin, and is an actin-activated ATPase that hydrolyzes ATP to form ADP and inorganic phosphate that yields the energy required for muscle contraction. The large myosin molecule consists of a number of polypeptide chains that are composed of light meromyosin filaments and a heavy meromyosin head. The heavy meromyosin generates the force during muscle contraction. It can be further divided into two components, a globular S1 that

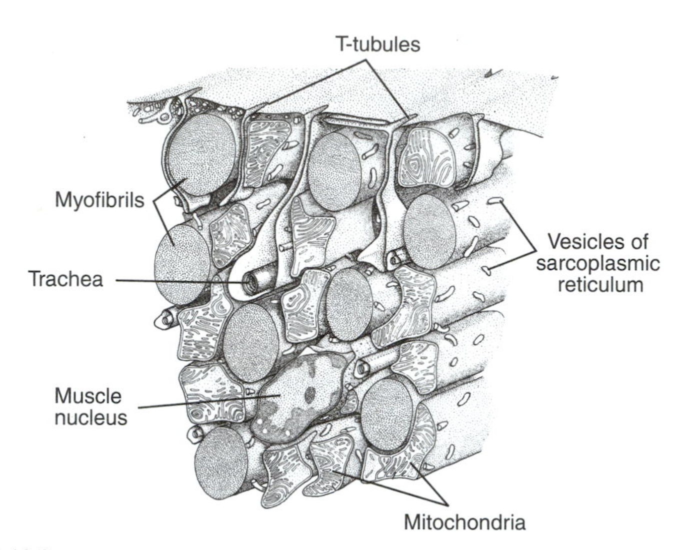

FIGURE 10.2 A cross section through a flight muscle fiber, showing the myofibrils in the cytoplasm and the invaginations of the T-tubules that permit membrane depolarizations to reach inside the cell. From Pringle (1975). Reprinted with permission.

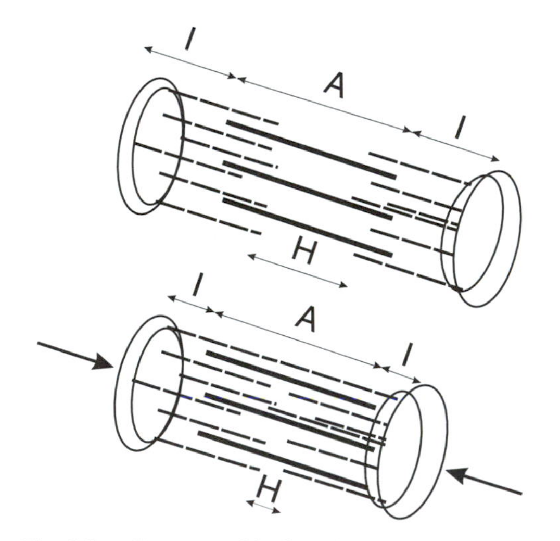

FIGURE 10.3 The sliding filament model of muscle contraction. The end plates of the sarcomere move closer together when the actin and myosin filaments slide past each other. The A band does not change in size, but both the I band and H zone decrease in size as the overlap between the myofilaments is altered.

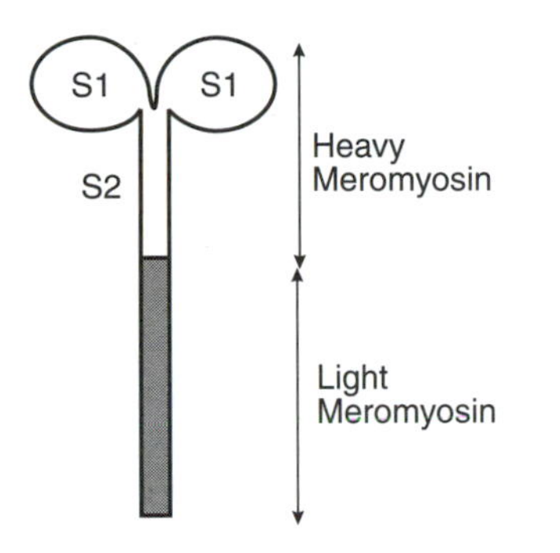

FIGURE 10.4 The structure of the myosin filament.

makes up the heads and has ATPase activity and a rod-shaped S2 region that links the heads with the filaments (Fig. 10.4).

Actin is a globular protein that exists in a complex with other proteins including the tropomyosin and troponin that control the interactions between actin and myosin (Fig. 10.5). The tropomyosin forms a two-stranded helical rod that runs parallel to the actin filament. The troponin complex is located at intervals along

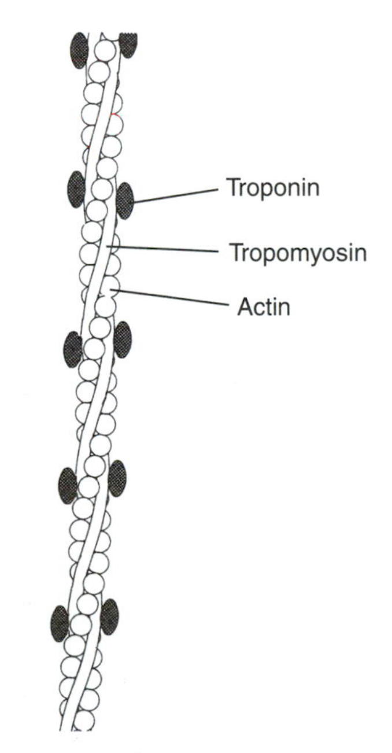

FIGURE 10.5 The composition of the actin filament, complexed with troponin and tropomyosin.

the actin filament, and consists of three subunits. One of the subunits binds to actin, another to tropomyosin, and a third to calcium.

Reaching deeply into the sarcoplasm midway between the Z line and the H zone, where the myofilaments overlap are the invaginations of the sarcolemma called the transverse tubules. The components of this T system are closely associated with the sarcoplasmic reticulum, the network of vesicles that surrounds the myofibrils and serves as a reservoir for calcium. A depolarization of the sarcolemma by a nervous impulse passes to the inside of the muscle via the transverse tubules and activates the sarcoplasmic reticulum to release calcium from its internal stores in the proximity of the myofilaments. A calcium pump restores the calcium to the sarcoplasmic reticulum (Fig. 10.6). When this calcium is reversibly bound by troponin it induces a conformational change in the troponin– tropomyosin complex that exposes binding sites on the actin filament and allows the S1 heads of the myosin cross-bridges to attach to actin (Fig. 10.7). Once the S1 myosin heads bind

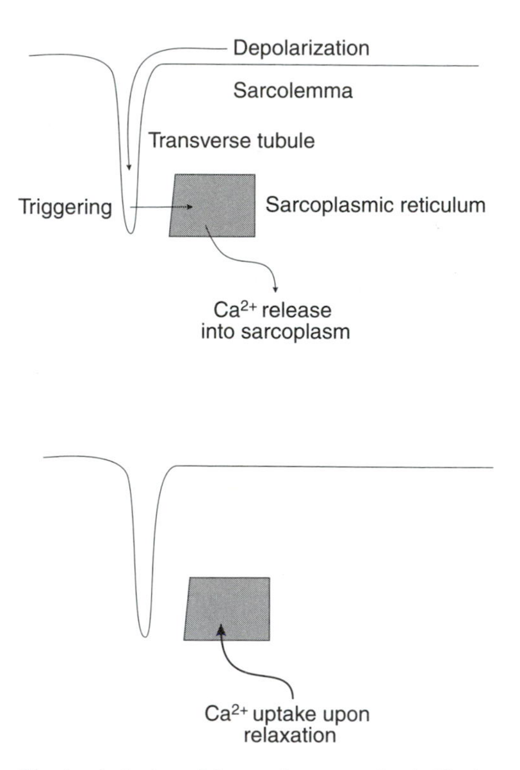

FIGURE 10.6 The depolarization of the sarcolemma reaches inside the muscle fiber via the transverse tubule, causing the sarcoplasmic reticulum to release calcium into the cytoplasm. The binding of the calcium by the actin complex triggers the contraction. The calcium is taken up by the sarcoplasmic reticulum during muscle relaxation. From Keynes and Aidley (1981). Reprinted with permission.

ATP, it is hydrolyzed into ADP and inorganic phosphate and the S1 heads change their shape to a high-energy configuration. The energized heads then bind to specific sites on actin as a cross-bridge. The binding causes a conformational change in the myosin heads that rotates them and produces the force that moves the actin filament toward the center of the sarcomere. They then revert to their low-energy state.

The attachment between the low-energy myosin and the actin is broken when a new molecule of ATP binds to the S1 heads. The free myosin heads are again able to bind to another area farther up the actin myofilament causing more contraction of the sarcomere. Thus, each myosin cross-bridge undergoes a cycle of attachment to actin, rotation, and detachment. The cross-bridge next reattaches to a new site on the actin and the cycle is repeated.

When the central nervous system ends the excitation, calcium is withdrawn from the sarcoplasm by the sarcoplasmic reticulum and the calcium-depleted troponin turns the muscle off by again covering the binding sites. The contraction of insect muscle is thus regulated by two events: the depolarization of the muscle by a nerve impulse and the elevated calcium concentration in the sarcoplasm. The nervous depolarization triggers contraction, and the elevated calcium determines when and for how long the contraction occurs.

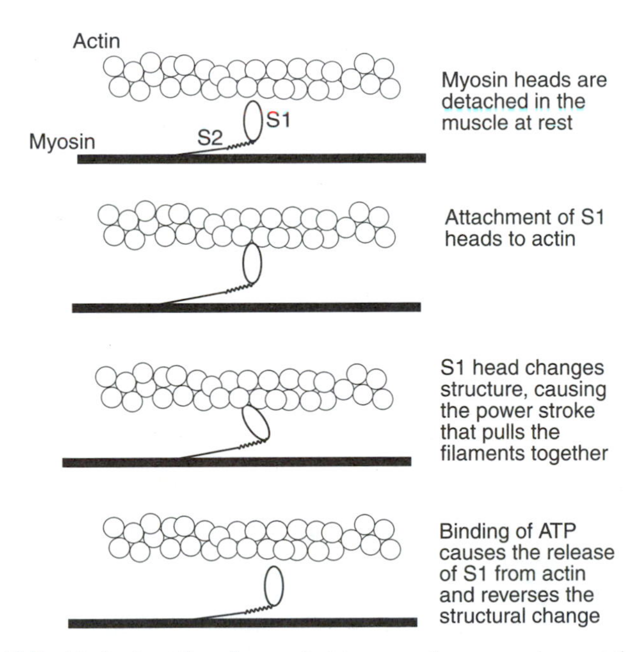

FIGURE 10.7 Mechanism of attachment, detachment, and movement between the myofilaments. Adapted from Huxley (1976). Reprinted with permission.

TYPES OF INSECT MUSCLES

There are two major types of insect muscle. **Visceral muscles** surround the viscera but do not attach to the body wall. **Skeletal muscles** are anchored to the exoskeleton at either end and move parts of the skeleton relative to each other. Skeletal muscles are further divided into **synchronous** and **asynchronous** muscles.

Synchronous Muscles

Most skeletal muscles in insects are called synchronous because they contract when they receive a nervous signal from the motor neurons that innervate them. Each nervous signal is followed by a single contraction of the muscle, and in the case of wing muscles, produce a wingbeat frequency in the range of 5–30 per second. The abundant sarcoplasmic reticulum in synchronous muscle cells enables a large amount of calcium to be released when the muscles are stimulated, and the sarcomeres undergo a maximal degree of contraction. The extensive sarcoplasmic reticulum also allows the calcium to be efficiently resequestered for muscle relaxation. Synchronous muscles are found in what are considered to be the relatively primitive flight muscles of orthopterans, lepidopterans, and odonates.

Synchronous skeletal muscles lie on a continuum, with **fast muscle fibers** at one extreme and **slow muscle fibers** at the other. Fast muscle fibers have shorter sarcomeres, more extensive sarcoplasmic reticulum, and a 3:1 ratio of actin to myosin filaments. They are innervated by excitatory fast neurons that release a large number of neurotransmitter packets to cause a strong contraction. Slow muscle fibers have less sarcoplasmic reticulum, longer sarcomeres, and a 6:1 ratio of actin to myosin filaments (Fig. 10.8). By virtue of their reduced sarcoplasmic reticulum they permit calcium to remain in the sarcoplasm longer and they consequently have a slower rate of relaxation. They are often innervated by slow neurons that release small packets of neurotransmitter at the neuromuscular junction and produce small depolarizations. Although a single impulse may not evoke a contraction, repeated firing of the slow neuron allows a summation of the depolarizations to occur. These differences in structure produce differences in function. Fast muscle fibers are associated with a rapid contraction, while slow muscle fibers undergo a slower, sustained contraction. Individual muscles may consist of all slow or fast fibers or mixed populations of slow, intermediate, and fast fibers (Fig. 10.9).

The contraction of most synchronous muscles is generally limited to a maximum shortening of about 50%, but there are special supercontracting and superextending muscles associated with structures that undergo an unusual degree of extension. For example, the intersegmental muscles that allow the elongation of the abdominal segments containing the ovipositor in some insects can stretch to over 10 times their normal length and shorten by as much as 90% (Fig. 10.10). This unusual change in length is due to the penetration of the Z disc by the

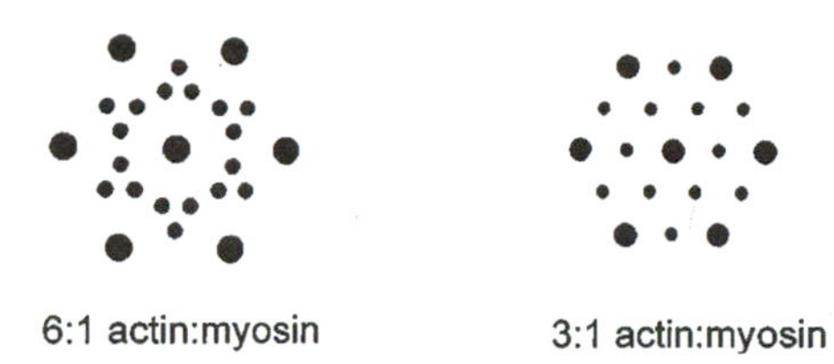

FIGURE 10.8 Cross sections of slow (left) and fast (right) muscle fibers. From Aidley (1985). Reprinted with permission.

myofilaments during supercontraction and by the breakup of the Z line into Z bodies during superextension.

Asynchronous Muscles

The insects in several more advanced orders, including dipterans, coleopterans, hymenopterans, and some hemipterans, have evolved smaller wings that enable them to occupy niches unavailable to those insects with large wings. In order for these smaller wings to achieve the necessary aerodynamic forces to support flight, they must beat much more rapidly. However, the higher wingbeat frequency that is necessary may exceed the rate at which the nervous impulses can reach the muscles. The wingbeat of synchronous muscles is limited to no more than 100 Hz because after the transmission of a nervous impulse, the neuron undergoes a brief refractory period during which time its membrane potentials are restored and no additional impulses can be transmitted. This structural design limits the rate at which a neuron can fire. If wingbeat were restricted by the speed of nervous transmission there would be a definite restriction on the lower size limit of flying insects, because flight by the smaller insects could only be possible if their smaller wings were able to beat faster. In the ceratopogonid midge, *Forcipomyia,* the wingbeat approximates 1000 per second, which is too fast for the control of flight muscles by individual nervous impulses. The rapid contraction of its flight

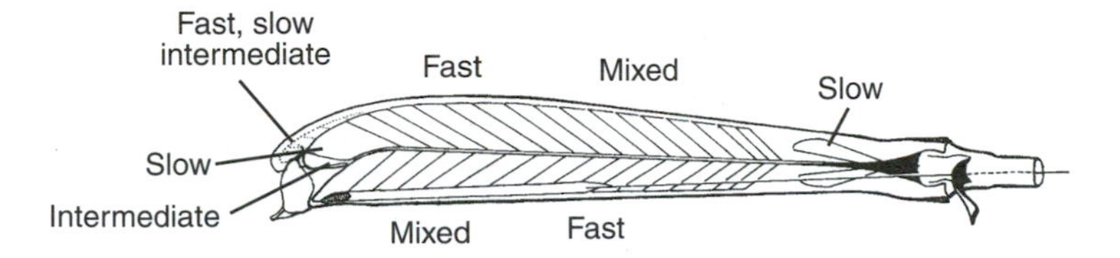

FIGURE 10.9 Populations of fast and slow muscle fibers in an insect leg. From Delcomyn (1985). Reprinted with permission.

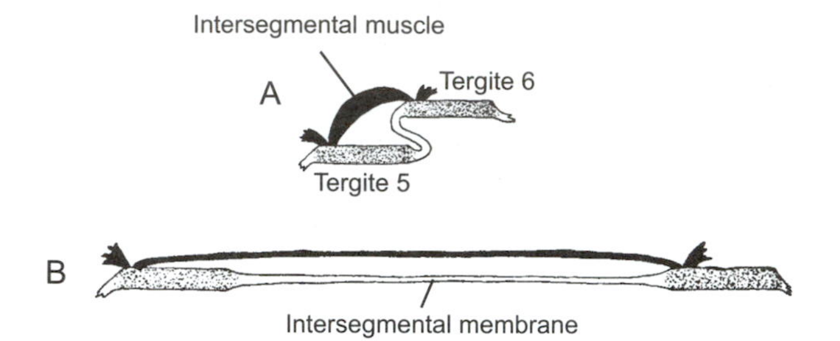

FIGURE 10.10 Intersegmental muscles allow the superextension (B) and supercontraction (A) of the body segments. Reprinted from *Journal of Insect Physiology*, Volume 29. Jorgensen, W.K. and M.J. Rice. Superextension and supercontraction in locust ovpositor muscles, pp. 437–448. Copyright 1983, with permission from Elsevier Science.

muscles is only possible because another type of muscle is present that does not require individual nerve impulses to stimulate their contraction.

These special muscles are called either **fibrillar,** owing to their large myofibrils, or **asynchronous,** because they contract without the 1:1 synchrony with electrical events. The frequency at which nervous signals activate these muscles is considerably less than the frequency of contractions. In contrast to the one nervous impulse per contraction in synchronous muscles, asynchronous muscles generally oscillate through 5–25 contractions for each nervous impulse (Fig. 10.11).

Asynchronous flight muscles appear to have evolved independently several times from synchronous muscle, but they show the same structure wherever they are found. Asynchronous flight muscles have a reduced sarcoplasmic reticulum that has a slower rate of sarcoplasmic calcium exchange. This provides a significant energy savings by not having to pump calcium in and out of the sarcoplasmic reticulum with each contraction. A constant level of calcium is maintained by a low frequency of nervous impulses, and self-oscillatory contractions are initiated by the characteristics of the myofibrils. Nervous excitation brings the contractile apparatus into a state of activation at which time it becomes sensitive to mechanical stretching. The stretching may expose more myosin heads, allowing them to contact the actin molecules more readily. There is also a greater overlap of actin and myosin filaments that allows an increased number of attachment points for the lower amplitude contractions. The contraction of asynchronous muscles is on the order of only 1–2% as a result of the shorter I bands than in synchronous muscle.

There are several other structural differences in asynchronous muscles. The protein **arthrin** is a form of actin that has been isolated from asynchronous flight muscles and may be involved in the regulation of muscle activation by stretch. Another protein, **projectin,** projects from the Z lines into the I bands of asynchronous muscle. **Flightin** is additional novel myofibrillar protein in asynchronous muscles that appears in different isoforms as the adult *Drosophila* matures its flight capabilities after emergence. A unique accessory protein on the actin myofila-

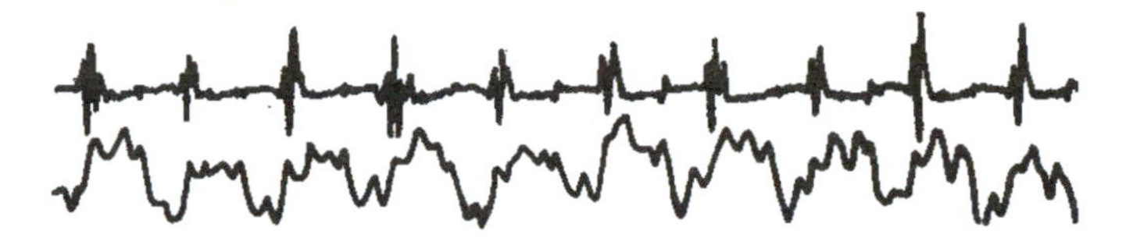

FIGURE 10.11 Nervous impulses (top in each row) and muscle contractions (bottom in each row). Traces for synchronous muscles, in which the nervous impulses cause a contraction, are shown in the top row. Asynchronous muscles, in which occasional nervous impulses maintain a state of contractibility, are shown at the bottom. From Pringle (1981). Reprinted with permission.

ments, **troponin-H,** is found only in asynchronous muscle. Troponin-H may modulate the mechanism of muscle activation by calcium and stretching.

Once activated, the stimulus for asynchronous flight muscles to contract is their stretching by antagonistic muscles. Much like the vibrations of a tuning fork, the oscillations of asynchronous flight muscle are based on their inertial load, with nervous impulses activating the muscles but not determining the frequency of their contraction. When the wings of synchronous fliers are trimmed, the wingbeat is not affected, but when the wings of asynchronous fliers are trimmed, the wing oscillations increase because the load on them has been reduced.

The storage of energy in the elasticity of the thorax is an essential property of insect asynchronous flight. Indirect flight muscles distort the thorax and the movement of wings follows. Instability in the configuration of the thorax was once considered to be a major component of wing movement, with a so-called click mechanism the result of the positions of reduced stability in the midrange of wing movement that made some wing configurations less energetically expensive than others. When the wings are moved, they encounter a resistance in the thorax movements, but once this resistance is overcome, they "click" into their next position. The muscles are therefore required to contract only to a certain point to bring the wings to the area of instability and release the energy stored in the thorax in order to make the wing move fully. Once the tension is released, the antagonistic muscles that were stretched begin to contract. Rather than move the wings throughout the entire range of aerodynamic effectiveness, the muscles have to move the wings only slightly and then the energy stored in elastic proteins in the thorax is used to complete the wing movement. The resistance of the indirect flight muscles to stretching may also allow the muscles themselves to store energy during their elongation.

Tracheal Supply to Muscles

The contraction of muscles requires energy and an adequate supply of oxygen. In spite of the enormous energy requirements of flight muscles, particularly during flight, their respiration is always aerobic. Tracheoles are in close contact with most insect muscles to provide the necessary oxygen, but in flight muscles, they indent the muscle membrane to become functionally intracellular. They penetrate to the interior of the muscle fibers by accompanying the invaginations of the T tubules and ensuring that the oxygen is carried directly to the point of its consumption, contacting or encircling the mitochondria. Surrounding these terminal tracheoles are abundant mitochondria that are able to utilize the oxygen carried by the tracheal system (Fig. 10.12).

There are large increases in tracheal ventilation that occur during flight. Air sacs in the tracheal system change their volume as a result of their compression by wing movements, increasing the convective ventilation. Also, abdominal contractions that are synchronized with spiracular openings can produce a pumping action that drives additional air through the tracheal system.

Neural Excitation and Modulation of Muscle Contraction

Muscle fibers in vertebrates bear single nervous innervations and are activated in an all-or-none fashion. The strength of a contraction depends on the number of total muscle fibers within the muscle that are recruited at one time, with fewer muscle fibers activated when smaller contractions are required. In the smaller insects, muscles often consist of only one or two fibers and the option of recruiting fewer muscle fibers for smaller muscle contractions may not exist. In insects, graded contractions must be accomplished by some mechanism other than the recruitment of additional muscle fibers.

The motor neurons that innervate insect skeletal muscles run along the muscle fibers and repeatedly synapse at intervals. One muscle fiber may receive multiple innervations from several motor neurons and these may consist of a combination of both fast and slow neurons. The **fast neurons** innervate all muscle fibers and cause a rapid muscle contraction (Fig. 10.13). The **slow neurons** innervate some skeletal muscles and cause small depolarizations and slight twitch muscle contractions. Repeated firing of the slow neurons causes a summation of depolarization effects and allows a muscle made of only a few fibers to engage in graded contractions rather than an all-or-nothing response. The jumping muscle of the locust hind leg is innervated by both fast and slow neurons. The slow neurons are used for ordinary movements, while the fast neurons are used for leaps.

The excitatory neurotransmitter at the insect neuromuscular junction is generally the amino acid L-glutamate. It is present in vesicles at the synapse and released at the muscle surface to cause a depolarization of the sarcolemma that is carried

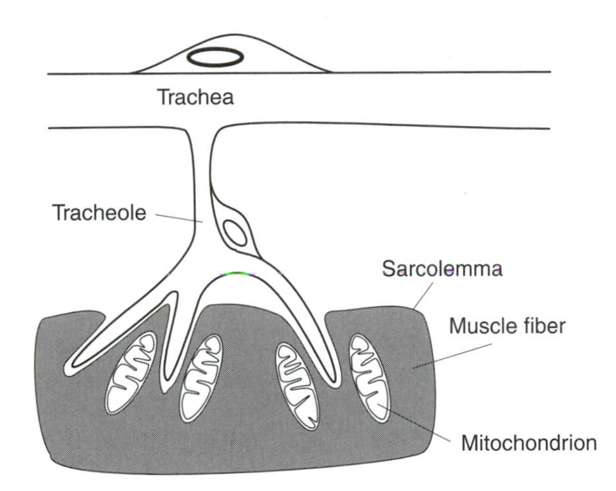

FIGURE 10.12 Penetration of the flight muscle fibers by tracheoles.

deep into the muscle by the transverse tubules. The muscle fibers of some muscles may have additional innervations by neurons that are able to inhibit membrane depolarization. These inhibitory neurons release the neurotransmitter γ-aminobutyric acid (GABA), which causes an influx of chloride ions that hyperpolarizes the membrane and prevents the activation of the muscle fiber.

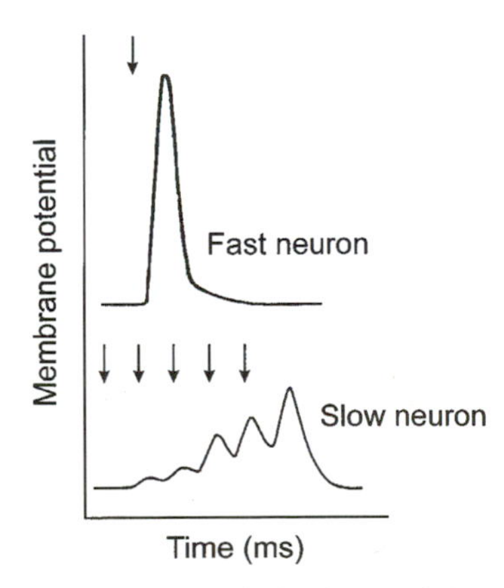

FIGURE 10.13 Changes in membrane potential after the stimulation of fast and slow neurons. In fast neurons (top), nervous stimulation (arrow) causes a sharp rise in membrane potential, causing a rapid muscle contraction. In slow neurons, each nervous impulse causes a small depolarization, but their effects can be summated. This allows a muscle that consists of only a few muscle fibers to engage in a graded contraction.

Some skeletal muscles may be innervated by other neurons identified as dorsal unpaired medial (DUM) neurons that release the neurotransmitter **octopamine,** which is able to modulate the effects of other neurotransmitters. Its presence may synergize the release of L-glutamate and elevate the level of cAMP within the muscle cells. In locusts, DUM neurons release octopamine shortly after flight is initiated and have a variety of effects on the interneurons involved with flight and wing proprioreception. The octopamine also increases the power output of flight muscles and stimulates the release of adipokinetic hormone to mobilize lipid from the fat body for flight energy. Other neurotransmitters such as 5-hydroxytryptamine, or serotonin, can modify the normal response of a muscle to excitatory transmitters and increase the rates of contraction and relaxation.

Myotropic Peptides

The contraction of both skeletal and visceral muscles can also be modulated by myotropic neuropeptides. The existence of these myoactive substances was first demonstrated with a bioassay that examined the contractions of a cockroach heart and later expanded to use other visceral muscles including those of the hindgut and oviduct. The neuropeptide **proctolin** was originally shown to cause contractions of the longitudinal muscles of the cockroach hindgut but has since been identified from both visceral and skeletal muscles in a wide variety of insects. It functions as a neuromodulator that works with glutamate as a cotransmitter at the neuromuscular junction. As mentioned in Chapter 7, there are a number of **cardioacceleratory peptides (CAP)** that control the muscles that produce the heartbeat. In newly emerged *Manduca* adults, the release of CAPs increases the heartbeat in order to expedite wing inflation. Small neuropeptides that have been grouped together into a family of **myokinins** have been identified from several insects. These are potent stimulators of hindgut and oviduct muscles. The insect **tachykinins** are related to vertebrate tachykinins, stimulating the contraction of the visceral muscles of the foregut and hindgut, and the muscles of the oviduct. Although many of these peptides are produced by neurosecretory cells within the insect, peptides produced by the male accessory glands and transferred to females during mating may also have myotropic effects. They are able to stimulate the contractions of the oviduct muscles when they are transferred from males along with sperm.

Evolution of Insect Wings

Insects were the first animals on the planet to fly. The modifications that led to flight probably occurred about 350 million years ago and are believed to have happened only once during the course of insect evolution, with all pterygote insects descending from a common ancestor. Abiotic factors may have been quite different at the time and these differences may have promoted the acquisition of flight. For example, the atmosphere was believed to be significantly denser during this period, with concentrations of oxygen as high as 35% compared to the current value of 21%. The density of the air is the major determinant of force production by an airfoil, be it biological or mechanical, and the increased density may have improved the chances that flight would evolve. Accompanying increased oxygen concentrations in the atmosphere were increases in tracheal diffusion and body size. Early flying insects were enormous compared to present-day insects; odonate ancestors living during the Carboniferous period are estimated to have wingspans greater than 70 cm.

Taking to the air was probably the most important evolutionary step in the entire history of insects, conferring significant advantages in finding food, locating mates, and escaping predation. Two major hypotheses attempt to explain the evolution of wings from unwinged ancestors. The first proposes that wings developed from thoracic winglets or **paranotal lobes** that might have been first used for thermoregulation, then as airfoils as they enlarged, and finally as flapping wings. The secondary wing articulation and musculature had to evolve accordingly. The second hypothesis suggests that wings evolved from winglets that originated as gills. In this second case, protowings could have first evolved in aquatic larvae and enlarged to allow improvements in underwater navigation. Once the ancestors left the water, the significance of the wings could have shifted to a second function as locomotor devices. This **articulated gill theory** has support in the behavior of some extant aquatic stoneflies that engage in surface skimming in which the wings are flapped but the insects remain in the water, a behavior intermediate between swimming and flying. The aquatic gills might have been first used to "row" within the water, and then became large enough for skimming and finally flight. The reduced muscle power and wing movements used in skimming are sufficient to propel them through the water and explain the initial evolutionary benefits of wings before insects actually left the ground.

Recent molecular data provides more evidence for the articulated gill theory. The ancestral arthropod limb segment is called a **podite,** Lobes along the inner (endite) and outer (exite) margins of the podite developed in some groups. The basal segment of the limb that is attached to the pleural wall is the **coxopodite** and its outer lobe is known as the **epipodite.** In crustaceans, the epipodites are often modified into gills (Fig. 10.14). Two ancient genes that have wing-specific functions in insects are also expressed in the epipodites of crustaceans, which are often modified to serve as gills. The expression of the same genes in the epipodites of crustaceans and the wings of insects suggests that insect wings evolved from these epipodites. The crustaceans consist of the only group that has been largely confined to the water or moist habitats. When terrestrialization occurred in the other groups, the epipodite gill was believed to have been modified into a wing in the line leading to pterygote insects, while it was lost in the other two groups (Fig. 10.15).

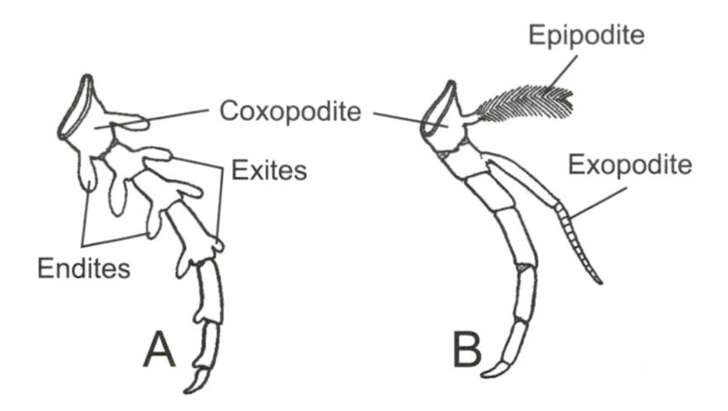

FIGURE 10.14 (A) Generalized trilobite limb. (B) A generalized crustacean limb. From Snodgrass (1935). Reprinted with permission.

Because insect wings arose not from walking legs as in birds and bats, but as outgrowths of the body wall, they develop as a sandwich of two epicuticular layers with hemolymph, nerves, and trachea in between (Fig. 10.16). Because procuticle is absent, chitin does not appear to be present in the membranous areas of the wings. Another difference between the wings of insects and those of other flying

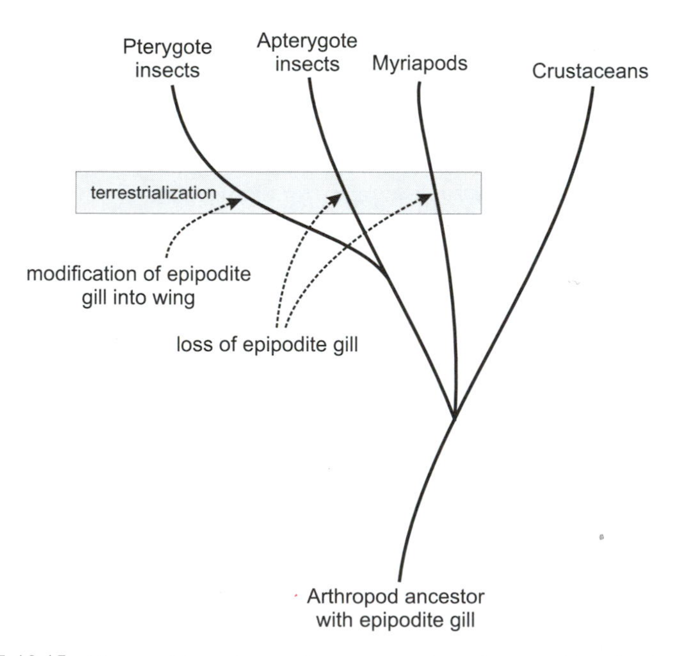

FIGURE 10.15 The possible evolution of insect wings from the epipodite gills of an arthropod ancestor. The epipodite gill remained in aquatic crustaceans, but after the terrestrialization of insects and myriapods, it evolved into the wing of pterygotes but was lost in apterygotes and myriapods. Adapted from Averof and Cohen (1997). Reprinted with permission.

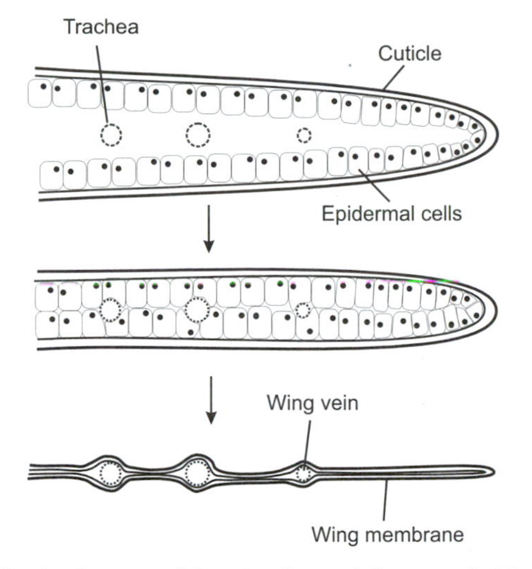

FIGURE 10.16 The development of the wing (bottom) from a sandwich of epidermal cells during pupal development (top).

animals is that there are no intrinsic muscles associated with the insect wing. Wing movement and changes in wing shape during flight must be controlled by muscles in the thorax, with some deformations of the wing resulting from the wing structure itself as it responds to the aerodynamic forces. Most of the epidermal cells in the insect wing degenerate after adult emergence, leaving a largely acellular membrane lined with veins, tracheae, and nerves. The loss of water as these cells degenerate makes the flapping movements of the wings more efficient. However, the absence of epidermal cells in the mature adult wing is also a reason that adult pterygotes no longer molt. Without functional epidermal cells, it is impossible to undergo the steps in molting and produce a new wing cuticle. Selection has favored the wing efficiency that comes with epidermal cell degeneration over the possible advantages of adult molting.

The development of flight was accompanied by an increased capacity for information processing by the insect central nervous system. The **flicker fusion frequency** is the maximum rate that individual light impulses can be resolved and is a measure of the speed of optical processing. In humans, the flicker fusion frequency is less than 20 Hz, while it can reach as high as 200–300 Hz in flying insects.

Basic Structure of the Insect Wing

Fully functional wings are found only in adult insects. The membranous wings are reinforced with rigid longitudinal and cross veins that may contain nerves and tracheae that are nourished by the hemolymph that flows through them. Although most of the epidermal cells have degenerated to leave only the epicuticular sandwich, a blood supply in the wings is essential to maintain their mechanical properties and prevent brittleness. Because wings arose only once in insect evolution, the pattern of veins present in a particular insect reflects its ancestry and evolutionary relationships among other insect groups and can be used diagnostically to classify insects. More advanced insects generally have smaller wings and fewer veins. The wing may be thrown into longitudinal corrugations behind the anterior margin to provide resistance to bending and the veins may be located on the fan-like pleats, with convex veins on the crest and convex veins in a trough (Fig. 10.17). A system of wing folds allows the deformation of the wing to occur during pronation and supination and permits the wings to fold at rest. Axillary sclerites connect the wing to the thorax and permit flexion of the wing over the abdomen.

A number of surface structures may appear on the wings. Scales are borne on the wings of lepidopterans and on the hind margin of the wings of mosquitoes. Spines may be present, as on the wings of odonates. Both tactile and proprioreceptive sensilla can also be found. Little is known about the roles of these accessory structures on flight. The scales may aid the escape of lepidopterans from spider webs.

Reynolds Number, Size, and Insect Movement

Whenever an object moves through a fluid medium such as air or water, forces are generated on the object by the elements of the fluid. The relationship between the viscous frictional forces from the fluid and the inertial force of the object can be quantified and expressed as a **Reynolds number** (**Re**). Re represents the ratio of the inertial and frictional forces. A low Re is produced by small size and/or low speed and indicates that frictional forces are large, while a high Re means that inertial forces are more important. For example, a toothpick with a low Re could be easily propelled in water but would be stopped by frictional forces from the viscosity of the water when the propulsion ceased. In contrast, a log floating in the water with a high Re would be initially difficult to move, but once it started moving, inertial forces would maintain its movement even when it was no longer pushed. The range of Reynolds numbers is quite large for living things, extending from approximately 10^{-6} for bacteria to 10^{7} for animals the size of whales. Insects have relatively small Reynolds numbers ranging between 10^{0} and 10^{4}, indicating that they are especially subject to the frictional forces in the water and air in which they must move. The effect of the viscous medium on such small animals has been compared to swimming in molasses. Among the implications of living at a low Reynolds number is that when frictional forces predominate, any movement is energetically expensive. Animals that are the size of insects encounter relatively

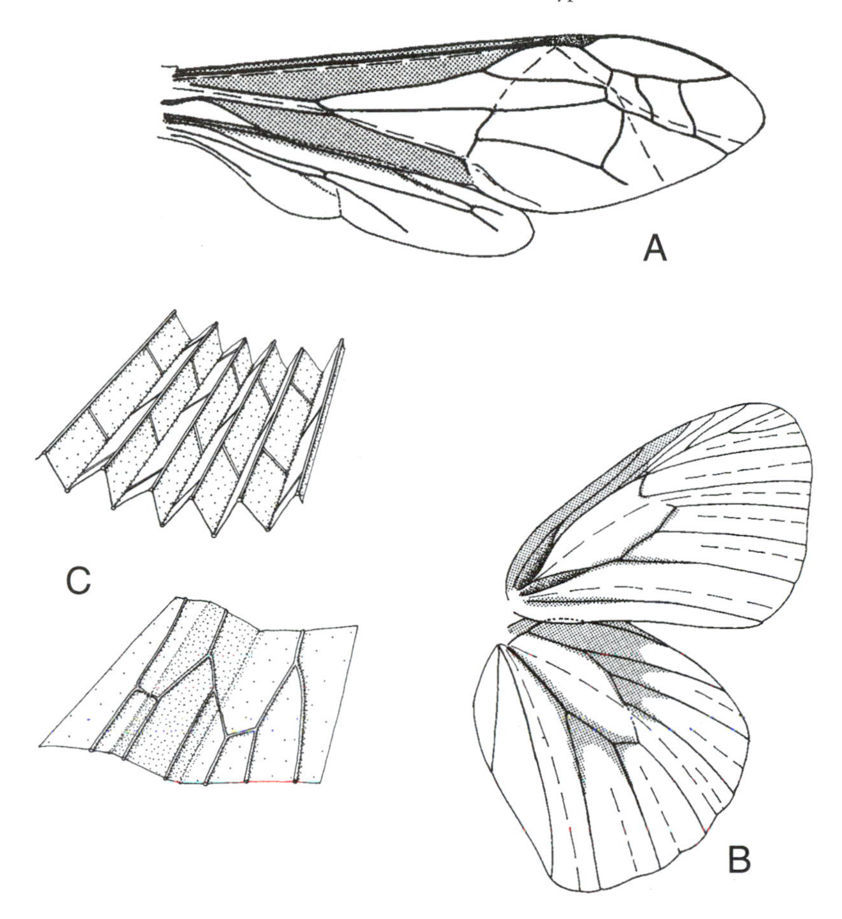

FIGURE 10.17 Corrugations and fold of the wings of hymenopterans (A) and lepidopterans (B). (C) Wing fluting over weak crossveins and lines of flexion. From Wooton, R.J. (1981). *Journal of Zoology* **193:** 447–468. Reprinted with the permission of Cambridge University Press.

huge energy costs to move about, and the aerodynamic strategies differ from those used by larger animals that fly. Metabolic rates during flight in some insects are as much as 100 times greater than those at rest, and the thoracic muscles of insects in flight exhibit the highest metabolic rates of any tissue. The metabolic rates of bumblebees in flight range between 60 and 65 ml of oxygen/g body mass/h, compared to about 45 ml/g/h for hummingbirds flying at the same speed. There is also a considerable inefficiency in the conversion of chemical energy to mechanical energy. As much as 80% or more of the energy used for insect flight is dissipated as heat.

The one advantage of small size is the relative strength it confers, but the apparent strength of insects lies not in any differences in their muscles, but in the relationship between the increase in body volume and muscle power that accompanies increases in body size. The power of a muscle varies with its cross-sectional area, which is a function of the square of its dimensions. The volume, and accompanying weight, of an animal that is powered by the larger muscle, increases with the cube of its dimensions. As animals get larger, their relative strength diminishes because their weight increases at a faster rate—a cubic function—than their muscular strength—a square function. Smaller insects benefit from their reduced size and lightweight exoskeleton.

Muscles Involved in Wing Movements

The flight muscles are usually the best developed of the muscles in the insect body and occupy most of the space in the thorax. There are three general categories of muscles that power insect flight: **direct, indirect,** and **accessory.** Direct flight muscles, consisting of the **basalar** and **subalar** muscles, insert directly at the base of the wing and provide the power for the downstroke in more primitive insects and also affect wing pronation and supination (Fig. 10.18). Another direct muscle, the **third axillary** muscle, inserts on the third axillary sclerite. It affects wing supination and is also responsible for wing flexion against the body wall when the wings are at rest.

In contrast, the indirect flight muscles move the wings not by moving them directly, but by changing the conformation of the thorax. These muscles include the **dorsoventral** group that extends from the tergum to the sternum. Because of the structural relationships among the sclerites that make up the thorax, in all insects, when the muscles pull the tergum down they indirectly cause the wings to raise and produce the upstroke (Fig. 10.18). Another indirect group, the **dorsal longitudinal** muscles, attach longitudinally between the two phragmata of each wing-bearing segment. When these contract, they shorten the segment and cause the tergum to elevate (Fig. 10.19). In more advanced insects, this deformation of the notum by the dorsal longitudinal muscles produces the power stroke by depressing the wing. These muscles are reduced in more primitive insects that instead use the direct muscles for the downstroke (Fig. 10.20).

There are also accessory muscles that insert into the thorax and influence its mechanical conformation. For example, the **pleurosternal** and **pleurotergal** muscles modulate the power output and the nature of the wingbeat by changing the orientation of the thoracic plates and the resonance of the thorax (Fig. 10.21).

Wing Movements during Flight

Unlike birds that can glide for sustained periods, most insects must constantly move their wings to generate the forces that will keep them in the air. Simply flap-

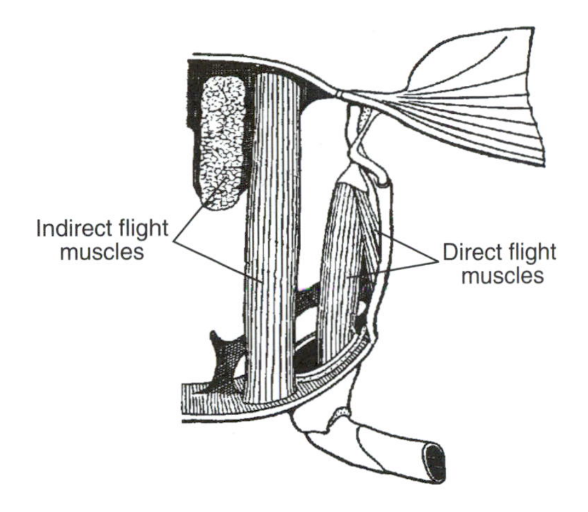

FIGURE 10.18 Cross section of a generalized insect thorax. The indirect flight muscles change the shape of the thoracic box. The direct flight muscles connect directly to the wing insertion. From Snodgrass (1935). Reprinted with permission.

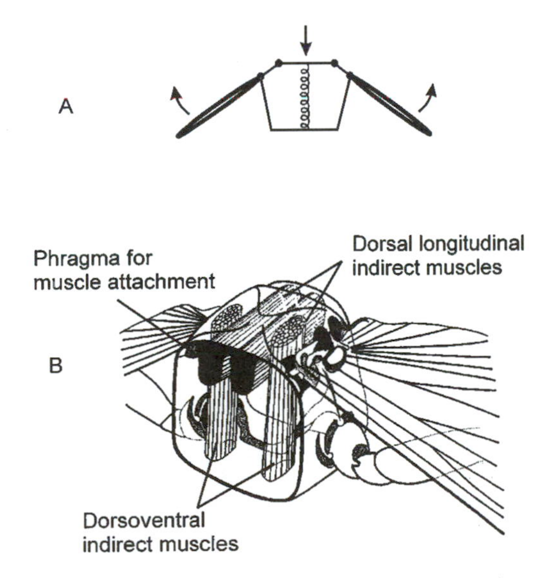

FIGURE 10.19 The mechanism of wing movement. (A) The downward movement of the thoracic tergum causes the wings to move upward. (B) The downward movement is mediated by the indirect dorsoventral muscles. The indirect dorsal longitudinal muscles attach to the phragmata at either end of the segment, causing it to shorten and move the wings downward. Reprinted from Brodsky, A.K. (1994). *The evolution of insect flight*. Copyright Andrei K. Brodsky 1994. By permission of Oxford Science Publications.

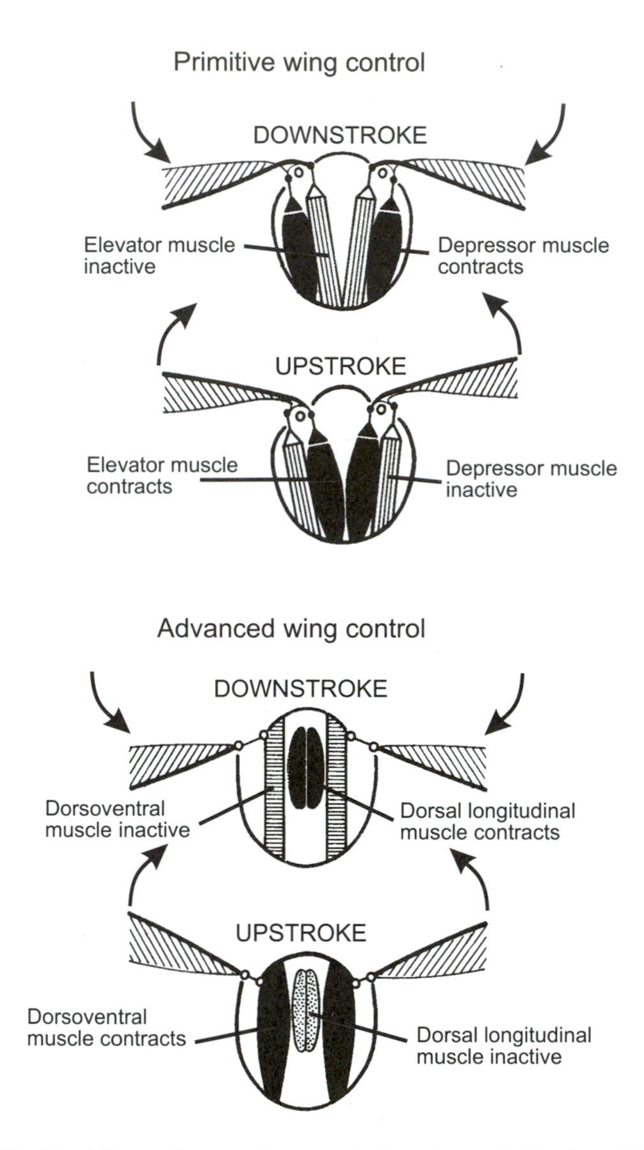

FIGURE 10.20 The differences between the more primitive wing control by direct flight muscles (left) and the more advanced wing control by indirect flight muscles (right). Reprinted from Nachtigall, W. (1985). *Insect Flight,* pp. 1–29. With permission from CRC Press.

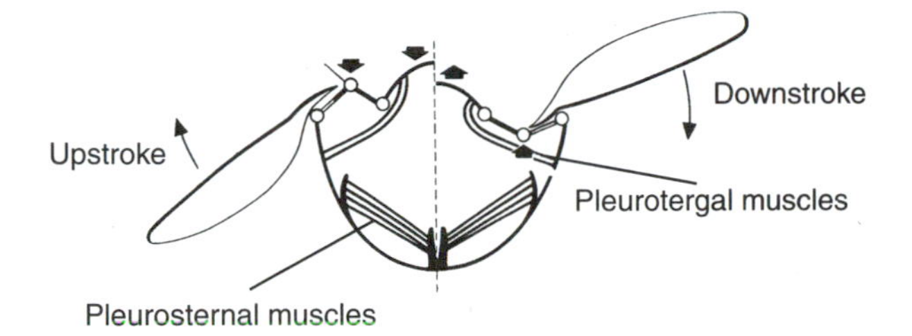

FIGURE 10.21 The pleurosternal and pleurotergal muscles are accessory muscles that change the shape of the thorax and modulate the output of the indirect flight muscles. Reprinted from Nachtigall, W. (1985). *Insect Flight,* pp. 1–29. With permission from CRC Press.

ping the wings alone would not create the necessary downward forces, as the downstrokes would be aerodynamically canceled out by the equivalent upstrokes. Deforming the wings during the flapping is therefore necessary in order to generate more upward forces than downward forces. There are several ways that insects can deform their wings, including twisting them to change the angle of attack during the stroke, and altering the curve of the wing from its leading edge to its tip, or its camber. Insects with two pairs of wings may vary the overlap between them during flight to change the total surface area during flapping.

Flight muscles not only power the wings but also modify their control. The power movements consist of an alternating upstroke and downstroke that is dependent on the alternating contraction of the elevator and depressor muscles. Separating these movements are the pronation and supination of the wings about their longitudinal axes as they reverse their up and down directions. The wing is pronated during the downstroke so that the leading edge faces downward and supinated during the upstroke so that the leading edge is up. In aircraft, birds, and insects, a wing produces lift only when a pressure gradient is created between the upper and lower surfaces of the wing as low pressure air flows over the upper surface. In *Drosophila,* the rotation of the wing produces a circulation of air in the opposite rotational direction and a wake capture that uses the vortex from the previous stroke to generate airflow around the wing. By capturing this lingering vortex wake, a strong force-generating airflow is created that supports flight. The air is forced backward and downward, moving the insect forward, generating thrust, and upward, generating lift.

Flying at such a small Reynolds number, the viscosity of the air is also an important factor to insects. Flight is controlled both by the frequency of wingbeat as well as the mechanical properties of the thorax, whose rigidity can be modified by accessory muscles. Larger insects may use the phenomenon of delayed stall in which the leading edge of the wing creates a vortex that remains closely bound to it and creates a low pressure area that generates lift.

Wing Coupling Mechanisms

The evolutionary trend in insects has been toward a reduction in wing size and number. The primitive condition is to have two pairs of wings that are minimally coordinated and beat independently, although their close association on neighboring thoracic segments certainly influences their movements. This is not the most efficient mechanism because the hind wings must operate in the area of turbulence generated by the forewings.

A more advanced approach to flight is the loss of one of the wing pairs or the coupling of the two wings so they serve as one functional unit. In many Coleoptera, the forewings are rigid and serve a protective function as a sheath when the wings are closed, but they are held open during flight while the hind wings beat. In other groups, notably the Hymenoptera and Lepidoptera, the two wings are mechanically coupled by lobes or spines at the base of the wing. In some Lepidoptera, a **jugal lobe** at the base of the forewing overlaps with the hind wing, causing them to beat together. In other Lepidoptera, a spine or **frenulum** at the base of the hind wing may engage a catch on the forewing. Many hymenopterans have a row of hooks, the **hamuli,** along the margin of the hind wing that catch along a sclerotized fold in the forewing (Fig. 10.22).

In dipterans, the second pair of wings has been eliminated rather than coupled. They are modified into knoblike halteres that beat antiphase to the forewings during flight (Fig. 10.23). The halteres monitor torque during flight by measuring the stress in the sensilla at their bases like a gyroscope. In the blowfly, *Calliphora,* there are more than 300 of these sensilla that monitor the insect's flight. Rotation of the body during flight causes a deflection of the halteres that is detected by the sensilla. The nervous signal from these sensilla is sent directly to the neurons that control the flight muscles and alter the wing beat accordingly to control steering (Fig. 10.24).

FLIGHT MUSCLE METABOLISM

Flight requires a relatively enormous amount of energy both because of the high cost of flying at a low Reynolds number and the inefficiency of the conversion of energy. Yet, insect flight muscles operate completely aerobically. Only about 10% of the chemical energy used for flight is translated into mechanical energy, with the rest being dissipated as heat. This heat can be used for thermoregulation in some insects to allow them to fly even when ambient temperatures are too low for optimal muscle activity (see Chapter 7, Circulatory Systems).

The substrates from which insect flight muscles draw their energy for contraction are located in several places (Fig. 10.25). Ideally, these substrates would be situated within the muscles themselves to be close to where they would be utilized, but the need to maintain high ratios of muscle power to weight limits the con-

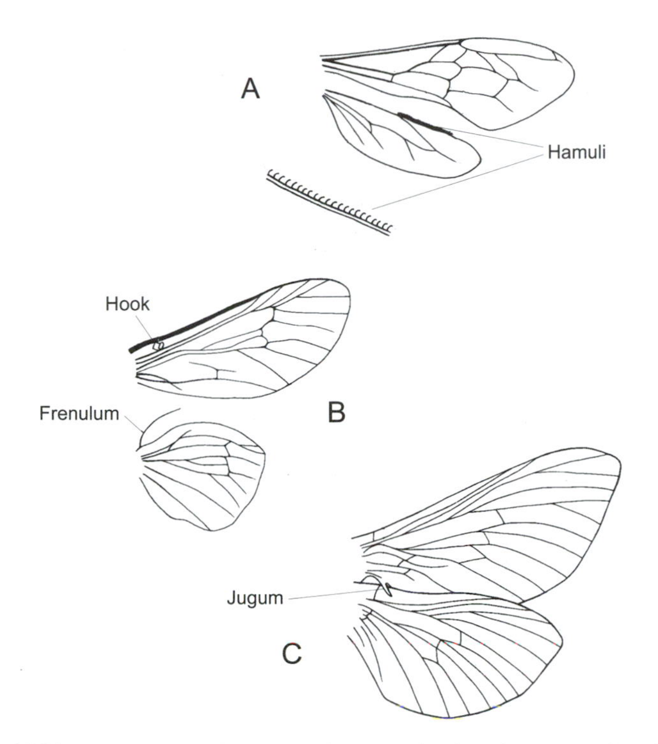

FIGURE 10.22 Mechanisms of wing overlap. (A) The hamuli, a row of small hooks in Hymenoptera. (B) The frenulum, a hair that protrudes from the posterior wing in some Lepidoptera. (C) The jugum, a lobe that projects backward from the anterior wing in some Lepidoptera. From Romoser and Stoffolano (1998). Reprinted with permission.

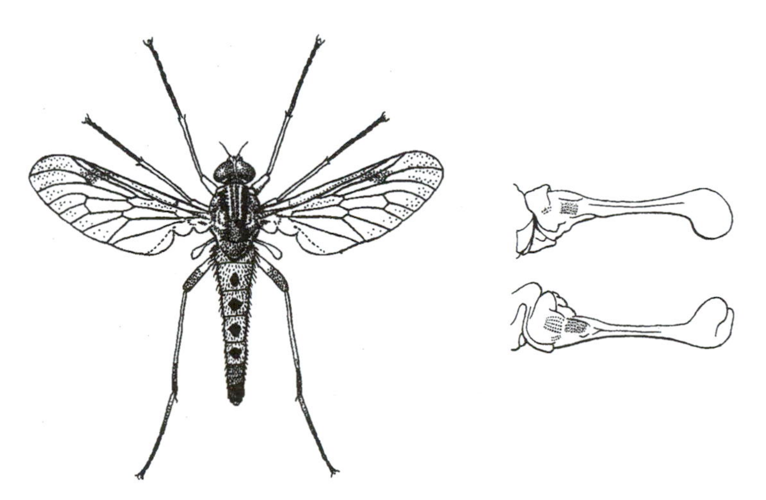

FIGURE 10.23 The second pair of wings has evolved into knob-like halteres in the Diptera. From Pringle (1975). Reprinted with permission.

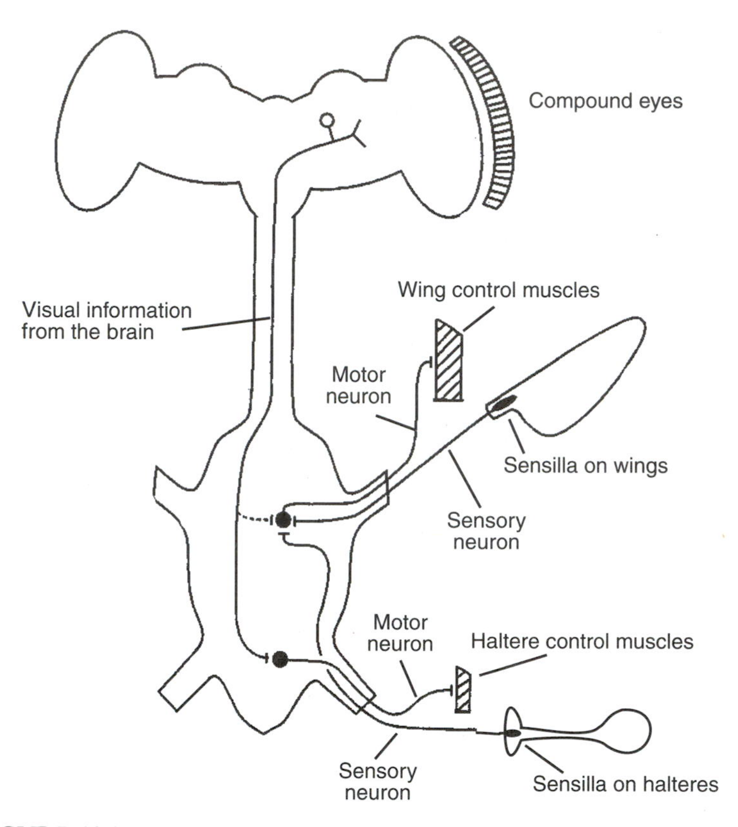

FIGURE 10.24 The mechanism of direct haltere control in the blowfly, *Calliphora*. The visual interneurons from the compound eyes activate the haltere control muscles. Twisting movements of the halteres activate their sensilla that feed to the wing muscle motor neurons and modulate their control. Reprinted with permission from Chan, W. P., F. Prete, and M. H. Dickinson. Visual input to the efferent control system of a fly's "gyroscope." *Science* **280:** 289–292. Copyright 1998. American Association for the Advancement of Science.

centrations they can attain there, necessitating that the bulk of the components be stored elsewhere. There are small amounts of substances located in the flight muscles themselves to power the initiation of flight, but the initial store of ATP in muscle cells is only sufficient for a few seconds of flight. It is replenished by the transfer of a phosphate group to ADP from **arginine phosphate**, providing an additional brief period of flight. This system is similar to vertebrate muscle where creatine phosphate is instead used as a reservoir for high-energy phosphoryl groups. The muscle may also store small amounts of other fuels including proline, glycogen, and triacylglycerol that are drawn upon during flight, but these resources are similarly limited.

Flight muscles draw the next most immediate source of energy from substrates in the hemolymph. The disaccharide trehalose is present in high concentrations as a

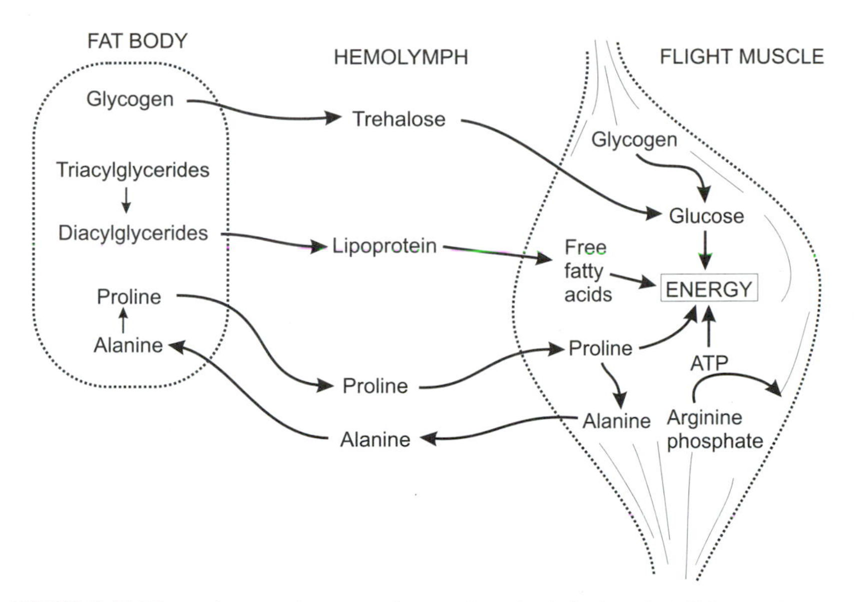

FIGURE 10.25 Utilization of various substrates from the fat body to fuel flight muscle contraction. Adapted from Wheeler (1989). Reprinted with permission.

circulating energy source that is used during the early phases of flight. Hemolymph diacylglycerol also bathes muscle cells and the amino acid proline is utilized in some insects for flight. These are mobilized from fat body reserves to maintain their levels in the hemolymph, but upper limits on the osmotic pressure of the hemolymph prevent the concentrations of these substances from getting too high.

The fuel for longer flights is stored in the fat body and transported to flight muscles through the hemolymph. This fuel use varies among insect orders. During long flights, dipterans and hymenopterans convert fat body glycogen into trehalose that is distributed to muscles through the hemolymph while migratory orthopterans utilize triacylglycerols in the fat body that are converted to diacylglycerols for transport.

Flight muscles completely oxidize carbohydrates to carbon dioxide and water in the absence of any anaerobic metabolism. Glycolysis in insect flight muscle occurs much like that in other animals, with a few additions. One is the presence of a glycerol phosphate shuttle, discussed in Chapter 6. With the conversion of glyceraldehyde phosphate to 1,3-diphosphoglycerate during glycolysis, NAD^+ acts as a hydrogen acceptor and is converted to NADH. Because the NAD^+ is present only in catalytic amounts, the reduced NADH must be reoxidized to maintain the biochemical pathway. In vertebrate muscle and insect muscles not involved in flight, NADH is oxidized during the conversion of pyruvate to lactic acid, and the liver or fat body rebuilds glucose as part of the oxygen debt. In insect flight muscle, there is no oxygen debt incurred because the NADH is reoxidized by the enzyme glycerol phosphate dehydrogenase with dihydroxyacetone phosphate as a substrate to form glycerol 3-phosphate. Unlike NADH, the glycerol 3-phosphate can pass into the mitochondrion where it is reoxidized to form dihydroxyacetone phosphate. The dihydroxyacetone phosphate then leaves the mitochondrion to be oxidized by the NADH. Thus, the glycerol 3-phosphate shuttles the hydrogen into the mitochondrion, where it can be harnessed for oxidative phosphorylation and also provide NAD^+ for glycolysis (Fig. 10.26).

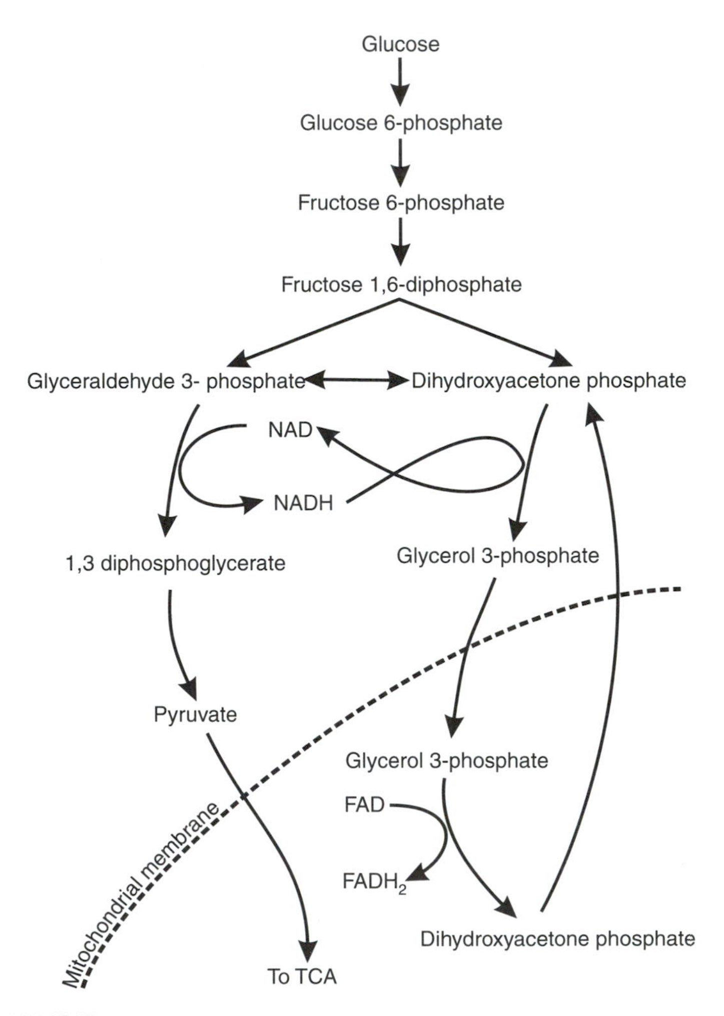

FIGURE 10.26 The glycerol 3-phosphate shuttle that operates in insect flight muscle.

Dipterans use proline as fuel for flight to different extents. The blowfly, *Phormia regina,* initially metabolizes proline to provide the tricarboxylic acid intermediates that are used for later flight. However, the tsetse fly, *Glossina,* also uses proline, but as the major fuel for flight. As a tsetse initiates flight, hemolymph proline decreases in concentration along with a concomitant increase in alanine concentrations and a brief increase in pyruvate during the first few seconds of flight. The pyruvate accumulates because it is produced faster than the rate of oxidation by mitochondria. These high levels of pyruvate initiate a series of biochemical steps beginning with the activation of the enzyme proline dehydrogenase that supplies tricarboxylic acid cycle intermediates to prime the cycle and speed up the rate of energy production. Glutamate is used to transaminate pyruvate so that proline enters the tricarboxylic acid cycle as α-ketoglutarate. The proline is not completely oxidized, losing two carbons to form alanine, which returns to the fat body and is reformed as proline by stored fatty acids. Thus, the use of proline is actually a mechanism for shuttling 2 carbon units from the fat body to the flight muscles (Fig. 10.27).

Lepidopterans may employ either lipids or carbohydrates, regulating their choice of fuels based on their feeding history. In general, insects that engage in long-range flights oxidize lipid, whereas those that use carbohydrate fly for only short periods. It has also been observed that insects with high wingbeat frequencies and asynchronous muscle tend to utilize carbohydrates, while those with synchronous flight muscles are more likely to utilize lipid. Lipid is the most concentrated form of energy storage. To fly 10 h each day, a migratory locust may use 70 mg of stored lipid. If glycogen were utilized instead, 500 mg, or 30% of the insect's weight, would have to be oxidized. Glycogen is considerably more hydrated than lipid and as a result, is also 8 times heavier for the same caloric potential.

In spite of its many advantages, there are at least two drawbacks to the use of lipid to power flight. It is energetically expensive to interconvert ingested carbohydrates into lipids for storage and then back to carbohydrates for utilization. The net energy yield from a given carbohydrate is 20% higher when it is utilized directly instead of converted first to lipids. For this reason, many moths use the sugar directly when they have recently fed on nectar and draw on their fat reserves when starved. The other major drawback to storing energy as lipids is that they must be transported to flight muscles for use, yet are insoluble in the aqueous transport medium of the hemolymph. A protein carrier molecule, **lipophorin,** is produced by the fat body to shuttle the diacylglycerols to the flight muscles (see Chapter 6).

In locusts, energy metabolism during flight is initiated by octopamine and regulated by **adipokinetic hormone** (**AKH**). Trehalose serves as the major fuel at the onset of flight, but as the hemolymph trehalose levels decline with activity, octopaminergic neurons within the corpus cardiacum stimulate the release of AKH. The AKH activates an adenylate cyclase that increases cAMP levels and subsequently

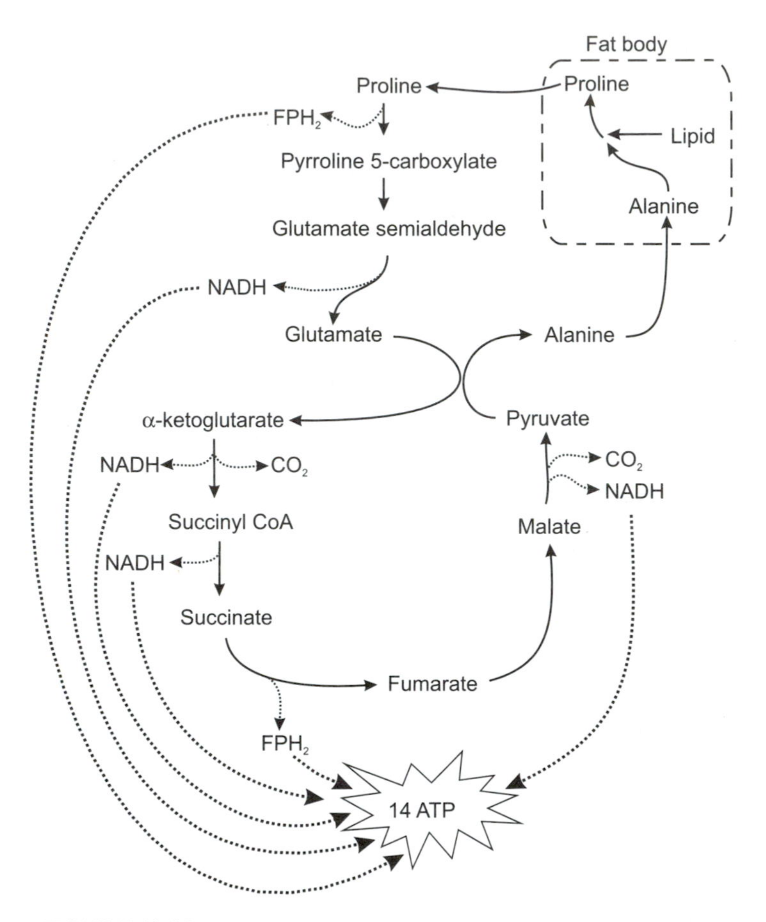

FIGURE 10.27 The utilization of the amino acid proline as a fuel for flight.

activates a protein kinase. The protein kinase then phosphorylates and activates a lipase that induces the release of diacylglycerols from the triacylglycerols stored in the fat body. AKH also induces the production of a lipoprotein carrier from the fat body that transports these diacylglycerols through the hemolymph to the flight muscles. The metabolism of carbohydrates that are stored in flight muscle during this lipid mobilization is also inhibited by AKH, so that the lipid reserves are used exclusively. Octopamine has several other effects on flight behavior, stimulating the interneurons involved in maintaining flight, the power output of the flight muscles themselves, and the proprioreceptors on the wing that monitor flight behavior. This amine may be the functional equivalent of "flight-or-fight" hormones in vertebrates, released during stress and causing an increase in the insect's arousal levels.

TABLE 10.1. Some Maximum Walking Speeds for Selected Insects (from Delcomyn, 1985)

Insect	Speed (cm/s)
Bristletails	21
Cockroaches	18–80
Grasshoppers	4–15
Earwigs	10
Flies	9–20
Beetles	11–58
Ants	1–2.6

TERRESTRIAL LOCOMOTION

Insects possess a rigid exoskeleton that limits any significant bending except in specialized areas of joints and cuticular membranes. Single joints of the insect leg allow movement to occur only in a single plane, but more complex three-dimensional movements are possible given the arrangement of joints within the appendage. The limited movements that are possible may simplify the neural and muscular control of each appendage. The presence of six walking legs provides more opportunities for balance during locomotion but also presents challenges for their coordination. Legs both suspend and support the insect off the ground and provide the forces necessary to propel it forward. Being suspended by the legs provides the body with a low center of gravity and an unusual stability, but muscular activity is always required to keep the body off the substrate (Fig. 10.28).

Six-legged insects walk using an alternating tripod gait in which three legs, the middle on one side and the anterior and posterior legs on the other, alternate their contact with the substrate (Fig. 10.29). This provides a tripod of stable support

FIGURE 10.28 A cross section of the generalized insect body, suspended by the legs with a low center of gravity. Reprinted from Delcomyn, F. *Comprehensive Insect Physiology, Biochemistry and Pharmacology* 1985. Volume 5, pp. 439–466. With permission from Elsevier Science.

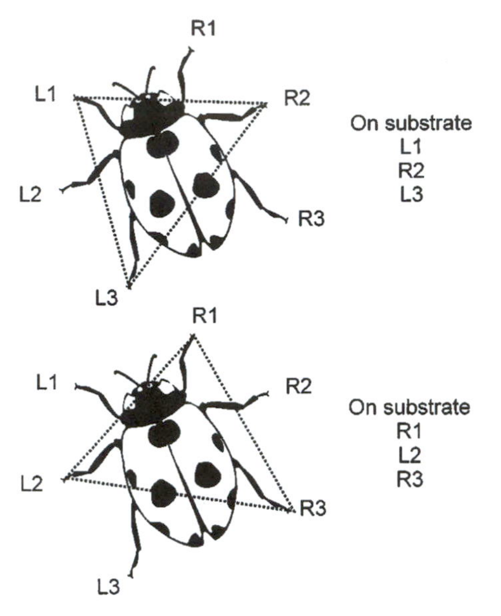

FIGURE 10.29 The alternating tripod gait of insects. The triangles show the legs in contact with the substrate.

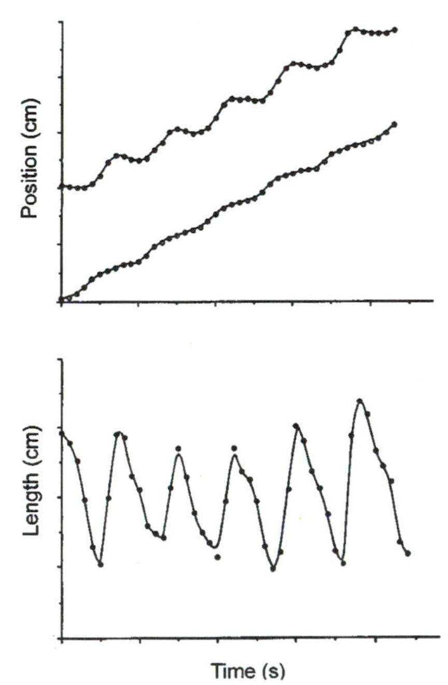

FIGURE 10.30 (Top) The movement of the head and rear of a larval *Drosophila* crawling in a straight line. (Bottom) The changes in body length of the same larva during locomotion. Reprinted from Berrigan, D. and D.J. Pepin (1995). *Journal of Insect Physiology* **41:** 329–337. With permission from Elsevier Science.

during walking, with the leg movements coordinated by the thoracic ganglia. Leg muscles are innervated by a mix of fast and slow neurons that can invoke a wide range of running speeds. Strictly speaking, insects do not run, if running is defined as movement in which all legs are no longer in contact with the surface, but a maximum progression of fast walking has been recorded for a number of insect species (Table 10.1).

The limbless locomotion that occurs in larval dipterans with hydrostatic skeletons that are maintained by hemolymph pressure is about 4 times more energetically expensive than for a legged caterpillar of the same mass. They crawl using a combination of telescoping segments and peristaltic contractions (Fig. 10.30).

ADDITIONAL REFERENCES

Averof, M., Cohen, S. M. 1997. Evolutionary origin of insect wings from ancestral gills. *Nature* **385:** 627–630.

Bailey, L. 1954. The respiratory currents of the tracheal system of the adult honeybee. *J. Exp. Biol.* **31:** 589–593.

Beenakkers, A. M. T., Van der Horst, D. J., Van Marrewijk, W. J. A. 1984. Insect flight metabolism. *Insect Biochem.* **14:** 243–260.

Berrigan, D., Pepin, D. J. 1995. How maggots move: Allometry and kinematics of crawling in larval Diptera. *J. Insect Physiol.* **41:** 329–337.

Bullard, B., Leornard, K., Larkins, A., Butcher, G., Karlik, C., Fyrberg, E. 1988. Troponin of asynchronous flight muscle. *J. Mol. Biol.* **204:** 621–637.

Chan, W. P., Dickinson, M. H. 1996. *In vivo* length oscillations of indirect flight muscles in the fruit fly *Drosophila virilis*. *J. Exp. Biol.* **199:** 2767–2774.

Chan, W. P., Prete, F., Dickinson, M. H. 1998. Visual input to the efferent control system of a fly's 'gyroscope.' *Science* **280:** 289–292.

Cruse, H. 1990. What mechanisms coordinate leg movement in walking arthropods? *Trends Neurosci.* **13:** 15–21.

Cruse, H., Bartling, C., Cymbalyuk, G., Dean, J., Dreifert, M. 1995. A modular artificial neural net for controlling a six-legged walking system. *Biol. Cybernet.* **72:** 421–430.

Daley, J., Southgate, R., Ayme-Southgate, A. 1998. Structure of the *Drosophila* projectin protein: Isoforms and implication for projectin filament assembly. *J. Mol. Biol.* **279:** 201–210.

Delcomyn, F. 1985. Walking and running. In *Comprehensive insect physiology biochemistry and toxicology* (Kerkut, L. E. Gilbert, Eds.), vol. 5, pp. 439–466. Pergamon Press, Oxford.

Delcomyn, F. 1985. Factors regulating insect walking. *Annu. Rev. Entomol.* **30:** 239–256.

Dickinson, M. 1994. The effects of wing rotation on unsteady aerodynamic performance at low Reynolds numbers. *J. Exp. Biol.* **192:** 179–206.

Dickinson, M. H., Farley, C.T., Full, R. J., Koehl, M. A., Kram, R., Lehman, S. 2000. How animals move: An integrative view. *Science* **288:** 100–106.

Dickinson, M. H., Gotz, K. G. 1996. The wake dynamics and flight forces of the fruit fly *Drosophila melanogaster*. *J. Exp. Bio. l* **199:** 2085–2104.

Dickinson, M. H., Hannaford, S., Palka, J. 1997. The evolution of insect wings and their sensory apparatus. *Brain Behav. Evol.* **50:** 13–24.

Dickinson, M. H., Lehmann, F. O., Sane, S. P. 1999. Wing rotation and the aerodynamic basis of insect flight. *Science* **284:** 1954–1960.

Dickinson, M. H., Lighton, J. R. 1995. Muscle efficiency and elastic storage in the flight motor of *Drosophila*. *Science* **268:** 87–90.

Dudley, R. 1998. Atmospheric oxygen, giant paleozoic insects and the evolution of aerial locomotor performance. *J. Exp. Biol.* **201:** 1043–1050.

Dudley, R. 2000. The evolutionary physiology of animal flight: Paleobiological and present perspectives. *Annu. Rev. Physiol.* **62:** 135–155.

Dudley, R. 2000. *The biomechanics of insect flight: Form function, evolution.* Princeton Univ. Press, Princeton, NJ.

Duve, H. 1975. Intracellular localization of trehalase in thoracic muscle of the blowfly, *Calliphora erythrocephala. Insect Biochem.* **5:** 299–311.

Eisenberg, E., Hill, T. L. 1985. Muscle contraction and free energy transduction in biological systems. *Science* **227:** 999–1006.

Elia, A. J., Money, T. G. A., Orchard, I. 1995. Flight and running induce elevated levels of FMRFamide-related peptides in the haemolymph of the cockroach, *Periplaneta americana* (L.). *J. Insect Physiol.* **41:** 565–570.

Ellington, C. P. 1985. Power and efficiency of insect flight muscle. *J. Exp. Biol.* **115:** 293–304.

Ellington, C. P. 1995. Unsteady aerodynamics of insect flight. *Symp. Soc. Exp. Biol.* **49:** 109–129.

Ellington, C. P. 1999. The novel aerodynamics of insect flight: Applications to micro-air vehicles. *J. Exp. Biol.* **202**(23): 3439–3448.

Fayyazuddin, A., Dickinson, M. H. 1996. Haltere afferents provide direct, electrotonic input to a steering motor neuron in the blowfly, *Calliphora. J. Neurosci.* **16:** 5225–5232.

Fraenkel, G., Pringle, J. W. S. 1938. Halteres of flies as gyroscopic organs of equilibrium. *Nature* **141:** 919–920.

Full, R. J., Koehl, M. A. R. 1993. Drag and lift on running insects. *J. Exp. Biol.* 176: 89–101.

Full, R. J., Tu, M. S. 1991. Mechanics of a rapid running insect: two-, four- and six-legged locomotion. *J. Exp. Biol.* **156:** 215–231.

Full, R. J., Tu, M. S. 1990. Mechanics of six-legged runners. *J. Exp. Biol.* **148:** 129–146.

Gans, C., Dudley, R., Aguilar, N. M., Graham, J. B. 1999. Late Paleozoic atmospheres and biotic evolution. *Histol. Biol.* **13:** 199–219.

Hardie, J., Hawes, C. 1982. The three-dimensional structure of the Z-disc in insect supercontracting muscle. *Tissue Cell* **14:** 309–317.

Harrison, J. F., Roberts, S. P. 2000. Flight respiration and energetics. *Annu. Rev. Physiol.* **62:** 179–205.

Jensen, M. 1956. Biology and physics of locust flight. III. The aerodynamics of locust flight. *Phil. Trans. R. Soc. London B* **239:** 511–552.

Jorgensen, W. K., Rice, M. J. 1983. Superextension and supercontraction in locust ovipositor muscles. *J. Insect Physiol.* **29:** 437–448.

Josephson, R., Ellington, C. 1997. Power output from a flight muscle of the bumblebee *Bombus terrestris*. I. Some features of the dorso-ventral flight muscle. *J. Exp. Biol.* **200:** 1215–1226.

Josephson, R. K., Malamud, J. G., Stokes, D. R. 2000. Asynchronous muscle: A primer. *J. Exp. Biol.* **203:** 2713–2722.

Keynes, R. D., Aidley, D. J. 1981. *Nerve and muscle.* Cambridge University Press, Cambridge, UK.

Kingsolver, J. G., Koehl, M. A. R. 1985. Aerodynamics, thermoregulation, and the evolution of insect wings: Differential scaling and evolutionary change. *Evolution* **39:** 488–504.

Kingsolver, J. G., Koehl, M. A. R. 1994. Selective factors in the evolution of insect wings. *Annu. Rev. Entomol.* **39:** 425–451.

Kukalova-Peck, J. 1978. Origin and evolution of insect wings and their relation to metamorphosis, as documented by the fossil record. *J. Morphol.* **156:** 53–126.

Lehmann, F. O., Dickinson, M. H. 1997. The changes in power requirements and muscle efficiency during elevated force production in the fruit fly *Drosophila melanogaster. J. Exp. Biol.* **200:** 1133–1143.

Maiorana, V. C. 1979. Why do adult insects not moult? *Biol. J. Linn. Soc.* **11:** 253–258.

Marden, J. H. 1989. Bodybuilding dragonflies: Costs and benefits of maximizing flight muscle. *Physiol. Zool.* **62:** 505–521.

Marden, J. H. 1998. From molecules to mating success: Integrative biology of muscle maturation in a dragonfly. *Am. Zool.* **38:** 528–544.

Marden, J. H. 2000. Variability in the size, composition, and function of insect flight muscles. *Annu. Rev. Physiol.* **62:** 157–178.

Marden, J. H., Kramer, M. G. 1994. Surface-skimming stoneflies: A possible intermediate stage in insect flight evolution. *Science* **266:** 427–430.

McMasters, J. H. 1989. The flight of the bumblebee and related myths of entomological engineering. *Am. Sci.* **77:** 164–169.

Miyan, J. A., Ewing, A. W. 1985. Is the 'click' mechanism of dipteran flight an artefact of CCl_4 anaesthesia? *J. Exp. Biol.* **116:** 313–322.

Molloy, J. E., Kyrtatas, V., Sparrow, J. C., White, D. C. S. 1987. Kinetics of flight muscles from insects with different wingbeat frequencies. *Nature* **328:** 449–451.

Nachtigall, W. 1985. Mechanics and aerodynamics of flight. In *Insect flight* (G. J. Goldsworthy, C. H. Wheeler, Eds.), pp. 1–29. CRC Press, Boca Raton, FL.

Orchard, I., Ramirez, J. M., Lange, A. B. 1993. A multifunctional role for octopamine in locust flight. *Annu. Rev. Entomol.* **38:** 227–249.

Osborne, M. P. 1967. Supercontraction in the muscles of the blowfly larva: An ultrastructural study. *J. Insect Physiol.* **13:** 1471–1482.

Paemen, L., Schoofs, L., Deloof, A. 1990. Presence of myotropic peptides in the male accessory reproductive glands of *Locusta migratoria*. *J. Insect Physiol.* 36: 861–867.

Peckham, M., Cripps, R., White, D., Bullard, B. 1992. Mechanics and protein content of insect flight muscles. *J. Exp. Biol.* **168:** 57–76.

Predel, R., Nachman, R. J., Gade, G. 2001. Myostimulatory neuropeptides in cockroaches: Structures, distribution, pharmacological activities, and mimetic analogs. *J. Insect Physiol.* **47:** 311–324.

Pringle, J. W. S. 1948. The gyroscopic mechanism of the halteres of Diptera. *Phil. Trans. R. Soc. London B* **233:** 347–384.

Pringle, J. W. S. 1968. Comparative physiology of the flight motor. *Adv. Insect Physiol.* **5:** 163–227.

Pringle, J. W. S. 1975. *Insect flight.* Oxford Biology Reader 52. Oxford University Press.

Pringle, J. W. S. 1981. The evolution of fibrillar muscle in insects. *J. Exp. Biol.* **94:** 1–14.

Rankin, M. A., Burchsted, J. C. A. 1992. The cost of migration in insects. *Annu. Rev. Entomol.* **37:** 533–559.

Roberts, S. P., Harrison, J. F. 1999. Mechanisms of thermal stability during flight in the honeybee *Apis mellifera*. *J. Exp. Biol.* **202:** 1523–1533.

Roff, D. A. 1994. Habitat persistence and the evolution of wing dimorphism in insects. *Am. Nat.* **144:** 772–798.

Roff, D. A., Fairbairn, D. J. 1991. Wing dimorphisms and the evolution of migratory polymorphisms among the Insecta. *Am. Zool.* **31:** 243–251.

Schoofs, L., Vanden Broeck, J., De Loof, A. 1993. The myotropic peptides of *Locusta migratoria*: Structures, distribution, functions and receptors. *Insect Biochem. Mol. Biol.* **23:** 859–881.

Smith, C. W., Herbert, R., Wootton, R. J., Evans, K. E. 2000. The hind wing of the desert locust (*Schistocerca gregaria* Forskal). II. Mechanical properties and functioning of the membrane. *J. Exp. Biol.* **203:** 2933–2943.

Srinivasan, M. V., Zhang, S., Altwein, M., Tautz, J. 2000. Honeybee navigation: Nature and calibration of the "Odometer." *Science* **287:** 851–853.

Srinivasan, M., Zhang, S., Lehrer, M., Collett, T. 1996. Honeybee navigation en route to the goal: Visual flight control and odometry. *J. Exp. Biol.* **199:** 237–244.

Tryba, A. K., Ritzmann, R. E. 2000. Multi-joint coordination during walking and foothold searching in the blaberus cockroach. I. Kinematics and electromyograms. *J. Neurophysiol.* **83:** 3323–3336.

Tu, M. S., Dickinson, M. H. 1996. The control of wing kinematics by two steering muscles of the blowfly (*Calliphora vicina*). *J. Comp. Physiol. A* **178:** 813–830.

Vigoreaux, J. O., Hernandez, C., Moore, J., Ayer, G., Maughan, D. 1998. A genetic deficiency that spans the flightin gene of *Drosophila melanogaster* affects the ultrastructure and function of the flight muscles. *J. Exp. Biol.* **201:** 2033–2044.

Weis-Fogh, T. 1967. Respiration and tracheal ventilation in locusts and other flying insects. *J. Exp. Biol.* **47:** 561–587.

Weis-Fogh, T., Jenson, M. 1956. Biology and physics of locust flight. I. Basic principles of insect flight: A critical review. *Phil. Trans. R. Soc. London B* **239:** 415–458.

Wendt, T., Guenebaut, V., Leonard, K. R. 1997. Structure of the *Lethocerus* troponin-tropomyosin complex as determined by electron microscopy. *J. Struct. Biol.* **118:** 1–8.

Wendt, T., Leonard, K. 1999. Structure of the insect troponin complex. *J. Mol. Biol.* **285:** 1845–1856.

Wheeler, C. H. 1989. Mobilization and transport of fuels to the flight muscles. In *Insect flight* (G. J. Goldsworthy, C. H. Wheeler, Eds.), pp. 273–303. CRC Press, Boca Raton, FL.

Wigglesworth, V. B. 1990. The direct transport of oxygen in insects by large tracheae. *Tissue Cell* **22:** 239–243.

Wigglesworth, V. B. 1990. The properties of the lining membrane of the insect tracheal system. *Tissue Cell* **22:** 231–238.

Wigglesworth, V. B., Lee, W. M. 1982. The supply of oxygen to the flight muscles of insects: A theory of tracheole physiology. *Tissue Cell* **14:** 501–518.

Wootton, R. J. 1981. Support and deformability in insect wings. *J. Zool. London* **193:** 447–468.

Wootton, R. J. 1986. The origin of insect flight: Where are we now? *Antenna* **10:** 82–86.

Wootton, R. J. 1990. The mechanical design of insect wings. *Sci. Am.* 263: 114–120.

Wootton, R. J. 1992. Functional morphology of insect wings. *Annu. Rev. Entomol.* **37:** 113–140.

Wootton, R. J., Evans, K. E., Herbert, R., Smith, C. W. 2000. The hind wing of the desert locust (*Schistocerca gregaria* Forskal). I. Functional morphology and mode of operation. *J. Exp. Biol.* **203:** 2921–2931.

Wootton, R. J., Kukalova-Peck, J. 2000. Flight adaptations in Palaeozoic Palaeoptera (Insecta). *Biol. Rev.* **75:** 129–167.

Wootton, R. J., Kukalova-Peck, J., Newman, D. J. S., Muzon, J. 1998. Smart engineering in the mid-Carboniferous: How well could Paleozoic dragonflies fly? *Science* **282:** 749–751.

Wray, J. S. 1979. Filament geometry and the activation of insect flight muscles. *Nature* **280:** 325–326.

Zera, A. J., Sall, J., Grudzinski, K. 1997. Flight-muscle polymorphism in the cricket *Gryllus firmus*: Muscle characteristics and their influence on the evolution of flightlessness. *Physiol. Zool.* **70:** 519–529.

Ziegler, C. 1994. Titin-related proteins in invertebrate muscles. *Comp. Biochem. Physiol.* **109:** 823–833.

Zhang, S. W., Lehrer, M., Srinivasan, M. V. 1999. Honeybee memory: Navigation by associative grouping and recall of visual stimuli. *Neurobiol. Learn. Mem.* **72:** 180–201.

CHAPTER 11

Nervous Systems

COMMUNICATION BY NERVOUS TRANSMISSION

One method to coordinate the activities of insect cells has already been described. Hormones, the chemical signals released by endocrine centers, can ultimately reach every cell by traveling through the circulatory system. However, the chemical messages are relatively slow in arriving because they depend on diffusion through the hemolymph for their transport. An alternative is the transmission of information within the insect by the electrical signals of the nervous system that provides a much more rapid signaling for the coordination of events.

If biological information is to be transferred in a timely manner, the nervous system is certainly the preferred route. Electrical signals can reach a distant part of the insect within milliseconds, compared to the minutes or hours that it might take a hormone to do the same. If food is discovered or the insect is threatened by a predator, the signals that are sent to the muscles that move the animal toward or away from the stimuli are best transmitted by nerves. By the time a hormonal signal was sent and received, the food might be gone and the predator might already have made a meal of the insect. An extensive hard-wired communication system takes up precious space, however, and a relatively small animal cannot afford to have too much space taken up by these lines to each effector organ. Insects use both the endocrine and nervous systems for information transfer between their cells, with each system having its own strengths and constraints. This chapter will deal with nervous transmission.

BASIC COMPONENTS OF THE NERVOUS SYSTEM

The Neuron

Nerve cells, or **neurons,** are the cellular building blocks that make up the nervous system. They may be distinguished, according to their function, as **sensory**

neurons, motor neurons, interneurons, or **neurosecretory neurons.** They are capable of integrating information, undergoing excitation, and transmitting the information by electrical and chemical signaling. The basic structure of a neuron consists of a cell body, or **perikaryon,** that contains the nucleus, and its projections that form the axons and dendrites (Fig. 11.1). All protein synthesis occurs in the perikaryon and metabolites synthesized there are transported to the neural processes. The perikaryon contains a large nucleus and an abundance of Golgi complexes and rough endoplasmic reticulum, whereas these organelles are generally absent from the axons, the processes that typically transmit information to other neurons. Axons often contain smooth endoplasmic reticulum and neurosecretory vesicles. Branches from the axons comprise the dendrites, regions that are

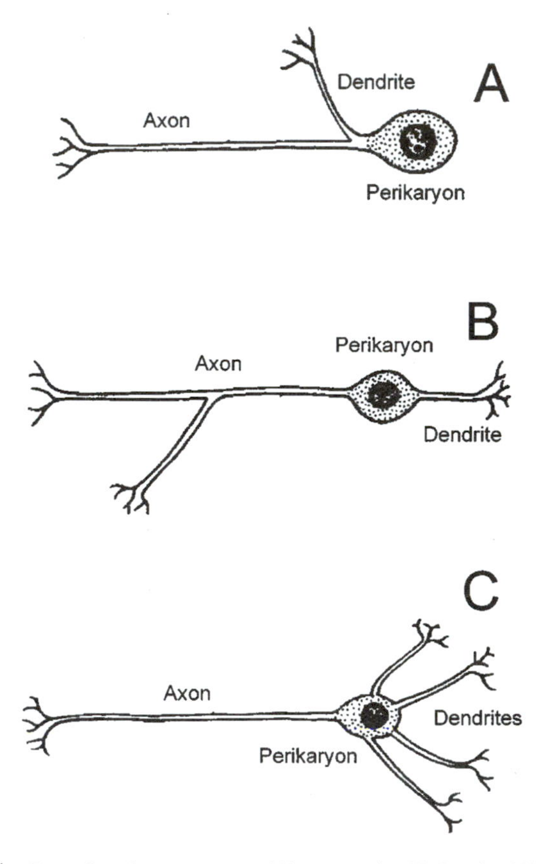

FIGURE 11.1 Examples of insect neurons. (A) Monopolar. (B) Bipolar. (C) Multipolar. Reprinted with permission from Romoser, W.S., J.G. Stoffolano Jr. (1998). *The Science of Entomology*, 4th edition. Copyright McGraw-Hill Education.

specialized for the receipt of information by the neuron. Most insect neurons are **monopolar,** unlike the case in vertebrates, bearing a single axon from the perikaryon, but those involved with peripheral receptors may be **bipolar,** and internal stretch receptors are often **multipolar** (Fig. 11.1). What are referred to as "nerves" are bundles of axons commonly surrounded by sheaths that arise from the perikarya located within the ganglia.

Sensory neurons carry messages from sensory receptors and motor neurons regulate the contraction of muscles. The connections between sensory and motor neurons are usually mediated by the interneurons that represent the bulk of the nerve cells in the central nervous system (Fig. 11.2). The aggregated perikarya of motor neurons and interneurons make up the **ganglia** of the insect central nervous system, with the perikarya of sensory neurons generally located not in the ganglion but near the receptor itself. These perikarya of motor neurons and interneurons are situated peripherally in the ganglion surrounding a central core, or **neuropil,** that also consists of the axons of sensory neurons, the axons and dendrites of interneurons, and the dendrites of motor neurons. The synapses, which are connections between the neurons, also lie within the neuropil.

Glial Cells

A system of barriers is present to maintain the chemical environment of the neurons separately from other tissues. A complex network of membranes and intercellular channels surrounds the nerve cells to maintain the ionic differences that are respon-

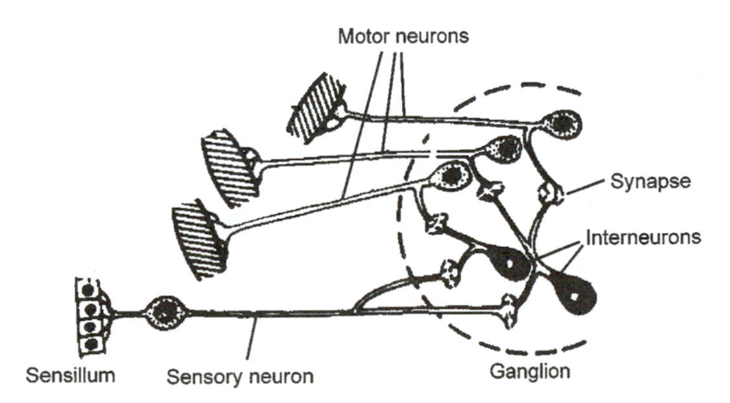

FIGURE 11.2 Interneurons bridge the connection between motor neurons and sensory neurons. Reprinted with permission from Romoser, W.S., J.G. Stoffolano Jr. (1998). *The Science of Entomology*, 4th edition. Copyright McGraw-Hill Education.

sible for the electrical potential that is required for the operation of the nervous system. Each neuron is almost completely surrounded by sheath material secreted by **glial cells** that insulates the neuron and that may provide it with nutrients. The sheaths are absent only at the synapse so as to allow the neurons to interact. The glial cells are instrumental in maintaining the controlled environment of the neuron through the presence of ion pumps that regulate the movement of sodium from the hemolymph to the extracellular fluid located in the channels formed by these cells. An additional outer layer of specialized glial cells forms the **perineurium** that secretes the outermost **neural lamella** and is largely responsible for the blood–brain barrier in insects. The perineurial layer and the lamella are collectively referred to as the **nerve sheath** (Fig. 11.3). The presence of glycogen within perineurial cells and the changes in its accumulations suggest that these cells are involved in the transfer of metabolites to the neurons, but direct evidence for the nutritive role of glial cells is lacking.

Maintenance of Electrical Potential and Nervous Transmission

All living cells actively engage in ion transport across the cell membranes, which results in an electrical potential difference that makes the inside of the cell more negative than the outside. Generally, potassium is taken up while sodium, magnesium, and calcium are pumped out. Neurons differ from other cells in that this electrical potential is able to vary substantially. The difference

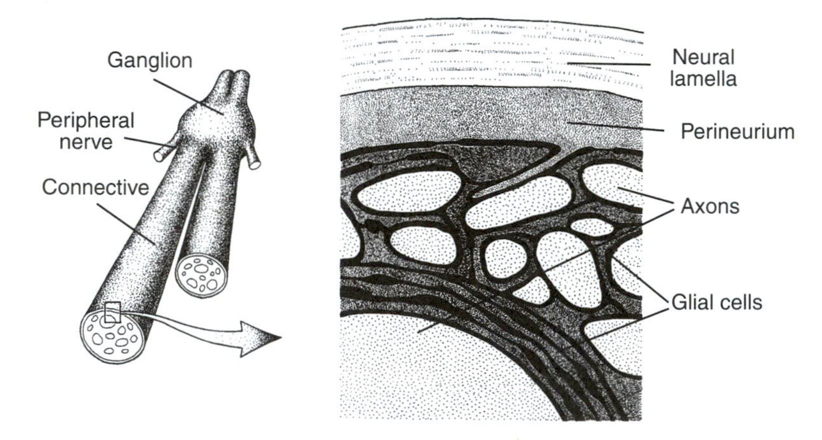

FIGURE 11.3 (Left) The connectives of the central nervous system of the cockroach. (Right) The axons are enclosed by the perineurium, a layer of glial cells, and overlaid by the neural lamella. Reprinted from Treherne, J.E. *Comprehensive Insect Physiology, Biochemistry and Pharmacology* 1985. Volume 5, pp. 115–137. With permission from Elsevier Science.

in electrical potential, or polarization, between the inside and outside of an unstimulated neuron is approximately –70 mV. This polarization, maintained by a sodium/potassium pump in the absence of any nervous transmission, is called the **resting potential.**

In the process of **sensory transduction,** sensory receptors convert energy from the environment into electrical energy. Light, mechanical deformation, or chemical signals cause the dendrites of the sensory receptors to undergo a depolarization that is proportional to the amount of stimulation. This variable change in electrical potential in the dendrite after stimulation is called the **receptor potential.** The generation of a receptor potential in turn causes the all-or-nothing depolarization along the axon known as an **action potential** that sweeps along the axon to the terminal arborizations at the synapse. The action potential in any one location in the axon lasts for 1–2 ms before the ion pumps restore it, but the depolarization moves down the axon until it reaches the synapse.

Events Occurring at the Synapse

At the terminal end of the axon, the electrical energy that was propagated must be converted to chemical energy before it can stimulate the neighboring neuron. When the action potential arrives at the presynaptic membrane, ion channels open that cause calcium to enter the neuron and stimulate synaptic vesicles to fuse with the membrane. The fusion of the vesicles causes the release of the neurotransmitters within them by exocytosis into the synaptic cleft, which varies between 200 and 500 μm in width. When more frequent depolarizations reach the presynaptic membrane, there is a greater fusion of vesicles and more release of the neurotransmitter. The neurotransmitter diffuses across the synaptic cleft, where it may reversibly bind to specific receptors on the postsynaptic membrane and induce a conformational change that alters the permeability of the membrane and induces its subsequent depolarization. This postsynaptic membrane is surrounded by a dendritic membrane that ultimately joins the axon and propagates this depolarization into another action potential that moves along the axon of the second neuron (Fig. 11.4).

A number of different neurotransmitters and neuromodulators are believed to be involved in nervous transmission at the insect synapse. The most common excitatory neurotransmitter is **acetylcholine,** released by interneurons in the neuropil and by sensory receptors. The amino acid **glutamate** appears to be the principal excitatory neurotransmitter at the neuromuscular junction. Some synapses contain neurotransmitters that inhibit rather than stimulate the postsynaptic membrane. **Gamma-aminobutyric acid (GABA)** is an inhibitory neurotransmitter in both the neuropil and the neuromuscular junction. Neuromodulators like GABA may affect either the release of a neurotransmitter from the presynaptic membrane or

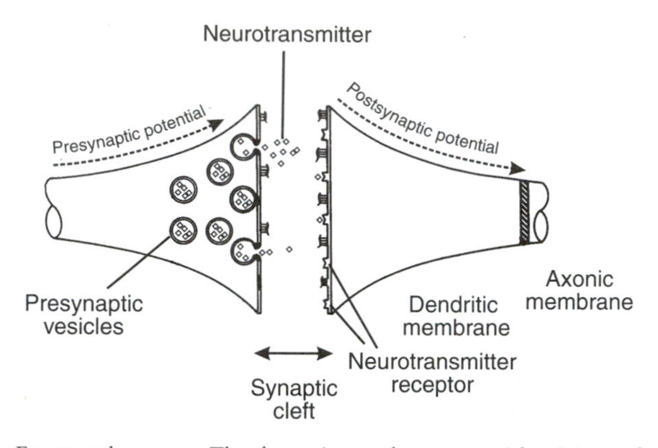

FIGURE 11.4 Events at the synapse. The change in membrane potential arriving at the presynapse causes the release of neurotransmitter stored in vesicles. The neurotransmitter binds to receptors on the postsynaptic membrane, leading to a change in membrane potential of the postsynaptic neuron. Reprinted from Shankland, D.L. and J.L. Frazier (1985). *Fundamentals of Insect Physiology*, pp. 253–286. This material is used by permission of John Wiley & Sons, Inc.

the response of the postsynaptic membrane to other released neurotransmitters. Several biogenic amines are involved in neuromodulation, including **histamine, serotonin,** and **octopamine.** Octopamine has a multifunctional role, capable of modulating the responses of flight muscles and sensory receptors and also acting as a hormone when it is released into the hemolymph.

Enzymes within the cleft degrade the neurotransmitters after they have been bound and allow some of these materials to be recycled to the presynaptic neuron for resynthesis of neurotransmitter in the vesicles. This degradation also prevents the neurons from being continually stimulated by a stimulus after it is no longer present and opens the receptor to stimulation by a subsequent release of neurotransmitter.

The synaptic connection of a single neuron with two or more neurons, or with a muscle cell that is innervated by multiple motor neurons in conjunction with neurotransmitters capable of exciting or inhibiting at the synapse, allows the nervous system to integrate nervous transmission in a much more complex fashion. In the cockroach cercus, a few hundred mechanoreceptors converge on an interneuron that mediates the escape response. When the neurotransmitter that is released by all the sensory cells is the same, an additive excitation or inhibition can be created. When excitatory and inhibitory connections are mixed, the response can be a total inhibition or excitation at the synapse or a partial effect. Divergent connections that involve a single neuron stimulating two or more different interneurons can take a message from a single receptor and spread it to several cells. There may also be loops that provide feedback to modulate the output of a motor or sensory neuron.

EVOLUTION AND STRUCTURE OF THE NERVOUS SYSTEM

The Central Nervous System

The insect central nervous system consists of the **brain** and **ventral nerve cord.** The system is composed of the individual neurons that are associated in functional groupings. These groupings evolved along with the reorganization that occurred during the evolution of the insect body from its primitive annelid-like ancestor. Neurons evolved from ectodermal cells that assumed the properties of irritability and conductivity, and were believed to have been ancestrally arranged in two lateral bands of ectodermal nerve cords that ultimately moved together and became fused laterally (Fig. 11.5). The primitive nerve mass at the anterior united the cords and served as the primitive brain or **archicerebrum.** The extensive reorganization of the nervous components that resided in the anterior body segments followed the consolidation of the anterior segments and their appendages into the head capsule and its mouthparts.

The insect brain is composed of a large grouping of neurons that lies above the esophagus, and for that reason has also been referred to as the **supraesophageal ganglion** (Fig. 11.6). It is a composite structure derived in part from the primitive archicerebrum of its annelid ancestors, but there is not total agreement as to the origin of its current form. The precise number of primitive segments that constitute the insect head is still a matter of controversy. The three sections of the brain, the **protocerebrum, deutocerebrum,** and **tritocerebrum** may have

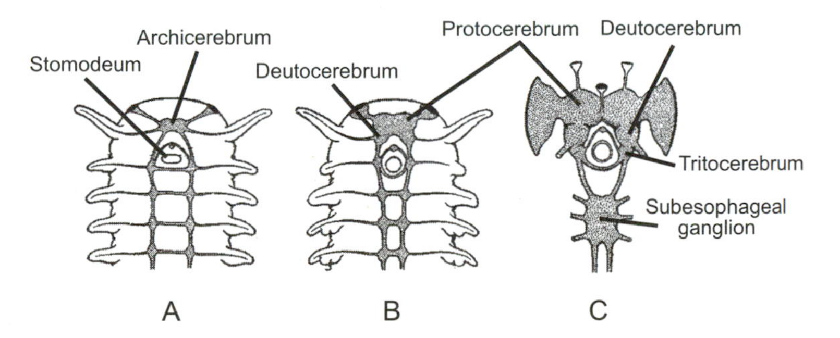

FIGURE 11.5 Possible evolution of the brain and central nervous system from an annelid-like ancestor. (A) The primitive archicerebrum originated as a mass of ganglia that united the two nerve cords. (B) The specialization of the protocerebrum and deutocerebrum into ocular and antennal centers. (C) The present-day insect brain, with its protocerebral, deutocerebral, and tritocerebral lobes, and the subesophageal ganglion that arose from the consolidation of the next three ganglia. From Snodgrass (1935). Reprinted with permission.

each originated from a different segment of the primitive head capsule, although some believe that the deutocerebrum may have evolved as an antennal lobe of the original protocerebrum.

The protocerebrum is associated with the compound eyes, ocelli, and some integumental sensory receptors. It consists of two large lateral lobes, and neurosecretory cells are present within the area known as the **pars intercerebralis** along the anteriodorsal midline (Fig. 11.7). The neurosecretory cells transport neurosecretory material to the corpora cardiaca and corpora allata, where it is released. There are several **glomeruli,** which are areas of the brain that are dense with synaptic contacts. Lateral to the pars intercerebralis are the **corpora pedunculata,** or **mushroom bodies,** which each consist of a dorsal **calyx,** or head, that rests on a stalk, the **pedunculus.** The degree of development of the corpora pedunculata was once thought to be associated with the presence of complex behaviors that require a strong visual and olfactory memory, such as in the social insects. The relative size of these areas tends to be correlated with the degree of behavioral complexity displayed by the insect, but they are sometimes also well developed in many insects such as cockroaches that may not be as "intelligent." Lateral to the corpora pedunculata are the **optic lobes,** the portion of the pro-

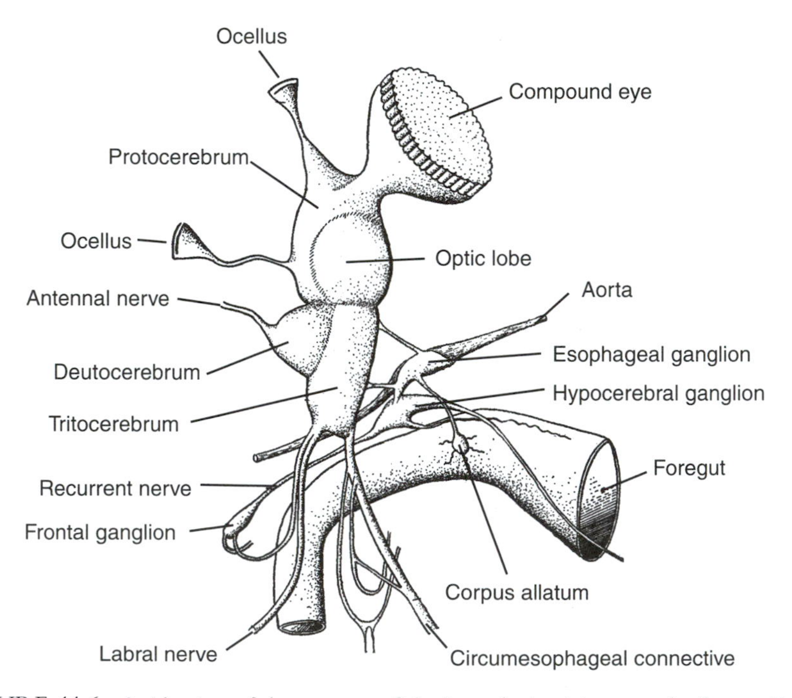

FIGURE 11.6 A side view of the structure of the insect brain, sitting atop the foregut. From Albrecht (1953). Reprinted with permission.

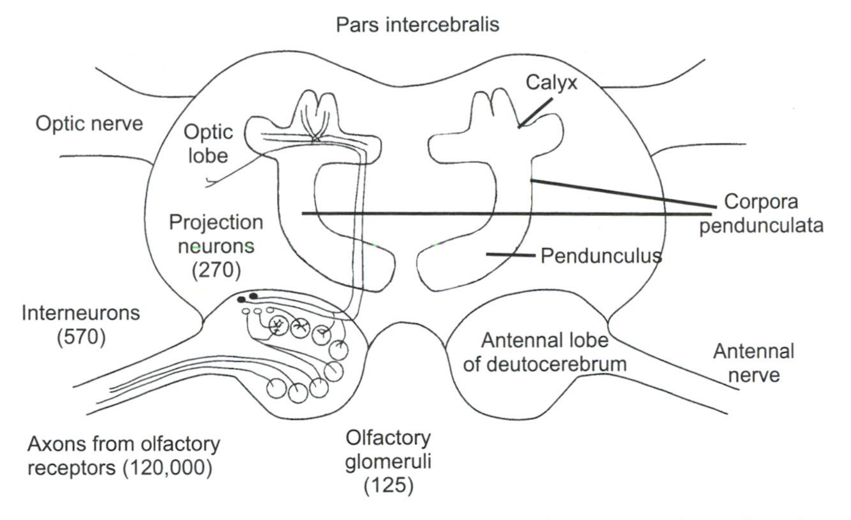

FIGURE 11.7 Glomeruli within the brain. Numbers in parentheses represent the numbers of neurons. From Lemon and Getz (1999). Reprinted with permission.

tocerebrum that extends to the compound eyes. Each consists of three groups of neuropils that process the sensory information from the compound eyes.

The deutocerebrum innervates the sensory receptors and muscles of the antennae. It, too, is divided into a series of glomeruli, the dense areas of neuropil surrounded by glial cells in which many synapses occur. These are areas of neural convergence where hundreds of thousands of receptor neurons synapse with several hundred interneurons, with these in turn synapsing with other neurons that project into the mushroom bodies of the protocerebrum (Fig. 11.7). The number of glomeruli varies among insects; locusts contain about 1000 glomeruli, and there are about 125 present in American cockroaches. Each glomerulus occupies a constant position and receives specific sensory neurons, suggesting that each may have a separate functional identity.

The tritocerebrum lies beneath the deutocerebrum and is the third and smallest part of the brain. It connects the central nervous system to the visceral nervous system through the frontal ganglion, and to the ventral nerve cord through the **circumesophageal** connectives. It also receives nerves from the labrum.

The **subesophageal ganglion** is the first ganglion of the ventral nerve cord and the only one in the head. It is a compound ganglion formed by the fusion of the ganglia of the mandibular, maxillary, and labial segments. It gives off paired nerves to each of the mouthpart appendages and innervates the salivary glands. The remainder of the ventral nerve cord consists of a series of paired ganglia, joined by connectives, which contain the perikarya of motor neurons and interneurons. The first three are in the thorax, serving as a locomotor center that controls the wings and the legs. In some insects, these ganglia are consolidated to form a compound **metathoracic ganglion.** Thysanurans have eight abdominal ganglia, but the evolutionary trend has been toward a reduction in the number of abdominal ganglia and most insects have fewer (Fig. 11.8). The metathoracic ganglion may be commonly fused with the first 1 to 3 abdominal ganglia, and abdominal ganglia 7 and 8 are commonly fused in insects to form the compound **terminal abdominal ganglion.** This terminal abdominal ganglion innervates the hindgut and reproductive structures and regulates oviposition behavior in females and the mating behavior of both sexes.

The **peripheral nervous system** involves all the nerves that radiate from the central nervous system. These include the nerves that innervate muscles, stretch receptors that may serve as proprioreceptors, innervations of the reproductive system and spiracles, and the various sensory receptors.

The Visceral Nervous System

The portion of the nervous system described so far is largely devoted to moving the insect around and mediating the interactions with the external environment. Another part of the insect nervous system is concerned with maintaining the internal environment and coordinating various internal functions. This element of the nervous system, termed the **visceral nervous system,** innervates the gut, heart, and endocrine glands. The major ganglion of the visceral nervous system is

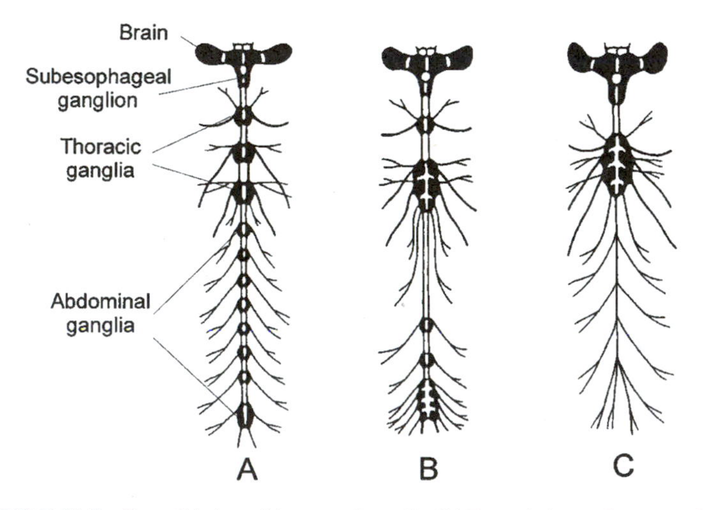

FIGURE 11.8 Consolidation of the ventral ganglia. (A) Example in a coleopteran. (B) Example in a hymenopteran. (C) Example in a dipteran. From Gullan and Cranston (2000). Reprinted with permission.

the **frontal ganglion** that arises from the frontal connectives issued by the tritocerebrum. The frontal ganglion innervates the foregut and controls crop emptying. It gives rise to the recurrent nerve that passes underneath the brain and expands into the **hypocerebral ganglion** that innervates the corpora allata, corpus cardiacum, and the fore- and midguts (Fig. 11.6).

SENSING THE ENVIRONMENT

The integument of insects provides an effective barrier against water loss and environmental assault. By separating the insect from its environment, it allows the proper internal conditions to be maintained. However, in order to operate effectively, the insect must also be able to detect changes in the environment and make responses that are biologically appropriate. Chemical messages from sources such as pheromones mediate mate location; volatiles from host plants and animals identify food and oviposition sites; and various signals from other individuals must trigger alarm and aggregation responses.

The nervous system, residing on the inside of the impenetrable integument, must be able to receive information across this barrier. To allow this, a compromise had to be struck between the need for the insect to conserve water and the need to sample the environment with a biological membrane that by necessity must contain a moist receptor surface. Insects have solved this problem by exposing these receptor surfaces only through extremely small pores. The pores are continuously open, but the small size of the pores minimizes the potential for water loss. Receptors are not distributed uniformly over the body but are concentrated on few areas that would be most likely to receive stimuli, such as the mouthparts, antennae, legs, and cerci, making most of the insect body insensitive to external stimuli.

Another unique property of the sensory receptors in insects is that they are all **primary sense cells** instead of the secondary sense cells that respond to taste, touch, and vision in vertebrates. A secondary sense cell is a cell of nonneural origin that is linked to a neuron. Vertebrate touch receptors in the skin are modified epidermal cells that produce a receptor potential and then relay that potential to a neuron. When the cell that receives environmental stimuli is a bipolar primary sense cell, it produces both a receptor potential and an action potential, and there is no need for the second neuron. (Fig. 11.9). One cell in insects performs the same function as two do in vertebrates: the primary sense cell and the neuron that innervates it.

Sensilla

The insect **sensillum,** or basic unit of sensory reception, originates from the same ectodermal tissues that give rise to the nervous system. The sensilla may be

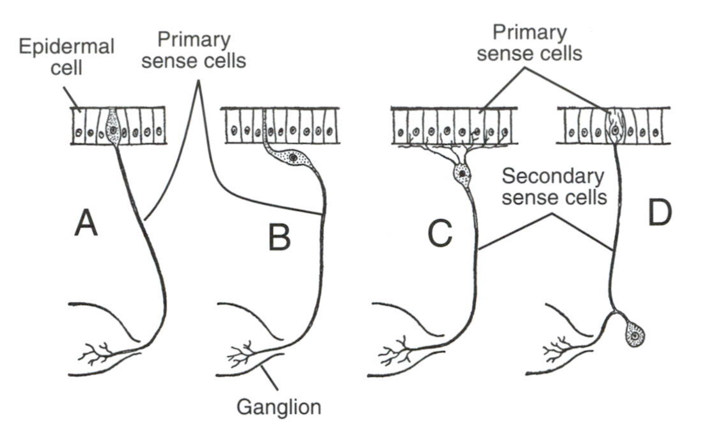

FIGURE 11.9 The differences between primary and secondary sense cells. (A) and (B) Primary sense cells directly receive the stimuli. (C) and (D) Secondary sense cells of nonneural origin that are linked to primary sense cells. From Snodgrass (1935). Reprinted with permission.

identified by their external morphology (Fig. 11.10), but their structure alone is not sufficient to determine their functions. A sensillum may contain several cuticular parts that differentiate from an epidermal mother cell into one or more cells, including the **sensory neurons,** a **tormogen cell** that creates the socket, a **trichogen cell** that creates the shaft of the hair, and a **thecogen** cell that produces the sheath component of the sensillum that isolates the axons from one another and provides the neuron with ions and nutrients (Fig. 11.11). The sensory neu-

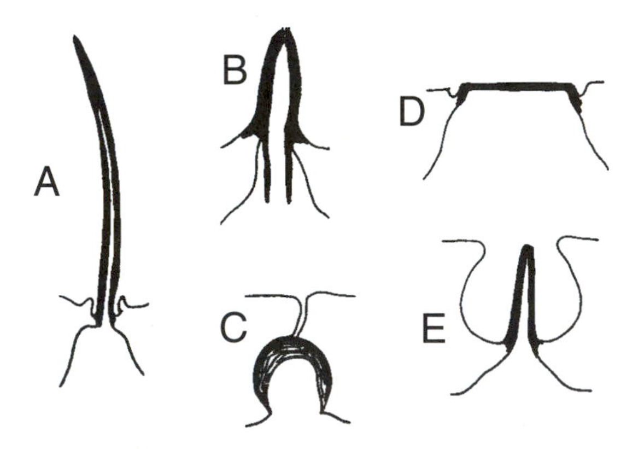

FIGURE 11.10 Examples of external morphologies of sensilla. (A) Trichoid sensillum. (B) Basiconic sensillum. (C) Campaniform sensillum. (D) Placoid sensillum. (E) Coeloconic sensillum. Reprinted from Zakaruk, R.Y. *Comprehensive Insect Physiology, Biochemistry and Pharmacology* 1985. Volume 6, pp. 1–69. With permission from Elsevier Science.

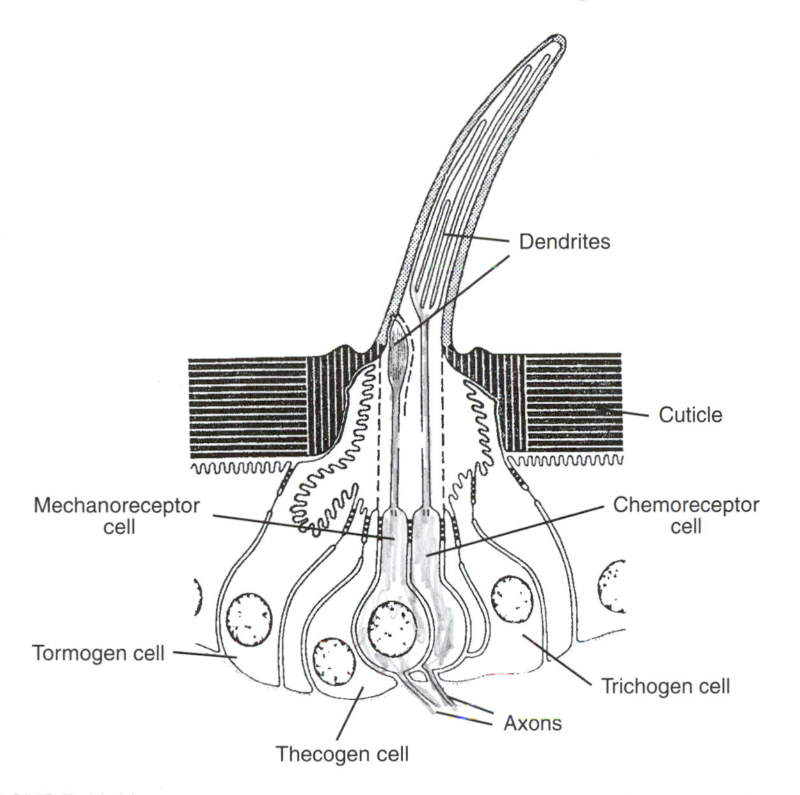

FIGURE 11.11 A cross section of a trichoid sensillum. Reprinted with permission from Altner, H. and L. Prillinger (1980). *International Review of Cytology* **67:** 69–139. Copyright Academic Press.

rons are bipolar, extending their dendrites into the cuticular portion and their axons to the central nervous system. The structure of the particular sensillum varies considerably depending on the sensory modality involved, and a description of these sensilla will be based on a functional classification according to what stimuli we believe the receptors may respond to.

Chemoreception

There are two general types of chemoreceptors that are capable of responding to chemical energy in the environment. **Contact chemoreceptors,** which function in what we commonly refer to as taste, respond to substances in solution at relatively high concentrations and a relatively close range. They are concentrated mostly on the mouthparts, legs, and ovipositor. The axons from these contact chemoreceptors generally connect with interneurons in the ganglia of the segment on which they appear. **Olfactory chemoreceptors** on the antennae and on the palps of the mouthparts mediate what is considered as smell, responding to substances traveling in air in relatively low concentrations and originating at greater distances. The axons from these olfactory receptors generally terminate in the deutocerebrum.

Although both contact and olfactory chemoreceptor sensilla often have the same outward appearance of a hair-like process or a small peg, they differ in the number of pores that are present on the shaft. For taste receptors, the perikarya of hair-like **trichoid sensilla** lie beneath the hair and the dendrites extend into the shaft where they reach a single pore at the tip (Fig. 11.12). A viscous fluid may be present at this single pore. Taste receptors on the legs may additionally have mechanoreceptive dendrites associated with them. Internal taste receptors can also be found within the pharynx of the digestive tract, but are usually subcuticular. Axons from most of the sensilla in the head run directly to the subesophageal ganglion without synapsing.

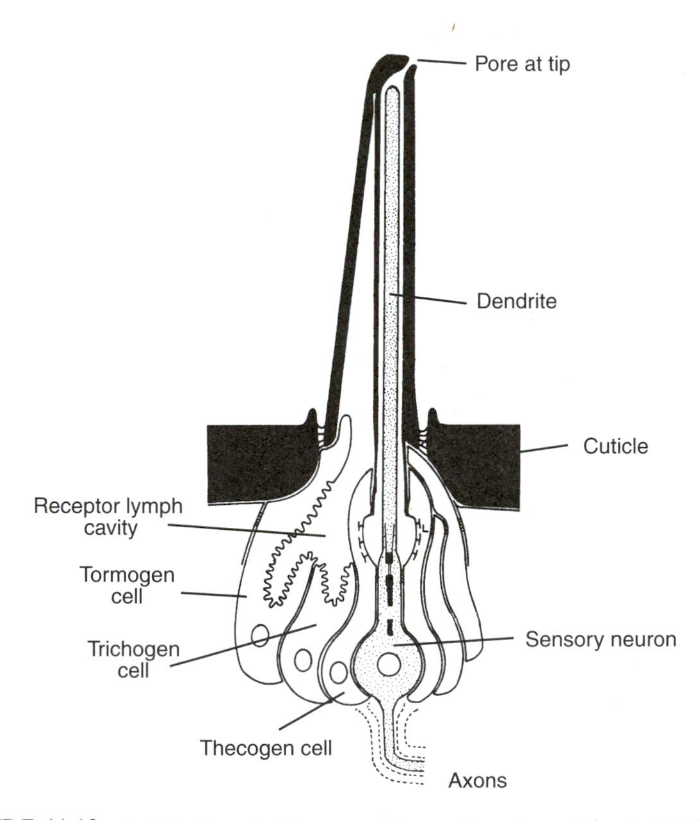

FIGURE 11.12 A trichoid contact chemoreceptor sensillum. From Zakaruk (1985). Reprinted with permission.

Olfactory chemoreceptors that are located on the antennae may consist of either basiconic sensilla, which are blunt hairs or short cones, or the hair-like trichoid sensilla. They differ from taste receptors in having numerous pores along their shafts, and lumens that are filled with multiple dendritic branches (Fig. 11.13). From one to three neurons may also be present that project their dendrites into the shaft. Coeloconic sensilla have double walls and numerous longitudinal grooves on the surface. The neurons in these sensilla can respond to thermal stimuli as well as odors. Chemical signals from the environment such as pheromones enter through the pores in the cuticle. The pores open into pore tubules that lead into the lumen of the sensillum and direct the odor molecules to sites on the dendrite. The axons of the olfactory sensilla extend to the deutocerebrum. In male *Antheraea polyphemus* moths, about 70% of the more than 60,000 sensilla present on the antennae are responsive to the female sex pheromone.

Stimulus molecules bind to specific receptors that are located on the dendrites within the sensillum lumen. The number of different receptor genes expressed by the neurons of the sensilla reflects the potential recognition of odor molecules. The

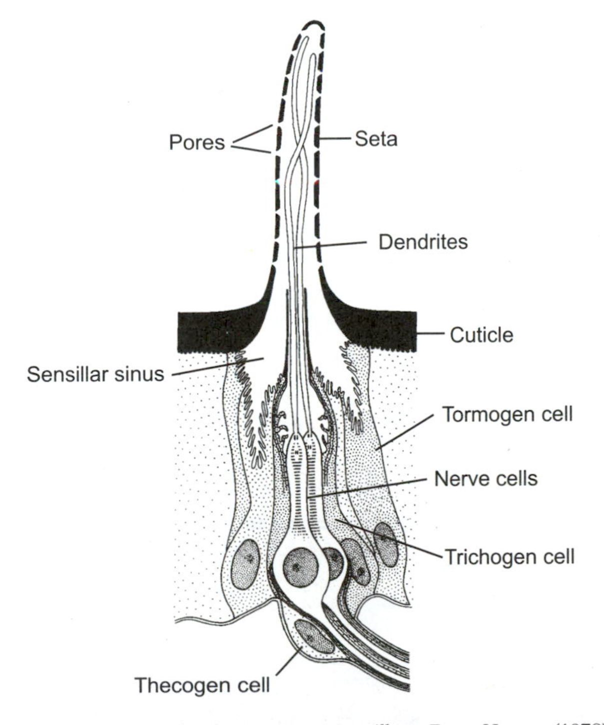

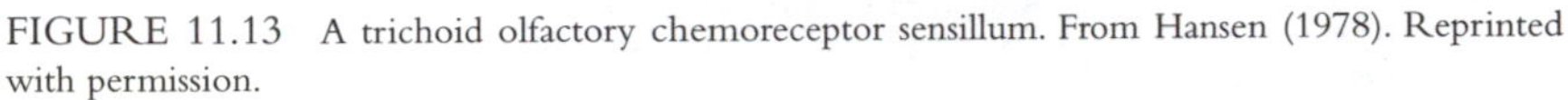
FIGURE 11.13 A trichoid olfactory chemoreceptor sensillum. From Hansen (1978). Reprinted with permission.

specificity of a particular receptor depends on the types of these specific receptors that are present. Specific sugars, feeding attractants and deterrents, host stimulants, and pheromones cause the dendrite to depolarize only when those molecules are able to bind to the receptors that are present. The binding causes the generation of a receptor potential that is transmitted to the perikaryon, with a subsequent action potential resulting in the axon. The presence of multiple receptors on sensory neurons differs from the vertebrate system in which only a single receptor is expressed by a given neuron.

A fluid, the **sensillum lymph,** surrounds the dendrites within the cavities of the shafts of both contact and olfactory chemoreceptors. The lymph bathes the dendrites to protect them from desiccation and transports stimulus molecules to the receptors. **Odorant binding proteins** that are contained within the lymph may provide another filter that is used to distinguish odor molecules. These binding proteins are largely responsible for the specificity and sensitivity of insect olfaction, enhancing the rate at which odor molecules are captured and partitioning these often hydrophobic molecules in the aqueous system that surrounds the dendrite (Fig. 11.14). The proteins are produced by the sensillar support cells and are secreted into the sensillum lymph. Three classes of odorant binding proteins have been identified in the antennae of adult lepidoptera: two classes of **general odor-**

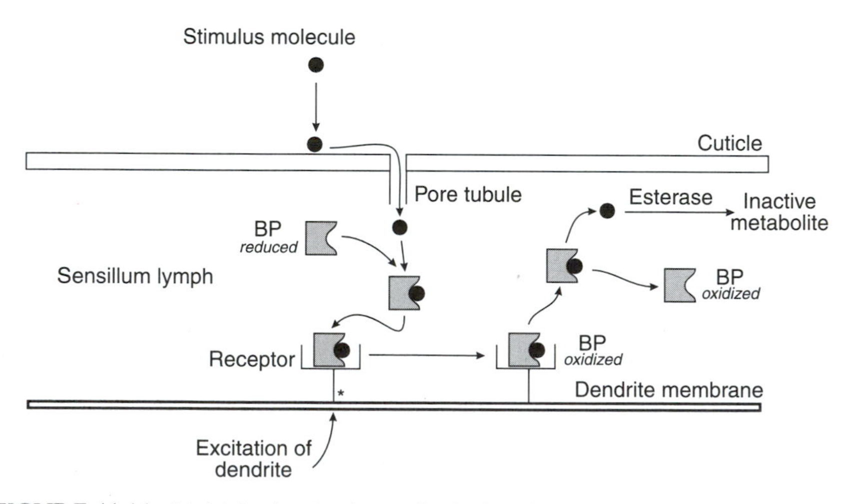

FIGURE 11.14 Model for the stimulation of a dendrite by a stimulus molecule. The molecule strikes the cuticle and diffuses across it to the pore tubule. It moves through the tubule into the sensillum lymph, where a reduced odorant binding protein (BP) takes up the stimulus molecule and delivers it to the receptor on the dendritic membrane. This causes the excitation of the dendrite and the oxidation of the BP, ultimately releasing the molecule into the lymph where it is degraded by esterases. From Ziegelberger (1995). Reprinted with permission.

ant binding proteins that are consistent with the distribution of general odorant receptors and a third class of **pheromone binding proteins** that predominate in the antennae of male moths that have a large number of sensilla that respond to the female sex pheromone. The different odorant binding proteins may recognize different odorants. If their distribution were restricted to different regions of the olfactory system, they could determine which odorant molecules bind to the receptor on the dendrite. The mechanism of transporting molecules from the outside to the surface of the dendrite can be efficient enough to allow a single odor molecule adsorbed on the cuticular surface to ultimately cause an action potential to be generated.

After the dendrite is stimulated, it must be rapidly inactivated to prevent its continuous stimulation and to allow another odor molecule to subsequently bind. An insect flying within an odor plume must make rapid assessments about the concentrations of odorants in the plume and respond accordingly. **Esterases** within the sensillum lymph as well as on the surface of the cuticle quickly inactivate the odorant molecules. Binding proteins in the lymph may also be involved in the degradation process. Those esterases that are present on the cuticle may act to prevent the entire surface of the insect from becoming a pheromone source for other insects when pheromones bind to it.

Sensory transduction refers to the ways in which the environmental energy is changed into electrical energy in the nervous system. There are several distinct steps in this process that have been identified (Fig. 11.15). In the first, **adsorption,** the odor molecule is adsorbed from the air onto the cuticular wall. The molecule then **diffuses** through the cuticle to the pores and pore tubules. **Binding** involves the specific binding of the odor molecules to receptor sites, followed by their conformational change and **activation.** The activated receptor induces a change in **membrane conductance** that leads to the generation of an action potential. Last, **inactivation** of the odor molecule occurs to make room for the next molecule. There are typically thousands of receptors present, many of which may show varying sensitivity to a given stimulus. Odor discrimination thus involves the brain discerning which of the receptors present has been activated and their degree of activation by an odor molecule.

The brain may receive this information from receptors in two general ways (Fig. 11.16). It may make use of specific receptors that react to only one compound. Many of the pheromone receptors fit into this category. When such a specialist receptor is stimulated in this **labeled line,** the brain "knows" that the particular compound must be present because it alone can cause the stimulation, and the behavior is directed accordingly. However, it would be impossible for the sensory system to accommodate every possible stimulus molecule with its own specific receptor. Many other receptors have a much broader response spectrum with differing sensitivities, and a particular molecule may activate a wider range of such receptors to greater or lesser degrees. The resulting code of neural activity from a field of generalist receptors is known as **across-fiber patterning.** This

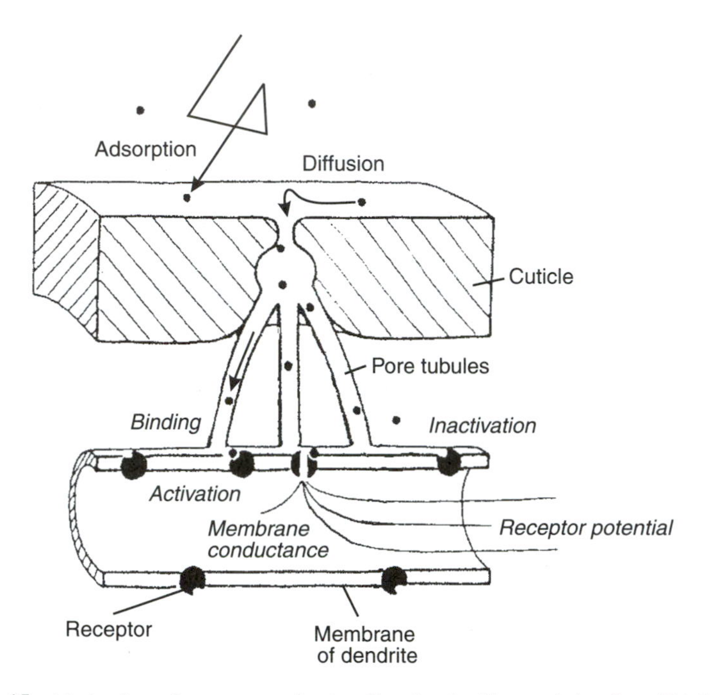

FIGURE 11.15 Mechanism of sensory transduction. Reprinted with permission from Kaissling, K.-E. (1974). *Biochemistry of sensory functions*, pp. 243–273. Copyright Springer-Verlag GmbH & Co. KG.

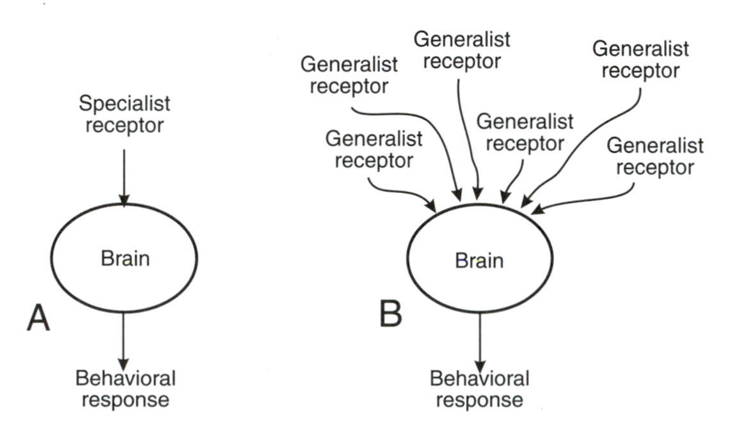

FIGURE 11.16 Production of a behavioral response from a labeled line (A) or an across-fiber pattern (B). In the labeled line, a specific receptor sends a specific message to the brain. In across-fiber patterning, the overall pattern from a number of generalist receptors is evaluated by the brain.

allows a large number of different stimuli to create different patterns that can be coded by the central nervous system without requiring large numbers of specific receptors for each molecule. Both strategies of sensory transduction have been identified. In several phytophagous insects, including the Colorado potato beetle and the locust, the quality of host plants is determined by a large number of contact chemoreceptors on the palps, without any single receptor being responsible for acceptance or rejection of the food. In other insects, such as the cabbage white butterfly, maxillary taste receptors are specific for glucosinolates and the stimulation of these receptors evokes feeding behavior. Similarly, feeding deterrents may act via specific labeled lines or by disrupting the normal across-fiber patterns that the insect requires for feeding.

The responses of sensilla can be measured electrophysiologically. The electrical potential change between the base of the antenna and its tip can be measured when an odor molecule is passed over it and the receptors are activated. This change in electrical potential can be monitored as an **electroantennogram,** which measures the summated receptor potentials in the whole antenna (Fig. 11.17). A more precise determination of the responses of receptors results from single cell recording, in which an electrode is implanted at the base of a single sensillum and the signals from the single receptors are measured. This technique is able to determine the sensitivity and specificity of individual sensilla. Both methods can provide preliminary information for screening potential attractants and deterrents, but neither can substitute for determining the behavioral responses of intact insects. The electrophysiological techniques fail to account for effects of physiological state and nervous integration on the activation of behavior.

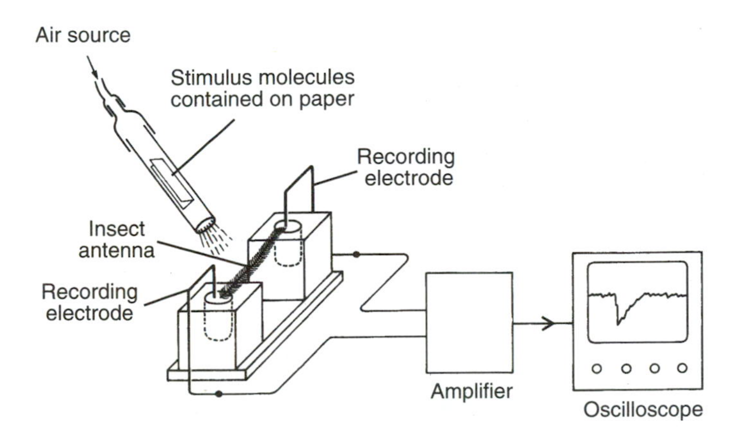

FIGURE 11.17 An example of an electroantenogram. The receptors on the whole excised antenna respond to the stimulus blown across it. The responses of all the sensory cells are amplified and displayed on an oscilloscope. From Gullan and Cranston (2000). Reprinted with permission.

Thermoreception and Hygroreception

The sensitivities to temperature and humidity are often found in the same antennal receptors. Combinations of cold, moist air and dry air receptors or cold and warm receptors are most common. These sensilla are typically small pegs located within cuticular pits (Fig. 11.18). Humidity may hydrate the cuticular peg, causing it to deform and function like a typical mechanoreceptor. The thermoreceptors that respond to cold increase their firing rates with falling temperatures.

Mechanoreception

Mechanoreceptors sense mechanical energy in the environment, such as pressure, gravity, vibration, and the internal forces generated by muscles, as they distort the body. The rigid exoskeleton is an ideal platform with which to detect these vibration stimuli and transfer them to sensory receptors. The various types of mechanoreceptors include cuticular structures, chordotonal organs, and stretch receptors.

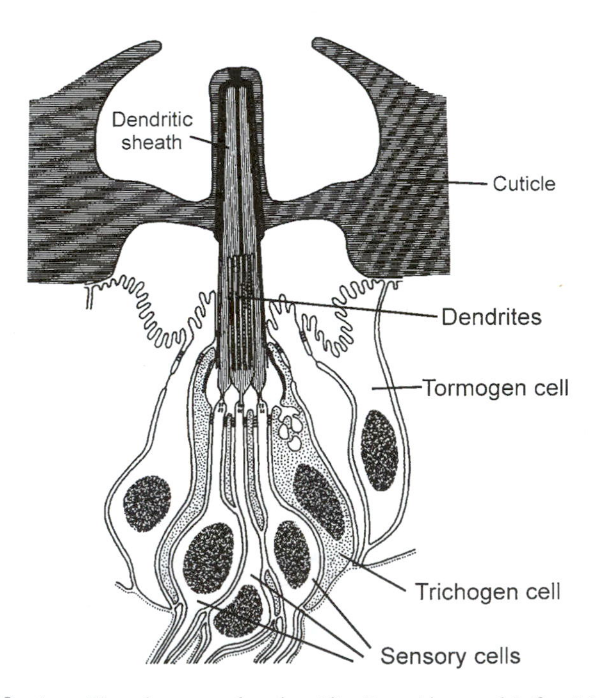

FIGURE 11.18 A sensillum that responds to humidity. From Altner and Loftus (1985). Reprinted with permission from the Annual Review of Entomology, Volume 30. ©1985 by Annual Reviews.

The cuticular structures consist of the hair-like external processes, commonly known as trichoid sensilla that respond to tactile stimulation. These setae resemble the chemoreceptor sensilla but have no pores and are innervated by a single neuron (Fig. 11.19). Some chemoreceptors may contain an additional mechanoreceptor at the base that operates independently of the chemoreceptor neurons. The tormogen, or socket cell, secretes a joint membrane that contains the elastic protein resilin to provide flexibility in its movement. Displacement of the seta deforms the tubular body of the dendrite and initiates a receptor potential that in turn triggers an action potential in the axon. Some trichoid sensilla display a **phasic response** that generates impulses only when deflected but not when constantly deformed. Other, usually more blunt sensilla, show a **tonic response** that is strong when initially deformed and steady but reduced under constant deformation.

Campaniform sensilla are dome-shaped structures usually located near joints or other structures such as the halteres that are subject to distortion and cuticular stress (Fig. 11.20). The **scolopale,** a cuticular cap, covers a single neuron underneath that is stimulated by its deformation when compressive forces are generated in the adjacent cuticle. The campaniform sensilla on the cerci of the cricket initiate a kicking response by the legs when they are deformed by movement of the cercal shaft.

Chordotonal organs consist of subcutaneous groupings of special sensilla known as **scolopidia,** which serve as mechanical transducers. An individual

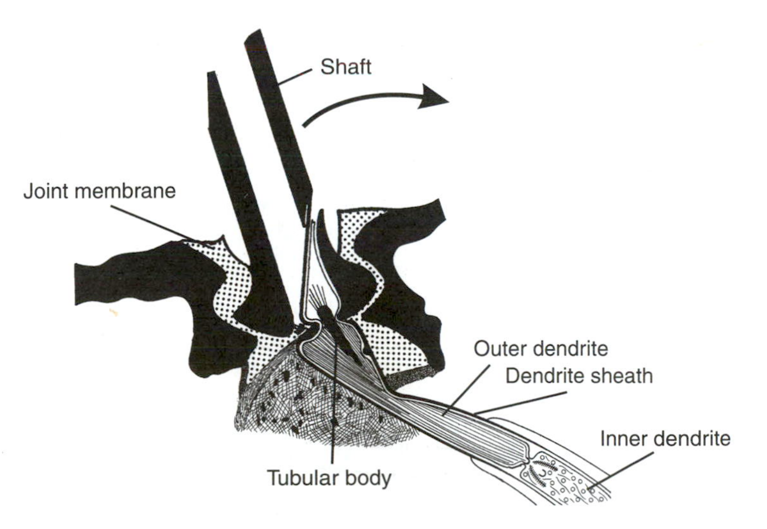

FIGURE 11.19 A mechanoreceptor that is activated by movement of its seta in the direction of the arrow. The deformation of the subcuticular tubular body initiates the receptor potential. Reprinted with permission from Barth, F.G. and R. Blickhan (1984). *Biology of the Integument.* Volume 1, pp. 554–582. Copyright Springer-Verlag GmbH & Co. KG.

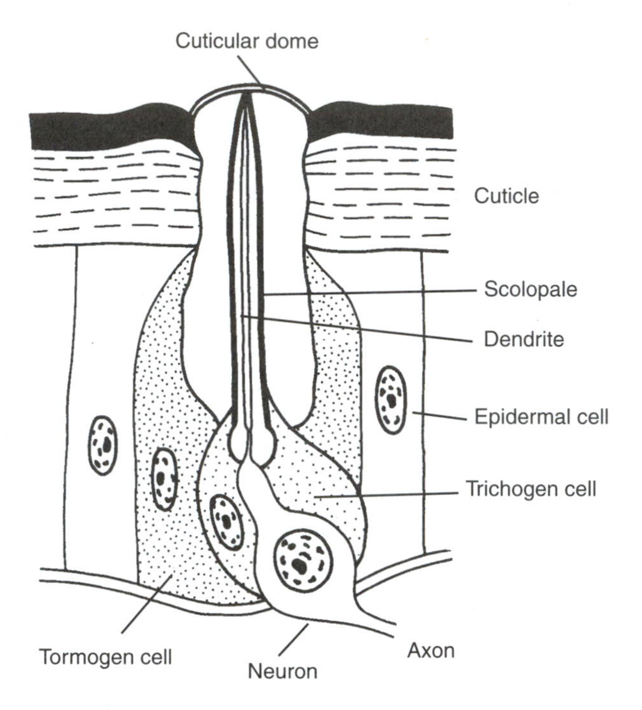

FIGURE 11.20 A campaniform sensillum. Sheer forces within the cuticle deform the cuticular dome. The resulting movement of the scolopale generates a receptor potential in the dendrite. From Chapman (1991). Reprinted with permission.

scolopidium is composed of cells that include one or more bipolar sensory cells, a glial cell, an attachment cell, and a scolopale (Fig. 11.21). It is usually attached to the cuticle at both ends and thus can measure the degree of cuticular distortion. The chordotonal organs are internal sense organs and are never associated with external cuticular processes. They are often used as **proprioreceptors** to measure relative body position in response to cuticular deformation by bending, gravity, airborne vibrations, or air movement.

A typical chordotonal organ is **Johnston's organ,** which is found in the antennal flagellum of many insect orders. Movement of the antennal shaft is transmitted to the groups of scolopidia below, allowing the insect to detect vibrations produced by individuals of the opposite sex and to receive feedback from air currents that deform the antennae during flight (Fig. 11.22). **Subgenual organs,** another type of chodotonal organ, have been identified in the tibia of insects, attached at one end to the cuticle and at the other to a trachea (Fig. 11.23). They respond to vibrations of the leg, perhaps as currents of hemolymph are created by the movements. The subgenual organs may be useful for predator avoidance or for responding to intraspecific signals.

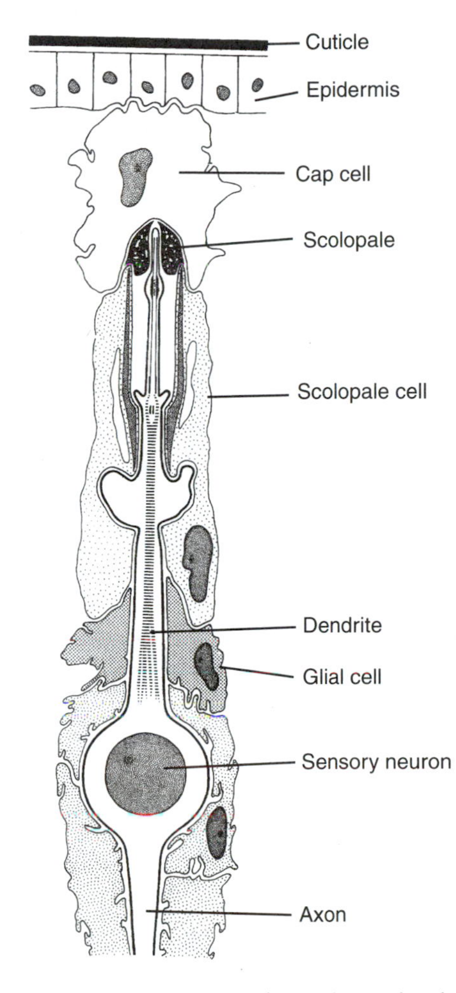

FIGURE 11.21 A subcuticular chordotonal organ. Attached to the cuticle at both ends, these sensilla can measure the degree of cuticular distortion. From Gullan and Cranston (2000). Reprinted with permission.

Tympanal organs are chordotonal structures that are able to detect sound vibrations. The tympanal organ consists of a thin drumlike piece of cuticle called a tympanum with a chordotonal organ with from one to several scolopidia attached on the inside that can monitor its movement (Fig. 11.24). Sounds cause the tympanal membrane to vibrate and the vibrations are detected by the scolopidia. The **tympanum** enables insects to respond to a broad range of frequencies from 100 Hz to as high as 140 kHz. They are often present in a paired configuration to allow the insect to detect the direction that the sound is coming from. In

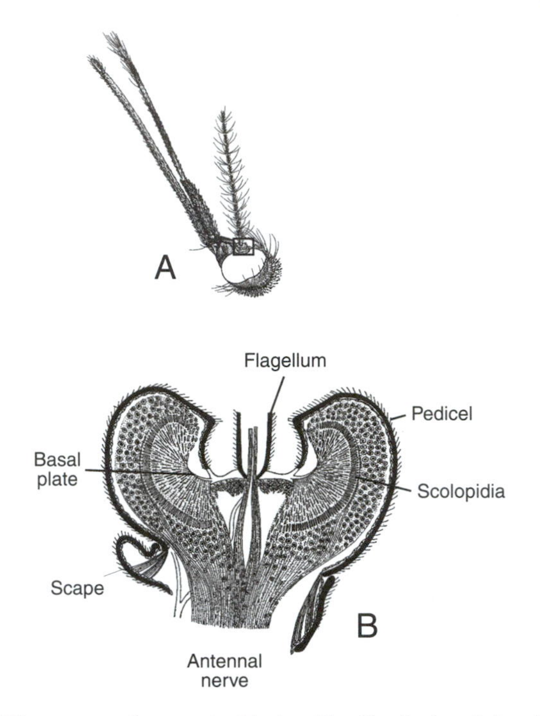

FIGURE 11.22 (A) The antenna of a mosquito. The box identifies the inset below. (B) A cross section of Johnston's organ at the base of the antenna. The basal plate deflects the scolopidia when it is displaced by movements of the flagellum. From Snodgrass (1935). Reprinted with permission.

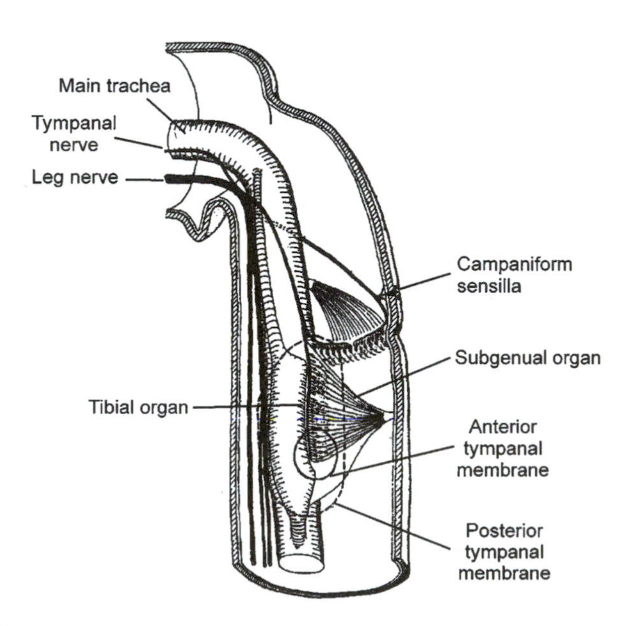

FIGURE 11.23 A subgenual organ within the tibia of an insect leg. One end is attached to the cuticle, with the other attached to the trachea. From Field and Matheson (1998). Reprinted with permission.

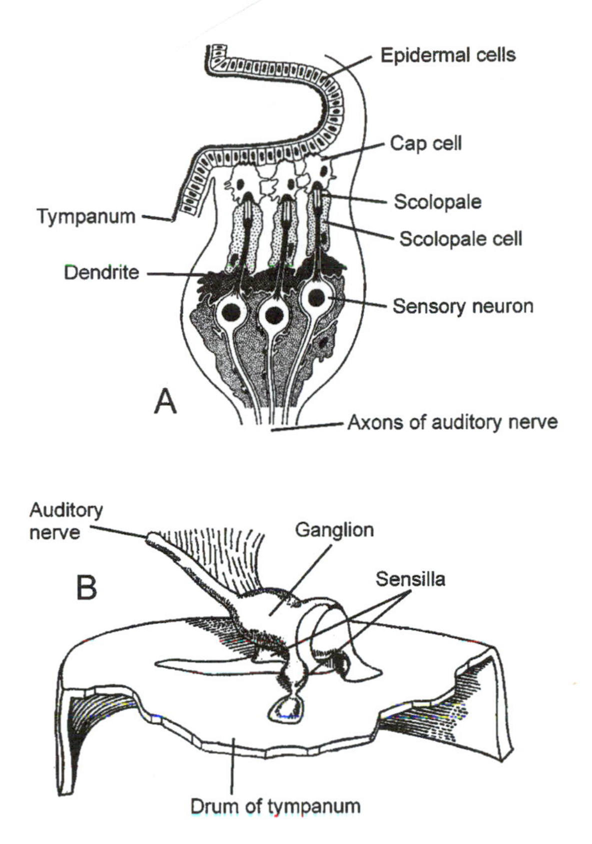

FIGURE 11.24 (A) A cross section of the sensillum underlying the tympanum. (B) The inner surface of a tympanal organ. Reprinted from Gray, E.G. (1960). *Philosophical Transactions of the Royal Society of London B* **243:** 75–94. With permission from The Royal Society.

crickets, tympanal organs are present on the fore tibia with their membranes backed against a large tracheal tube. In some grasshoppers, the organ is located on either side of the first abdominal segment. Night-flying noctuid moths have two scolopidia associated with each abdominal tympanal organ. Each has a differing sensitivity and allows the moth to respond to the ultrasonic cries of bats. The cell with the greater sensitivity responds to the far-away cries, while the less sensitive cell informs the moth that the bat is nearby and causes it to engage immediately in an evasive dive.

Stretch receptors are multipolar neurons that are found throughout the internal organs. They consist largely of dendritic endings embedded at several points in muscle or connective tissue or within the basement membrane that surrounds an organ (Fig. 11.25). They provide feedback for proprioreception when the structure that is monitored changes in shape. The foregut is often innervated by multipolar neurons that can respond to gut movement and distention. Within the

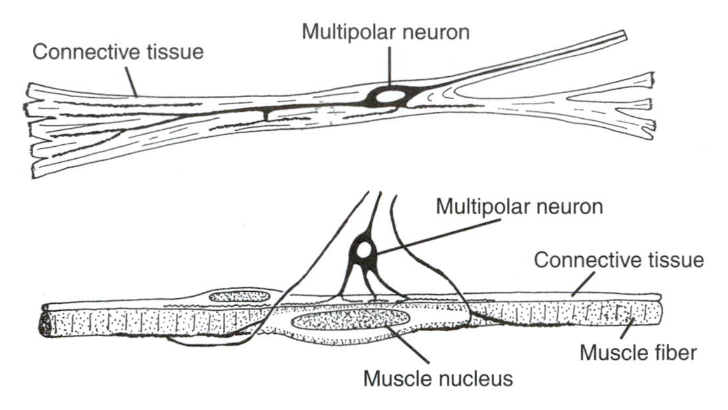

FIGURE 11.25 Two stretch receptors. (Top) A multipolar neuron associated with the connective tissue of a cockroach. (Bottom) A multipolar neuron associated with the muscle fiber and a strand of connective tissue in a moth. The dendrites of the multipolar neuron extend into the fiber tract. From Osborne (1969). Reprinted with permission.

abdomen, dorsal longitudinal stretch receptors may be present in each segment that extend to the intersegmental regions.

Visual Receptors

The sensitivity to visual stimuli is possible through a number of different receptors in insects. Even infrared irradiation can be detected by paired thoracic pit organs in some buprestid beetles that breed only in trees that have recently been killed by fire. The beetles are able to respond to the infrared emitted from forest fires in the wavelength range of 2 to 4 μm using dome-shaped sensilla that are exposed during flight and activated specifically by the infrared wavelengths.

Dorsal Ocelli

Of the receptors responding to light in the visual spectrum, **dorsal ocelli,** or simple eyes, are the least complex of the visual structures. They often appear in a triangular pattern on the head between the compound eyes in the winged adults of most orders and in the larvae of hemimetabola, but may also occur singly. A typical ocellus consists of a convex transparent corneal lens with an aggregation of hundreds of light-sensitive **retinula cells** below (Fig. 11.26). A region of each of the retinula cells known as the **rhabdom** contains the visual pigment **rhodopsin** that absorbs light and initiates the receptor potentials. The visual pigments and the mechanism of light perception in insects will be described in more detail below. A lower layer of cells

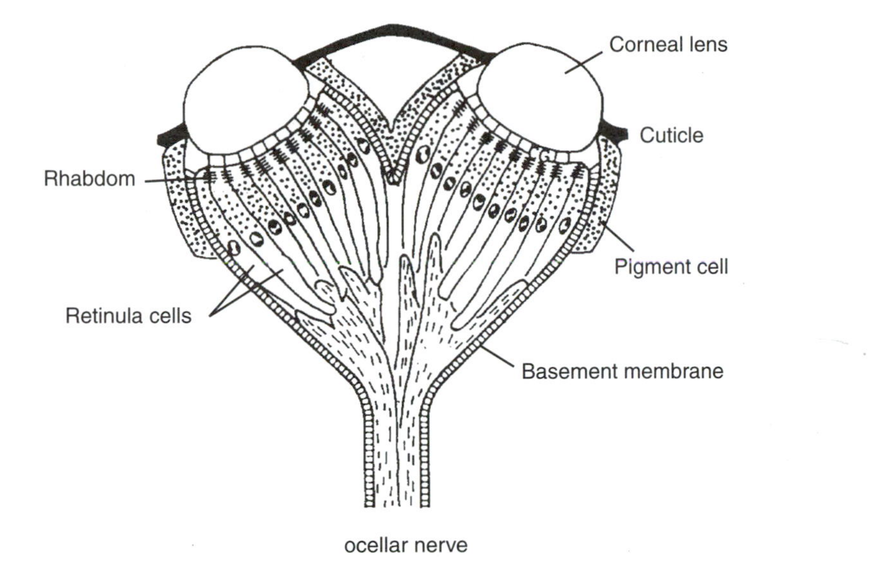

FIGURE 11.26 The generalized structure of a pair of ocelli. From Toh and Tateda (1991). Reprinted with permission.

that contain urate crystals or closely packed tracheae may function as a reflecting **tapetum.** The axons from these retinula cells merge with interneurons so that fewer axons enter the brain than the number of receptor cells present. In flies, the three ocelli are connected to the protocerebrum by a single ocellar nerve. The ocelli appear to be poorly designed for image perception and may be able only to scan the horizon for light intensity to provide general information for navigation. They appear to interact with the processing of stimuli from the compound eyes to facilitate locomotion that is based on the perception of visual images.

Lateral Ocelli

Lateral ocelli, or **stemmata,** are also relatively simple eyes that are the only photoreceptors present in the larvae of holometabolous insects (Fig. 11.27). Despite the name, their structures are more similar to the compound eyes than to the dorsal ocelli, but they fail to meet the image quality that is possible with compound eyes. The number of receptors associated with stemmata is usually too low to allow the formation of anything but a coarse mosaic of the environment. The larval holometabolous insects that bear these stemmata generally crawl and have no need for a visual system with the higher performance of the compound eyes in adults, where flight and mate identification require better image processing. Typically, a stemma bears a cuticular **corneal lens** above a **crystalline cone** that serve as the

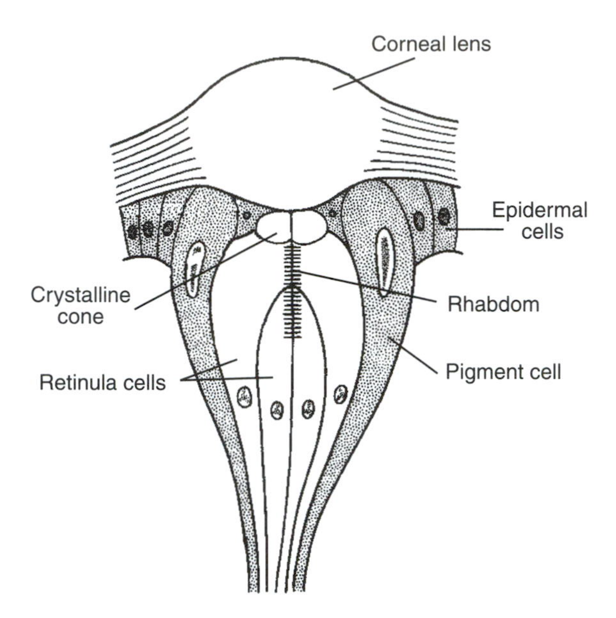

FIGURE 11.27 A lateral ocellus, or stemma. From Gullan and Cranston, (2000). Reprinted with permission.

optical elements. As in the dorsal ocelli, a portion of the plasma membrane of the retinula cells is specialized as a **rhabdomere** to contain a large number of microvilli that contain the visual pigment. The increase in surface area that the rhabdomere provides allows those neurons to pack abundant visual pigment into each cell. These may contain multiple photopigment systems, suggesting that color discrimination is possible. Two or more rhodopsins that are tuned to different regions of the visual spectrum can allow the insect to discriminate visual stimuli on the basis of their wavelengths.

Depending on the species, there may be from as few as 3 to more than 5000 retinula cells below that are grouped around a central rhabdom consisting of the rhabdomeres of the retinula cell, and several of these rhabdoms may be present within a single stemma. The light from the lenses is focused on one or more of the rhabdoms. Stemmata may be present in insects either in groups or singly and provide a coarse mosaic of the environment but with far better resolution than the dorsal ocelli. The stemmata may also be sensitive to polarized light.

Compound Eyes

Compound eyes are the primary visual receptors of adult insects and larval hemimetabola. They consist of a few to several thousand groups of **ommatidia,**

the individual optical units (Fig. 11.28). Dragonflies may have as many as 10,000 ommatidia, a worker honeybee has about 5500, *Drosophila* has 800, and subterranean insects may only have a few. In the primitive proturans and diplurans, compound eyes are completely absent. The ommatidia of some insects may be of different sizes in different areas of the compound eyes. For example, the aquatic beetle *Gyrinus* has a dorsal pair of eyes that looks above the water and a ventral pair of eyes that looks below into the water. Not all the facets that face the environment may respond similarly; different ommatidia in the compound eyes of honeybees process visual information differently, with shape detection and pattern and color discrimination processed best in the ventral part of the frontal visual field.

Eyes are receptors that evaluate the three-dimensional world by projecting images on a two-dimensional field, and the nature of that field in insects differs from that in vertebrates. In vertebrates, the receptor surface is concave (Fig. 11.29A), allowing a single lens in front of the retina to produce an image. In arthro-

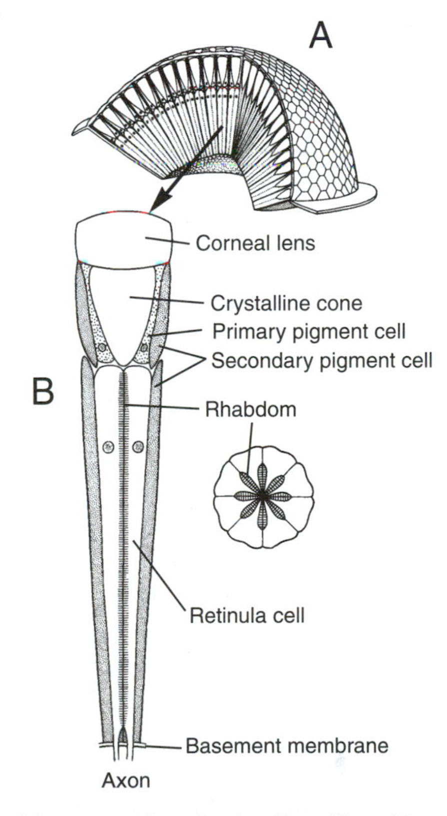

FIGURE 11.28 (A) A section of the compound eye showing the position of the ommatidia. (B) A cross section of a single ommatidium. From Gullan and Cranston (2000). Reprinted with permission.

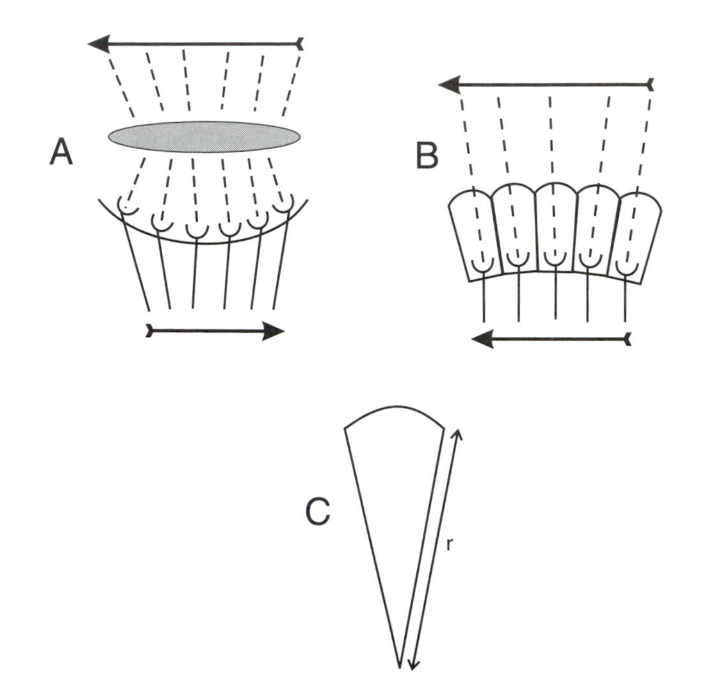

FIGURE 11.29 (A) The lens of vertebrates projects a reversed image on the concave receptor surface. (B) The lens system of the insect compound eye projects an unreversed image on a convex receptor surface. (C) The resolution of the ommatidium is dependent on the radius of its curvature (*r*). From Goldsmith (1990). Reprinted with permission.

pods, the receptor surface is convex, and an image is formed only when the individual ommatidia allow a narrow cone of light to enter that is perpendicular to the receptor surface (Fig. 11.29B). This design has implications for image resolution, as it is primarily dependent on the radius of the eye's curvature (Fig. 11.29C). A larger radius provides an increased resolution, but brings with it larger facets and there are limits to the size of the eye that can be accommodated on an insect's head. A compromise in some insects is the development of a **foveal region** where the radius of curvature of some of the ommatidia is large and visual acuity in the region is greater. Figure 11.30 shows an approximation of what a human would resemble if a compound eye, with the same resolution of a lens eye, was present.

The compound eyes provide a panoramic view of the world with a large field of vision. Even ants, with a relatively small number of facets on either side of the head, can perceive almost the entire visual field above and below the horizon except for a blind area of about 10% of the total field that lies below the thorax and abdomen. In the design of the insect eye, visual acuity is sacrificed for a panoramic view of the world.

Compound eyes may be placed in one of two categories: the **apposition eyes** of most day-active insects or the **superposition eyes** of nocturnal insects that sacrifice resolution for increases in sensitivity. The optical portion of the apposition eye consists of a modified cuticle that forms a corneal lens (Fig. 11.31A). The lens is produced by two modified epidermal cells, the **corneagen cells,** which move to the outer edge of the ommatidium later in development and produce the masking pigments. Bristles or conical protuberances may be present on or at the edge of the cornea to increase the visual capacity of the eye and reduce reflection to also provide camouflage. **Semper cells** produce a second lens, the crystalline cone, which is surrounded by pigment cells. The corneal lens and crystalline cone together focus the light on the optically active receptor of the ommatidium and are isolated from neighboring units by the pigment cells. Elongated retinula cells in groups of 8–12 are arranged around the longitudinal axis of the ommatidium. Their cell membranes along the ommitidial axis contain dense microvillar borders that comprise the rhabdomere of each cell. The rhabdomeres of the packed retinula cells are closely apressed to form the rhabdom, onto which the light is focused. As in the stemmata, they contain the visual pigment rhodopsin that becomes activated when it absorbs light.

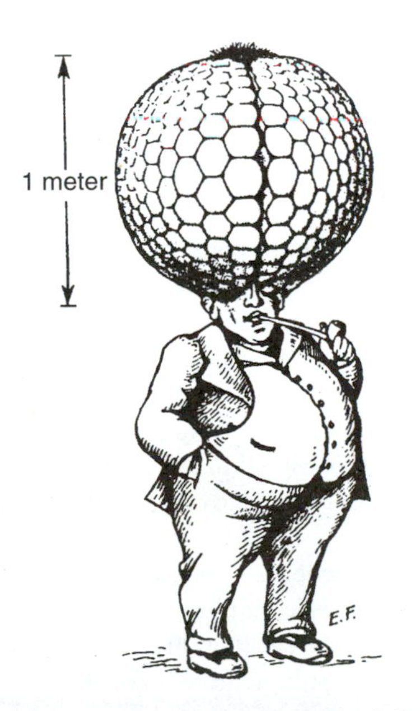

FIGURE 11.30 The approximate size of a compound eye in humans if the same resolution were to be obtained as for a lens eye. From Kirschfield (1976). Reprinted with permission.

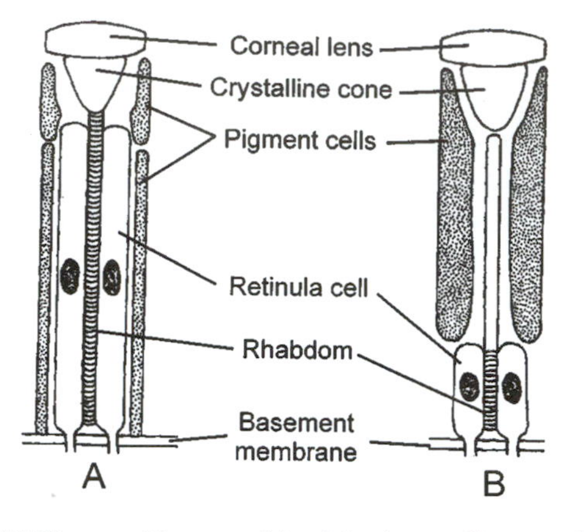

FIGURE 11.31 (A) The apposition eye of day-flying insects. Pigment cells isolate the individual ommatidia. (B) The superposition eye of night-flying insects. The pigment within the pigment cells moves downward when the eye is dark-adapted, allowing light from neighboring ommatidia to strike the rhabdom. Reprinted with permission from Romoser, W.S., J.G. Stoffolano Jr. (1998). *The Science of Entomology*, 4th edition. Copyright McGraw-Hill Education.

Visual Pigments

All visual systems that have been examined in living organisms share a structurally similar optical receptor molecule, suggesting that the evolution of visual receptors is an ancient event. Opsins are the membrane proteins that are localized in the photoreceptor sensilla that bind with vitamin A–derived retinoids to form the visual pigment **rhodopsin.** There are different opsins with characteristic absorption peaks that determine which wavelengths of light may be absorbed, and the presence of these different opsins in different photoreceptor cells determines the degree of color vision possible. The absorption of light by the pigment causes an isomerization of the molecule from *cis* to *trans* configuration (Fig. 11.32), and these changes in opsin conformation initiate a biochemical cascade that ultimately results in the excitation of the sensillum. When the active metarhodopsin that results catalyzes G-protein activation, the activated G-protein activates phospholipase C (PLC). PLC catalyzes the breakdown of the membrane phospholipid phosphatidyl 4,5-biphosphate (PIP_2) into inositol trisphosphate (IP_3) and diacylglycerol (DAG), which causes membrane channels to open and a depolarizing receptor potential to be generated (see Chapter 1). Unlike the visual cycle in vertebrates, where the absorption of light is only required for activation, the insect cycle requires light absorption for both activation and regeneration of the active rhodopsin (Fig. 11.33).

Drosophila has eight photoreceptor neurons in each ommatidium that can be divided into three classes of spectral sensitivity. There are six outer photoreceptors that are sensitive to blue wavelengths of light, and the remaining two central neu-

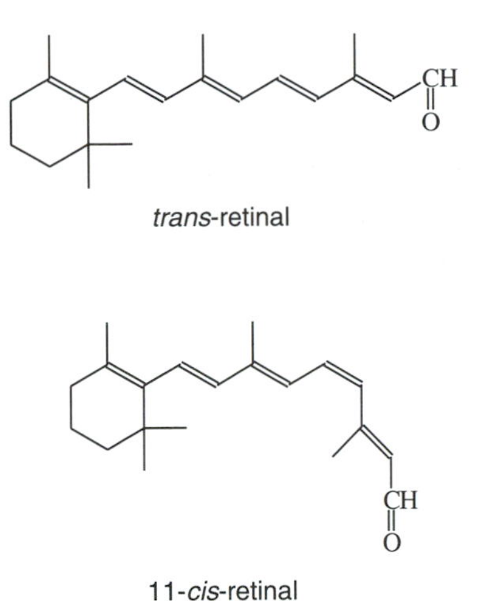

FIGURE 11.32. The two forms of rhodopsin. The 11-*cis* retinal is an unstable form that is converted to its *trans* form when it absorbs a photon of light.

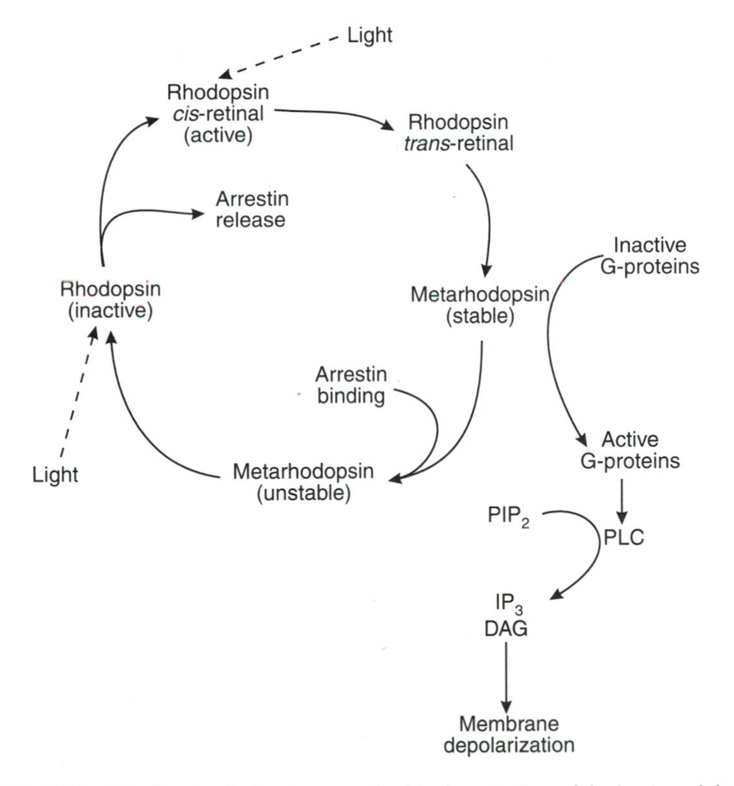

FIGURE 11.33 The biochemical pathway involved in the activation of rhodopsin and the resulting membrane depolarization. When the unstable *cis* retinal absorbs a photon of light, it converts it to the *trans* form, which is transformed to the metarhodopsin in its stable form that is capable of activating G-proteins. This pathway ultimately leads to membrane depolarization. The metarhodopsin loses its ability to activate the G-proteins after it binds with the protein arrestin and becomes an unstable form. Light converts this metarhodopsin to an inactive rhodopsin that releasees its arrestin and again becomes capable of responding to light.

rons are sensitive to either blue-green light or ultraviolet wavelengths. Some ommatidia in regions of the compound eye may lack some of these pigments. The dorsal ommatidia of most insects lack green receptors; in the honeybee, these regions have receptors only for blue and UV light, suggesting that they might be best adapted for detecting objects against the open sky or the sky itself. The spatial arrangements of the visual pigments can also modify their spectral sensitivity. When the rhabdoms are in a **fused** arrangement, the rhabdomeres that bear different pigments act as lateral filters for each other. When the rhabdoms are **tiered,** distal receptors filter the light that reaches more proximal receptors. Most insects have a combination of both fused and tiered rhabdoms.

Superposition Compound Eyes

The rhabdom of apposition compound eyes extends from the basement membrane of the ommatidium up to the crystalline cone, with screening pigments within each surrounding pigment cell that prevent light from the other ommatidia from striking the rhabdom and the pigments of apposition eyes but that do not move within the cells. In contrast, the rhabdom of the **superposition** eye is separated from the crystalline cone by a larger distance (Fig. 11.31B). In darkness, the pigments move downward and reduce the isolation between neighboring ommatidia, with complete adaptation occurring in 15 to 60 min. A dark-adapted superposition eye can spread the light from as many as 30 ommatidia, increasing the sensitivity but decreasing the resolution (Fig. 11.34). In some *Drosophila* mutants that lack eye pigments, their ability to resolve images is reduced because the light they receive is spread over a wider visual field.

Perception of Polarized Light

Although the light originating from the sun is unpolarized and vibrates in all directions, as it travels through the earth's atmosphere the particles it encounters cause it to become polarized and vibrate in a specific direction. The degree of polarization changes with regard to the position of the sun and the orientation of the observer, making it possible to determine the sun's position even when clouds obscure it, as long as a portion of the sky is visible. Many insects can make use of this phenomenon for navigation by detecting polarized light. This ability is present in many social Hymenoptera such as bees, ants, and wasps that must orient to find food and return to their nests.

The ability to perceive polarized light lies largely within the orientation of the visual pigment within the rhabdomere. Some of the elongated rhabdomeres contain uniformly oriented rhodopsin within their microvilli (Fig. 11.35) and

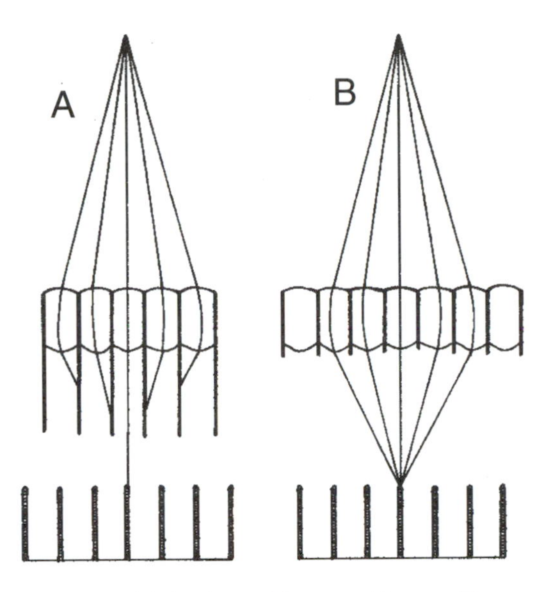

FIGURE 11.34 Light- (A) and dark-adapted (B) superposition eyes. The movement of screening pigments allows the light from neighboring ommatidia to strike the rhabdom. From Richards and Davies (1977). Reprinted with permission.

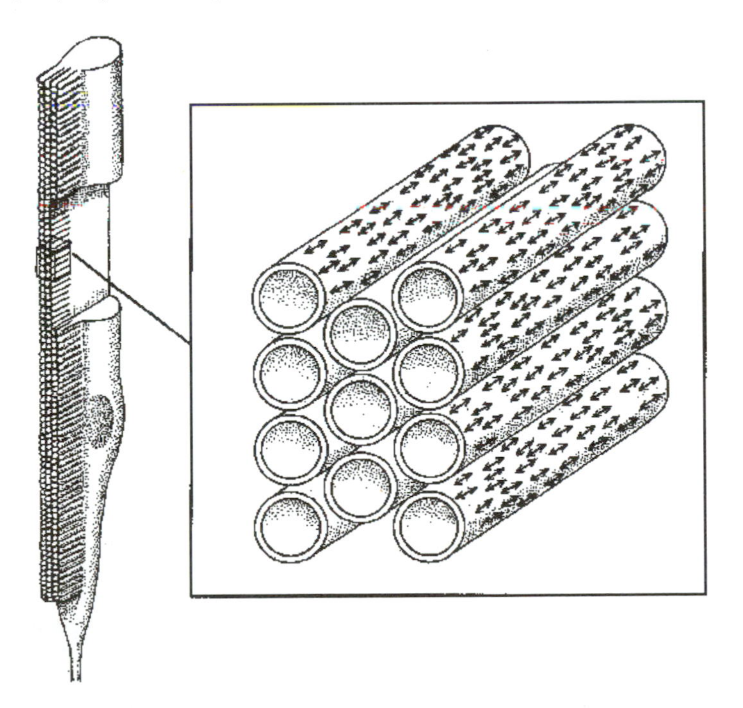

FIGURE 11.35 (Left) An elongated rhabdomere. (Right) The orientation of rhodopsin within the microvilli of the rhabdomere that allows the reception of polarized light. From Wehner (1976). Reprinted with permission.

the absorption of light is maximal when the light is polarized in the same direction as the pigment is oriented. If the receptors are moved as the insect rotates about its vertical axis, the output of the receptors is modulated as the microvilli become parallel to the light. Other receptors held at different angles record different responses to the rotation within the field of polarization. By scanning the sky, it is believed that the insect can record the degree of polarization from the pattern registered in its receptors and can then later orient by matching the pattern in the sky with its recorded pattern in memory. Water can polarize reflected light, and aquatic insects can use their polarized light detection to identify the water surface.

Variations in the Compound Eye Structure

Some Diptera and Hemiptera possess a variant of the apposition eye called a **neural superposition eye.** Its optics are the same as an apposition eye, but the receptors are organized in a different way. Rather than having a typical rhabdom that consists of the fused individual rhabdomeres of the retinula cells, the rhabdomeres are unfused and receive and transmit separate images that are reconstructed neurally so that the rhabdomeres of different ommatidia that view the same point converge on the same synapse, making them superposition by neural configuration rather than optical design (Fig. 11.36).

The hummingbird hawkmoth, *Macroglossum stellatarum*, is a day-active moth with superposition eyes that diverge from the usual superposition structure, possessing both high sensitivity and high spatial resolution. There are up to four rhabdoms under the facets of some of the ommatidia, which differs significantly from the usual superposition ommatidium. Gradients of resolution and sensitivity are also present, although these were not previously considered to be possible under the principles governing the operation of the superposition eye.

Extraocular Photoreception

Genital photoreceptors are present in many butterflies. They develop during the pupal stage and are present only in adults. In the female, these receptors are involved in oviposition behavior, informing the female that the ovipositor is extended far enough to successfully lay eggs. They play a role in copulation in the male, providing information that the vagina of the female is aligned properly for penile insertion. Two pairs of genital receptors, P1 and P2 (Fig. 11.37), mediate these behaviors in both sexes. Their structures resemble that of a **phaosome,** a primitive photoreceptor first identified from the epidermis of earthworms. The presence of photoreceptor pigments within the genital photoreceptors has yet to be confirmed.

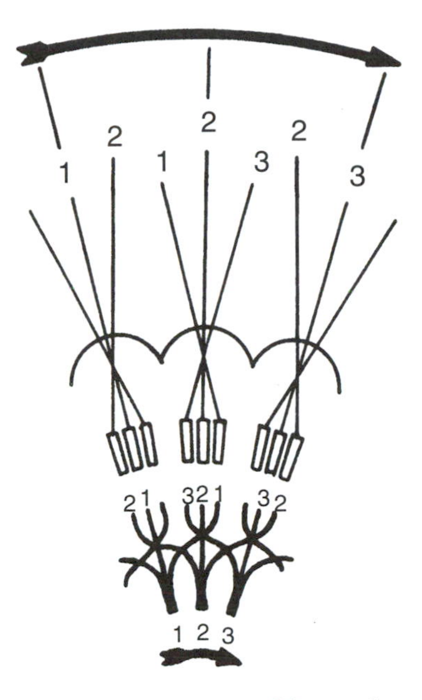

FIGURE 11.36 The optical and nervous arrangements of the neural superposition eye of Diptera. The pattern of the arrow above is transmitted to the optic nerves below as optically apposition, but neurally superposition. From Land (1985). Reprinted with permission.

Magnetic Sensitivity

Some insects, including migrating monarch butterflies, honeybees, and ants, show a sensitivity to geomagnetic fields that allows them to orient based on the field strength. Monarchs appear to have a "compass" that somehow leads them to their

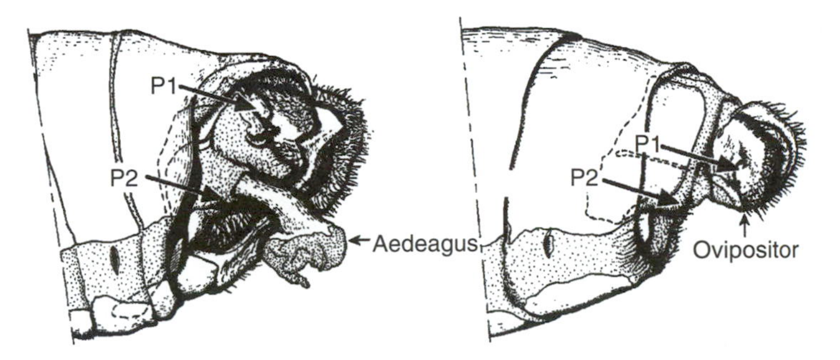

FIGURE 11.37 The genital photoreceptors in male (left) and female (right) Lepidoptera. From Arikawa (2001). Reprinted with permission.

overwintering sites in Mexico, and when exposed to a brief magnetic pulse in the laboratory, they become disoriented. Bees use magnetic fields to supplement other information when attempting to locate foraging sites. Iron granules are deposited in the trophocytes of the fat body, where expansion and contraction of the granule from magnetic fields may initiate a neural response in the workers. The migratory ant, *Pachycondyla marginata*, contains magnetic iron oxide particles in the head and abdomen that may explain their abilities to orient along magnetic lines.

ADDITIONAL REFERENCES

Nervous System

Anton, S., Hansson, B. S. 1996. Antennal lobe interneurons in the desert locust *Schistocerca gregaria* (Forskal): Processing of aggregation pheromones in adult males and females. *J. Comp. Neurol.* **370:** 85–96.

Anton, S., Hansson, B. S. 1999. Functional significance of olfactory glomeruli in a moth. *Proc. R. Soc. London B* **266:** 1813–1820.

Callec, J. J. 1985. Synaptic transmission in the nervous system. In *Comprehensive insect physiology, biochemistry and pharmacology* (G. A. Kerkut, L. I. Gilbert, Eds.), vol. 5, pp. 139–179. Pergamon Press, Oxford.

Carlson, S. D., Juang, J. L., Hilgers, S. L., Garment, M. B. 2000. Blood barriers of the insect. *Annu. Rev. Entomol.* **45:** 151–174.

Carlson, S. D., Saint Marie, R. L. 1990. Structure and function of insect glia. *Annu. Rev. Entomol.* **35:** 597–621.

De Bruyne, M., Clyne, P. J., Carlson, J. R. 1999. Odor coding in a model olfactory organ: The *Drosophila* maxillary palp. *J. Neurosci.* **19:** 4520–4532.

Gilbert, C., Strausfeld, N. J. 1991. The functional organization of male-specific visual neurons in flies. *J. Comp. Physiol. A* **169:** 395–411.

Hansson, B. S., Anton, S. 2000. Function and morphology of the antennal lobe: New developments. *Annu. Rev. Entomol.* **45:** 203–231.

Hartenstein, V. 1997. Development of the insect stomatogastric nervous system. *Trends Neurosci.* **20:** 421–427.

Hildebrand, J. G. 1996. Olfactory control of behavior in moths: Central processing of odor information and the functional significance of olfactory glomeruli. *J. Comp. Physiol. A* **178:** 5–19.

Hildebrand, J. G., Montague, R. A. 1986. Functional organization of olfactory pathways in the central nervous system of *Manduca sexta*. In *Mechanisms in insect olfaction* (T. L. Payne, M. C. Birch, C. E. J. Kennedy, Eds.), pp. 277–285. Clarendon Press, Oxford.

Homberg, U., Christensen, T. A., Hildebrand, J. G. 1989. Structure and function of the deutocerebrum in insects. *Annu. Rev. Entomol.* **34:** 477–501.

Homberg, U., Montague, R. A., Hildebrand, J. G. 1988. Anatomy of antenno-cerebral pathways in the brain of the sphinx moth *Manduca sexta*. *Cell Tissue Res.* **254:** 255–281.

Kanzaki, R., Arbas, E. A., Hildebrand, J. G. 1991. Physiology and morphology of protocerebral olfactory neurons in the male moth *Manduca sexta*. *J. Comp. Physiol.* A **168:** 281–298.

Kloppenburg, P., Camazine, S. M., Sun, X. J., Randolph, P., Hildebrand, J. G. 1997. Organization of the antennal motor system in the sphinx moth *Manduca sexta*. *Cell Tissue Res.* **287:** 425–433.

Kutsch, W., Breidbach, O. 1994. Homologous structures in the nervous system of arthropods. *Adv. Insect Physiol.* **24:** 1–113.

Lane, N. J. 1985. Structure of components of the nervous system. In *Comprehensive insect physiology biochemistry and pharmacology* (G. A. Kerkut, L. I. Gilbert, Eds.), pp. 1–47. Pergamon Press, Oxford.

Lemon, W. C., Getz, W. M. 1999. Neural coding of general odors in insects. *Ann. Entomol. Soc. Am.* **92:** 861–872.

Masson, C., Mustaparta, H. 1990. Chemical information processing in the olfactory system of insects. *Physiol. Rev.* **70:** 199–245.

May, M. 1991. Aerial defense tactics of flying insects. *Am. Sci.* **79:** 316–328.

Mobbs, P. G. 1985. Brain structure. In *Comprehensive insect physiology, biochemistry and pharmacology* (G. A. Kerkut, L. I. Gilbert, Eds.), vol. 5, pp. 299–370. Pergamon Press, Oxford.

Treherne, J. E. 1985. Blood–brain barrier. In *Comprehensive insect physiology, biochemistry and pharmacology* (G. A. Kerkut, L. I. Gilbert, Eds.), vol. 5, pp. 115–137. Pergamon Press, Oxford.

Treherne, J. E., Schofield, P. K. 1981. Mechanisms of ionic homeostasis in the central nervous system of an insect. *J. Exp. Biol.* **95:** 61–73.

Truman, J. W. 1996. Metamorphosis of the insect nervous system. In *Metamorphosis: Postembryonic reprogramming of gene expression in amphibian and insect cells.* (L. I. Gilbert, J. R. Tata, B. G. Atkinson, Eds.), pp. 283–320. Academic Press, San Diego.

Usherwood, P. N. R. 1994. Insect glutamate receptors. *Adv. Insect Physiol.* **24:** 309–341.

van den Berg, M. J., Ziegelberger, G. 1991. On the function of the pheromone binding protein in the olfactory hairs of *Antheraea polyphemus*. *J. Insect Physiol.* **37:** 79–85.

Visser, J. H. 1986. Host odor perception in phytophagous insects. *Ann. Rev. Entomol.* **31:** 121–144.

Vogt, R. G., Prestwich, G. D., Lerner, M. R. 1991. Odorant-binding-protein subfamilies associate with distinct classes of olfactory receptor neurons in insects. *J. Neurobiol.* **22:** 74–84.

Vogt, R. G., Riddiford, L. M. 1981. Pheromone binding and inactivation by moth antennae. *Nature* **293:** 161–163.

Vogt, R. G., Riddiford, L. M. 1986. Pheromone reception: A kinetic equilibrium. In *Mechanisms in insect olfaction* (T. L. Payne, M. C. Birch, C. E. J. Kennedy, Eds.), pp. 201–208. Clarendon Press, Oxford.

Vogt, R. G., Rybczynski, R., Lerner, M. R. 1991. Molecular cloning and sequencing of general odorant-binding proteins GOBP1 and GOBP2 from the tobacco hawk moth *Manduca sexta*: Comparisons with other insect OBPs and their signal peptides. *J. Neurosci.* **11:** 2972.

Vosshall, L. B., Amrein, H., Morozov, P. S., Rzhetsky, A., Axel, R. 1999. A spatial map of olfactory receptor expression in the *Drosophila* antenna. *Cell* **96:** 725–736.

Vosshall, L. B., Wong, A. M., Axel, R. 2000. An olfactory sensory map in the fly brain. *Cell* **102:** 147–159.

Warren, J. T., Dai, J. D., Gilbert, L. I. 1999. Can the insect nervous system synthesize ecdysteroids? *Insect Biochem. Mol. Biol.* **29:** 571–579.

Weevers, R. de G. 1985. The insect ganglia. In *Comprehensive insect physiology biochemistry and pharmacology* (G. A. Kerkut, L. I. Gilbert, Eds.), vol. 5, pp. 213–297. Pergamon Press, Oxford.

Wolf, R., Heisenberg, M. 1990. Visual control of straight flight in *Drosophila melanogaster*. *J. Comp. Physiol.* A **167:** 269–283.

Sensory Receptors

Acosta-Avalos, D., Wajnberg, E., Oliveira, P. S., Leal, I., Farina, M., Esquivel, D. M. 1999. Isolation of magnetic nanoparticles from *Pachycondyla marginata* ants. *J. Exp. Biol.* **202:** 2687–2692.

Altner, H., Prillinger, L. 1980. Ultrastructure of invertebrate chemo- thermo- and hygroreceptors and its functional significance. *Int. Rev. Cytol.* **67:** 69–139.

Altner, H., Loftus, R. 1985. Ultrastructure and function of insect thermo- and hygroreceptors. *Annu. Rev. Entomol.* **30:** 273–295.

Angeli, S., Ceron, F., Scaloni, A., Monti, M., Monteforti, G., Minnocci, A., Petacchi, R., Pelosi, P. 1999. Purification, structural characterization, cloning and immunocytochemical localization of chemoreception proteins from *Schistocerca gregaria*. *Eur. J. Biochem.* **262:** 745–754.

Angioy, A. M., Liscia, A., Pietra, P. 1981. Some functional aspects of the wing chemosensilla in *Phormia regina* (Meig.) (Diptera Calliphoridae). *Monitore Zool. Ital.* **15:** 221–228.

Arikawa, K. 1993. Valva-opening response induced by the light stimulation of the genital photoreceptors of male butterflies. *Naturwissenschaften* **80:** 326–328.

Arikawa, K. 2001. Hindsight of butterflies. *BioScience* **51:** 219–225.

Arikawa, K., Aoki, K. 1982. Response characteristics and occurrence of extraocular photoreceptors on lepidopteran genitalia. *J. Comp. Physiol.* A **148:** 483–489.

Barth, F. G., Blickhan, R. 1984. Mechanoreception. In *Biology of the integument* (J. Bereiter-Hahn, A. G. Matoltsy, K. S. Richards, Eds.), vol. 1, pp. 554–582. Springer-Verlag, Berlin.

Briscoe, A. D., Chittka, L. 2001. The evolution of color vision in insects. *Annu. Rev. Entomol.* **46:** 471–510.

Callahan, P. S. 1975. Insect antennae with special reference to the mechanism of scent detection and the evolution of the sensilla. *Int. J. Insect Morphol. Embryol.* **4:** 381–430.

Carlson, S. D., Chi, C. 1979. The functional morphology of the insect photoreceptor. *Annu. Rev. Entomol.* **24:** 379–416.

Chapman, R. F. 1982. Chemoreception: The significance of receptor numbers. *Adv. Insect Physiol.* **16:** 247–356.

Chapman, R. F., Ascoli-Christensen, A. 1999. Sensory coding in the grasshopper (Orthoptera: Acrididae) gustatory system. *Ann. Entomol. Soc. Am.* **92:** 873–879.

Clyne, P. J., Warr, C. G., Carlson, J. R. 2000. Candidate taste receptors in *Drosophila*. *Science* **287:** 1830–1834.

Dethier, V. G. 1963. *The physiology of insect senses.* Wiley, New York.

Dickinson, M. J. 1990. Comparison of encoding properties of campaniform sensilla on the fly wing. *J. Exp. Biol.* **151:** 245–261.

Eguchi, E., Watanabe, K., Hariyama, T., Yamamoto, K. 1982. A comparison of electrophysiologically determined spectral responses in 35 species of lepidoptera. *J. Insect Physiol.* **28:** 675–682.

Etheredge, J. A., Perez, S. M., Taylor, O. R., Jander, R. 1999. Monarch butterflies (*Danaus plexippus* L.) use a magnetic compass for navigation. *Proc. Natl. Acad. Sci. USA* **96:** 13,845–13,846.

Feng, L., Prestwich, G. D. 1997. Expression and characterization of a lepidopteran general odorant binding protein. *Insect Biochem. Mol. Biol.* **27:** 405–412.

Fent, K., Wehner, R. 1985. Ocelli: A celestial compass in the desert ant *Cataglyphis*. *Science* **228:** 192–194.

Field, L. H., Matheson, T. 1998. Chordotonal organs of insects. *Adv. Insect Physiol.* **27:** 1–228.

Fullard, J. H., Yack, J. E. 1993. The evolutionary biology of insect hearing. *TREE* **8:** 248–252.

Gao, Q., Chess, A. 1999. Identification of candidate *Drosophila* olfactory receptors from genomic DNA sequence. *Genomics* **60:** 31–39.

Giger, A., Srinivasan, M. 1997. Honeybee vision: Analysis of orientation and colour in the lateral, dorsal and ventral fields of view. *J. Exp. Biol.* **200:** 1271–1280.

Gilbert, C. 1994. Form and function of stemmata in larvae of holometabolous insects. *Annu. Rev. Entomol.* **39:** 323–349.

Giurfa, M., Menzel, R. 1997. Insect visual perception: Complex abilities of simple nervous systems. *Curr. Opin. Neurobiol.* **7:** 505–513.

Gleadall, I. G., Hariyama, T., Tsukahara, Y. 1989. The visual pigment chromophores in the retina of insect compound eyes, with special reference to the Coleoptera. *J. Insect Physiol.* **35:** 787–795.

Glendinning, J. I., Ensslen, S., Eisenberg, M. E., Weiskopf, P. 1999. Diet-induced plasticity in the taste system of an insect: Localization to a single transduction pathway in an identified taste cell. *J. Exp. Biol.* **202:** 2091–2102.

Goldsmith, T. H. 1990. Optimization, constraint, and history in the evolution of eyes. *Q. Rev. Biol.* **65:** 281–322.

Goldsmith, T. H., Bernard, G. D. 1974. The visual system of insects. *Physiol. Insecta* **2:** 165–272.

Gribakin , F. G. 1979. Cellular mechanisms of insect photoreception. *Int. Rev. Cytol.* **57:** 127–184.

Gray, E. G. 1960. The fine structure of the insect ear. *Phil. Trans. R. Soc. London B* **243:** 75–94.

Hansen, K. 1978. Insect chemoreception. In *Taxis and behavior* (G. L. Hazelbauer, Ed.), vol. 5, pp. 233–292. Chapman and Hall, London.

Hansson, B. S. 1995. Olfaction in lepidoptera. *Experientia* **51:** 1003–1027.

Hekmat-Scafe, D. S., Steinbrecht, R. A., Carlson, J. R. 1997. Coexpression of two odorant-binding protein homologs in *Drosophila*: Implications for olfactory coding. *J. Neurosci.* **17:** 1616–1624.

Hekmat-Scafe, D. S., Steinbrecht, R. A., Carlson, J. R. 1998. Olfactory coding in a compound nose. Coexpression of odorant-binding proteins in *Drosophila*. *Ann. NY Acad. Sci.* **855:** 311–315.

Horridge, G. A. 1977. The compound eye of insects. *Sci. Am.* **237:** 108–120.

Horridge, G. A. 1992. What can engineers learn from insect vision? *Phil. Trans. R. Soc. London B* **337:** 271–282.

Hsu, C. Y., Li, C. W. 1994. Magnetoreception in honeybees. *Science* **256:** 95–97.

Ioannides, A. C., Horridge, G. A. 1975. The organization of visual fields in the hemipteran acone eye. *Proc. R. Soc. London B* **190:** 373–391.

Jacobs, G. A. 1995. Detection and analysis of air currents by crickets. *BioScience* **45:** 776–785.

Kastberger, G. K., Schuhmann, K. 1993. Ocellar occlusion effect on the flight behavior of homing honeybees. *J. Insect Physiol.* **39:** 589–600.

Kirschfield, K. 1976. The resolution of lens and compound eyes. In *Neural principles of vision* (F. Zettler, R. Weiler, Eds.), vol. 5, pp. 354–370. Springer-Verlag, Berlin.

Kiselev, A., Subramanium, S. 1994. Activation and regeneration of rhodopsin in the insect visual cycle. *Science* **266:** 1369–1374.

Krieger, J., Von Nickisch-Rosennegk, E., Mameli, M., Pelosi, P., Breer, H. 1996. Binding proteins from the antennae of *Bombyx mori*. *Insect Biochem. Mol. Biol.* **26:** 297–307.

Lall, A. B., Lord, E. T., Trouth, C. O. 1985. Electrophysiology of the visual system in the cricket *Gryllus firmus* (Orthoptera: Gryllidae): Spectral sensitivity of the compound eyes. *J. Insect Physiol.* **31:** 353–357.

Land, M. F. 1985. The eye: Optics. In *Comprehensive insect physiology, biochemistry and pharmacology* (G. A. Kerkut, L. I. Gilbert, Eds.), vol. 6, pp. 225–275. Pergamon Press, Oxford.

Lehrer, M. 1998. Looking all around: Honeybees use different cues in different eye regions. *J. Exp. Biol.* **201:** 3275–3292.

Lessing, D., Carlson, J. R. 1999. Chemosensory behavior: The path from stimulus to response. *Curr. Opin. Neurobiol.* **9:** 766–771.

Maida, R., Krieger, J., Gebauer, T., Lange, U., Ziegelberger, G. 2000. Three pheromone-binding proteins in olfactory sensilla of the two silkmoth species *Antheraea polyphemus* and *Antheraea pernyi*. *Eur. J. Biochem.* **267:** 2899–2908.

McIver, S. B. 1975. Structure of cuticular mechanoreceptors of arthropods. *Annu. Rev. Entomol.* **20:** 381–397.

McIver, S. B. 1985. Mechanoreception. In *Comprehensive insect physiology biochemistry and pharmacology* (G. A. Kerkut, L. I. Gilbert, Eds.), vol. 6, pp. 71–132. Pergamon Press, Oxford.

Menzel, R., Ventura, D. F., Hertel, H., de Souza, J., Greggers, U. 1986. Spectral sensitivity of photoreceptors in insect compound eyes: Comparison of species and methods. *J. Comp. Physiol.* A **158:** 165–177.

Michelsen, A. 1979. Insect ears as mechanical systems. *Am. Sci.* **67:** 696–705.

Michelsen, A., Fink, F., Gogala, M., Traue, D. 1982. Plants as transmission channels for insect vibrational songs. *Behav. Ecol. Sociobiol.* **11:** 269–281.

Michelsen, A., Larson, O. N. 1983. Strategies for acoustic communication in complex environments. In *Neuroethology and behavioral physiology* (F. Huber, H. Markl, Eds.), pp. 321–331. Springer-Verlag, Berlin.

Michelsen, A., Nocke, H. 1974. Biophysical aspects of sound communication in insects. *Adv. Insect Physiol.* **10:** 247–296.

Miyako, Y., Arikawa, K., Eguchi, E. K. 1993. Ultrastructure of the extraocular photoreceptor in the genitalia of a butterfly, *Papilio xuthus*. *J. Comp. Neurol.* **327:** 458–468.

Miyako, Y., Arikawa, K., Eguchi, E. 1995. Morphogenesis of the photoreceptive site and development of the electrical responses in the butterfly genital photoreceptors during the pupal period. *J. Comp. Neurol.* **363:** 296–306.

Muir, L. E., Thorne, M. J., Kay, B. H. 1992. *Aedes aegypti* (Diptera: Culicidae) vision: Spectral sensitivity and other perceptual parameters of the female eye. *J. Med. Entomol.* **29:** 278–281.

Nordstrom, P., Warrant, E. J. 2000. Temperature-induced pupil movements in insect superposition eyes. *J. Exp. Biol.* **203:** 685–692.

Osborne, M. P. 1969. Structure and function of neuromuscular junctions and stretch receptors. In *Insect ultrastructure* (A. C. Neville, Ed.), vol. 5. Royal Entomological Society, London.

Pollock, J. A., Benzer, S. 1988. Transcript localization of four opsin genes in the three visual organs of *Drosophila:* RH2 is ocellus specific. *Nature* **333:** 779–782.

Roeder, K. D. 1965. Moths and ultrasound. *Sci. Am.* **212:** 94–102.

Roeder, K. D., Treat, A. E. 1957. Ultrasonic reception by the tympanic organ of noctuid moths. *J. Exp. Zool.* **134:** 127–157.

Rossel, S., Wehner, R. 1986. Polarization vision in bees. *Nature* **323:** 128–131.

Schmitz, H., Bleckmann, H., Murtz, M. 1997. Infrared detection in a beetle. *Nature* **386:** 773–774.

Schwartzkopff, J. 1974. Mechanoreception. *The physiology of insecta* **2:** 273–352.

Shankland, D. L., Frazier, J. L. 1985. Nervous system: Electrical events. In *Fundamentals of insect physiology* (M. Blum, Ed.), pp. 253–286. Wiley, New York.

Srinivasan, M., Lehrer, M. 1985. Temporal resolution of colour vision in the honeybee. *J. Comp. Physiol. A* **157:** 579–586.

Srinivasan, M., Lehrer, M., Wehner, R. 1987. Bees perceive illusory colours induced by movement. *Vision Res.* **27:** 1285–1289.

Steinbrecht, R. A., Laue, M., Maida, R., Ziegelberger, G. 1996. Odorant-binding proteins and their role in the detection of plant odours. Entomol. *Exp. Appl.* **80:** 15–18.

Steinbrecht, R. A. 1996. Are odorant-binding proteins involved in odorant discrimination? *Chem. Senses* **21:** 719–727.

Steinbrecht, R. A., Laue, M., Zhang, S. -G., Ziegelberger, G. 1994. Immunocytochemistry of odorant-binding proteins. In *Olfaction and taste XI.* pp. 804–807.

Steinbrecht, R. A., Laue, M., Ziegelberger, G. 1995. Immunolocalization of pheromone-binding protein and general odorant-binding protein in olfactory sensilla of the silk moths *Antheraea* and *Bombyx. Cell Tissue Res.* **282:** 203–217.

Steinbrecht, R. A., Stankiewicz, B. A. 1999. Molecular composition of the wall of insect olfactory sensilla—The chitin question. *J. Insect Physiol.* **45:** 785–790.

Stengl, M., Hatt, H., Breer, H. 1992. Peripheral processes in insect olfaction. *Annu. Rev. Physiol.* **54:** 665–681.

Stocker, R. F. 1994. The organization of the chemosensory system in *Drosophila melanogaster:* A rewiew [sic]. *Cell Tissue Res.* **275:** 3–26.

Strausfeld, N. J., Hildebrand, J. G. 1999. Olfactory systems: common design, uncommon origins? *Curr. Opin. Neurobiol.* **9:** 634–639.

Stumpner, A., Von Helversen, D. 2001. Evolution and function of auditory systems in insects. *Naturwissenschaften* **88:** 159–170.

Toh, Y., Tateda, H. 1991. Structure and function of the insect ocellus. *Zool. Sci.* **8:** 395–413.

Tomlinson, A. 1988. Cellular interactions in the developing *Drosophila* eye. *Development* **104:** 183–193.

Warrant, E., Bartsch, K., Günther, C. 1999. Physiological optics in the hummingbird hawkmoth: A compound eye without ommatidia. *J. Exp. Biol.* **202:** 497–511.

Weevers, R. de G. 1966. The physiology of a lepidopteran muscle receptor I. The sensory response to stretching. *J. Exp. Biol.* **44:** 177–194.

Wehner, R. 1976. Polarized-light navigation by insects. *Sci. Am.* **235:** 106–114.

Wehner, R. 1981. Spatial vision in arthropods. *Handb. of Sensory Physiology* VII/**6C:** 287–616.

Wehner, R. 1983. The perception of polarised light. *Symp. Soc. Exp. Biol.* **36:** 331–369.

Wehner, R. 1989. Neurobiology of polarization vision. *Trends Neurosci.* **12:** 353–359.

Wehner, R. 1994. Insect vision: Exploring the third dimension. *Ethol. Ecol. Evol.* **6:** 395–401.

Wehner, R., Michel, B., Antonsen, P. 1996. Visual navigation in insects: Coupling of egocentric and geocentric information. *J. Exp. Biol.* **199:** 129–140.

Zacharuk, R. Y. 1980. Ultrastructure and function of insect chemosensilla. *Annu. Rev. Entomol.* **25:** 27–47.

Zackaruk, R. Y. 1985. Antennae and sensilla. In *Comprehensive insect physiology, biochemistry and pharmacology* (G. A. Kerkut, L. I. Gilbert, Eds.), vol. 6, pp. 1–69. Pergamon Press, Oxford.

Zacharuk, R. Y., Shields, V. D. 1991. Sensilla of immature insects. *Annu. Rev. Entomol.* **36:** 331–354.

Ziegelberger, G. 1995. Redox-shift of the pheromone-binding protein in the silkmoth *Antheraea polyphemus*. *Eur. J. Biochem.* **232:** 706–711.

Ziegelberger, G. 1996. The multiple role of the pheromone-binding protein in olfactory transduction. *Ciba Found. Symp.* **200:** 267–275.

Zufall, F., Schmitt, M., Menzel, R. 1989. Spectral and polarized light sensitivity of photoreceptors in the compound eye of the cricket (*Gryllus bimaculatus*). *J. Comp. Physiol. A* **164:** 597–608.

Zuker, C. S. 1996. The biology of vision of *Drosophila*. *Proc. Natl. Acad. Sci. USA* **93:** 571–576.

CHAPTER 12

Communication Systems

TYPES OF COMMUNICATIONS

Communication occurs not only between the cells of multicellular organisms, as described in preceding chapters, but also between two or more whole organisms. Intraspecific and interspecific communications are absolutely essential for survival and reproduction. In order to reproduce, an individual must find and accurately identify the opposite sex so that time and energy are not wasted on mating attempts with stray objects in the environment that would fail to result in the production of offspring. Insects are frequently the prey of other animals, and there has been intense selective pressure for the evolution of defensive communications that minimize this loss. Also, because the competition for resources is most vigorous among the members of one's own species that occupy an identical ecological niche, communication with others is necessary to recognize this competition so that individuals might be identified and cooperated with, avoided, or possibly forced away from an area.

Many definitions of communication have been proposed, but none has been completely satisfactory to everyone. Animals produce many signals, not all of which are considered to be communications. Some definitions require that the interchange between two individuals must be in some sort of code whose structure has been forged through the process of natural selection, in order for a signal that is produced to be classified as a communication. The reception and interpretation of this code, whether visual, auditory, or chemical, must be mutually beneficial by this criterion and result in the increased fitness of the participants. However, insect communication will be considered here in its broadest sense; if one individual gives off any signal that produces a change in the behavior of another individual, it will be considered to have been a communication. This chapter will review some of the ways in which insects produce these signals.

One way to classify the systems of communication is according to the receptors that are involved in receiving the stimuli. We can identify various chemical, tactile, acoustical, and visual signals as being the primary means of communication in insects, each with its own advantages and disadvantages (Table 12.1). The reliance of the species on one signal over another is primarily a function of the ecological context in which the insect must function. For example, with the exception of those species that produce their own light, insects that are active at night are generally less dependent on visual cues for communication. Although signaling systems will be described individually, it must be emphasized that the expression of a behavior in natural situations often requires the involvement of signals using several sensory modalities. Tactile communication, for example, often elicits behavioral changes only in conjunction with other visual and chemical cues.

VISUAL COMMUNICATION

Insects use visual signals primarily to identify food and mates and to orient themselves in the environment. The compound eyes are most important in receiving the visual signals because they have the best resolution of the optical receptors present. Visual signals are often releasers of the highly ritualized and stereotyped insect behavioral sequences.

Visual Tracking

Another important function of visual communication is the identification and tracking of food using the visual signals from prey. Of the insects that communicate using vision for obtaining food, the visual system of the praying mantis may be the best understood. To capture prey, the praying mantis uses its two compound eyes to determine the distance of its prey on the background by triangulation before it strikes. Binocular trangulation only works at the close ranges of about 25 mm at which the insect can strike with its raptorial forelegs. At farther prey dis-

TABLE 12.1 The characteristics of some signals used in insect communication.

Characteristic	Type of signal			
	Visual	Acoustical	Tactile	Chemical
Range	Medium	Long	Short	Long
Rate of signal change	Fast	Fast	Fast	Slow
Circumvention of obstacles	Poor	Good	Poor	Good
Energetic cost	Low	High	Low	Low

Source: Lewis (1984).

tances, binocular vision is no longer effective because of the small distance between the eyes. At these greater distances, the mantis uses **saccadic tracking** to maintain the image of the prey on the **foveal region** of the compound eyes, a group of ommatidia in the center of each eye that is capable of high spatial resolution. The greater resolution in the foveas results from the larger facets and smaller angles between the neighboring visual axes. The saccades consist of rapid head movements that center the image of the prey on the foveas, followed by periods of no movement. By comparing coordinates of visual information from both the foveas as well as the ommatidia that surround the foveas of each eye, the mantis can still estimate prey distance by binocular triangulation at greater distances.

Visual Defense

Insects also use visual communication for defense. Many lepidopterans bear conspicuous eyespots on their wings, which are the circular patterns that resemble the eyes of vertebrates. The lepidopteran, *Nymphalis io,* responds to the presence of birds by lowering its wings to expose its eyespots, and the sight of the eyespots releases escape behavior in the birds, often preventing the butterflies from being eaten. Butterflies that have their eyespots removed experimentally are much more susceptible to bird predation than those that retain their eyespots. The noctuid moth, *Catocala,* has cryptic forewings and brightly colored hindwings that are usually concealed. When grasped by birds, the moth exposes its hindwings and startles the bird sufficiently that it momentarily releases its beak to allow the moth to escape.

Bioluminescence

A prime reason for communicating visually is to be seen by a member of the opposite sex of the same species. An insect must be relatively conspicuous against the complex environmental background if it is to be visually identified by another individual. However, being conspicuous during the day also has its drawbacks, because a conspicuous visual signal is also one that is more apparent to predators. Some insects have consequently evolved the ability to communicate visually at night, when predators are less active. This capacity to produce light has evolved in at least eight insect families within four orders: the Collembola, Homoptera, Diptera, and Coleoptera, but bioluminescence occurs in more beetles, largely in the families Elateridae and Lampyridae, than in any other group. The function of luminescence in collembolans, which produce a light from their whole bodies, has never been determined but may be for defense rather than for mating. In the firefly beetles (Lampyridae), rapid bursts of yellow-green light of wavelength between 550 and 580 nm are produced from a light organ at the ventral tip of the abdomen, and the insects use this visual signal to find a mate. These lampyrids may also use the light they produce to illuminate landing sites.

There are two strategies that are used by fireflies to bring the sexes together. In some species, it is the female that produces specific flashes to attract males of the same species. In other species, the males initiate the flash signals and when a receptive female is nearby, she returns the signal. Each species has its own flash pattern that serves as an isolating mechanism when species distributions overlap.

Flashing begins at twilight once the insects perceive a decrease in the intensity of ambient light to below a certain threshold. The fireflies are able to receive the flash signals with their superposition eyes that function well in dim light after they have been dark-adapted. Both the flash duration as well as its spectral qualities are important to a receiver; the green and yellow sensitivities of the eyes are matched to the emission spectra of the light produced. As an example, the males of the firefly *Photinus pyralis* begin patrolling open fields beginning at dusk, flying slowly and producing a 500-ms flash every 6 s. The male then pauses for 2 s, and if a female responds with a flash during this period, the male flies toward her and flashes again. The male ultimately reaches the female after a series of these flash exchanges.

A rather dishonest approach is used by fireflies in the genus *Photuris* that employs visual communication for another purpose. In a strategy known as **aggressive mimicry,** they are able to mimic the flash patterns of up to five other species, attracting the naïve males that respond to the false signals and instead of mating with them, they consume the males that respond. After their adult emergence, the *Photuris* females attract males of their own species, but once they mate their behavior changes and only then begin to attract the males of different species in order to prey on them.

Light Production

The light organ, or lantern, of adult fireflies is derived from fat body cells that are modified during development. Specialized light-emitting **photocytes** lie just beneath the transparent epidermis and cuticle and contain large numbers of photocyte granules, or **peroxisomes,** in their cytoplasm along with numerous mitochondria localized near the tracheoles. It has been assumed that the chemical reactions that produce the luminescence occur within these granules, but their functions are not known. The photocytes are arranged in rosettes that are separated by central cylinders containing the nerves and tracheal branches that run vertically through the photocyte layers (Fig. 12.1). The tracheal system ends in a tracheal cell that in turn envelops a tracheolar cell and the nerve ending that innervates it. The nerves innervate only the tracheoles and not the photocytes. Flashing is initiated by a nerve impulse that releases the neurotransmitter octopamine at the tracheolar cell. The message then travels to the photocyte by the release of the gas nitric oxide (NO) that diffuses through the 17-μm distance and briefly increases the level of oxygen in the photo-

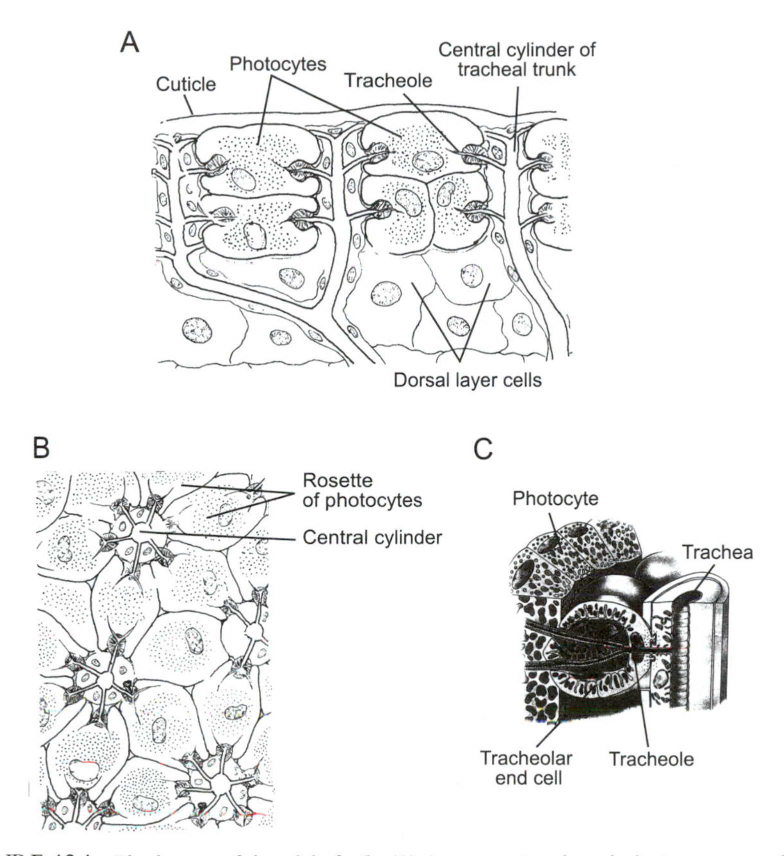

FIGURE 12.1 The lantern of the adult firefly. (A) A cross-section through the integument of the lantern showing the cuticle at the top and the photocytes below. (B) A ventral view just below the cuticle showing the arrangement of photocytes in rosettes around a central cylinder of tracheae. From Peterson and Buck (1968). Reprinted with permission. (C) The tracheal branching into tracheoles that regulates the passage of oxygen to the photocytes. Reprinted from Peterson, M.K. and J. Buck (1968). *Biological Bulletin* **135**: 335–348. With permission from the Marine Biological Laboratory.

cyte. The NO appears to inhibit the respiration of the mitochondria, thereby making more oxygen available to the peroxisomes for the light reactions. **Dorsal layer cells** underlying the photocytes contain white urate crystals and provide a reflecting surface for the light that is emitted.

Biochemistry of Luminescence

Light production results from the oxidation of the substrate **luciferin** that is catalyzed by the enzyme **luciferase** to yield oxyluciferin in an electronically excited state. The process involves the conversion of chemical energy into an electronic state transition that is large enough to yield a photon of visible light. Firefly luciferin is a benzothiazoyl-thiazole derivative; luciferase is a 62-kDa protein that has been frequently cloned and expressed in bacteria and eukaryotes. Altering the amino acid composition of luciferase by substitution results in a shift in emission peaks of the light produced, and these slight variations in luciferin structure represent the mechanism by which different species can produce a variety of wavelengths.

Luciferase catalyzes the condensation of luciferin with ATP in the presence of magnesium, and the resulting adenylate, still bound to the enzyme, reacts with oxygen to form a dioxetane, a four-member peroxide ring (Fig. 12.2). The ring decomposes across the peroxide bond to yield oxyluciferin in an excited state for the production of a photon of light. The luminescent reaction is very efficient, with a yield of about 90% relative to the consumption of the luciferin substrate.

ACOUSTICAL COMMUNICATION

Insects produce sound and vibrational signals through the air, water, or solid substrates, and receive and interpret these signals by what we may infer to be hearing. Demonstrations of insect hearing come from mostly the orders Orthoptera, Diptera, Coleoptera, Lepidoptera, and Homoptera, but the production of sounds can be easily missed if only the human ear, with its relatively poor performance, is used as an analyzing instrument. Thus, there are undoubtedly many more instances of acoustical communication in insects that we may be unaware of. Vibrations may be produced by a variety of mechanisms and through a variety of substrates to ultimately reach another insect and change its behavior.

Sound Production by Percussion

Sounds can be produced by **percussion,** involving the bringing together of two rigid structures. The male Australian whistling moth, *Hecatesia thyridion,* has the costa of its forewings modified into small knobs called **castanets.** As the wings come together at the top of the wing stroke, these castanets strike each other and produce sounds that resonate through the wings and attract females.

The production of sound can also occur by the striking of a body part against the substrate. Pscopterans and plecopterans can tap their abdomens against the ground to make drumming sounds; in plecopterans, a peg or hammer is located on the seventh to ninth abdominal segments that strikes the substrate. Using their heads, the social ants and termites can drum on the ground to warn colony members of danger. Males of the katydid, *Meconema thalassinum,* produce a drumming sound by tapping one of their hind tarsi on the ground.

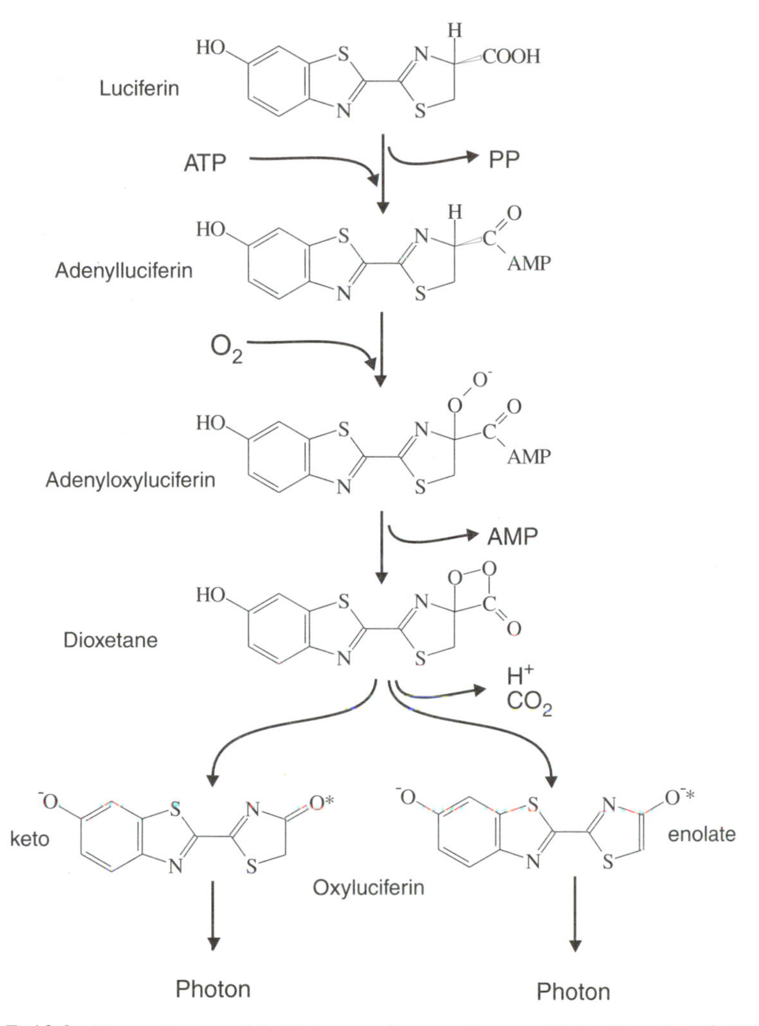

FIGURE 12.2 The oxidation of luciferin to release a photon of light. From Wood (1995). Reprinted with permission.

Sound Production by Vibration

Sounds can be produced by using muscles to directly vibrate a membrane. The special membrane is called a **tymbal,** and functions much like the lid of a tin can when it is pressed and released with a finger. The tymbal mechanism is present in many homopterans and is most utilized in male cicadas where they appear as paired structures that are borne on the first abdominal segment. Some arctiid moths bear a tymbal on either side of the metathorax.

The chitinous membrane comprising the tymbal is normally bowed outward and is surrounded by a sclerotized ring. A tymbal muscle is attached to the center of the tymbal, and when the muscle contracts, it pulls the membrane rapidly inward and then relaxes to release it, in the process moving the air and producing a sound as it moves in either direction (Fig. 12.3). Tracheal air sacs may be present behind the tymbal to produce a sympathetic resonance and prevent other internal tissues from damping the sounds. In those pentatomid species that bear tymbals lacking these sacs, much softer sounds are issued. Some species of cicada produce double pulses from the inward movement of the tympanum that is due to the buckling of the tymbal in two stages. The two tymbal muscles on either side contract alternately at a frequency of 120 Hz, producing a final signal of about 240 Hz. Accessory muscles that attach to the rim of the tymbal and increase its convex shape are able to modulate the volume of the sounds that are produced. By altering the curvature of the tymbal when the muscle contracts, the energy required to move the tymbal and the intensity of the sound produced are increased. One meter away, the male cicada cry has been recorded at 100 decibels, but only about 30 decibels are required to stimulate a female.

Insectivorous bats use ultrasonic pulses to find their food at night by echolocation. Bats can first detect a moth about 6 m away, but with its tympanum, the moth can register the bat's approach at the greater distance of about 35 m and execute evasive maneuvers to escape predation. In addition to the evasive behavior, some arctiid and ctenuchid moths produce ultrasonic clicks from their metathoracic tymbals that either may signal the moth's distastefulness to the bat or jam the bat's ultrasonic detection system and interfere with its echolocation.

Many insects produce vibratory sounds by **stridulation,** the rubbing of two body parts against each other. One body part that contains a series of teeth or corrugations is referred to as the **file** or **pars stridens,** and the opposing part may

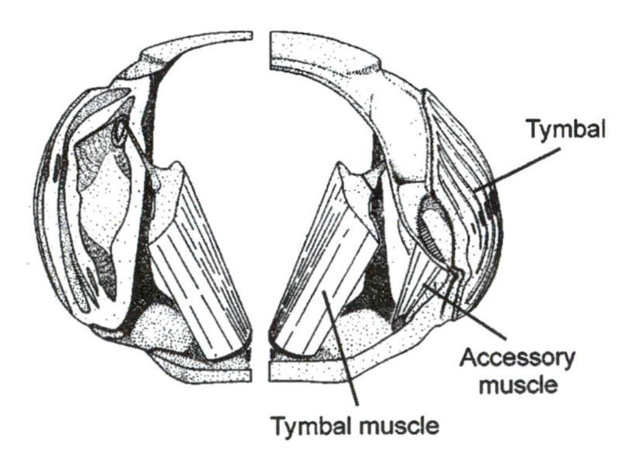

FIGURE 12.3 The tymbal on the outside body wall and its musculature. From Simmons and Young (1978). Reprinted with permission.

consist of a peg or tooth that is known as the **scraper** or **plectrum,** and is dragged across the file to create the sounds. Crickets have a row of chitinous teeth on the ventral side of a wing vein that acts as the file, with the scraper formed from the inner edge of the elytron (Fig. 12.4). Sounds are produced during the closing of the wings when the scraper on one side of the wings is moved against the file on the other side of the wings. A wing area known as the **harp** is corrugated and is driven to vibration by the stridulation, radiating the sounds that are generated. A **mirror** may also be present, an area consisting of thin cuticle that is supported by surrounding wing veins and vibrates when it is distorted by the scraper–file interaction.

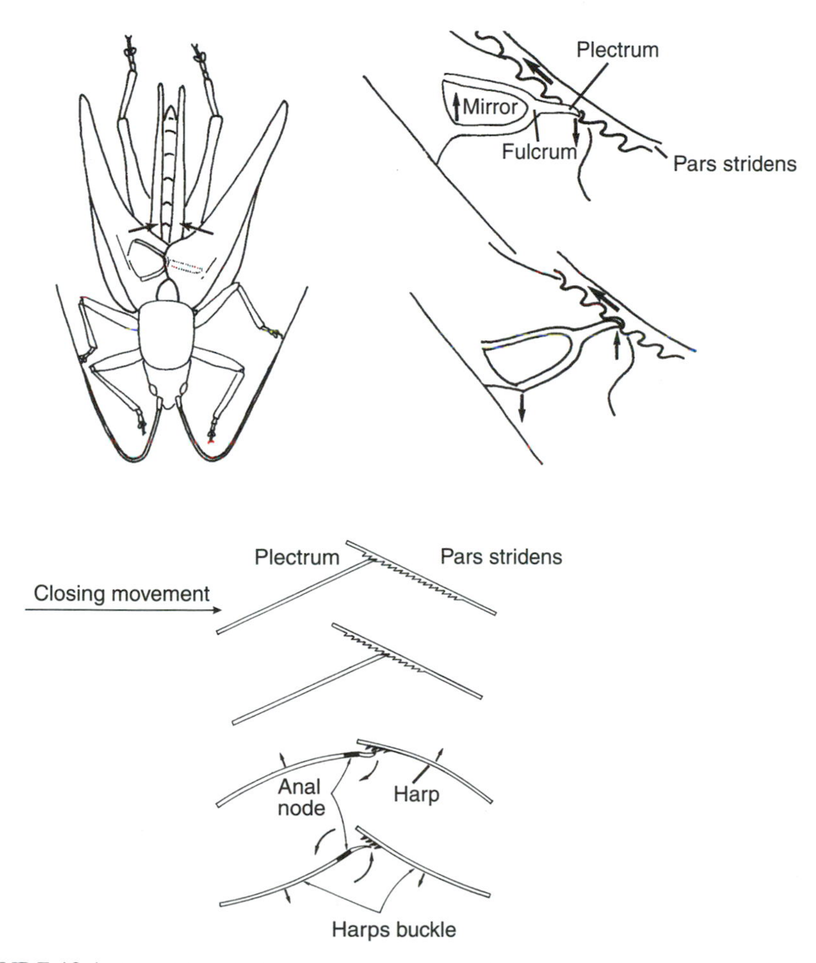

FIGURE 12.4 (Top) The stridulatory apparatus of a katydid. The mirror frame vibrates as the file on one elytron is moved across the plectrum. (Bottom) The stridulatory apparatus of a cricket. From Ewing (1989). Reprinted with permission.

Mole crickets modify their burrows into acoustical chambers, creating a horn that increases the efficiency of their harps' vibrations (Fig. 12.5). A song produced when the cricket is outside its burrow is significantly less intense than one produced while in the burrow. The song produced outside is only about 4% of the intensity of one produced within the burrow, which is amplified by the chamber and can be heard by humans up to 600 m away.

Beetles have a variety of mechanisms with which to stridulate. Some have a file at the top of their heads that is scraped by a ridge under the anterior of the pronotum. In others, the file is located under the head and the scraper located on the anterior margin of the prosternum. The file and scraper may both be located on the thorax or abdomen, or the file alone on the thorax or abdomen and the scraper on an appendage (Fig. 12.6). Leg stridulation occurs in some grasshopper

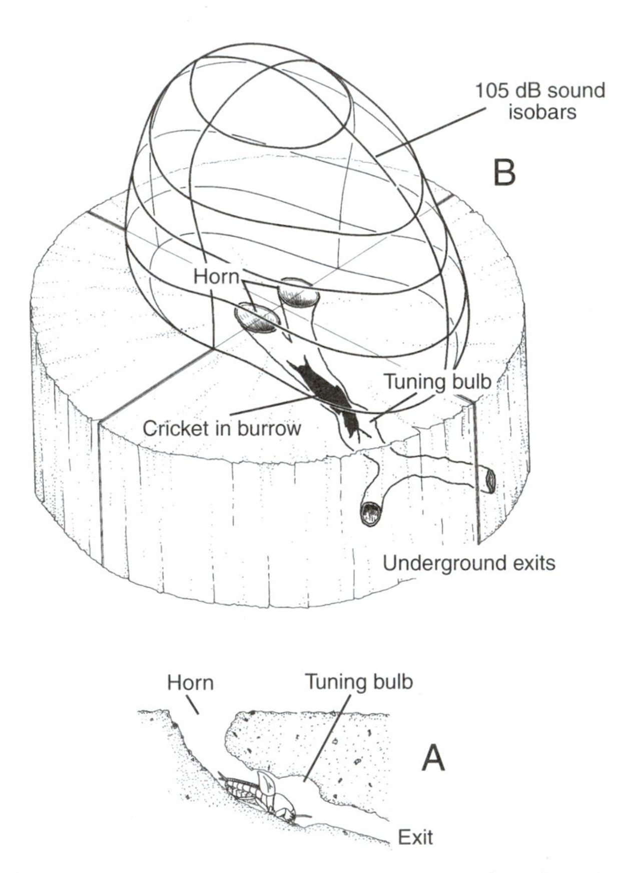

FIGURE 12.5 (A) A cross section of a mole cricket burrow. (B) A three-dimensional view of the acoustical chamber that amplifies the stridulations. From Bennett-Clark (1989). Reprinted with permission.

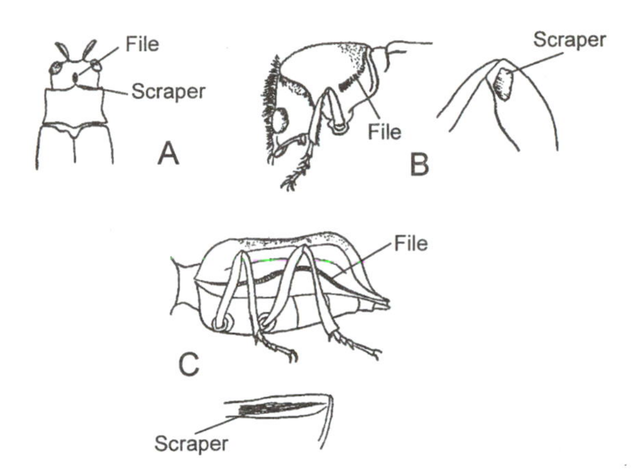

FIGURE 12.6 Various stridulatory structures on coleopterans. From Haskell (1961). Reprinted with permission.

families. The hind femora bear stridulatory pegs that are rubbed against a sharpened wing vein (Fig. 12.7).

There are tachinid flies that are parasitoids of male crickets and use the songs of their cricket hosts to locate them. Rather than possessing a typical Johnston's organ that is used to detect vibration in many other dipterans, these flies have evolved a tympanum much like those of the cricket. With this tympanum tuned to the songs of its host, the female fly is able to locate the male cricket as it calls and deposit its eggs on or near him. Although not an example of a signal originating with an insect, the calling song of *Hyla* tree frogs attracts mosquitoes in the genus *Corethrella* for a blood meal.

The displacement of air during the rapid movement of the wings during flight creates sounds, and some insects have used this mechanism as a means of communication. In those mosquito species that swarm, the males hover over a conspicuous

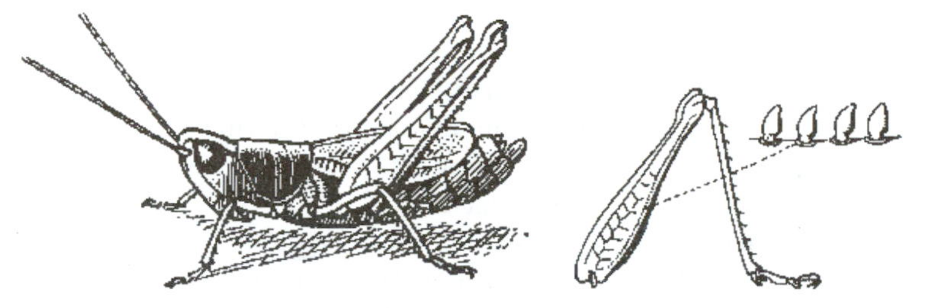

FIGURE 12.7 (Right) The stridulatory pegs on the legs of a grasshopper. From Haskell (1961). Reprinted with permission.

marker at sunset and when females enter the swarm, the males are attracted to them by the tone produced by the female's wing beat frequency. *Drosophila* males and females produce pulses and bursts of sound from their incomplete wing beats while they walk. The male's song stimulates the female to engage in the courtship ritual. Honeybees also produce pulses by vibrating their wings during the waggle dance to inform other workers about the location of resources.

Sound Production by the Expulsion of Air

Cockroaches in the genus *Gromphadorhina* expel air through their tracheal systems in response to disturbances and during courtship. The fourth abdominal spiracle is modified with an inner constriction and a large opening to the outside, and as the air exits, the spiracle creates a hissing sound. All spiracles but the fourth are closed when the abdominal muscles contract so that air is vigorously forced out of the single open pair. The tsetse, *Glossina,* has been observed making ultrasonic sounds during feeding and mating that have been attributed to the release of air through the tracheal system when the tracheae supplying air to flight muscles are contracted.

TACTILE COMMUNICATION

Tactile signals are used for short-range communication, generally for aggressive or sexual encounters. Aphids engage in violent kicking when they are crowded, physically displacing other aphids in order to gain more space for themselves. Many insects must touch their mates before copulation can begin, receiving close-range chemical signals that release the required stepwise behaviors. Contact chemoreception is especially important in dipteran mating systems, with female cuticular hydrocarbons that are perceived by touch serving as essential behavioral releasers. Males of both the house fly, *Musca domestica,* and the tsetse, *Glossina morisitans,* use visual stimuli as initial attractants, but as they approach the females, cuticular hydrocarbons act as attractants that are necessary for the close-range stimulation of the males. The male cuticles of both these dipterans contain substances, the **abstinons,** that have anti-homosexual activity when contacted, deterring the males from attempts to copulate with other males. The crane fly, *Tipula oleracea,* requires a rigid sequence of tactile communications in order to mate, including stereotyped leg movements. Courtship can be broken off if any of the tactile stimuli are perceived at the wrong time or missing altogether.

The social ants and termites live in darkened underground galleries and must use cues other than visual ones to recognize nestmates. Individuals of both groups produce colony-specific hydrocarbons that are detected by a sweep of the antennae or palps across the cuticular surface of another individual. Each colony has a chemical signature that is read by touch and that triggers aggressive behaviors if it fails to match the signature that is expected.

CHEMICAL COMMUNICATION

Insects make an extensive use of chemicals that are released into the surrounding medium for communication. They employ chemicals for communicating to find a mate, gather together, provide others of their own species with the location of food, identify nestmates, and defend themselves against predators. Insects live in a complex chemical world, with chemical cues governing much of their behavior and most of their interactions. The classification of the chemicals that insects use to communicate is based on the functional roles of those chemicals in the interactions they mediate.

The chemicals that mediate physiological or behavioral processes can be categorized as either **hormones** or **semiochemicals.** Hormones, as we have seen, are produced by one organism and mediate physiological reactions within that producing organism. In contrast, any chemical that mediates an interaction between two organisms, whether of the same or different species, is a semiochemical (Greek: *semeon,* a signal). Semiochemicals are further divided into two categories based on whether the use of the chemical is between members of the same or different species. **Pheromones** are semiochemicals that mediate intraspecific interactions, and **allelochemicals** are those that mediate interspecific interactions (Fig. 12.8).

PHEROMONES

Chemical communication between single-celled protozoa was certainly well established before multicellular organisms evolved. The hormones that organize and coordinate the development of multicellular animals may thus be the intracellular equivalent of the pheromones that the unicellular organisms employ, suggesting that pheromones may be the chemical ancestors of hormones. Pheromones are produced from **exocrine glands,** which are modified epidermal cells, and transmitted to another individual of the same species.

The first insect pheromone, bombykol, was isolated in the mid 1950s from over 300,000 *Bombyx mori* silkworm moths. From this large amount of biomass, only 5.3 mg of the active product was obtained. To identify the pheromone components active in the boll weevil, *Anthonomus grandis,* 4 million insects were processed along with 55 kg of their fecal material. With improvements in analytical instrumentation, it was no longer necessary to use such a large number of insects, and

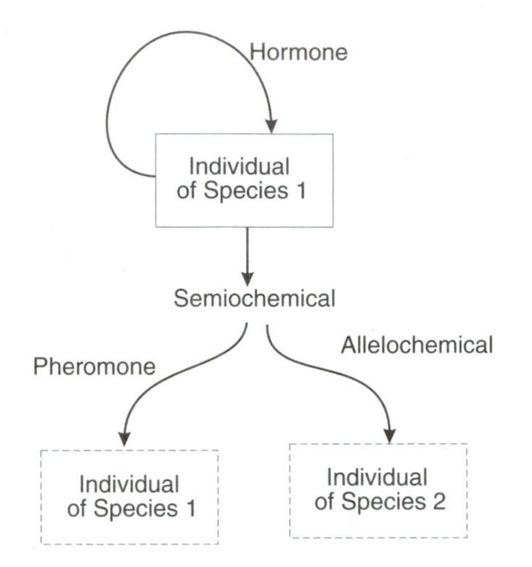

FIGURE 12.8 The actions of hormones and semiochemicals. Hormones are produced by an individual and act on cells within its body. Semiochemicals are chemicals that mediate the interactions between two organisms. If the organisms are of the same species, the semiochemical is called a pheromone. If the organisms are a different species, the semiochemical is termed an allelochemical.

the isolation and identification of insect pheromones have proceeded at an ever-increasing rate. The identification of the sex pheromones for over 1600 insect species from more than 90 families has been accomplished, and a number of **para-pheromones,** which are analogs and mimics of the natural products, have been synthesized for use as tools in pest management.

Pheromones were first called ectohormones because they were secreted by glands and had the physiological effects of a hormone, but unlike hormones were produced outside of the organism that they affected. Hormones are substances secreted internally, so the name ectohormone was thus a contradiction in terms and the name pheromone (Greek: *phereum,* to carry; *horman,* to excite) was adopted instead. Pheromones are chemicals produced by specialized glands that are secreted to the outside by one animal and have a specific effect on another individual of the same species. They are typically active in extremely small concentrations, and usually as a mixture of compounds in a species-specific pheromone blend. The individual components in the blends are often common to several species, with the precise proportions of the individual components conferring the species specificity. It is not uncommon for geographic variants of a species to produce significantly different proportions of the pheromone components. There is one compound, (*Z*)-7-dodecen-1-yl acetate, that is used by over 126 insect species as a component in their pheromone blends, and incredibly is also a component of the pheromone that female Asian elephants produce in their urine to broadcast their readiness to mate to male elephants.

For a chemical signal to be distinct in the chemically noisy environment, one strategy is to use one very complex component that would never appear otherwise, and, indeed, several species appear to employ only a single component in their pheromones. The American cockroach, *Periplaneta americana,* uses periplanone B alone to attract males from a distance. It is a bit unfortunate that bombykol, the first pheromone to be identified, appeared to consist of a single component. Using this as a model for much of the future research on insect pheromones, many subsequent investigations looked for single-component pheromones in other species. However, many of these single components failed to show the same activity as an intact insect did in bioassays, and as the pheromones from more insects were identified, it became apparent that many of the same chemical components were present in different species. When electroantennograms showed responses of receptors to the single components, but the whole insects perceiving them failed to display the behavior, it was eventually realized that most pheromones consisted of blends of different compounds, and the species specificity that is present in these pheromones results from the specific pattern of the blended components that the receiver collects. Even what was initially believed to be a single component in bombykol ultimately turned out to be a multiple-component system. Pheromones produced by two species of the leaf-rolling moth, *Archips argyrospilus* and *A. mortuanus,* are good examples of these blends. Both species produce a pheromone with the same four components, but with respective proportional blends of 60:40:4:200 and 90:10:1:20. The ratio of the first two components is the most important for maintaining the reproductive isolation between these species. The specificity of most pheromones lies in their blends.

As with hormones, pheromones can be broadly subdivided according to their mode of influence. **Releaser pheromones** stimulate an immediate and reversible behavioral response that is mediated by the nervous system soon after the receiver perceives it. There are a large number of insect behaviors that are released by pheromones. In contrast, **primer pheromones** mediate a fundamental physiological change in the receiver that reprograms it for an altered response, acting directly on the nervous system or some other physiological system. The response may not be immediate as for a releaser, and it may be a novel response that was not expressed previously for that same stimulus (Fig. 12.9). Primers are utilized mostly by social insects to regulate a variety of social interactions.

RELEASER PHEROMONES

Releaser pheromones make up a diverse group of chemicals, and are best subdivided on the basis of the functions they serve. The most common functional

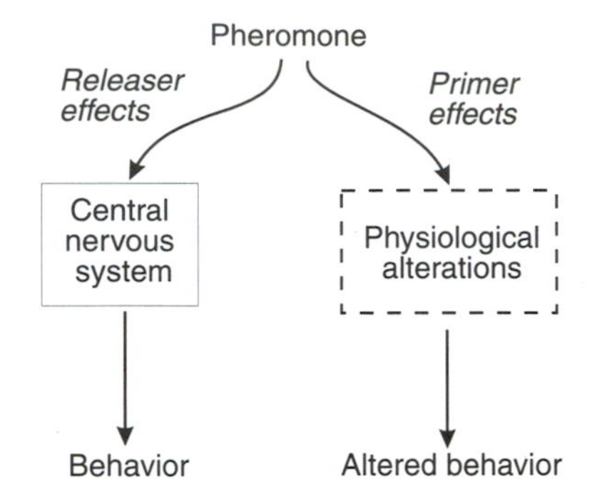

FIGURE 12.9 The releaser and primer effects of pheromones. A releaser pheromone acts on the central nervous system and causes the immediate release of a behavior. A primer pheromone alters the physiological state of the receiver, causing another behavior to be expressed.

categories include **sex pheromones, aggregation pheromones, alarm pheromones, trail pheromones,** and **dispersal,** or **spacing pheromones.**

Sex Pheromones

Sex pheromones are chemicals produced by insects of one sex that elicit a behavioral response in members of the opposite sex. They advertise the presence of an individual and lure others of the species for mating. The mechanisms by which the pheromones are produced and received vary considerably among insects. If the perception of the pheromone releases a long-range searching behavior, it is called a **sex attractant.** If it facilitates closer-range courtship behavior or copulation, it may be referred to as an **aphrodisiac.** The pheromone may be produced by either the male or the female of the species, and, in some cases, mating behaviors are released by pheromones that are produced by both of the sexes. Volatile sex pheromones are usually released at certain times during the insect's life cycle and certain periods of the day rather than continuously during an individual's life, and their release may be terminated once the individual mates. **Calling behavior,** which consists of a particular posture and the eversion of pheromone glands to allow the pheromone to evaporate, occurs in many sexually mature insects.

The use of sex pheromones by moths has probably been studied the most intensively of any insect. Typically, the female releases a long-distance attractant that arouses the male, which becomes airborne and flies upwind into the chemical plume. The plume is not a continuous swirl of increasing concentration of pheromone molecules, but rather is more like the smoke from a chimney in a gentle breeze. Filaments of the plume drift downward interspersed with air that is free of any pheromone molecules, so the male encounters a discontinuous stimulus trail. The pulsed signals are necessary to find the female; with a continuous strong signal, the pheromone receptors become adapted and the male terminates his flight. He maintains a zigzag flight pattern in and out of the plume as he enters and leaves the trail until reaching the female. After landing, the male may release a short-distance aphrodisiac pheromone to attract the female, and copulation follows.

Aggregation Pheromones

Unlike sex pheromones that act on only one sex, **aggregation pheromones** bring many individuals of both sexes together. Their activity may resemble that of sex pheromones because they often increase the probability that copulation will occur in the population. They are produced mainly by coleopterans as a defense against predators and to overwhelm the resistance of a host tree. For example, female bark beetles are the first to bore into a tree and release an aggregation pheromone that along with the host terpene attracts both males and females. The sexes are brought together for mating but the large number of individuals present also benefits the beetles by overcoming the tree's defenses. Coccinelid beetles produce an aggregation pheromone that attracts large numbers of males and females to overwintering sites, and their aposematic coloration is enhanced and predators discouraged when the brightly colored insects are aggregated. The honeybee queen produces an aggregation pheromone from her mandibular glands that is responsible for the retinue of workers that attend to her, and also stabilizes the colony around the queen when it swarms.

Alarm Pheromones

Alarm pheromones are produced mostly by social insects to warn other colony members of danger and to recruit for colony defense. It is certainly more adaptive and effective for a species to mount a collective response to some traumatic stimulus rather than an individual response. In the honeybee, the release of *iso*-amyl acetate by alarmed workers releases a frenzied attack by other workers. Ants produce their alarm pheromones in their mandibular glands and release them when attacking prey to recruit other colony members. Aphids and treehoppers also secrete alarm pheromones that cause them to fall off the plant and escape possible predation. The green peach aphid, *Myzus persicae,* secretes the alarm pheromone (*E*)-β-farnesene from its cornicles when attacked by predators. Alarm pheromones generally consist of low-molecular-weight, highly volatile compounds that easily spread throughout a colony yet evaporate quickly to terminate the aggression when the danger no longer exists.

Trail Pheromones

Trail pheromones are also found mostly in social insects, including the ants, termites, bees, and wasps. When a worker locates a resource, she lays down a trail when returning to the colony that other workers can use to find the resource. Flying insects use trail pheromones to stimulate colony members to enter the hive. Bees mark the nest entrance with products from the Nasonov gland that induce workers to enter. There is also evidence for a pheromone deposited from the tarsi of bees and wasps that may serve as a trail pheromone at the nest entrance.

Venomous ants use their sting to lay down a product of the poison glands as a trail pheromone, while nonvenomous ants synthesize their pheromone in Dufour's gland or in the gut. These terrestrial trail pheromones serve as a sensitive index to the amount of food present at a distant location, because each worker returning from the resource adds to the trail's intensity. Once the food is exhausted, returning workers no longer lay down a trail and it soon dissipates. Trail pheromones tend to be more stable than alarm pheromones, but they are still relatively volatile, a necessity if trails that are no longer informative are to be avoided.

Trail pheromones appear to have arisen as metabolic by-products that were eventually adapted as signals, and may be exceptions to the rule that pheromones exist as specific component blends. A single component, 3-ethyl-2,5-dimethylpyrazine, is present in the poison glands of several species of *Myrmica* and is able to induce trail-following in all those species, but there are some other species that do indeed use multiple components.

Epideictic Pheromones

Spacing pheromones are also known as **epideictic pheromones.** They maintain the densities of individuals attempting to exploit an exhaustible resource to numbers that are below its carrying capacity. Female tephritid fruit flies oviposit on the flesh of fruits where the larvae develop and mature. Some of the fruits are only large enough to support a single larva. Immediately after she oviposits, the female circles the fruit as she trails her ovipositor on its surface and deposits a pheromone that deters other females from laying eggs on that fruit. Female bark beetles are the first to attack a tree and produce an aggregation pheromone that attracts both males and females. Along with the increase in the number of aggregating males comes an increase in the epideictic pheromone they produce, and this prevents subsequent males and females from landing on that resource so that it is not overpopulated.

Funeral Pheromones

So-called **funeral pheromones** are produced in dead ants that stimulate other live colony members to remove them to a pile outside the nest. When an object is covered with an extract of the saturated fatty acids that are thought to be responsible, it is treated as if it were a dead ant.

PRIMER PHEROMONES

The effects of pheromones acting as primers are at the same time subtle yet physiologically profound. They alter the physiology of the receiver so that it displays a modified response pattern to future stimuli. Unlike releasers, there have been too few primer pheromones identified for them to be further subdivided categorically. Because the effects of primer pheromones are so subtle, possible bioassays for their actions are relatively difficult to perform and often take months to complete. The lack of suitable bioassays for primer pheromones has been an obstacle in their study and is certainly a major reason that the chemical nature of only one primer pheromone has been determined.

Primer pheromones are most often used by social insects for the regulation of colony activities. In honeybee colonies, there is only one fertile queen along with several thousand reproductively sterile workers. The queen produces a multicomponent pheromone from her paired mandibular glands, the only primer pheromone that has been completely identified that, along with substances produced by the workers, inhibits the ovarian development of workers and maintains them as nonreproductives. The major compound of the queen mandibular pheromone, 9-oxo-2-decenoic acid, acts on the endocrine system of workers to suppress their synthesis of JH.

By regulating the levels of JH in workers, the queen mandibular pheromone also affects the ontogeny of worker polyethisms that affects the rate at which workers change their behavior as they age from inside activities of brood rearing to outside activities of foraging. Low levels of JH maintain the workers in the nest, while higher levels trigger the onset of foraging behavior. Among the releaser functions of this queen mandibular pheromone is as a queen recognition factor that is responsible for attracting and maintaining the retinue of workers that constantly attend to her, and regulating the swarm behavior of the colony.

If the queen is removed, levels of the queen mandibular pheromone decline within the colony and the workers soon become agitated and begin preparations for rearing a replacement queen. They elongate existing cells that already contain newly hatched larvae and feed them royal jelly, a special secretion from their mandibular glands that, acting as a primer pheromone, brings the female larvae onto the developmental pathway to become queens rather than workers. The ingestion of royal jelly by the developing larvae affects their release of JH, and it is the presence of JH during a JH-sensitive period that ultimately determines which developmental polyphenism the larvae will assume (see Chapter 2). Shortly after the last larval molt, larvae destined to be either workers or queens have identical ovaries, but during the next 24 h most of the ovarioles of the workers degenerate. The royal jelly also contains substances that preserve the food that is stored in the cells of the hive.

There are other primer pheromones in honeybees that originate with the brood. The larval cuticle of the brood contains a blend of 10 fatty acid esters and stimulates the development and synthesis of proteins by the hypopharyngeal glands of the workers.

The queen fire ant, *Solenopsis invicta,* produces a primer pheromone in its poison sac and possibly in another gland that has yet to be identified. Like the primer pheromone in honeybees, it prevents unmated queen ants that may be present in the colony from developing their ovaries and shedding their wings, and acts as a releaser pheromone to attract workers to the queen. It acts indirectly on caste determination by affecting the behavior of the workers toward developing larvae. In response to primer pheromone, workers restrict the quantity of food given to female larvae that results in their development as workers.

There is also evidence for primer pheromones in nonsocial insects. Adult male *Schistocerca gregaria* locusts produce a primer pheromone on the surface of the cuticles that accelerates the growth of male and female immatures.

PHEROMONE RELEASE AND SYNTHESIS

The synthesis of pheromones may occur throughout the adult insect's life, but release of the synthesized pheromones generally occurs only during certain environmental and physiological circumstances. Bark beetles release their pheromones only during the day, while nocturnal moths engage in calling behavior and pheromone release only at night. The larvae of the moth, *Antheraea polyphemus,* feed only on oak leaves, and adults require the presence of *trans*-2-hexanol from the oak leaves in order to call. Females that mate more than once release pheromones periodically, but those females that only mate once usually terminate their pheromone release after mating.

Pheromones are generally produced by modified epidermal cells that can be found in various places throughout the body. These are often clustered into groups designated as **exocrine glands,** secretory glands that direct their products to the outside of the organism. Exocrine glands have been generally classified into two major types depending on their structural organization. Type I glands appear to be most immediately derived from epidermal cells, formed by a simple epithelial layer or lining an internal reservoir that temporarily stores the secretions. Type II glands consist of both secretory and duct cells, with the secretory components far removed from the cuticular epidermal cells.

The pheromone produced by adult male *Schistocerca gregaria* locusts affects the rate of maturation of other nearby males and females and originates in secretory cells dispersed throughout the epidermis. Many female lepidopterans release pheromones from type I glands that are modified epidermal cells located within the intersegmental membrane, whose surface area is increased by longitudinal folds. These glands may be deeply invaginated, but when the female displays calling behavior, the glands are everted from the body cavity and the volatile secretions are allowed to evaporate to the outside. Eversible brush structures on the sternites at the tip of the abdomen of male noctuid moths spread out fan-like when the male engages in courtship. These so-called **hair pencils** consist of hypertrophied trichogen cells accompanied by long hairs. Some glands cells are internal, such as Dufour's gland in ants and the frontal gland of termites that produce alarm pheromones. The gland cells produce pheromones that are stored in a reservoir that is lined with cuticle. The type II mandibular glands of bees produce alarm pheromones, sex pheromones, and the queen substance that inhibits ovarian development in the workers of a colony. In some stinging ants, the poison gland of the sting apparatus produces a trail pheromone.

An extensive variety of chemical compounds have been identified as components of insect pheromones. In general, signals that require a rapid dispersal, such as the alarm pheromones, utilize smaller molecules, and signals requiring a more persistent exposure, such as sex pheromones, utilize larger molecules. Their activity is determined by the molecule's shape, size, and functional groups that are present, and degree of unsaturation. Most work on mechanisms of pheromone synthesis has been focused on lepidopteran sex pheromones. The components of the blends are closely related to the fatty acids, generally with 10, 12, 14, 16, or 18 carbons in a straight chain, alcohols, acetates or aldehydes as single functional groups, and one, two, or three double bonds.

The sex pheromone of the cabbage looper, *Trichoplusia ni,* consists of a blend of saturated and unsaturated acetate esters of fatty acids. Beginning with fatty acids, the six components of the blend are produced by the action of a Δ-11 desaturase, an enzyme unique to lepidopteran sex pheromone glands, along with chain-shortening reactions, reduction, and acetylation (Fig. 12.10). A complex blend of products is thus produced by employing only a few recurring steps.

The release of sex pheromones by most moths occurs during the evening hours and is controlled both by an endogenous circadian rhythmicity and physiological factors such as mating status. Pheromone synthesis is regulated hormonally by a neuropeptide from the subesophageal ganglion, the **pheromone biosynthesis activating neuropeptide** (**PBAN**), a 33- to 34-amino-acid peptide that regulates pheromone production by activating enzymes involved in the synthesis of the final product. In the few moths that have been studied, PBAN controls pheromone production by regulating a step in fatty acid biosynthesis. In the moth *Sesamia nonagrioides,* PBAN activates an acetyl transferase that produces the major component, (*Z*)-11-hexadecenyl acetate (Fig. 12.11). Some female

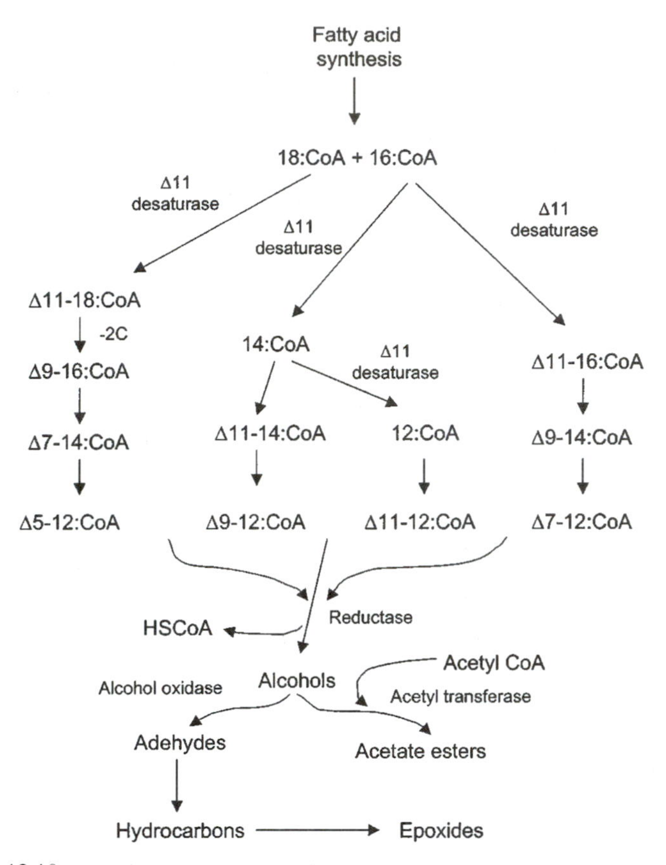

FIGURE 12.10 Sex pheromone synthesis from fatty acids in the cabbage looper, *T. ni*. A small number of recurring steps produce the variety of hydrocarbons used as pheromones. From Roelofs 1995; Tillman *et al.* (1999). Reprinted with permission.

moths terminate their production of sex pheromone once they mate because the release of PBAN from the subesophageal ganglion is suppressed after mating. The signal that mating has occurred is transmitted from a spermatheca full of sperm to the subesophageal ganglion by a neural signal in some species and by a humoral signal from the spermathophore in others.

Juvenile hormone has been implicated in the control of pheromone biosynthesis in coleopterans. The aggregation of bark beetles on coniferous host trees is released by the pheromones produced by the attacking beetles in their defecations. In the bark beetle, *Ips,* feeding on the *Pinus* host activates the corpora allata to produce JH III, and this activates two key enzymes: 3-hydroxy-3-methylglutaryl-CoA (HMG-CoA) synthase and an HMG-CoA reductase (Fig. 12.12) that in turn activate the isoprenoid pathway toward the synthesis of the sex and aggregation pheromones ipsdienol and ipsenol.

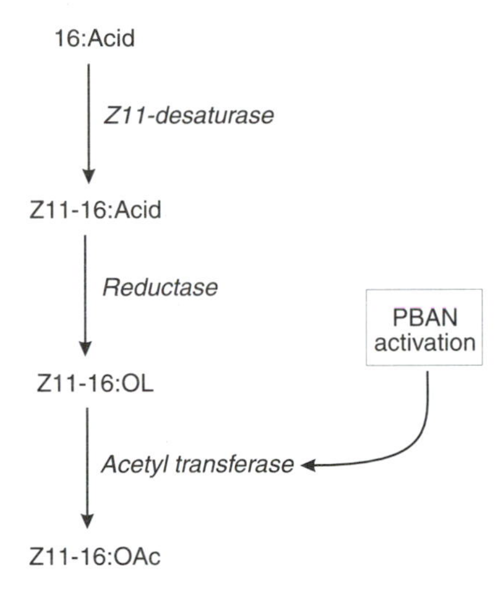

FIGURE 12.11 Pheromone biosynthesis activating neuropeptide (PBAN) activates the enzyme acetyl transferase responsible for pheromone production in moths. From Mas *et al.* (2000). Reprinted with permission.

ALLELOCHEMICALS

The other major category of semiochemicals consists of the allelochemicals (Greek: *allelon,* of one another), which mediate interspecific interactions. These affect species other than the ones that are producing them and may adversely affect either the emitter or the receiver. If the signal is adaptively favorable to the emitter but not to the receiver, the substance is considered to be an **allomone.** If the signal is adaptively favorable to the receiver but not to the emitter, the substance is classified as a **kairomone.** If both receiver and emitter benefit, the substance is a **synomone** (Fig. 12.13).

Allomones

Allomones are chemical countermeasures that are used primarily for defense. They are thus allelochemicals that are adaptive to the sender but not the receiver. A wide variety of chemical defenses can be mounted against potential predators, including oral and anal discharges, toxic components in the hemolymph made available by reflex bleeding, glandular discharges, and bites and stings that are supplemented with poisons.

Many orthopterans and the larvae of lepidopterans commonly discharge oral secretions consisting of gut contents mixed with salivary secretions when they are disturbed. In response to predation, some coleopterans engage in **autohemor-**

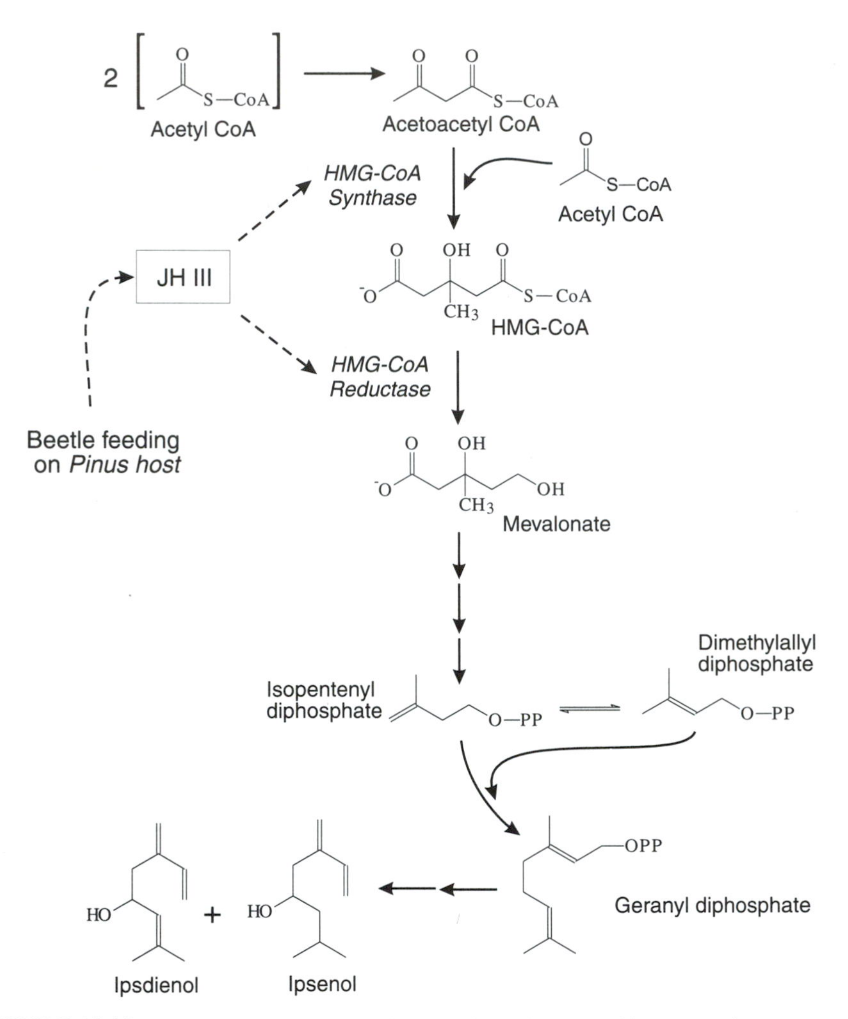

FIGURE 12.12 Pheromone biosynthesis in bark beetles and its control by JH. Two key enzymes involved in the pathway toward the synthesis of ipsenol and ipsdienol are activated by JH. From Tillman *et al.* (1999). Reprinted with permission.

rhage, in which allomone-fortified hemolymph that deters predators is released through the intersegmental membranes. Termite soldiers of some species supplement their large mandibles with toxic monoterpene hydrocarbons that are squirted on enemies as they fight. In the termite subfamily Nasutitermitinae, the soldier caste is composed of specialized **nasutes,** lacking mandibles for defense but are capable of squirting a gummy latex substance from the frontal gland of their pointy heads (Fig. 12.14). Bombardier beetles in the genus *Brachynus* discharge and accurately aim hot quinines at attackers. Synthesized as a binary weapon, hydrogen peroxide, hydroquinones, peroxidases, and catalases are separately stored in two

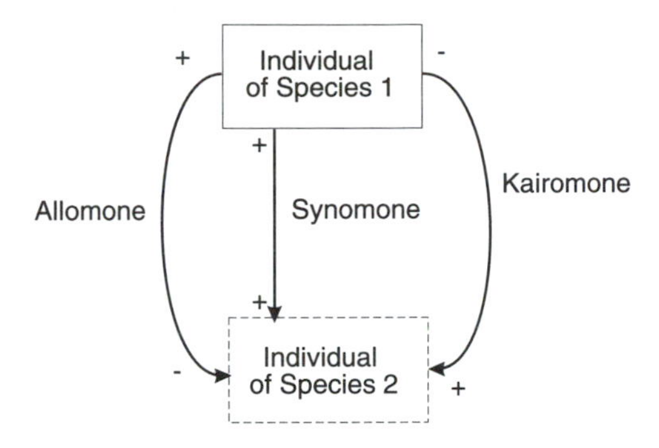

FIGURE 12.13 Allelochemicals, which mediate interspecific interactions. Allomones are allelochemicals that are adaptively favorable to the emitter but not the receiver. Kairomones are adaptively favorable to the receiver and not the emitter. Synomones are adaptively favorable to both the emitter and the receiver.

chambers and mixed together in a reaction vessel to produce an explosive discharge at 100°C (Fig. 12.15). Social Hymenoptera have modified their ovipositors into a sting apparatus that is used to effectively inject toxic and painful secretions under the skin of potential predators. Dytiscid beetles synthesize a mixture of steroids that are identical to the vertebrate hormones and discharge them into the water as a defense against vertebrate predators.

Given the extensive use of chemicals as pheromones to communicate between members of the same species, it is not surprising that some insects have evolved the role of illegitimate signalers that exploit these intraspecific chemicals and use them as allomones. Wild potato plants produce the alarm pheromone of the aphid *M. persicae,* (*E*)-β-farnesene, and prevent the aphids from feeding by releasing their alarm response. The bolas spider produces a drop of moth sex pheromone at the end of a silken cord that it twirls around to entice the male moths to their deaths. A staphylinid beetle takes up residence in termite colonies and is able to acquire food from the termite workers because it has mimicked the cuticular hydrocarbons of the termite. In what has been called a "wolf-in-sheep's clothing" strategy, *Chrysopa sloaaonae* lacewing larvae pluck the wax from the wooly alder aphids on which they feed and place the wax on their own dorsum. With this disguise, the

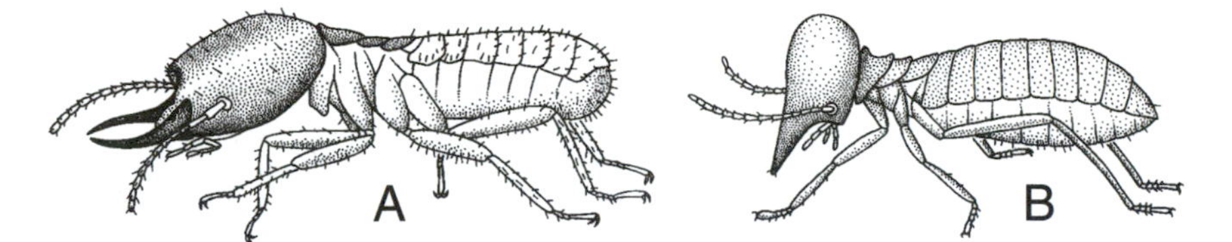

FIGURE 12.14 (A) A typical member of the termite soldier caste of *Coptotermes*. (B) A specialized nasute of *Nasutitermes*. From Gullan and Cranston (2000). Reprinted with permission.

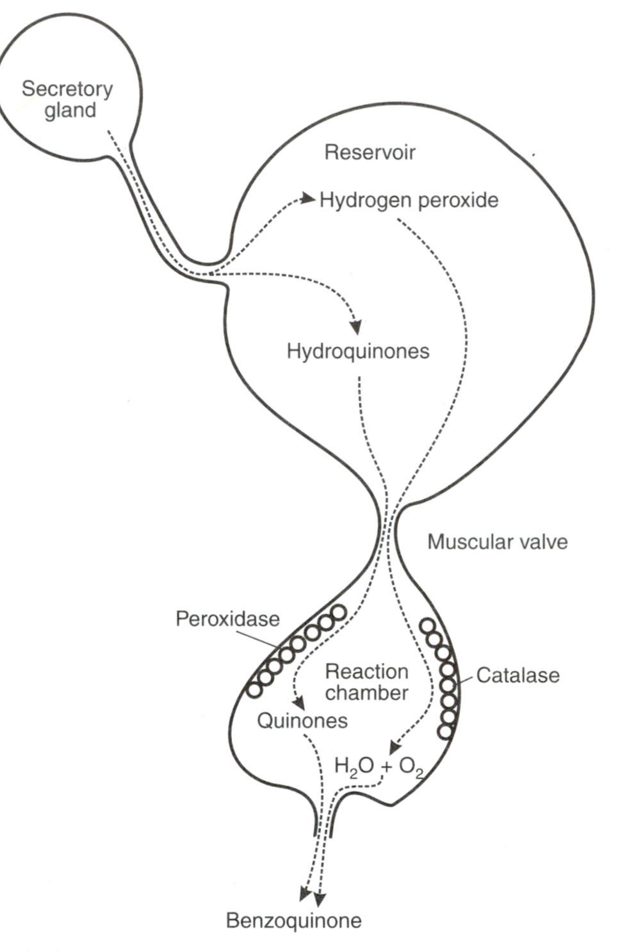

FIGURE 12.15 The production of hot benzoquinones by the bombardier beetle. The secretory gland produces hydrogen peroxide and hydroquinones into the reservoir. They are released into the reaction chamber, where the catalase and peroxidase reside, to produce an explosive reaction.

lacewings are immune from assault by the ants that normally protect the aphids. This use of the cuticular hydrocarbon that establishes the identity of the aphids is unique as an allomone in that the lacewing does not synthesize it but physically coats itself with the secretions from the aphid.

Kairomones

Kairomones (Greek: *kairos,* opportunistic) benefit the receiver rather than the emitter. They have been described as pheromones and allomones that have evolutionarily backfired and may not represent a distinct class of chemical signals themselves. Kairomones may be hormones, pheromones, or allomones that are normally used by one organism but exploited by an illegitimate receiver. They may be normal products of metabolism of one species that another now uses to locate its host. For example, many phytophagous insects find their host plants by the chemical fingerprint of secondary plant substances that they use as kairomones. Mosquitoes locate vertebrate hosts for blood meals by using as kairomones the carbon dioxide and other chemicals that are produced during normal vertebrate metabolism. The parasitic mite, *Varroa jacobsoni,* is attracted to honeybee drone larvae by fatty acid esters present in the bee larvae, and the rabbit flea uses the hormones of its rabbit hosts as a kairomone to locate the rabbit and also to mature its reproductive system.

Illegitimate receivers have also evolved the ability to exploit the pheromones of other species and use them as kairomones. Predatory beetles and flies can recognize the aggregation pheromones of bark beetles to locate them as prey. A braconid parasitoid uses the epideictic pheromone produced by the apple maggot fly to find the eggs to parasitize. Some beetles, cockroaches, and mites can exploit trail pheromones that have been laid by foraging ants.

Synomones

Synomones are chemicals that are adaptive to both the sender and the receiver. These include floral scents that attract pollinators and thus benefit both insect and plant. The hymenopterous parasites *Trichogramma* are attracted to tomato plants where they may find suitable hosts to parasitize. Damaged pine trees produce terpenes that bark beetles use as kairomones to find the trees, but the same chemicals are acting as synomones when they attract pteromalid hymenopterous parasites that then parasitize the beetles, benefiting both the parasites and the tree.

THE MULTICOMPONENT NATURE OF BEE COMMUNICATION

Humans have been aware of the ability of bees to communicate the location of resources to others in the hive at least since Aristotle wrote about the phenomenon of honeybee recruitment over 2000 years ago. The method by which this communication occurs was first discovered by Von Frisch, who described how a foraging worker returns to the hive and communicates both the distance and direction of the resources by performing a dance for other workers using a symbolic language (Fig. 12.16). If the food is close to the hive, the returning forager performs a round dance that gives no information about the direction of the food, only that the food is nearby. If the food is distant, the forager performs a waggle dance that imparts

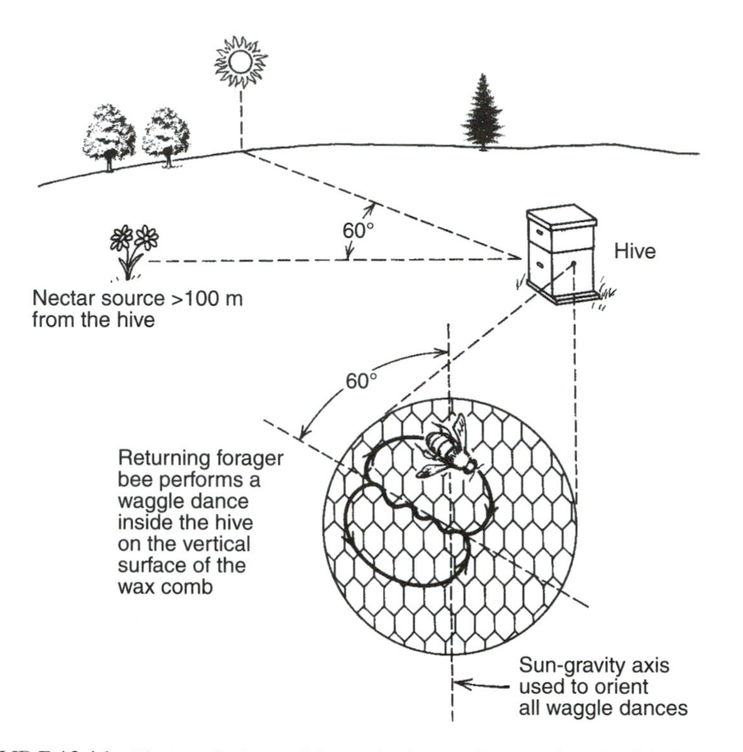

FIGURE 12.16 The waggle dance of the worker bee on the vertical comb of the hive is performed at an angle equal to the angle of the nectar source from the hive. From Matthews and Matthews (1978). Reprinted with permission.

both direction and distance. The waggle dance is performed on the vertical surface of the combs within the hive, based on the distance of the food and its angular direction with respect to the sun. The bees are able to compensate for the movement of the sun during the intervening period. Accompanying the dance are sounds produced by both the forager and the other workers in response. The forager produces an airborne sound by beating her wings as she dances, and an observing worker responds by producing sounds as she presses her thorax against the comb. These comb vibrations are cues that the forager uses to stop dancing and regurgitate food samples to give the workers additional information about the quality of the food source. Most races of honeybees have their own dialects of the dance language that may reflect the differences in the ecological ranges in which they forage, but they all communicate with vision, sound, and tactile stimuli.

ADDITIONAL REFERENCES

Acoustical Communication

Aidley, D. J. 1969. Sound production in a Brazilian cicada. *J. Exp. Biol.* **51:** 325–337.

Bailey, W. J. 1978. Resonant wing systems in the Australian whistling moth *Hecatesia* (Agarasidae, Lepidoptera). *Nature* **272:** 444–446.

Bennet-Clark, H. 1989. Songs and the physics of sound production. In *Cricket behavior and neurobiology* (F. Huber, T. Moore, W. Loher, Eds.), pp. 227–261. Comstock, Ithaca.

Bennet-Clark, H. 1997. Tymbal mechanics and the control of song frequency in the cicada *Cyclochila australasiae*. *J. Exp. Biol.* **200:** 1681–1694.

Bennet-Clark, H. 1998. Sound radiation by the bladder cicada *Cystosoma saundersii*. *J. Exp. Biol.* **201:** 701–715.

Bennet-Clark, H. C. 1995. Insect sound production: Transduction mechanisms and impedance matching. *Symp. Soc. Exp. Biol.* **49:** 199–218.

Bennet-Clark, H. C. 1998. How cicadas make their noise. *Sci. Am.* **278:** 58–61.

Bennet-Clark, H. C. 1999. Resonators in insect sound production: How insects produce loud pure-tone songs. *J. Exp. Biol.* **202:** 3347–3357.

Bennet-Clark, H. C., Daws, A. G. 1999. Transduction of mechanical energy into sound energy in the cicada *Cyclochila australasiae*. *J. Exp. Biol.* **202:** 1803–1817.

Blest, A. D., Collett, T. S., Pye, J. D. 1963. The generation of ultrasonic signals by a New World arctiid moth. *Proc. R. Soc. London B* **158:** 196–207.

Cade, W. 1975. Acoustically orienting parasitoids: Fly phonotaxis to cricket song. *Science* **190:** 1312–1313.

Cocroft, R. B., Tieu, T. D., Hoy, R. R., Miles, R. N. 2000. Directionality in the mechanical response to substrate vibration in a treehopper (Hemiptera: Membracidae: *Umbonia crassicornis*). *J. Comp. Physiol. A* **186:** 695–705.

Elsner, N. 1983. A neuroethological approach to the phylogeny of leg stridulation in gomphocerine grasshoppers. In *Neuroethology and behavioral physiology* (F. Huber, H. Markl, Eds.), pp. 54–68. Springer-Verlag, Berlin.

Ewing, A. W. 1989. *Arthropod bioacoustics*. Comstock, Ithaca.

Fenton, M. B., Fullard, J. H. 1981. Moth hearing and the feeding strategies of bats. *Am. Sci.* **69:** 266–275.

Fuchs, S. 1976. The response to vibrations of the substrate and reactions to the specific drumming in colonies of carpenter ants (*Camponotus,* Formicidae, Hymenoptera). *Behav. Ecol. Sociobiol.* **1:** 155–184.

Fullard, J. H., Fenton, M. B., Simmons, J. A. 1979. Jamming bat echolocation: The clicks of arctiid moths. *Canad. J. Zool.* **57:** 647–649.

Fullard, J. H., Yack, J. E. 1993. The evolutionary biology of insect hearing. *TREE* **8:** 248–252.

Gopfert, M. C., Wasserthal, L. T. 1999. Hearing with the mouthparts: Behavioural responses and the structural basis of ultrasound perception in acherontiine hawkmoths. *J. Exp. Biol.* **202:** 909–918.

Haskell, P. T. 1961. *Insect sounds*. Quadrangle Books, Chicago.

Huber, F. 1974. Sensory and neuronal mechanisms underlying acoustic communication in Orthopteran insects. *Adv. Behav. Biol.* **15:** 55–97.

Huber, F., Thorson, J. 1985. Cricket auditory communication. *Sci. Am.* **253:** 60–68.

McIver, S. B. 1985. Mechanoreception. In *Comprehensive insect physiology biochemistry and pharmacology* (G. A. Kerkut, L. I. Gilbert, Eds.), vol. 6, pp. 71–132. Pergamon Press, Oxford.

McKeever, S. 1977. Observations of *Corethrella* feeding on tree frogs. *Mosq. News* **37:** 522–523.

Michelsen, A., Fink, F., Gogala, M., Traue, D. 1982. Plants as transmission channels for insect vibrational songs. *Behav. Ecol. Sociobiol.* **11:** 269–281.

Michelsen, A., Nocke, H. 1974. Biophysical aspects of sound communication in insects. *Adv. Insect Physiol.* **10:** 247–296.

Nelson, M. C. 1979. Sound production in the cockroach *Gromphadorhina portentosa*: The sound-producing apparatus. *J. Comp. Physiol. A* **132:** 27–38.

Nelson, M. C., Fraser, J. 1980. Sound production in the cockroach *Gromphadorhina portentosa*: Evidence for communication by hissing. *Behav. Ecol. Sociobiol.* **6:** 305–314.

Pearman, J. V. 1928. On sound production in the Psocoptera and on a presumed stridulatory organ. *Entomol. Mon. Mag.* **14:** 179–186.

Robert, D., Amoroso, J., Hoy, R. R. 1992. The evolutionary convergence of hearing in a parasitoid fly and its cricket host. *Science* **258:** 1135–1137.

Romer, H., Bailey, W. J. 1990. Insect hearing in the field. *Comp. Biochem. Physiol. A* **97:** 443–447.

Roth, L. M. 1948. A study of mosquito behavior. An experimental laboratory study of the sexual behavior of *Aedes aegypti* (Linnaeus). *Am. Midl. Nat.* **40:** 265–352.

Roth, L. M., Hartmann, H. B. 1967. Sound production and its evolutionary significance in Blattaria. *Ann. Entomol. Soc. Am.* **60:** 740–752.

Saini, R. K. 1985. Sound production associated with sexual behaviour of the tsetse, *Glossina morsitans morsitans*. *Insect Sci. Appl.* **6:** 637–644.

Sickmann, T., Kalmring, K., Muller, A. 1997. The auditory-vibratory system of the bushcricket *Polysarcus denticauda* (Phaneropterinae, Tettigoniidae). I. Morphology of the complex tibial organs. *Hear. Res.* **104:** 155–166.

Simmons, P., Young, D. 1978. The tymbal mechanism and song patterns of the bladder cicada, *Cystosoma saundersii*. *J. Exp. Biol.* **76:** 27–45.

Sismondo, E. 1980. Physical characteristics of the drumming of *Meconema thalassinum*. *J. Insect Physiol.* **26:** 209–212.

Skals, N., Surlykke, A. 1999. Sound production by abdominal tymbal organs in two moth species: The green silver-line and the scarce silver-line (Noctuoidea: Nolidae: Chloephorinae). *J. Exp. Biol.* **202:** 2937–2949.

Stumpner, A., Von Helversen, D. 2001. Evolution and function of auditory systems in insects. *Naturwissenschaften* **88:** 159–170.

Young, D., H. Bennet-Clark, C. 1995. The role of the tymbal in cicada sound production. *J. Exp. Biol.* **198:** 1001–1019.

Visual Communication

Blest, A. D. 1957. The function of eyespot patterns in the Lepidoptera. *Behavior* **11:** 209–256.

Buck, J. 1966. Unit activity in the firefly lantern. In *Bioluminescence in progress* (F. J. Johnson, Y. Haneda, Ed.), pp. 459–474. Princeton University Press, Princeton.

Buck, J. B. 1948. The anatomy and physiology of the light organ in fireflies. *Ann. NY Acad. Sci.* **49:** 397–482.

Burkhardt, D. 1983. Wavelength perception and colour vision. *Symp. Soc. Exp. Biol.* **36:** 371–397.

Case, J. F. 1984. Vision in mating behaviour of fireflies. In *Insect communication* (T. Lewis, Ed.), pp. 195–222. Academic Press, London.

Chase, M. R., Bennett, R. R., White, R. H. 1997. Three opsin-encoding cDNAs from the compound eye of *Manduca sexta*. *J. Exp. Biol.* **200:** 2469–2478.

Corrette, B. J. 1990. Prey capture in the praying mantis *Tenodera aridifolia* sinensis: Coordination of the capture sequence and strike movements. *J. Exp. Biol.* **148:** 147–180.

Ghiradella, H. 1977. Fine structure of the tracheoles of the lantern of a photurid firefly. *J. Morphol.* **153:** 187–204.

Goldsmith, T. H., Fernandez, H. R. 1968. The sensitivity of housefly photoreceptors in the mid-ultraviolet and the limits of the visible spectrum. *J. Exp. Biol.* **49:** 669–677.

Kral, K., Vernik, M., Devetak, D. 2000. The visually controlled prey-capture behaviour of the European mantispid *Mantispa styriaca. J. Exp. Biol.* **203:** 2117–2123.

Lall, A. B., Lord, E. T., Trouth, C. O. 1985. Electrophysiology of the visual system in the cricket *Gryllus firmus* (Orthoptera: Gryllidae): Spectral sensitivity of the compound eyes. *J. Insect Physiol.* **31:** 353–357.

Lall, A. B., Worthy, K. M. 2000. Action spectra of the female's response in the firefly *Photinus pyralis* (Coleoptera: Lampyridae): Evidence for an achromatic detection of the bioluminescent optical signal. *J. Insect Physiol.* **46:** 965–968.

Land, M. F. 1992. Visual tracking and pursuit: Humans and arthropods compared. *J. Insect Physiol.* **38:** 939–951.

Land, M. F. 1997. Visual acuity in insects. *Annu. Rev. Entomol.* **42:** 147–177.

Lewis, T. 1984. The elements and frontiers of insect communication. In *Insect communication* (T. Lewis, Ed.), pp. 1–27. Academic Press, London.

Lloyd, J. E. 1983. Bioluminescence and communication in insects. *Annu. Rev. Entomol.* **28:** 131–160.

Mager, H. I. X., Tu, S.-C. 1995. Chemical aspects of bioluminescence. Photochem. *Photobiol.* **62:** 607–614.

McElroy, W. D., DeLuca, M. 1985. Biochemistry of insect luminescence. In *Comprehensive insect physiology biochemistry and pharmacology* (G. A. Kerkut, L. I. Gilbert, Eds.), vol. 4, pp. 553–563. Pergamon Press, Oxford.

McElroy, W. D., DeLuca, M. A. 1983. Firefly and bacterial luminescence: Basic science and applications. *J. Appl. Biochem.* **5:** 197–209.

Muir, L. E., Kay, B. H., Thorne, M. J. 1992. *Aedes aegypti* (Diptera: Culicidae) vision: Response to stimuli from the optical environment. *J. Med. Entomol.* **29:** 445–450.

Peterson, M. K., Buck, J. 1968. Light organ fine structure in certain asiatic fireflies. *Biol. Bull.* **135:** 335–348.

Prokopy, R. J., Owens, E. D. 1983. Visual detection of plants by herbivorous insects. *Annu. Rev. Entomol.* **28:** 337–364.

Rossel, S. 1980. Foveal fixation and tracking in the praying mantis. *J. Comp. Physiol. A* **139:** 307–331.

Rossel, S. 1986. Binocular spatial localization in the praying mantis. *J. Exp. Biol.* **120:** 265–281.

Smith, D. F. 1963. The organization and innervation of the luminescent organ in a firefly *Photuris pennsylvanica* (Coleoptera). *J. Cell Biol.* **16:** 323–359.

Trimmer, B. A., Aprille, J. R., Dudzinski, D. M., Lagace, C. J., Lewis, S. M., Michel, T., Qazi, S., Zayas, R. M. 2001. Nitric oxide and the control of firefly flashing. *Science* **292:** 2486–2488.

Wilson, T., Hastings, J. W. 1998. Bioluminescence. *Annu. Rev. Cell Dev. Biol.* **14:** 197–230.

Wood, K.V. 1995. The chemical mechanism and evolutionary development of beetle bioluminescence. Photochem. *Photobiol.* **62:** 662–673.

Yamawaki, Y. 2000. Saccadic tracking of a light grey target in the mantis, *Tenodera aridifolia. J. Insect Physiol.* **46:** 203–210.

Chemical Communication

Albert, S., Bhattacharya, D., Klaudiny, J., Schmitzova, J., Simuth, J. 1999. The family of major royal jelly proteins and its evolution. *J. Mol. Evol.* **49:** 290–297.

Ali, M. F., Morgan, E. D. 1990. Chemical communication in insect communities: A guide to insect pheromones with special emphasis on social insects. *Biol. Rev.* **65:** 227–247.

Ando, T., Kasuga, K., Yajima, Y., Kataoka, H., Suzuki, A. 1996. Termination of sex pheromone production in mated females of the silkworm moth. *Arch. Insect Biochem. Physiol.* **31:** 207–218.

Averill, A. L., Prokopy, R. J. 1987. Residual activity of oviposition-deterring pheromone in *Rhagoletis pomonella* (Diptera: Tephritidae) and female response to infested fruit. *J. Chem. Ecol.* **13:** 167–177.

Billen, J. 1991. Ultrastructural organization of the exocrine glands in ants. *Ethol. Ecol. Evol.* **1:** 67–73.

Billen, J., Morgan, E. D. 1998. Pheromone communication in social insects: Sources and secretions. In *Pheromone communication in social insects* (R. K. Vander Meer, M. D. Breed, K. E. Espelie, M. L. Winston, Eds.), pp. 3–33. Westview Press, Boulder, CO.

Birch, M. C., Poppy, G. M. 1990. Scents and eversible scent structures of male moths. *Annu. Rev. Entomol.* **35:** 25–58.

Blum, M. S. 1969. Alarm pheromones. *Annu. Rev. Entomol.* **14:** 57–80.

Blum, M. S. 1987. Biosynthesis of arthropod exocrine compounds. *Annu. Rev. Entomol.* **32:** 381–414.

Breed, M. D. 1981. Individual recognition and learning of queen odors by worker honeybees. *Proc. Natl. Acad. Sci. USA* **78:** 2635–2637.

Breed, M. D. 1998. Recognition pheromones of the honey bee. *BioScience* **48:** 463–470.

Butler, C. G., Fletcher, D. J. C., Watler, D. 1969. Nest entrance marking with pheromones by the honeybee *Apis mellifera* L., and by a wasp, *Vespula vulgaris L. Anim. Behav.* **17:** 142–147.

Cardé, R. T. 1984. Chemo-orientation in flying insects. In *Chemical ecology of insects* (W. J. Bell, R. T. Cardé, Eds.), pp. 111–124. Chapman and Hall, London.

Cardé, R. T., Baker, T. C., Roelofs, W. L. 1975. Behavioral role of individual components of a multichemical attractant system in the Oriental fruit moth. *Nature* **253:** 348–349.

Cardé, R. T., Cardé, A. M., Hill, A. S., Roelofs, W. C. 1977. Sex pheromone specificity as a reproductive isolating mechanism among the sibling species *Archips argyrospilus* and *A. mortuanus* and other sympatric tortricine moths (Lepidoptera: Tortricidae). *J. Chem. Ecol.* **8:** 1207–1215.

Conner, W. E., Boada, R., Schroeder, F. C., Gonzalez, A., Meinwald, J., Eisner, T. 2000. Chemical defense: Bestowal of a nuptial alkaloidal garment by a male moth on its mate. *Proc. Natl. Acad. Sci. USA* **97:** 14,406–14,411.

Dettner, K., Liepert, C. 1994. Chemical mimicry and camouflage. *Annu. Rev. Entomol.* **39:** 129–154.

Dussourd, D. E., Ubik, K., Harvis, C., Resch, J., Meinwald, J., Eisner, T. 1988. Biparental defensive endowment of eggs with acquired plant alkaloid in the moth *Utetheisa ornatrix. Proc. Natl. Acad. Sci. USA* **85:** 5992–5996.

Eisner, T. 1958. The protective role of the spray mechanism of the bombardier beetle, *Brachynus ballistarius* lec. *J. Insect Physiol.* **2:** 215–220.

Eisner, T., Aneshansley, D. J., Eisner, M., Attygalle, A. B., Alsop, D. W., Meinwald, J. 2000. Spray mechanism of the most primitive bombardier beetle (*Metrius contractus*). *J. Exp. Biol.* **203:** 1265–1275.

Eisner, T., Eisner, M., Rossini, C., Iyengar, V. K., Roach, B. L., Benedikt, E., Meinwald, J. 2000. Chemical defense against predation in an insect egg. *Proc. Natl. Acad. Sci. USA* **97:** 1634–1639.

Eisner, T., Goetz, M. A., Hill, D. E., Smedley, S. R., Meinwald, J. 1997. Firefly "femmes fatales" acquire defensive steroids (lucibufagins) from their firefly prey. *Proc. Natl. Acad. Sci. USA* **94:** 9723–9728.

Eisner, T., Hicks, K., Eisner, M., Robson, D. S. 1978."Wolf-in-sheep's clothing" strategy of a predaceous insect larva. *Science* **199:** 790–794.

Eisner, T., Meinwald, J. 1995. The chemistry of sexual selection. *Proc. Natl. Acad. Sci. USA* **92:** 50–55.

Eisner, T., Ziegler, R., Mccormick, J. L., Eisner, M., Hoebeke, E. R., Meinwald, J. 1994. Defensive use of an acquired substance (carminic acid) by predaceous insect larvae. *Experientia* **50:** 610–615.

Gibson, R. W., Pickett, J. A. 1983. Wild potato repels aphids by release of aphid alarm pheromone. *Nature* **302:** 608–609.

Hölldobler, B., Morgan, E. D., Oldham, N. J., Liebig, J. 2001. Recruitment pheromone in the harvester ant genus *Pogonomyrmex. J. Insect Physiol.* **47:** 369–374.

Howard, R. W., Akre, R. D. 1995. Propaganda, crypsis, and slave-making. In *Chemical ecology of insects 2* (R. T. Cardé, W. J. Bell, Eds.), pp. 364–424. Chapman and Hall, New York.

Jacobson, M. 1974. Insect pheromones. In *The physiology of insecta* (M. Rockstein, Ed.), vol. 3, pp. 229–276. Academic Press, New York.

Karlson, P., Butenandt, A. 1959. Pheromones (Ectohormones) in insects. *Annu. Rev. Entomol.* **4:** 39–58.

Le Conte, Y., Arnold, G., Trouiller, J., Masson, C., Chappe, B., Ourisson, G. 1989. Attraction of the parasitic mite *Varroa* to the drone larvae of honey bees by simple aliphatic esters. *Science* **245:** 638–639.

Lenoir, A., D'ettorre, P., Errard, C., Hefetz, A. 2001. Chemical ecology and social parasitism in ants. *Annu. Rev. Entomol.* **46:** 573–599.

Mas, E., Lloria, J., Quero, C., Camps, F., Fabrias, G. 2000. Control of the biosynthetic pathway of *Sesamia nonagrioides* sex pheromone by the pheromone biosynthesis activating neuropeptide. *Insect Biochem. Mol. Biol.* **30:** 455–459.

Masler, E. P., Raina, A. K., Wagner, R. M., Kochansky, J. P. 1994. Isolation and identification of a pheromonotropic neuropeptide from the brain-suboesophageal ganglion complex of *Lymantria dispar.* A new member of the PBAN family. *Insect Biochem. Mol. Biol.* **24:** 829–836.

Meinwald, J., Eisner, T. 1995. The chemistry of phyletic dominance. *Proc. Natl. Acad. Sci. USA* **92:** 14–18.

Mohammedi, A., Crauser, D., Paris, A., Le Conte, Y. 1996. Effect of a brood pheromone on honeybee hypopharyngeal glands. *C. R. Acad. Sci. III* **319:** 769–772.

Nordlund, D. A., Lewis, W. J. 1976. Terminology of chemical releasing stimuli in intraspecific and interspecific interactions. *J. Chem. Ecol.* **2:** 211–220.

Norris, M. J. 1964. Accelerating and inhibiting effects of crowding on sexual maturation in two species of locusts. *Nature* **203:** 784–785.

Oldham, N. J., Morgan, E. D., Gobin, B., Billen, J. 1994. First identification of a trail pheromone of an army ant (*Aenictus* species). *Experientia* **50:** 763–765.

Pankiw, T., Winston, M. L., Slessor, K. N. 1994. Variation in worker response to honey bee (*Apis mellifera* L.) queen mandibular pheromone. *J. Insect Behav.* **7:** 1–15.

Plettner, E., Otis, G. W., Wimalaratne, P. D. C., Winston, M. L., Slessor, K. N., Pankiw, T., Punchihewa, P. W. K. 1997. Species- and caste-determined mandibular gland signals in honeybees (*Apis*). *J. Chem. Ecol.* **23:** 363–377.

Prestwich, G. D. 1985. Communication in insects. II. Molecular communication of insects. *Q. Rev. Biol.* **60:** 437–456.

Prestwich, G. D. 1993. Chemical studies of pheromone receptors in insects. *Arch. Insect Biochem. Physiol.* **22:** 75–86.

Raina, A. K. 1993. Neuroendocrine control of sex pheromone biosynthesis in Lepidoptera. *Annu. Rev. Entomol.* **38:** 329–349.

Raina, A. K., Menn, J. J. 1993. Pheromone biosynthesis activating neuropeptide: From discovery to current status. *Arch. Insect Biochem. Physiol.* **22:** 141–151.

Ramaswamy, S. B., Jurenka, R. A., Linn, C. E., Roelofs, W. L. 1995. Evidence for the presence of a pheromonotropic factor in hemolymph and regulation of sex pheromone production in *Helicoverpa zea. J. Insect Physiol.* **41:** 501–508.

Ramaswamy, S. B., Mbata, G. N., Cohen, N. E., Moore, A., Cox, N. M. 1994. Pheromonotropic and pheromonostatic activity in moths. *Arch. Insect Biochem. Physiol.* **25:** 301–315.

Rasmussen, L. E., Lee, T. D., Roelofs, W. L., Zhang, A., Daves, Jr., G. D. 1996. Insect pheromone in elephants. *Nature* **379:** 684.

Renou, M., Guerrero, A. 2000. Insect parapheromones in olfaction research and semiochemical-based pest control strategies. *Annu. Rev. Entomol.* **45:** 605–630.

Renwick, J. A. A., Dickens, J. C. 1979. Control of pheromone production in the bark beetle, *ips cembrae. Physiol. Entomol.* **4:** 377–381.

Roelofs, W. L. 1995. Chemistry of sex attraction. *Proc. Natl. Acad. Sci. USA* **92:** 44–49.

Roelofs, W. L., Hill, A., Cardé, R. T. 1975. Sex pheromone components of the redbanded leafroller, *Argyrotaenia velutinana. J. Chem. Ecol.* **1:** 83–89.

Rothschild, M. 1965. The rabbit flea and hormones. *Endeavor* **24:** 162–167.

Schildknecht, H. 1970. The defensive chemistry of land and water beetles. *Angew. Chem. Int. Ed. Engl.* **9:** 1–9.

Schlein, Y., Galun, R., Ben-Eliahu, M. N. 1981. Abstinons. Male-produced deterrents of mating in flies. *J. Chem. Ecol.* **7:** 285–290.

Schmitzová, J., Klaudiny, J., Albert, S., Schröder, W., Schreckengost, W., Hanes, J., Júdova, J., Simúth, J. 1998. A family of major royal jelly proteins of the honeybee *Apis mellifera* L. *Cell. Mol. Life Sci.* **54:** 1020–1030.

Schröder, F. C., Farmer, J. J., Attygalle, A. B., Smedley, S. R., Eisner, T., Meinwald, J. 1998. Combinatorial chemistry in insects: A library of defensive macrocyclic polyamines. *Science* **281:** 428–431.

Slessor, K. N., Kaminski, L. A., King, G. G. S., Borden, J. H., Winston, M. L. 1988. Semiochemical basis of the retinue response to queen honey bees. *Nature* **332:** 354–356.

Teal, P. E. A., Davis, N. T., Meredith, J. A., Christensen, T. A., Hildebrand, J. G. 1999. Role of the ventral nerve cord and terminal abdominal ganglion in the regulation of sex pheromone production in the tobacco budworm (Lepidoptera: Noctuidae). *Ann. Entomol. Soc. Am.* **92:** 891–901.

Tillman, J. A., Seybold, S. J., Jurenka, R. A., Blomquist, G. J. 1999. Insect pheromones—An overview of biosynthesis and endocrine regulation. *Insect Biochem. Mol. Biol.* **29:** 481–514.

Vargo, E. L., Hulsey, C. D. 2000. Multiple glandular origins of queen pheromones in the fire ant *Solenopsis invicta*. *J. Insect Physiol.* **46:** 1151–1159.

Whittaker, R. H., Feeny, P. P. 1971. Allelochemics: chemical interactions between species. *Science* **171:** 757–770.

Wilson, E. O. 1965. Chemical communication in the social insects. *Science* **149:** 1064–1071.

Wilson, E. O., Bossert, W. H. 1963. Chemical communication among animals. *Recent Progr. Horm. Res.* **19:** 673–716.

Wilson, E. O., Durlach, N., Reth, L. M. 1958. Chemical releasers of necrophoric behaviour in ants. *Psyche* **65:** 108–114.

Yeargan, K. V. 1994. Biology of bolas spiders. *Annu. Rev. Entomol.* **39:** 81–99.

Physiology of Communication

Alcock, J. 1979. *Animal behavior: An evolutionary approach.* Sinauer, Sunderland, MA.

Bloch, G., Hefetz, A., Hartfelder, K. 2000. Ecdysteroid titer, ovary status, and dominance in adult worker and queen bumble bees (*Bombus terrestris*). *J. Insect Physiol.* **46:** 1033–1040.

Breed, M. D., Gamboa, G. J. 1977. Behavioral control of workers by queens in primitively eusocial bees. *Science* **195:** 694–696.

Evans, J. D., Wheeler, D. E. 1999. Differential gene expression between developing queens and workers in the honey bee, *Apis mellifera. Proc. Natl. Acad. Sci. USA* **96:** 5575–5580.

Fletcher, D. J. C., Ross, K. G. 1985. Regulation of reproduction in eusocial Hymenoptera. *Annu. Rev. Entomol.* **30:** 319–343.

Giray, T., Robinson, G. E. 1996. Common endocrine and genetic mechanisms of behavioral development in male and worker honey bees and the evolution of division of labor. *Proc. Natl. Acad. Sci. USA* **93:** 11,718–11,722.

Hartfelder, K., Engels, W. 1998. Social insect polymorphism: Hormonal regulation of plasticity in development and reproduction in the honeybee. *Curr. Top. Dev. Biol.* **40:** 45–77.

Haynes, K. F., Yeargan, K. V. 1999. Exploitation of intraspecific communication systems: Illicit signalers and receivers. *Ann. Entomol. Soc. Am.* **92:** 960–970.

Hölldobler, B. 1999. Multimodal signals in ant communication. *J. Comp. Physiol. A* **184:** 129–141.

Kirchner, W. H., Towne, W. F. 1994. The sensory basis of the honeybee's language. *Sci. Am.* **270:** 74–80.

Lewis, T. 1984. The elements and frontiers of insect communication. In *Insect communication* (T. Lewis, Ed.), pp. 1–27. Academic Press, London.

Nieh, J. C. 1999. Stingless-bee communication. *Am. Sci.* **87:** 428–435.

Nijhout, H. F. 1999. Control mechanisms of polyphenic development in insects. *BioScience* **49:** 181–192.

Robinson, G. E. 1992. Regulation of division of labor in insect societies. *Annu. Rev. Entomol.* **37:** 637–665.

Robinson, G. E., Vargo, E. L. 1997. Juvenile hormone in adult eusocial Hymenoptera: Gonadotropin and behavioral pacemaker. *Arch. Insect Biochem. Physiol.* **35:** 559–583.

Stich, H. F. 1963. An experimental analysis of the courtship pattern of *Tipula oleracea* (diptera). *Canad. J. Zool.* **41:** 99–109.

Tumlinson, J. H., Teal, P. E. A. 1982. The sophisticated language of insect chemical communication. *J. Ga. Entomol. Soc.* **17:** 11–23.

Glossary

A band The dark band within a muscle sarcomere that is formed from the actin proteins.

Accessory gland Glands that are part of the male and female reproductive systems. In the female, they secrete cement, venoms and lubricants. In the male, they produce the seminal fluid, spermatophore, and active peptides that affect the female.

Accessory pulsatile organ An accessory heart at the base of the wings, legs, and antennae that supplements the movement of hemolymph from the body cavity.

Acetylcholine A neurotransmitter that is released from the presynaptic neuron and binds to receptors on the postsynaptic neuron of the synapse to perpetuate the action potential across neurons.

Acetylcholinesterase The enzyme present in the synapse that breaks down acetylcholine to acetate and choline, making the acetylcholine receptors available for the next release of neurotransmitter.

Acrosome The organelle at the tip of the sperm that is derived from the Golgi apparatus and breaks down the egg membrane to allow the sperm to penetrate for fertilization.

Actin A protein that makes up the muscle myofibrils. Together with myosin, it mediates muscle contraction.

Action potential A change of constant amplitude in the membrane potential of a neuron that is triggered by a depolarization of the dendrite. The action potential sweeps down the axon to the synapse.

Acyl carrier protein A protein linked to intermediates in the synthesis of fatty acids.

Adipohemocyte A type of insect hemocyte characterized by a relatively small nucleus and a cytoplasm containing abundant fat globules.

Adipokinetic hormone A hormone often produced by the corpora cardiacum that acts on the fat body to cause it to release stored lipids into the hemolymph.

Aedeagus The intromittent organ of copulation in the male.

Aeropyle A modification of the egg chorion that permits gaseous exchange.

Alarm pheromone A chemical produced by members of a species that induces a behavioral alarm or alertness in other individuals of that species.

Alary muscle Paired muscles that support the heart within the body cavity.

Allatostatin A hormone produced by the brain that inhibits juvenile hormone production by the corpus allatum.

Allatotropin A hormone produced by the brain that stimulates juvenile hormone production by the corpus allatum.

Ametabolous A type of development without any metamorphosis. Found in apterygotes where, except for size, adults and immatures closely resemble each other and have similar ecological habits.

Amniotic cavity A cavity formed by the amniotic folds during development. It encloses the germ band and frees it from the rest of the embryonic serosa.

Anabolism The metabolic production of protein, carbohydrates, and fats from ingested food.

Anthropomorphism The tendency to attribute human characteristics to nonhuman animals.

Apneustic A tracheal system that lacks spiracles.

Apolysis The separation of the epidermis from the overlying cuticle. Apolysis marks the beginning of the molt.

Apomictic parthenogenesis A form of reproduction in which the oocyte does not undergo any reduction division but remains diploid and the embryo develops as a clone of the mother.

Apposition eye A type of compound eye used by day-active insects in which each ommatidium is surrounded by pigment cells and isolated from the light that enters neighboring ommatidia.

Apyrene sperm Sperm that lack a nucleus.

Archicerebrum The primitive ganglionic mass that gave rise to the brain.

Arthrodial membrane The intersegmental membrane between two sclerotized plates that bestows flexibility to the cuticle. The arthrodial membrane typically consists of endocuticle covered by epicuticle, with little exocuticle.

Asynchronous muscle A specialized flight muscle that requires only periodic activation by the central nervous system. Once activated, the stretching by antagonistic muscles stimulates its contraction.

Atrium The chamber immediately below the spiracle.

Automictic parthenogenesis A type of reproduction that does not require fertilization by sperm. The egg is haploid but the diploid number is restored by fusion of the egg nucleus with a polar body.

Axoneme The organelle that propels the sperm by causing the flagellum to move. The axoneme consists of a series of microtubules.

Basement membrane The innermost layer of the integument that is secreted by hemocytes, forming a continuous layer of connective tissue that separates the body cavity from the integument.

Blastoderm A continuous layer of cells that surrounds the egg early during embryogenesis, derived from the energids that migrate to the periphery.

Blastokinesis A displacement of the embryo within the egg that enables it to utilize more yolk.

Bombyxin An insulin-like peptide that was first characterized as a "small" PTTH, but is not believed to have PTTH activity.

Brachial chamber A region within the rectum of larval dragonflies that contains tracheal gills that can extract oxygen from the water taken up through the anus.

Bursa copulatrix A modified pouch in the genital tract of some female insects into which sperm are first deposited before they migrate to the spermatheca.

Bursicon A hormone that mediates cuticular sclerotization.

Calyx The pedicel of each ovariole that leads to the lateral oviduct.

Campaniform sensillum A dome-shaped sensory receptor that measures cuticular distortion.

Cardioacceleratory peptide A peptide that regulates the rate of contraction of the dorsal vessel.

Carnitine A carrier molecule for acetyl CoA entering the mitochondrion from lipid metabolism.

Catabolism The metabolic processes that break down ingested organic molecules and release energy, captured by ATP, and waste products.

Cecropin A family of hemolymph proteins that mediate a humoral response to foreign invaders.

Cement layer The outermost layer of the epicuticle, produced by dermal glands and consisting of a shellac-like coating. The cement layer that seals the wax layer and provides protection from physical abrasion.

Chitin A polysaccharide consisting of linked *N*-acetyl-glucosamine residues. Chitin is insoluble in water, dilute alkali, alcohol, and organic solvents, but soluble in concentrated acids and hot alkali solutions. The chains of chitin are associated with protein, providing a framework for the stabilization of the cuticular proteins.

Chordotonal organ A subcuticular sensillum composed of units called scolopidia, attached to the cuticle at one or both ends. External movement distorts the scolopidium and causes the dendrite to depolarize.

Choriogenesis The synthesis and production of the chorion, or egg shell.

Chorion The outer shell of the insect egg. Its synthesis is one of the last acts of the follicular epithelium that surrounds the oocyte before it degenerates.

Chromosome puff Areas on the chromosome visible under light microscopy that are evidence of mRNA transcription.

Cibarium The area within the preoral cavity where food remains before it enters the mouth and the true digestive tract.

Circumesophageal connectives Nervous connections between the brain and the ventral nerve cord.

Citric acid cycle A series of aerobic biochemical reactions that ultimately degrade carbohydrates, fatty acids, and proteins to carbon dioxide and water and capture the energy differences in ATP.

Cleavage A series of mitotic divisions that transform the zygote into an embryo.

Corneagen cell Modified epidermal cells that produce the cornea of the compound eye.

Corneal lens The lens of the compound eye.

Corpora allata Ductless glands, usually found in pairs (singular: corpus allatum), that synthesize and release juvenile hormone.

Corpora cardiaca The neurohemal organ for the neurosecretory cells of the brain, also containing secretory cells of its own. The usually paired structures (singular: corpus cardiacum) store and release a number of neurohormones synthesized in the brain.

Corpora penduncluata The mushroom bodies of the protocerebral lobes of the brain. They contain abundant nerve cell perikarya and interneurons. Its size is correlated with behavioral complexity and is most highly developed in social Hymenoptera that display complex behaviors.

Countercurrent heat exchange The thermoregulatory mechanism that maintains a warm thorax. Cooler hemolymph from the abdomen is moved to the thorax at the same time that warm hemolymph from the thorax is moving backward. The warm hemolymph is used to preheat the cooler hemolymph before it enters the thorax.

Cryptonephridium A modification of the excretory system in insects that live in dry habitats. The distal ends of the Malpighian tubules are directly in contact with the rectum and remove water and salts before they are excreted.

Crystalline cone A hard transparent refractive structure of the compound eye.

Cutaneous respiration The ability to take up oxygen through the integument.

Cuticulin A layer of the epicuticle consisting of waxes.

Cystocyte A cell that arises from the division of a cystoblast.

Defensin A peptide that mediates humoral immunity.

Dermal gland A modified epidermal cell that produces the cement layer, as well as defensive secretions and pheromones.

Determination The nonvisible commitment of an undifferentiated cell to its differentiated state.

Deutocerebrum The middle portion of the insect brain that produces neurons that innervate the antennae.

Differentiation The visible physiological commitment of a cell to a specific developmental pathway.

Direct flight muscles Thoracic muscles that connect directly with the wing.

Diuretic hormone A hormone that increases the activity of the Malpighian tubules.

Diverticulum An invagination of the alimentary canal that produces a blind sac for the storage of ingested food.

Dorsal closure A stage of embryogenesis in which the embryonic ectoderm grows over the dorsal side of the embryo.

Dorsal diaphragm Connective tissue and muscle that form a dorsal compartment that separates the perivisceral and pericardial sinuses.

Dorsal ocelli The simple eyes usually found on the vertex of the head between the compound eyes.

Dorsal vessel The structure used for pumping hemolymph forward from the abdomen into the head. The dorsal vessel is divided into the posterior heart and anterior aorta.

Ecdysial line A programmed area of weakness in the cuticle that allows the insect to emerge when the endocuticle is sufficiently digested.

Ecdysiotropin A hormone that stimulates the production of ecdysteroids.

Ecdysis The process of shedding the old cuticle at the end of molting.

Ecdysis triggering hormone A hormone released from epitracheal glands that acts on the cells of the ventral nerve cord and causes the initial expression of preecdysis behavior and the release of eclosion hormone.

Ecdysteroid The collective term for derivatives of ecdysone. Ecdysteroids trigger mainly apolysis and vitellogenin production by affecting gene expression.

Eclosion The term used to describe the ecdysis of adult insects and the hatching of first instar larvae from the egg.

Eclosion hormone A peptide that initiates the stereotyped behaviors associated with ecdysis.

Ectoperiptrophic space The compartment created by the peritrophic membrane in the midgut.

Ectothermy Body temperature regulation using external sources, usually the sun.

Ejaculatory duct The common ectodermal duct uniting the vasa deferentia of the male reproductive system. Sperm travel from the vas deferens to the ejaculatory duct and into the female.

Embryonic primordium The thickened portion of the blastoderm that will develop into the embryo.

Endochorion A middle layer of the chorion that can be divided into an inner endochorion with a network of pillars and the outer endochorion that creates a roof network.

Endocuticle The innermost layer of the cuticle secreted by epidermal cells. It is unsclerotized and capable of being resorbed during the molting process.

Endocrine gland A gland that produces and secretes hormones within the body.

Endoperitrophic space A space within the gut that is created by the peritrophic membrane.

Endopterygote Insects that undergo a metamorphosis with their wings developing on the inside of the body.

Endoskeleton An invagination of the body wall that provides rigidity and attachment sites for muscles.

Endothermy The use of heat generated from flight muscles to warm the body.

Energids The daughter nuclei from the mitotic division of zygote nucleus during meroblastic cleavage, surrounded by an island of cytoplasm.

Epicuticle The thin, top layer of the cuticle, consisting of the inner and outer epicuticles, the wax layer, and the cement layer.

Epideictic pheromone A chemical substance produced by one individual that maintains the proper density among other individuals of that species.

Epidermis The single layer of cells that secretes the cuticle.

Epitracheal gland A group of neurosecretory cells associated with the trachea that produce the ecdysis-triggering hormone.

Esophagus A portion of the foregut that is differentiated as a simple tube that leads to the crop.

Esterase An enzyme that hydrolyzes esters.

Eupyrene sperm A type of sperm in Lepidoptera that contains a nucleus and is used to fertilize eggs.

Exochorion The proteinaceous outermost layer of the chorion.

Exocrine gland A gland that secretes chemicals to the outside of the body.

Exocuticle The outer layer of the procuticle that is sclerotized and incapable of resorption.

Exopterygote Insects that undergo a metamorphosis in which the wings develop as buds on the outside of the body wall.

Exoskeleton The hardened body wall of an insect.

Exuviae The old skin that is cast during the molting process. Always used as a plural noun.

Exuvial space The area created between the epidermal cells and the old cuticle by the digestion of the old endocuticle during the molting process.

Fast muscle fiber A type of muscle cell characterized by short sarcomeres, more extensive sarcoplasmic reticulum, and a ratio of actin:myosin filaments of 3:1. They produce a strong contraction when stimulated by excitatory fast neurons.

Fast neuron A type of neuron that releases a large number of neurotransmitter packets to cause a strong muscle contraction.

Fixed action pattern An innate stereotyped motor program.

Follicular epithelium The cells that surround the oocyte and sequester yolk proteins and synthesize the chorion.

Foregut The first of the three principal divisions of the insect alimentary canal. The foregut begins at the mouth and continues to the proventriculus.

Fovea A group of ommatidia in the center region of the compound eyes that are capable of higher spatial resolution.

Freeze tolerant Those insect species able to survive the formation of extracellular ice crystals by synthesizing ice-nucleating proteins that raise the supercooling point of body fluids and serve as catalysts for the extracellular nucleation of ice.

G proteins A family of signal-coupling proteins that act as intermediaries between activated membrane receptors and cellular effectors.

Gastric caecae Sac-like diverticulae at the anterior region of the midgut that are involved with water absorption and the creation of a flow within the midgut.

Gastrulation During embryonic development, the formation of the gastrula from endoderm.

Germ band The region of thickened blastoderm cells that becomes the embryo later in development.

Germarium The anterior regions of the male and female reproductive systems that produce oocytes from oogonia and spermatocytes from spermatogonia.

Glial cell An accessory cell that surrounds neurons and provides them with nourishment and insulation.

Glycerol 3-phosphate shuttle A biochemical pathway that regenerates NAD^+ from NADH by transferring hydrogen atoms between the cytoplasm and mitochondria.

Goblet cell A cell present in the midgut of larval lepidopterans that secretes potassium into the lumen and creates a flow of water in the gut.

H zone The area visible in the center of the A band of muscles consisting of myosin filaments without any overlapping actin.

Haploid parthenogenesis A type of reproduction in which the oocyte divides meiotically to form a haploid gamete that may develop with or without fertilization.

Hemimetabolous Insects that have a gradual or incomplete metamorphosis and the development of wings on the outside of the body wall.

Hemocoel The body cavity containing the internal organs and hemolymph.

Hemocyte One of a variety of different blood cells of insects.

Hindgut The posterior region of the alimentary canal derived from the proctodeal invagination.

Holoblastic A type of cleavage during embryogenesis in which the entire egg cell is divided.

Holometabolous The type of metamorphosis in which a complex change occurs between larvae and adults, involving a pupal stage. Wings develop on the outside of the body wall.

Holopneustic The condition of the respiratory system in which all spiracles are open and functional, usually involving two thoracic and eight abdominal spiracles.

Hormone A chemical produced by specialized tissues and released into the blood that affects target tissues elsewhere in the body.

Hypermetamorphosis Endopterygote insects whose larvae change form prior to pupation, producing different forms during the larval stage.

Hypocerebral ganglion A part of the stomodeal nervous system; a group of neurons that begins at the frontal ganglion and continues rearward to innervate the gut.

Hypopharynx A tongue-like structure that develops as a internal lobe of the head and often divides the preoral cavity into a cibarium and a salivarium.

I band The muscle band seen in longitudinal section that consists of only the actin filaments.

Indirect flight muscles The dorsoventral and dorsal longitudinal muscles in the thorax change the conformation of the thorax and indirectly move the wings.

Inner chorionic layer A thin proteinaceous inner layer of the chorion that is secreted above the vitelline envelope.

Integument The components of the exoskeleton, consisting of the basement membrane, epidermis, and cuticle.

Interneuron A neuron that lies entirely within a ganglion and serves as a switch to direct the stimuli from other neurons.

JH binding protein A protein that binds to JH and protects it from degradation while circulating in the hemolymph.

Johnston's organ A sensillum found in the second antennal segment that monitors antennal deflection.

Juvenile hormone A sesquiterpene that is produced by the corpus allatum and has a wide range of effects on metamorphosis and reproduction.

Kairomone An interspecific chemical message that benefits the receiver but not the emitter.

Kinesis A nondirectional response of an organism to a stimulus.

Lateral ocelli The simple eyes found in holometabolous larvae.

Lateral oviduct The tube that extends from the calyx of the ovary to the median oviduct, usually derived from embryonic mesoderm.

Lipophorin A protein carrier that binds to lipid molecules in the hemolymph, transporting them from the fat body to target tissues.

Luciferase The enzyme that acts on the substrate luciferin in the presence of oxygen and ATP to produce light.

Luciferin The molecule that is oxidized by luciferase to produce light.

Malpighian tubules The excretory organs of insects, arising at the junction of the midgut and hindgut.

Median oviduct The tube that connects the lateral oviducts and expels eggs out of the body of the female. The median oviduct is derived from ectodermal invaginations and may be modified into a genital pouch for the incubation of eggs.

Melanin A dark, organic pigment deposited by the epidermal cells.

Meroblastic A type of embryonic cleavage that divides only the nucleus but not the cytoplasm, ultimately producing a synctium of blastodermal cells.

Meroistic The type of ovariole that contains trophocytes, or nurse cells, to nourish the oocyte.

Mesocuticle A transitional layer of the cuticle between the endocuticle and the exocuticle.

Mesoderm The embryonic layer of tissue that is derived from the interaction between the ectoderm and endoderm.

Metabolism The process by which food is transformed into body tissues and waste products.

Metapneustic The type of respiratory system in which only the spiracles on the last abdominal segment are functional.

Micropyle A pore in the chorion of the egg through which sperm can enter for fertilization.

Midgut The central portion of the alimentary canal between the foregut and the hindgut where most digestion occurs. The midgut is derived from endodermal tissues.

Molting gel An enzyme produced by epidermal cells that digests the old endocuticle during the molting process. It is first secreted in an inactive form and activated once the new cuticle begins to be formed.

Motor neuron A neuron that innervates muscles.

Muscle fiber A muscle cell, often multinucleate and containing myofibrils within its cytoplasm.

Myofibril Contractile proteins embedded in the cytoplasm of muscle fibers that consist of the myofilaments, myosin and actin.

Myofilament The individual myosin or actin molecules that comprise the myofibrils in muscle.

Myogenic Muscle contractions that occur independently of nervous stimulation.

Myosin A large globular protein that comprises the thick myofilaments of muscle cells.

Myotropin A peptide that causes muscle contraction.

Nephrocyte Cells found mostly in the pericardial sinus that take up nonparticulate colloids and release metabolites into the hemolymph.

Neural superposition eye A type of compound eye that is optically apposition, but whose neurons are configured so that light from neighboring ommatidia make it functionally superpositioned.

Neuroblast Cells that give rise to neurons.

Neurohormone A hormone produced by neurosecretory cells and released from a neurohemal organ into the hemolymph.

Neuromodulator A neurotransmitter released at the synapse that modifies the conditions under which other nerve impulses are transmitted and received.

Neuropil The mass of neurons within a ganglion.

Neurosecretory cell A specialized neuron that produces hormones that are released into the hemocoel.

Neurotransmitter A chemical released at the neural synapse that enables the nervous activation to pass to an adjacent neuron.

Nidi Groups of regenerative cells in the midgut (singular, nidus).

Nodule A mechanism of cellular defense in which small foreign objects are surrounded by hemocytes and sequestered.

Odorant binding protein A peptide in the receptor lymph that binds to odorant molecules that enter through the pore tubule and shuttles the molecules to receptor sites.

Oenocyte A specialized hemocyte that usually resides between the basement membrane and the epidermal cells. Oenocytes appear to synthesize wax and possibly ecdysteroids.

Oenocytoid A hemocyte that resembles an oenocyte.

Oocyte The female gamete that differentiates from oogonia.

Oogonia The first stage in the differentiation of an oocyte from a primary female germ cell.

Ootheca The protein synthesized to cover an egg mass.

Operculum An area of programmed weakness on the insect chorion that allows the first instar larva to escape from the egg.

Optic lobes The lateral lobes of the protocerebrum that supply nerves to the compound eyes.

Ostia The slit-like valves in the heart that allow hemolymph to enter (Singular, ostium).

Ovariole The tubes that comprise the ovary and contain the oocytes as they develop into eggs.

Ovary The paired female reproductive organs.

Oviparity The most common form of insect development, in which the egg is fertilized as it is oviposited and develops outside the female's body.

Oviposition The passage of the egg to the outside of the female.

Ovoviparity A specialized form of development in which the egg is retained and incubated within the common oviduct. The egg hatches shortly after it is oviposited.

Ovulation The passage of the egg out of the ovariole and into the lateral oviduct.

Oxidation The energy releasing process of removing electrons from a substrate.

Paedogenesis The specialized form of reproduction in which the ovaries become mature during the immature stages and the larva produces eggs that develop parthenogenetically.

Panoistic An ovariole that lacks nurse cells and nourishes the oocyte through the follicular epithelium.

Pars intercerebralis The medial portion of the protocerebrum of the brain that often contains neurosecretory cell bodies.

Parthenogenesis Egg development that occurs without fertilization by the male gamete.

Patency The opening between follicle cells that permits the uptake of vitellogenin from the hemolymph.

PBAN Pheromone biosynthesis-activating neuropeptide, responsible for the activation of enzymes involved in pheromone synthesis.

Pericardial cell Specialized cells that lie at the side of the heart and filter the blood.

Pericardial sinus The sinus formed by the dorsal diaphragm that consists of the hemolymph cavity around the heart.

Perikaryon A nerve cell body that contains the nucleus.

Perineural sinus The sinus formed by the ventral diaphragm that consists of the hemolymph cavity around the ventral nerve cord.

Perineurium A sheath that surrounds a group of neurons.

Peripheral nervous system The system of nerves consisting largely of sensory receptors.

Periplasm The region of the oocyte cytoplasm that lies just below the vitelline membrane.

Peritreme The cuticular sclerite that surrounds a spiracle.

Peritrophic matrix A delicate chitinous membrane secreted by midgut cells that protects the cells and forms digestive compartments within the gut.

Perivisceral sinus The sinus formed by the dorsal and ventral diaphragms that consists of the hemolymph cavity around the digestive tract.

Phagocyte A blood cell that engulfs and consumes foreign bodies in the hemolymph.

Pharate The state of the instar after apolysis but before ecdysis; the instar concealed by the old cuticle.

Phasic response A response of a neuron or whole sensillum in which an action potential is generated only when the sensillum is distorted.

Pheromone A chemical produced by an individual of one species that mediates behavior of another individual of the same species.

Phragma An epidermal invagination producing an internal ridge for muscle attachment.

Phytoecdysteroid Ecdysteroids that are produced and derived from plants.

Plasma The noncellular liquid portion of the hemolymph.

Plasmatocyte The basic form of insect hemocyte. Large, abundant insect blood cells.

Plastron A physical gill found in aquatic insects in which a bubble of air is held in place by hydrofuge hairs and allow oxygen to be extracted from the water.

Plectrum A part of a stridulatory apparatus consisting of a structure that is rubbed against a membrane.

Pole cells Energids that migrate to the posterior pole of the egg early during development and ultimately differentiate into the germ cells of the adult.

Polyembryony The production of several embryos from a single egg, a form of reproduction occurring in some parasitic hymenopterans and Strepsiptera.

Polyphenism The presence of several different phenotypes in a population that are determined by environmental factors and not genotype.

Polytrophic meroistic A type of ovariole in which nurse cells accompany the oocytes within the follicle and supply it with nutrients.

Pore canal Small channels that extend from epidermal cells through the cuticle up to the epicuticle and carry substances such as waxes.

Pre-oral cavity The space enclosed by the mouthparts, forming a cavity in front of the true mouth, which is the beginning of the alimentary canal.

Primary sense cell A sensillum that is directly innervated by the nervous system without synapsing first with an interneuron.

Primary urine The waste products produced by the Malpighian tubules and emptied into the hindgut. The primary urine is modified by the rectum and excreted as secondary urine.

Primer pheromone A chemical produced by an individual of one species that has a fundamental physiological effect on another individual of the same species.

Proctodeum The tissue formed by the epidermal invagination at the posterior end of the embryo that produces the hindgut.

Procuticle The undifferentiated chitinous cuticle that develops into the endocuticle and exocuticle.

Prohemocyte Small, round hemocytes that give rise to other hemocyte types.

Propneustic The configuration of the tracheal system in which only the anterior spiracles are open and functional.

Proprioreception The perception of where an organism's body parts are immediately located.

Prothoracic gland The endocrine gland that synthesizes and secretes ecdysteroids.

Prothoracicotropic hormone The hormone produced by neurosecretory cells in the brain that activate the prothoracic gland to synthesize ecdysteroids.

Protocerebrum The anterior and most complex lobe of the brain, subdivided into protocerebral and optic lobes.

Proventriculus A specialized area of the posterior region of the crop that serves as a valve for the passage of food into the midgut, and can be modified into a gizzard for grinding food prior to its digestion.

Receptor potential The change in the membrane potential of a sensillum after stimulation. The potential change is proportional to the strength of the stimulus.

Rectal pad Specialized cells in the rectum involved in the uptake of materials from the lumen to the hemolymph.

Rectum The posterior portion of the hindgut, modified to resorb fluids from the lumen and return some of the needed components to the hemolymph.

Reduction The energy-storing process that involves the addition of electrons to a substance.

Reflexive bleeding The defensive release of hemolymph that contains distasteful substances through intersegmental membranes to avoid predation.

Releaser pheromone A chemical released by an individual of one species that has an immediate effect on releasing some behavior of other individuals of that same species.

Resting potential The normal steady membrane potential in an unstimulated neuron.

Retinula cell A cell comprising the ommatidia of the compound eye that contains optically active pigments and is stimulated upon the reception of light.

Reynolds number A number that expresses the relationship between the size of an organism and the physical forces acting upon it. The Reynolds number is a ratio of inertial and frictional forces.

Rhabdom The optically active structure of a light receptive sensillum, consisting of the individual rhabdomeres of adjacent retinula cells.

Rhodopsin A protein pigment capable of absorbing a photon of light and transferring the energy to other biological molecules.

Ring gland The composite endocrine gland of larval dipterans consisting of the prothoracic gland, corpus allatum, and corpus cardiacum.

Salivarium The posterior portion of the preoral cavity into which the salivary duct empties.

Sarcolemma The cell membrane of a muscle cell.

Sarcomere A unit of muscular contraction between the two Z lines, containing the myofibrils.

Sarcoplasm The cytoplasm of a muscle cell.

Scolopale A rod-like capsule that covers the distal end of a sense cell.

Second messenger A substance that acts inside the cell to alter the rate of biological processes once it is activated by a hormone that acts on the outside of the cell.

Secondary urine The urine after it is processed by the rectum and eliminated from the body.

Seminal fluid The product of the male accessory glands that serves as a transport medium for sperm and contains physiologically active substances that can affect the physiology of the female that receives it.

Seminal vesicle The enlarged area of the vas deferens of the male reproductive system that serves as a storage reservoir for sperm.

Semiochemical A chemical that is involved in the communication between two organisms.

Semper cell A cell of the ommatidium that produces the crystalline cone.

Sensillum A sense organ.

Sensillum lymph The hemolymph that bathes the dendrite of a sensillum.

Sensory neuron The nerve cells that innervate sensory organs.

Sensory transduction The process of transforming environmental energy into biological energy.

Sex pheromone A chemical released from a gland of an individual of one species and causes members of the opposite sex of that species to orient towards it for mating.

Slow muscle fiber A type of skeletal muscle cell that contains reduced sarcoplasmic reticulum, longer sarcomeres, and a 6:1 ratio of actin:myosin filaments. They have a relatively slow rate of relaxation because they permit calcium to remain in the sarcoplasm longer.

Somatic mesoderm One of the mesodermal bands that develops during embryogenesis and gives rise to skeletal muscles, fat body, and a portion of the reproductive system.

Spermatheca A special sac in the female reproductive system that receives, stores, and releases sperm.

Spermatid A male reproductive cell arising from the division of spermatocytes that contains the haploid chromosome number. The spermatid ultimately differentiates into a spermatozoan.

Spermatocytes A male reproductive cell that arises from the division of a male germ cell, the spermatogonium.

Spermatogonia The primoridial male germ cell.

Spermatophore A secretion of the male accessory glands that surrounds the sperm and protects it during transit to the female.

Spiracle The usually paired aperature in the integument that serves as the opening of the tracheal system to allow gaseous exchange.

Spiracular gill An extension of the cuticle that surrounds a spiracle, forming a gill that allows both aquatic and terrestrial respiration to occur.

Splanchic mesoderm One of the two mesodermal bands that develop during embryogenesis, giving rise to the visceral muscles.

Stomodeum The anterior epidermal invagination during embryogenesis that produces the foregut of the alimentary tract.

Storage excretion The ability to sequester metabolic wastes in the body.

Storage hexamerin One of a family of proteins that act primarily as storage proteins to provide amino acids required for protein synthesis in those developmental phases that do not feed.

Stylet sheath The salivary secretion of phytophagous hemipterans that seals the mouthparts around the plant surface.

Subgenual organ A chordotonal organ usually located in the tibia, attached at one end to the cuticle and at the other to a trachea.

Subesophageal ganglion The first ganglion of the ventral nerve cord, consisting of the fused ganglia of the mandibular, maxillary and labial segments.

Superposition eye A variety of compound eye found in insects that maneuver under low light conditions. The movement of screening pigments allows light from neighboring ommatidia to stimulate other ommatidia.

Synchronous muscle A primitive wing muscle that requires nervous stimulation for each contraction.

Syngamy The fusion of male and female gametes to form a zygote.

Synomone A chemical produced by an individual of one species and received by another that has an adaptive effect on both.

Tachykinin A peptide that stimulates the contraction of visceral muscles, homologous with the vertebrate peptides.

Taenidia The spiral thickenings of the tracheal epicuticle that prevent the collapse of the tracheal tubes from air pressure.

Taxis An oriented response to a stimulus.

Temporal polyethism The changes in behavior that accompany aging.

Testis The male reproductive organ that produces sperm, usually consisting of a pair of testes.

Thecogen cell An accessory cell that produces the sheath of the sensillum that isolates the axons from one another and provides them with nutrients.

Tokus A hemolymph compartment at the tip of the abdomen containing tufts of aerating trachea that oxygenate hemocytes as they circulate through the hemocoel.

Tonic response The response of a sensillum that is strong when initially deformed and steady but reduced under constant deformation.

Tormogen cell The modified epidermal cell that produces the socket of a seta or sensillum.

Tracheal gills Evaginations of the integument that are covered by a thin cuticle and richly supplied with tracheae to allow aquatic respiration.

Tracheoblast A cell that gives rise to tracheoles.

Tracheole A tracheal end cell that transfers oxygen from the tracheal system to body tissues.

Trail pheromone A volatile chemical that is laid down by foraging members of one species and is used by others of that species to locate the resources.

Transverse tubule Invaginations of the sarcolemma to form an internal system of membranes that carries a depolarization deep within the muscle cell to the sarcoplasmic reticulum.

Trehalose A disaccharide that is the principal hemolymph sugar.

Triacylglycerol Uncharged esters of glycerol that serve as the storage form of fatty acids.

Trichogen cell A modified epidermal cell that produces a seta.

Trophocyte A nurse cell that provides the developing oocyte with nutrients.

Tropomyosin A regulatory peptide complexed with the actin myofilaments.

Troponin A regulatory peptide complexed with the actin myofilaments.

Tunica propria The noncellular envelope that completely surrounds the ovariole.

Tymbal A sound-producing organ.

Tympanum An auditory organ consisting of thinly stretched integumental membrane with a group of chordotonal sensilla below.

Urea A primary waste product of mammals, reptiles, and birds.

Uric acid The primary waste product of insects and some vertebrates.

Vas deferens The duct of the male reproductive system that connects the vas efferens with the ejaculatory duct. It may be enlarged into a seminal vesicle for the storage of sperm.

Vas efferens A tube that connects each testicular follicle with the vas deferens.

Ventral diaphragm An internal septum that divides the body cavity into compartments, located between the ventral nerve cord and the gut.

Ventral nerve cord The chain of interconnected ventral ganglia, connecting to the tritocerebrum by the circumesophageal connectives and extending to the end of the abdomen.

Visceral nervous system The portion of the nervous system that innervates and controls the gut and endocrine organs.

Vitellarium The area of the ovariole in which oocytes deposit yolk during vitellogenesis.

Vitellin The vitellogenins that have been deposited in the cytoplasm after modification by the follicular epithelium.

Vitelline envelope The membrane surrounding the yolk within an egg that forms the innermost layer of the chorion.

Vitellogenesis The process by which yolk produced by the fat body is taken up by the oocyte and deposited in its cytoplasm.

Vitellogenin A female-specific yolk protein that is synthesized by the fat body and taken up by the oocyte through receptor mediated endocytosis.

Vitellophage An extraembryonic cell that digests the yolk stored within the egg that is used for embryogenesis.

Viviparity A method of reproduction in which a female gives birth to live offspring that have hatched within her body.

Wandering behavior A stereotyped set of behaviors that are initiated prior to pupation and place the last instar larva in a suitable environment to pupate.

Wax layer A lipid-containing layer of the epicuticle that serves as a major barrier to water loss.

Z line The disc of protein within a muscle cell that separates sarcomeres.

Index